AF614797

METHODS IN MOLECULAR BIOLOGY™

*Series Editor*
John M. Walker
School of Life Sciences
University of Hertfordshire
Hatfield, Hertfordshire, AL10 9AB, UK

For further volumes:
http://www.springer.com/series/7651

# Spectroscopic Methods of Analysis

## Methods and Protocols

Edited by

**Wlodek M. Bujalowski**

*The University of Texas Medical Branch at Galveston, Galveston, TX, USA*

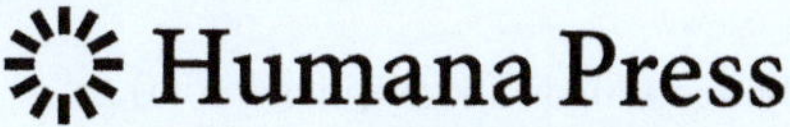

*Editor*
Wlodek M. Bujalowski, Ph.D.
The University of Texas Medical Branch at Galveston
Galveston, TX, USA

ISSN 1064-3745 ISSN 1940-6029 (electronic)
ISBN 978-1-61779-805-4 ISBN 978-1-61779-806-1 (eBook)
DOI 10.1007/978-1-61779-806-1
Springer New York Heidelberg Dordrecht London

Library of Congress Control Number: 2012934692



Printed on acid-free paper

Humana Press is a brand of Springer
Springer is part of Springer Science+Business Media (www.springer.com)

# Preface

Quantitative elucidation of structural, energetic, and dynamic aspects of macromolecular interactions is indispensable for understanding the functional activities of biomolecules. In this regard, solution studies are of paramount importance, as they address the actual or/and potential behaviors of biological systems in the environment very close to the environment in the living cell. Among other aspects, solution studies focus on such primary questions as: how many, how strong, how fast, which part/domain, where? Optical spectroscopic methods are a major part of current research methodologies used to examine function–structure relationships of biological systems. The sensitivity of the methods makes the examination of the systems of interest possible at concentration ranges where the limiting laws of thermodynamics apply. This feature provides a firm, physical basis for interpretation of the obtained results. In turn, it allows the experimenter to venture into more complex environments, informing how far he/she can go, still preserving the predictive power of thermodynamics.

The optical spectroscopic methods are not confined to small molecules and permit the studies of even the largest biological systems in their full splendor, including the living cell. Technological advances allow the experimenter to place reporter groups almost at will in different, strategically relevant locations, providing access to energetic, kinetic, and structural data, unavailable by high-resolution structural techniques. Different branches of optical spectroscopy are specifically suited for such quantitative, solution analyses. Technological development in, e.g., laser optics, computer analyses, instrumentation, and theory, make access to new optical spectroscopic properties of molecules, or novel applications of the well-known features. The developments tremendously enrich the available research repertoire and provide access to new, previously uncharted phenomena. These technological advances are more and more common in typical biochemical/biophysical laboratories, as the equipment begins to be within the reach of an average research group. Although strictly structural analyses, using optical spectroscopic methods, are of limited resolution, their resolution of energetic and dynamic aspects of interactions and structure, in solution, has no rival. This is particularly true in terms of the time range and the size of the examined system.

The presented book considers a range of important and timely biological problems, as predominantly tackled by optical spectroscopic methods. The leading experts in their fields contribute different chapters of the book with intention to provide an account of a given approach.

The first three chapters address experimental and theoretical methodology of the fluorescence properties, including fluorescence lifetime and fluorescence resonance energy transfer which can be applied to analyze properties of a fluorophore/molecule alone, or when placed in an examined biological system. Chapters 4–10 focus on application of the fluorescence spectroscopy to examine a large spectrum of biological activities of macromolecules, ranging from RNA dynamics, motor protein activities, large protein-nucleic acid complexes, fast kinetic analysis of macromolecule–ligand interactions, allosteric regulation, immunoassays, and biosensors. Chapters 11–14, provide account of current methodologies as applied to large biological systems, including mammalian cells, bacterial pathogens, and light-harvesting pigment-protein complex in plants. Single-molecule approaches are the subjects of Chaps. 15–18. The scopes of the discussed research encompass RNA folding, ribonucleoprotein assembly, mechanics of polycystic kidney disease proteins, kinetics of a single cross-bridge during contraction of muscle, and unique methodology to carry out spatially controlled, repeatable measurements of single molecules.

Although fluorescence spectroscopy dominates current optical spectroscopic applications, it is by no means the only branch of optical spectroscopy presently used to address function/activities of biological systems. Unique information about the behavior of the examined macromolecules can be obtained by examining electric filed effect and properties of natural reporting groups within the system of interest in carefully selected experimental conditions. An additional feature of these approaches is the fact that they can be applied without introducing any reporting labels. Thus, Chaps. 19 and 20 discussed applications of electro-optical analysis of macromolecular structure and dynamics, using electric dichroism and birefringence, and cryoradiolysis and cryospectroscopy of heme-oxygen intermediates in cytochromes P450.

The discussions are presented in a very practical manner, introducing the research approaches, as applied in the laboratory environment, to a larger audience. The volume is intended for all individuals interested in getting acquainted with the application of the considered methodologies; yet, it should be valuable and enough penetrating for established researchers.

*Galveston, TX, USA* *Wlodek M. Bujalowski, Ph.D.*

# Contents

## Contributors

IRINA AKOPOVA · *Department of Molecular Biology and Immunology, Center for Commercialization of Fluorescence Technologies, University of North Texas Health Science Center, Fort Worth, TX, USA*
HENRIQUE T.M.C.M. BALTAR · *Macquarie University, North Ryde, NSW, Australia*
P. BOJARSKI · *Molecular Spectroscopy Division, Institute of Experimental Physics, University of Gdansk, Gdansk, Poland*
JULIAN BOREJDO · *Department of Molecular Biology and Immunology, Center for Commercialization of Fluorescence Technologies, University of North Texas Health Science Center, Fort Worth, TX, USA*
WLODEK M. BUJALOWSKI · *Department of Biochemistry and Molecular Biology, Sealy Center for Structural Biology, Sealy Center for Cancer Cell Biology, The University of Texas Medical Branch at Galveston, Galveston, TX, USA; Department of Obstetrics and Gynecology, Sealy Center for Structural Biology, Sealy Center for Cancer Cell Biology, The University of Texas Medical Branch at Galveston, Galveston, TX, USA*
PHILIP BUTTERWORTH · *Macquarie University, North Ryde, NSW, Australia*
YI-CHUN CHEN · *Bioengineering Department, University of Illinois at Urbana-Champaign, Urbana, IL, USA*
ROBERT M. CLEGG · *Loomis Laboratory of Physics, Department of Physics, University of Illinois at Urbana-Champaign, Urbana, IL, USA; Bioengineering Department, University of Illinois at Urbana-Champaign, Urbana, IL, USA; Center for Biophysics and Computational Biology, University of Illinois at Urbana-Champaign, Urbana, IL, USA*
SABATO D'AURIA · *Laboratory for Molecular Sensing, IBP-CNR, Naples, Italy*
I.G. DENISOV · *Department of Biochemistry, School of Molecular and Cellular Biology, University of Illinois, Urbana, IL, USA*
CHRISTOPHER J. FISHER · *Department of Physics and Astronomy, Department of Molecular Biosciences, University of Kansas, Lawrence, KS, USA*
DAVID P. GIEDROC · *Department of Chemistry, Indiana University, Bloomington, IN, USA*
EWA M. GOLDYS · *Australian School of Advanced Medicine, Macquarie University, North Ryde, NSW, Australia*
Y.V. GRINKOVA · *Department of Biochemistry, School of Molecular and Cellular Biology, University of Illinois, Urbana, IL, USA*
B. GROBELNA · *Faculty of Chemistry, University of Gdansk, Gdansk, Poland*
NICHOLAS E. GROSSOEHME · *Winthrop University, Rock Hill, SC, USA*
WOJCIECH I. GRUDZINSKI · *Department of Biophysics, Institute of Physics, Maria Curie-Sklodowska University, Lublin, Poland*

WIESLAW I. GRUSZECKI · *Department of Biophysics, Institute of Physics, Maria Curie-Sklodowska University, Lublin, Poland*
IGNACY GRYCZYNSKI · *Center for Commercialization of Fluorescence Technologies, University of North Texas Health Science Center, Fort Worth, TX, USA*
ZYGMUNT GRYCZYNSKI · *Center for Commercialization of Fluorescence Technologies, University of North Texas Health Science Center, Fort Worth, TX, USA*
KATHLEEN B. HALL · *Department of Biochemistry and Molecular Biophysics, Washington University School of Medicine, St. Louis, MO, USA*
MARIA J. JEZEWSKA · *Department of Biochemistry and Molecular Biology, Sealy Center for Structural Biology, Sealy Center for Cancer Cell Biology, The University of Texas Medical Branch at Galveston, Galveston, TX, USA*
MARTIN KRATZMEIER · *Liquid Phase Separation Division, Agilent Technologies Deutschland GmbH, Waldbronn, Germany*
A. KUBICKI · *Molecular Spectroscopy Division, Institute of Experimental Physics, University of Gdansk, Gdansk, Poland*
L. KUŁAK · *Department of Theoretical Physics and Quantum Informatics, Gdansk University of Technology, Faculty of Applied Physics and Mathematics, Gdansk, Poland*
TULLIO LABELLA · *Laboratory for Molecular Sensing, IBP-CNR, Naples, Italy*
TIMOTHY M. LOHMAN · *Department of Biochemistry and Molecular Biophysics, Washington University School of Medicine, St. Louis, MO, USA*
RAFAL LUCHOWSKI · *Department of Biophysics, Institute of Physics, Maria Curie-Sklodowska University, Lublin, Poland*
MANDY SIU YU LUNG · *Australian School of Advanced Medicine, Macquarie University, North Ryde, NSW, Australia*
LIANG MA · *Department of Neuroscience and Cell Biology, University of Texas Medical Branch, Galveston, TX, USA*
EVGENIA G. MATVEEVA · *Department of Molecular Biology and Immunology, Center for Commercialization of Fluorescence Technologies, University of North Texas Health Science Center, Fort Worth, TX, USA*
PRASAD METTICOLLA · *Department of Molecular Biology and Immunology, Center for Commercialization of Fluorescence Technologies, University of North Texas Health Science Center, Fort Worth, TX, USA*
DAVID MILLAR · *Department of Molecular Biology, The Scripps Research Institute, La Jolla, CA, USA*
PRIYA MUTHU · *Department of Molecular Biology and Immunology, Center for Commercialization of Fluorescence Technologies, University of North Texas Health Science Center, Fort Worth, TX, USA*
BEATA MYSLIWA-KURDZIEL · *Department of Plant Physiology and Biochemistry, Faculty of Biochemistry, Biophysics and Biotechnology, Jagiellonian Uniwersity, Cracow, Poland*
ANDRES F. OBERHAUSER · *Department of Neuroscience and Cell Biology, Sealy Center for Structural Biology and Molecular Biophysics, The University of Texas Medical Branch at Galveston, Galveston, TX, USA;*

*Department of Biochemistry, Sealy Center for Structural Biology and Molecular Biophysics, The University of Texas Medical Branch at Galveston, Galveston, TX, USA*

D. Hern Paik · *JILA, National Institute of Standards and Technology and University of Colorado, Boulder, CO, USA; Department of Molecular, Cellular and Developmental Biology, University of Colorado, Boulder, CO, USA*

Claudio Pellecchia · *Department of Chemistry, University of Salerno, Salerno, Italy*

Thomas T. Perkins · *JILA, National Institute of Standards and Technology and University of Colorado, Boulder, CO, USA; Department of Molecular, Cellular and Developmental Biology, University of Colorado, Boulder, CO, USA*

Paul Pilowsky · *Australian School of Advanced Medicine, Macquarie University, North Ryde, NSW, Australia*

Goran Pljevaljčić · *Department of Molecular Biology, The Scripps Research Institute, La Jolla, CA, USA*

Dietmar Porschke · *Research Group Biomolecular Dynamics, Max Planck Institute for Biophysical Chemistry, Göttingen, Germany*

S. Rangelowa-Jankowska · *Molecular Spectroscopy Division, Institute of Experimental Physics, University of Gdansk, Gdansk, Poland*

Rae Robertson-Anderson · *Department of Molecular Biology, The Scripps Research Institute, La Jolla, CA, USA*

Giuseppe Ruggiero · *Laboratory for Molecular Sensing, IBP-CNR, Naples, Italy*

Pabak Sarkar · *Department of Molecular Biology and Immunology, Center for Commercialization of Fluorescence Technologies, University of North Texas Health Science Center, Fort Worth, TX, USA*

Edwin van der Schans · *Department of Molecular Biology, The Scripps Research Institute, La Jolla, CA, USA*

Tanya Shtoyko · *University of Texas at Tyler, Tyler, TX, USA*

S.G. Sligar · *Department of Biochemistry, School of Molecular and Cellular Biology, University of Illinois, Urbana, IL, USA*

Bryan Q. Spring · *Center for Biophysics and Computational Biology, University of Illinois at Urbana-Champaign, Urbana, IL, USA*

Maria Staiano · *Laboratory for Molecular Sensing, IBP-CNR, Naples, Italy*

Anna Stecka · *Department of Plant Physiology and Biochemistry, Faculty of Biochemistry, Biophysics and Biotechnology, Jagiellonian University, Cracow, Poland*

Maria Strianese · *Department of Chemistry, University of Salerno, Salerno, Italy*

Kazimierz Strzalka · *Department of Plant Physiology and Biochemistry, Faculty of Biochemistry, Biophysics and Biotechnology, Jagiellonian University, Cracow, Poland*

A. Synak · *Molecular Spectroscopy Division, Institute of Experimental Physics, University of Gdansk, Gdansk, Poland*

Danuta Szczesna-Cordary · *University of Miami Miller School of Medicine, Miami, FL, USA*

Eric J. Tomko · *Department of Biochemistry and Molecular Biophysics, Washington University School of Medicine, St. Louis, MO, USA*
Meixiang Xu · *Department of Neuroscience and Cell Biology, University of Texas Medical Branch, Galveston, TX, USA*
Piotr Wasko · *Department of Biophysics, Institute of Physics, Maria Curie-Sklodowska University, Lublin, Poland*
Colin G. Wu · *Department of Biochemistry and Molecular Biophysics, Washington University School of Medicine, St. Louis, MO, USA*

# Chapter 1

## Fluorescence Lifetime Imaging Comes of Age How to Do It and How to Interpret It

**Yi-Chun Chen, Bryan Q. Spring, and Robert M. Clegg**


### Abstract

Fluorescence lifetime imaging (FLI) has been used widely for measuring biomedical samples. Practical guidelines on taking successful FLI data are provided to avoid common errors that arise during the measurement. Several methods for analyzing and interpreting FLI results are also introduced; e.g., a model-free data analysis method called the polar plot allows visualization and analysis of FLI data without iterative fitting, and an image denoising algorithm called variance-stabilizing-transform TI Haar helps to elucidate the information of a complex biomedical sample. The instrument considerations and data analysis of Spectral-FLI are also discussed.


**Key words:** FLIM, FLI, Fluorescence lifetime imaging microscopy, Polar plot, Chebyshev transform, TI Haar denoising, Spectral FLIM

## 1. Introduction

Fluorescence lifetimes have important characteristics critical for a detailed analysis of fluorescence signals. The lifetimes are sensitive to the local environment of fluorophores and they are often the most reliable way to measure the efficiency of FRET (Förster Resonance Energy Transfer). Barring aggregation, binding to—or interaction with—other molecular components, the lifetimes are independent of fluorophore concentrations. Lifetimes are especially advantageous when studying complex cellular physiological and biological structures. For this reason, fluorescence lifetime imaging (FLI; when specifically referring to microscopy, this is called FLIM) has become popular for in vivo experiments (1, 2). This chapter introduces both routine and novel methods for carrying out FLI experiments and analyzing FLI data. In particular, we discuss issues that are typically encountered during in vivo experiments. The topics will be demonstrated with our frequency-domain FLI setup (3, 4), but there are several ways to carry out FLIM measurements (1, 2).

Wlodek M. Bujalowski (ed.), *Spectroscopic Methods of Analysis: Methods and Protocols*, Methods in Molecular Biology, vol. 875, DOI 10.1007/978-1-61779-806-1_1, © Springer Science+Business Media New York 2012

We also present a model-free data analysis method called the polar plot (4–9); this method of analysis and display is becoming more prevalent in the literature, and has also been incorporated into commercial packages of FLIM. Polar plots offer immediate visualization and aid the interpretation of FLI results. Prior knowledge of a lifetime model is not required, and recursive fitting processes are bypassed. Polar plots are especially suitable for comparing large data sets and complicated cellular dynamics. Details on how to interpret data via polar plots is summarized.

Another image analysis procedure we have found to be valuable for interpreting FLI data is a denoising algorithm called variance-stabilizing-transform TI Haar (VST-TI Haar). TI Haar was developed by Willett and Nowak for photon-limited medical images having Poisson or signal-independent Gaussian noise (10, 11). Because the detectors used for our FLI instrumentation have different signal-to-noise characteristics, the TI Haar algorithm can only be applied to denoise FLI images (12) after applying a variance-stabilizing transform. Image denoising is very helpful when biological samples with complex morphologies or heterogeneities are encountered. Therefore, VST-TI Haar denoising provides an important asset in FLI data analysis.

The third method, Spectral-FLI, is growing in popularity (13–40). A spectrograph is easily integrated into most current FLI setups. In Spectral-FLI, the spectrum containing the fluorescence signals from all fluorophores present in every pixel of the image is acquired at the same time as the lifetime data. When lifetime-resolved spectra are available at every pixel of the image, complex mixtures of fluorescence components can be distinguished (unmixed) that were otherwise difficult to separate.

Before discussing these new methods for carrying out FLI experiments, a brief introduction of frequency-domain FLI will be presented. Although FLI is becoming more common, it is necessary to familiarize readers with the fundamental concepts of FLI, as well as the terminology. For more detailed discussions about FLI techniques, several good review articles and two new books are available (1, 2, 41–46). Some practical guidelines for successfully acquiring FLI data with in vivo samples will also be discussed. We do not present detailed derivations of the equations; only the basic expressions and final results are shown. The origin of the equations can be found in the cited references.

## 2. Methods

### *2.1. Fluorescence Lifetimes*

When a fluorophore is excited to a higher energy electronic state by a short pulse of light, it can return to the ground state by emitting a photon, or the excited molecule can pass through several alternative pathways of de-excitation; e.g., dynamic quenching, FRET, internal conversion (vibrational relaxation), intersystem crossing to and from the triplet state, and photolysis (photodestruction of the molecule). The average rate of decay from an ensemble of excited molecules can be measured conveniently by detecting the time course of fluorescence decay. In the case of a single component, the decay is a decreasing exponential (Eq. 1) and this defines the fluorescence lifetime $\tau$. The fluorescence response $I(t)$ of a single component excited by a short pulse can be expressed as

$$I(t) = I_0 \, e^{-t/\tau}. \tag{1}$$

where $I_0$ is the amplitude of the exponential fluorescence decay, which is related to the fluorophore concentration and the kinetic rate constant of the natural radiative pathway. The *natural rate of radiation* is the rate of de-excitation provided that the *only pathway for de-excitation is radiative* (i.e., the fluorescence pathway); the natural rate of radiation can theoretically be calculated quantum mechanically. The *measured rate of decay*, $1/\tau$ of (Eq. 1), is equal to the *sum of all rates of all pathways* for exiting the excited state.

### 2.2. Measuring Lifetimes in the Frequency Domain

There are two general methods to acquire fluorescence lifetime information: the time-domain and frequency-domain. In the time-domain method, fluorophores are (usually) excited by ultra-short pulses, and the exponential decay of fluorescence signal is recorded directly in time and analyzed numerically to calculate the lifetime (see (Eq. 1)). For the frequency-domain technique, the excitation light intensity is modulated repetitively at a high frequency of 10–200 MHz, (Eq. 2). The emitted fluorescence signal, (Eq. 3), is modulated repetitively at the same frequency as the excitation, but with a different modulation depth and phase delay relative to the excitation. The modulation and phase of the fluorescence relative to the excitation light depend on the frequency of modulation, and the value(s) of the fluorescence lifetime(s). The lifetime information is derived from the differences of the modulation depth and phase of the fluorescence signal relative to the excitation.

Figure 1 shows the instrument setup of a frequency-domain homodyne FLI system constructed in our laboratory. Homodyne FLI is a wide-field, rapid data acquisition system. As shown in the figure, the light source intensity is modulated repetitively by a Pockels cell, which is an electro-optic modulator, at 10–200 MHz frequency $f (f = \omega/2\pi)$. The fundamental frequency component of the light intensity modulation can be expressed as

$$E = E_0(1 + M_E \cos(\omega t + \phi_E)), \tag{2}$$

where $M_E$ and $\phi_E$ are the modulation depth (fraction of the average light intensity that is modulated) and phase of the excitation, respectively. The repetitive modulation of the excitation light

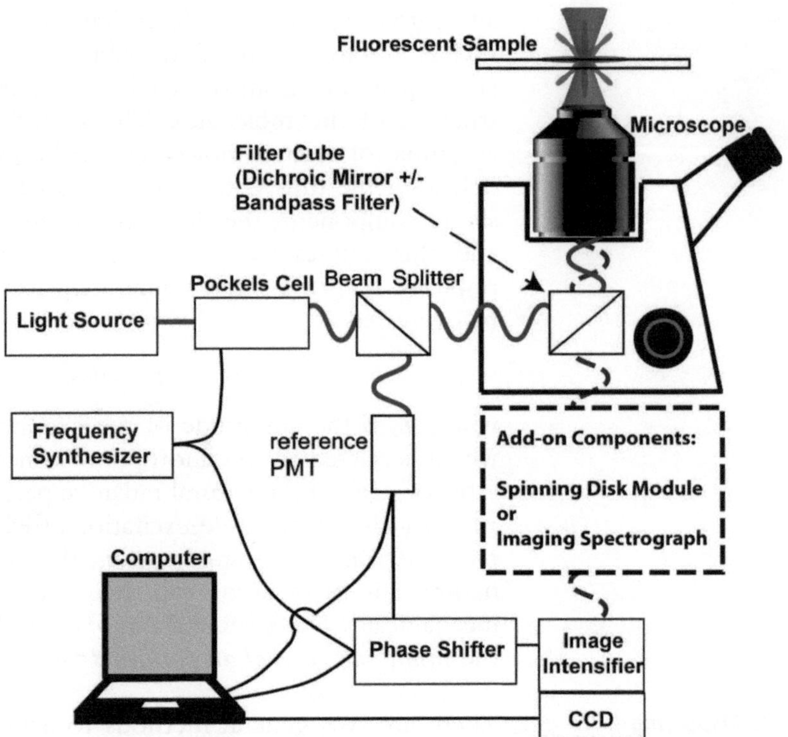


Fig. 1. The instrument setup of a full-field frequency-domain FLI. The light source is first modulated at 10–100 MHz by a Pockels cell, then the excitation light is directed to the sample. Part of the excitation signal is monitored by a reference PMT. The fluorescence emission of the sample is then demodulated by an image intensifier, which is modulated at the same MHz frequency as the Pockels cell but with a series of different phase shifts. Then the image is received by the CCD. With add-on components, such as a spinning disc module or a spectrograph, FLI is capable of achieving confocality or resolving spectral information. The frequency synthesizer and phase shifter control the modulation/demodulation frequency and the relative phase between the excitation and modulation of the intensifier.

(and thereby also that of the fluorescence intensity) does not have to be a pure sinusoid. The data are analyzed with a digital Fourier decomposition of the recorded signal at each pixel, and the component of the Fourier analysis of interest is usually the fundamental Fourier component (i.e., the frequency component at the repetition frequency, (Eq. 2)). It is possible to analyze the higher Fourier components. Each Fourier component can be analyzed separately, also on the polar plot (see below for a description of the polar plot). However, for simplicity, we will assume that the excitation modulation, (Eq. 2), the fluorescence emission, (Eq. 3), and the resulting homodyne signal, (Eq. 4), are all pure sinusoids. See the literature for a discussion of the analysis of multiple Fourier components (2, 47).

The emitted fluorescence signal from a sample contains the lifetime information in both the fractional modulation depth and the phase delay, relative to the excitation. Multiple fluorescing species (components), each with a separate lifetime, contribute additively to the total recorded fluorescence signal as in (Eq. 3).

$$F = \sum_i A_i \tau_i E_0 (1 + M_E M_{\tau_i} \cos(\omega t + \phi_E + \phi_{\tau_i})). \qquad (3)$$

Here, the lifetime of component $i$, $\tau_i$, is related to the modulation depth $M_{\tau_i} = 1/\sqrt{1 + \omega^2 {\tau_i}^2}$ and phase lag $\phi \tau_i = \tan^{-1}(\omega \tau_i)$, $A_i$ is the fluorescence *intensity* of fluorophore $i$, which depends on the fluorophore concentration and its corresponding radiative rate constant.

In the homodyne technique, the amplification level of the image intensifier is modulated at exactly the same frequency as the Pockels cell (Fig. 1); the two modulation wave forms must be phase-locked, meaning that the repetitive frequencies must be identical, and their relative phases (phase difference between them) must remain constant during the measurement. In effect, assuming for simplicity that the modulated excitation light is a pure sinusoid (at frequency $\omega$), the fluorescence signal, $F$ of (Eq. 3), is multiplied with the modulated amplification of the intensifier (again assumed to be a pure sinusoid), producing a steady-state time-invariant signal that depends on the phase difference between the fluorescence light falling on the intensifier cathode and the modulated amplification of the intensifier. The recorded signal at each pixel on the CCD is the homodyne, steady-state fluorescence intensity, $G$ of (Eq. 4):

$$G = Q \sum_i A_i \tau_i E_0 G_0 (1 + M_E M_G M_{\tau_i} \cos(\phi_E + \phi_G + \phi_{\tau_i})). \quad (4)$$

The additional terms $M_G$ and $\phi_G$ are the modulation depth of the intensifier amplification and the phase lag introduced by the image intensifier. $G_0$ is the time-averaged amplification applied to the intensifier. $Q$ is a factor to account for all the instrument factors. As one sees in (Eq. 4), the recorded signal is a steady-state intensity (i.e., does not depend on time). The phase of the voltage amplification of the image intensifier (relative to the excitation) is shifted in steps from $0 \rightarrow 2\pi$ to reconstruct a complete period of the cosine function of (Eq. 4). Knowing $M_E \cdot M_G$, $\phi_E$ and $\phi_G$, the modulation and phase of the sample fluorescence can be determined. A minimum of three data points at different phases is needed for fitting a sinusoidal function, because there are three unknowns: DC, AC, and the phase. Fluorophores of known lifetime are used to calibrate the instrument parameters ($M_E \cdot M_G$, $\phi_E$, etc.). In this way, the modulation depth (AC/DC) and phase from the fluorescence sample are corrected for the instrument parameters in order to determine the true phase and modulation parameters of the fluorescence signal. In the rest of the chapter, the modulation and phase refer to the true values of the modulation and phase of the fluorescence after the analysis; these are the modulation and phase values of the fluorescence signal relative to the excitation, and they are the parameters used to make a polar plot.

These are the experimental basics and terminology of frequency-domain FLI. In the next section, we describe the FLI instrument. We discuss several factors that are crucial for acquiring reliable FLI data.

### *2.3. Instrument Considerations*

It is always better to avoid noise contributions during the data acquisition as much as possible, rather than correcting the results afterward. Many of these noise sources can be minimized by careful consideration before taking FLI data. However, some sources of noise are unavoidable and are intrinsic to the measurement and instrument. Thorough discussions on the types of errors in FLI measurement can also be found in the literature (48). Important considerations regarding the instrument are summarized in the following subsections.

#### *2.3.1. Laser and Pockels Cell Stability*

A reference photomultiplier tube (Fig. 1, reference PMT) records a fraction of the modulated excitation light passing through the Pockels cell. Because both the laser and Pockels cell can undergo thermal drift over time with temperature fluctuations, the reference PMT signal is used to calibrate the instrument continuously in real time. We also use the homodyne technique to acquire the PMT signal. The modulation depth and phase shift from the reference PMT signal (which represents the excitation light parameters) are initially acquired with a FLI sample of a known lifetime standard in order to calibrate the excitation parameters. By knowing the $M_\tau$ and $\phi_\tau$ of the standard, we can determine the $M_E$ and $\phi_E$ of (Eq. 3). The identical measurement is then made on the sample, and the actual phase and modulation can be determined by using the standard calibration. This corrects for delays due to the speed of light as well as changes in modulation depth as the light passes through the microscope. In addition, since the PMT and the FLI data are acquired simultaneously, fluctuations in light intensity are also corrected.

Random noise can be reduced by averaging the homodyne acquired data over an extended time, and also by taking several consecutive average measurements.

#### *2.3.2. Modulation and Demodulation Frequency*

The modulation and phase of the fluorescence signal are functions of both the sample's lifetime and the modulation frequency. Therefore, for every lifetime value, there exists an optimal frequency that provides the most reliable modulation and phase values, or in other words the best lifetime resolution, (Eq. 5), (7).

$$\omega_{op}{}^2 = \frac{1+\sqrt{3}}{2\cdot\tau^2}. \quad (5)$$

$\omega_{op}$ represents the optimal angular frequency ($\omega_{op} = 2\pi f_{op}$) for a sample with lifetime $\tau$. For example, 100 MHz is the optimal modulation frequency for a 1.86-ns lifetime, and 40 MHz gives the best lifetime resolution for a 4.66-ns lifetime fluorophore. While the modulation frequency sets the lifetime resolution, the highest frequency the instrument can reliably operate depends on the electronic components. For example, the phase shifter and Pockels cell are both capacitive, and have the response times that work best in certain frequency ranges. The typical response times of the phase shifter or Pockels cell are less than 100 ps, which is more than enough for most fluorescence lifetime measurements. But measuring lifetimes below 50 ps requires faster electronics; although, because we operate in the frequency mode, faster measurements can still be made with reduced signal-to-noise ratios.

*2.3.3. Excitation Light Intensity*

Often, higher excitation intensities result in better signal-to-noise ratios. But especially for biomedical samples, care must be taken to avoid photo damage and photo bleaching. Sometimes underlying reaction mechanisms that are triggered by light affect the fluorescence response; and the reactions often respond differently to different levels of light intensity. Photosynthesis is an example. In the case of photosynthesis, different fluorescence quenching pathways are induced by different levels of light intensities. This is related to the mechanisms by which plants balance their photophysical response between the light-harvesting reaction and protection mechanisms against light damage. The fluorophore's concentration is another parameter that influences the signal intensity and thereby the signal-to-noise ratio. But when the signal is derived from intrinsic fluorescence, the concentration of the fluorophore is beyond the experimenter's control. In such cases, the exposure time is the major variable parameter to obtain a satisfactory and reliable signal-to-noise ratio.

For wide-field imaging, the laser beam is expanded to illuminate the full observed field of the sample. The laser beam has a Gaussian intensity profile, and the light intensity of even an expanded beam is not identical over the entire image. To avoid as much as possible the induction of different reaction activities mentioned in the previous paragraph at different locations of the image, an expander with diffuser is recommended. That is, one should try to keep the excitation light intensity the same throughout the image. Then, at least, any changes in time are similar throughout the image. Note that, barring intermolecular interactions and the sort of light-triggered reactions discussed above, the fluorescence lifetime is generally independent of the fluorophore concentration and excitation light intensity. This is also the case when there are multiple fluorophores or multiple deactivation pathways. However, if particular deactivation pathways respond differently to different levels of excitation, the laser intensity can be a factor influencing the measured fluorescence response.

*2.3.4. Optics Alignment*

Some components are especially sensitive to the alignment of the optics. For photomultiplier tubes, the angle and the area that photons impinge on the photocathode can influence the efficiency of the photoelectron collection onto the first dynode, and therefore influence the PMT output (this effect depends on the type of tube). For the Pockels cell, which uses the birefringence property of the crystal to modulate the light intensity, a correct alignment is necessary to produce the deepest modulation depth and the highest light intensity. To align the optics, the laser beam must be measured at different points along the optical path. Mounts with four degrees of freedom are desirable. For accurate Pockels cell alignment, one can also check for Lissajous figures using a diffusing element (49). By placing a light diffuser in front of the Pockels cell, which produces a beam with the wave-vector of the laser light oriented in all directions, an isogyro pattern can be seen by placing a white screen right after the Pockels cell (image not shown here, see Fig. 4 in (49)). If the Pockels cell is well-aligned, when the diffuser is not in the beam, then the outgoing laser beam will be in the center of the isogyro pattern when a diffuser was in the beam. The isogyro pattern is a simple and accurate way to align a Pockels cell.

*2.3.5. Fluorescence Lifetime Standard*

As mentioned above, frequency-domain FLI requires a fluorescence lifetime standard to calibrate the instrument parameters. Common fluorophore lifetimes can be found in the several references (50, 51). Fluorescein is one of the most common lifetime standards, because of its high quantum yield, and the excitation and emission wavelengths are convenient and similar to many dyes. Because the measured lifetime of the sample depends on how well the instrument is calibrated, any error in the measurement of the lifetime standard will degrade the accuracy of the sample measurement. Therefore, it is important to carefully characterize and calibrate the instrument with a known and trusted reference fluorophore. A second lifetime standard can be used to iteratively cross-calibrate the instrument. Also, a lifetime standard that has a similar lifetime value as the sample is preferable, as they are in the same optimal modulation frequency range. For many applications, it is not necessary that the reference fluorophore have a single lifetime; however, if an accurate measurement of the lifetime is desired, one should know the values of the lifetime components. This is because one is calibrating directly the modulation depth and the phase delay, see (Eq. 4). It is only important to be able to determine the modulation and phase parameters of the excitation light. And it is most important that the standard be chemically stable and that the phase and modulation characteristics be highly reproducible.

### 2.4. Model-Free Representation of Data: Polar Plot

In the polar plot analysis, the modulation and the phase of the fluorescence are used to construct a model-free plot. Model-free means the plot only requires the measured modulation depth and the phase of the fluorescence. Figure 2a shows how to make a polar plot from the modulation and phase of the sample data. The measured values of $M_{\text{tot}}$ and $\phi_{\text{tot}}$ are used to define new coordinates as in (Eq. 6) and (Eq. 7) (4–9):

$$x_{\text{tot}} = M_{\text{tot}} \cos \phi_{\text{tot}} = \frac{\sum_i A_i \tau_i M_{\tau_i} \cos \phi_{\tau_i}}{\sum_i A_i \tau_i} \tag{6}$$

and

$$y_{\text{tot}} = M_{\text{tot}} \sin \phi_{\text{tot}} = \frac{\sum_i A_i \tau_i M_{\tau_i} \sin \phi_{\tau_i}}{\sum_i A_i \tau_i}. \tag{7}$$

For emphasis, we have written the subscript "*tot*" in (Eq. 6) and (Eq. 7) in order to indicate that the signal is the overall measured signal. Equations (Eq. 6) and (Eq. 7) show how $M_{\text{tot}}$ and $\phi_{\text{tot}}$ are related theoretically to the corresponding parameters of individual fluorescence components in the case of multiple lifetime components. When there is single lifetime, the $(x, y)$ points will fall on a universal semicircle centered at $(x, y) = (0.5, 0)$. When the measured fluorescence signal is due to several lifetime components, the $(x, y)$ positions on the polar plot will lie inside of the semicircle, and the locations of the $(x, y)$ points are determined by the weighted intensity fractions of the contributing individual lifetime components (see Fig. 2b and the following discussion). As $x$ and $y$ are also functions of the modulation frequency, data points corresponding to the same lifetime measured at different modulation frequencies have different locations on the polar plot. Polar

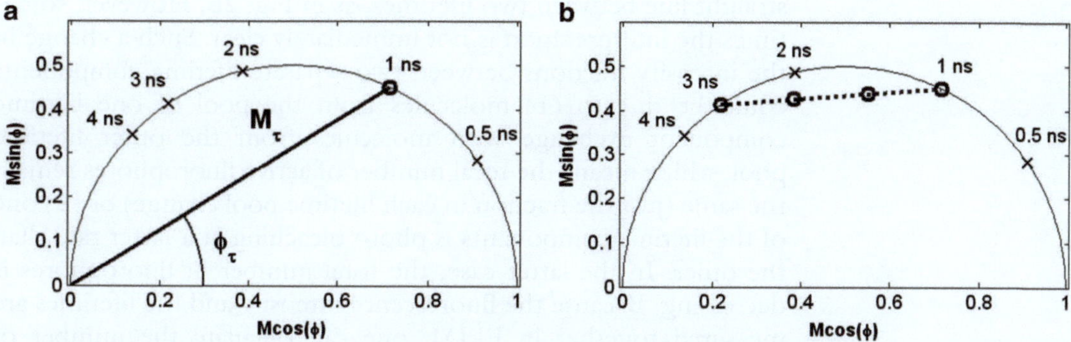


Fig. 2. Polar plot representation of lifetime data acquired at 100 MHz. (**a**) The measured modulation $M_{\text{tot}}$ and phase $\phi_{\text{tot}}$ are used to define the axes of the new axis system of polar plot, which has $x = M_{\text{tot}} \cdot \cos(\phi_{\text{tot}})$ and $y = M_{\text{tot}} \cdot \sin(\phi_{\text{tot}})$. (**b**) On the polar plot, single lifetime data falls on the universal semicircle, and multiple lifetimes data locate inside of the semicircle. The location depends on the relative contribution from each lifetime components.

plots are useful for interpreting complicated cellular dynamics; the following discussion presents more details.

*2.4.1. Data Points on the Semicircle*

In Fig. 2, it is shown that the locations on a polar plot of simulated single-lifetime data of a 100-MHz measurement lie on the universal semicircle. Longer lifetimes are located more counter-clockwise on the semicircle, and shorter lifetimes are more clockwise. That is, for a single frequency of modulation, the phase angle increases and the modulation depth decreases as the single lifetimes become longer. If there is light scattering in the recorded data, which is like having a fluorescence lifetime of 0 ns, the corresponding polar plot scattering component is at the most clockwise point: $(x, y) = 1, 0$. It was mentioned earlier that 100 MHz gives the best lifetime resolution of 1.86 ns lifetime. The reason for this is easy to see from the distribution of points on the polar plot. Lifetime values ranging from 1 to 2 ns have the best resolution of the modulation and phase when modulating at 100 MHz.

*2.4.2. Data Points Inside of the Semicircle*

If there are multiple lifetime components, the points on the polar plot using the measured modulation depth and phase will fall inside of the semicircle. As shown in Fig. 2b, if the signal is composed of 3 and 1 ns, then the polar plot points lie on a straight line connecting the points on the semicircle corresponding to the two single lifetime positions (2, 7, 8). The relative distances on the straight line from the measured point to the corresponding single-lifetime points on the semicircle are proportional to the intensity contributions of these single-lifetime components. If the 3-ns component contributes higher intensity, then the point is closer to 3 ns, and vice versa; that is, the intensity contribution of each component is proportional to the fractional distance on the opposite side of the straight line from the measured point. When measuring FLI during active cellular dynamics that interconvert between two fluorescence components, the measured polar plot points progress along the straight line between two lifetimes, as in Fig. 2b. However, sometimes the interpretation is not immediately clear. Such a change in the intensity fractions between two separate lifetime components could be due to (1) molecules from the pool of one lifetime component exchange with molecules from the other lifetime pool, which means the total number of active fluorophores remain the same (just the fraction in each lifetime pool change) or (2) one of the lifetime components is photo bleaching at a faster rate than the other. In the latter case, the total number of fluorophores is decreasing. Because the fluorescence intensity and the lifetimes are measured together in FLIM, one can calculate the number of molecules of each component. The individual lifetimes can be determined from the intersection of the extension of the straight line. Thus, the molecular mechanism responsible for the phase and modulation changes can often easily be determined. A derivation of the relation between the lifetime data and fluorescence intensities, which allows the calculation of the concentrations of the different lifetime components, is given in the literature (5).

*2.4.3. Data Points outside of the Semicircle*

Assuming that the instrument is calibrated carefully, there are generally two reasons if the polar plot data fall outside of the semicircle: (1) either Förster resonant energy transfer (FRET) is taking place (or some other excited state reaction) or (2) photo bleaching is happening during illumination. To tell which one is the real cause, one can simply check the data to see whether the average of the recorded sinusoidally modulated signal is a decaying with time.

If the fluorescence signal comes *solely from the acceptor* of a FRET pair, the modulation $M_{AD}$ and phase $\phi_{AD}$ (after the initial analysis and corrections) are given as (2):

$$M_{AD} = M_{DA} \cdot M_A \tag{8}$$

and

$$\phi_{AD} = \phi_{DA} + \phi_A. \tag{9}$$

Here, $M_{DA}$ and $\phi_{DA}$ are the modulation and phase of the donor in the presence of FRET, and $M_A$ and $\phi_A$ are the modulation and phase of the acceptor without any FRET (i.e., what would be measured if the acceptor were directly excited by excitation light). A calculation using (Eq. 8) and (Eq. 9) shows that the polar plot points, if just the acceptor fluorescence is detected, lie outside the semicircle. According to (Eq. 8) and (Eq. 9), if the acceptor lifetime in the absence of FRET is known, the donor $M_{DA}$ and $\phi_{DA}$ in the presence of FRET can be calculated and the FRET efficiency can also be determined.

In the case of a decaying sinusoidal (for instance, with very rapid photolysis taking place during the data acquisition), one can fit the data with Chebyshev polynomials to evaluate the true modulation, phase, and decay constant (for instance, the decay rate of the fluorescence intensity due to photolysis). See (52) for a detailed discussion. In the Chebyshev algorithm, the data set is operated on by a bounded Chebyshev transform to obtain Chebyshev polynomial coefficients. Then algebraic equations involving recursion relations of the Chebyshev coefficients are solved algebraically to derive the parameters of the corresponding fluorescence parameters (phases, modulation values, and the time constant of the photolysis decay). The method has the advantage of avoiding recursive iterative fitting procedures, which would require starting parameters and is very time-consuming. Because the Chebyshev method is a single-pass procedure, and the parameters are determined by numerical analysis of an algebraic equation, the Chebyshev fitting procedure is rapid (orders of magnitude faster than iterative fitting) and suitable for in vivo FLI, especially when there is large amount of data (e.g., a high number of pixels).

### 2.5. Denoising FLI Images: VST-TI Haar Algorithm

As FLI instrumentation has matured significantly, it has become popular for in vivo experiments. Denoising algorithms have been shown to be very promising for FLI image analysis (12). The intrinsic and unavoidable noise of the measurement is due mainly to photon noise and contributions from the intensifier and CCD. If one knows the characteristics of the noise, one can apply an appropriate denoising algorithm in order to improve the construction of the lifetime-resolved image with greater precision. Quantitative measurements are more useful in evaluating complicated cellular dynamics.

The usefulness of denoising in image processing is demonstrated in Fig. 3. Wild-type *Chlamydomonas reinhardtii* cells grown in a liquid media were deposited onto a nitrocellulose filter paper, which has a pore size of 1.2 μm (5). The cells were excited with a 488-nm laser and the fluorescence emission of chlorophyll a was measured. The modulation frequency was 100 MHz, and a 40× objective lens was used in the FLI microscope. Figure 3 is one

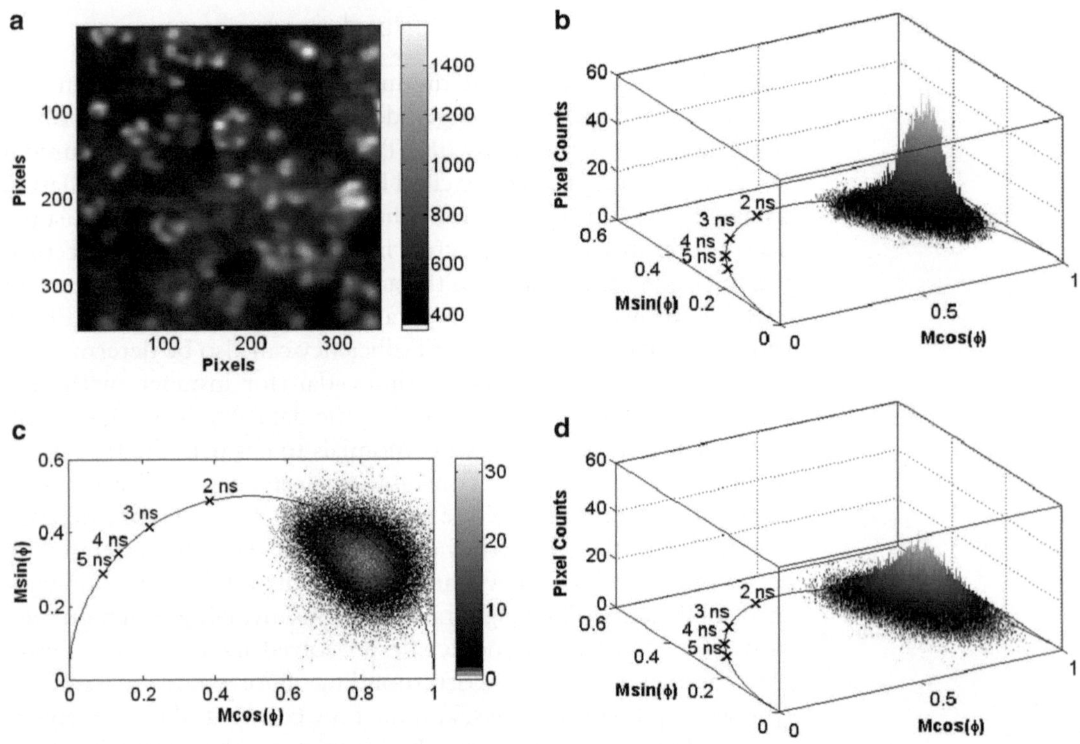


Fig. 3. VST-TI Harr denoising of *Chlamydomonas reinhardtii* (Chlamy) cells FLI data. Chlamy cells were deposited on a nitrocellulose filter paper (with pore size 1.2 μm) for the FLI measurement. The cells were excited by a continuous 488 nm laser modulated at 100 MHz and imaged by a 40× objective lens. The modulation frequency is 100 MHz. (**a**) The fluorescence intensity image of the Chlamy cells calculated from the FLI data. The gray-scale bar is fluorescence intensity in arbitrary unit (A. U.). (**b**) The histogram of VST-TI Harr denoised FLIM data on the polar plot. The data is composed of a major lifetime pool, and a minor fraction of signal at longer lifetime value. (**c**) and (**d**) are the polar plot of the raw data (before denoising) in 2D and 3D, respectively. The gray-scale bar and the *z*-axis represent the pixel counts. Compared with the denoised data in (**b**), the longer lifetime pool is not as noticeable in (**c**) and (**d**).

of the FLI data sets taken during the Kautsky transient (53, 54). Figure 3a is the fluorescence intensity image of the cells. The grayscale bar indicates the florescence intensity in arbitrary units. After VST-TI Haar denoising, the fluorescence lifetime analysis of every pixel from Fig. 3a is shown on the 3D polar plot (Fig. 3b). The *z*-axis represents the number of pixels corresponding to a particular point on the polar plot. It is clearly shown that the cells' lifetimes are not from a single component. Besides the major population, some pixels show longer lifetimes. Figure 3c, d, are polar plots in 2D and 3D before denoising of the same data as analyzed in Fig. 3a, b (after denoising). The centroid of the lifetime pool remains the same after the denoising; however, the width of the lifetime distribution is noticeably tighter. Therefore, denoising the image can considerably reduce the spread of the lifetime data. Only noise components due to random variance of the measurement are removed; the variance due to distributions of lifetimes is not affected. This is especially relevant in cases when it is important to study the morphology of the sample and the statistics of lifetime heterogeneity.

Originally, the TI Haar was a sophisticated algorithm designed by Nowak and Willett for denoising Poisson noise from photon counting data. VST-TI Haar has extended the use of this powerful method to full-field FLI, or any modality where photoelectron amplification degrades the Poisson statistics of the photon counts (e.g., intensified CCD cameras, electron-multiplying CCD cameras, and PMTs operating in charge accumulation mode). Taken together, TI Haar and VST-TI Haar can be applied to any imaging modality where the analysis involves fitting a portion of the image (e.g., super-resolution imaging by fitting the PSF of single-fluorophores) or where a fit is carried out at each pixel in a sequential series of images (e.g., FLIM and linear spectral unmixing analysis of hyperspectral images). These denoising analyses are especially desirable for applications where frame averaging mitigates visualization of dynamic biological events; thus, denoising can be used to compensate for the loss of SNR at faster frame rates. The steps for denoising FLI data are summarized in the following subsections.

#### 2.5.1. Find the Variance-vs.-Mean Signal Characteristics

Regular FLI data is acquired at different fluorescence intensities. The exposure time is kept constant and the light intensity is adjusted using a gradient neutral density filter. It is important to diffuse the excitation intensity over the entire region of interest in the image. Select a small homogenous region in each image (e.g., a $50 \times 50$ pixels region in the center of the image), and calculate the variance and the mean from the data. The plotted data will look as shown in Fig. 4, which shows data measured from two different image intensifier gains at 100 MHz. The variance and mean of the data represent the value of the noise and signal, respectively.

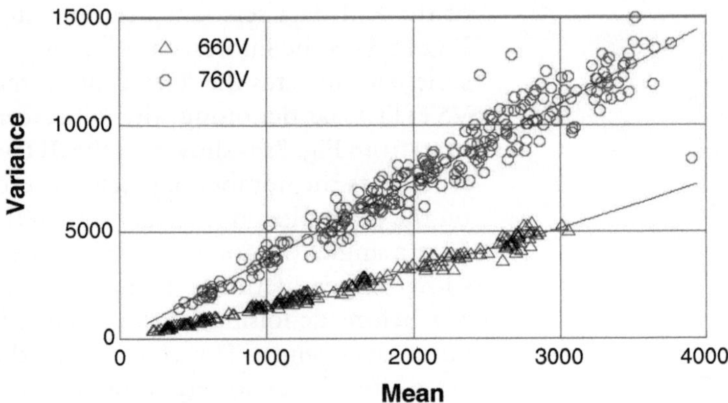


Fig. 4. The mean-variance curves of the ICCD at two different settings, 760 and 660 V. From the fitting of the curve, the noise characteristic information is used for VST-TI Harr calculation.

By fitting the data, the variance $\sigma(x)^2$ can be expressed as a function of the signal intensity $x$:

$$\sigma(x)^2 = f(x). \tag{10}$$

*2.5.2. Calculating the Variance-Stabilizing Transform*

As photon noise is Poissonian, the variance-vs.-mean characteristic from just a CCD is a linear curve, and has a slope of 1, after taking into account the CCD gain (the number of photoelectrons per gray-scale level, which is determined by the analog-to-digital converter) to rescale the data into units of photoelectrons. But with an image intensifier, new noise is introduced into the image. The recorded signal has larger variance that cannot be characterized simply as Poissonian. The noise distribution of the intensified CCD (ICCD) has a signal-dependent Gaussian distribution (this is not just the Gaussian corresponding to the limit at high number of photons of the original Poissonian noise; there are additional effects introduced by the intensifier (12)). In order to use TI-Haar denoising of the signal-independent Gaussian noise, a variance-stabilizing transform has to first be applied. From (Eq. 10), the variance-stabilized data is

$$y(x) = VST(x) = \int \frac{\mathrm{d}x}{\sqrt{f(x)}}. \tag{11}$$

*2.5.3. Translation-Invariant Haar Denoising (TI Haar)*

TI Haar is a multiscale image estimation method for data with Poisson and Gaussian noise. The algorithm has several important properties that make it useful for FLI data. This multiscale estimate can remove the noise without blurring the edges and boundaries in the image, and the image can be reconstructed with high accuracy. The algorithm developed by R.M. Willett and R.D. Nowak (10, 11) also has the advantages of reduced computational time. For frequency-domain FLI, every phase-shifted image should be denoised before calculating the modulation and phase values. The basic TI-Haar algorithm is available on R. Willett's homepage, and the transformed variance-vs.-mean characteristic is required as an input parameter.

#### 2.5.4. Inverse Variance-Stabilizing Transform

After applying TI-Haar denoising, the data must undergo an inverse VST to recover the actual denoised image, with the true signal-to-noise dependence. In (Eq. 11), the VST algorithm calculates $y$ as a function of $x$ ($y$ are the normalized values of the original $x$-data). In inverse VST, one simply expresses $x$ as a function of $y$, and then applies the normal analysis to the denoised data.

### 2.6. Spectral-FLI

A fluorescence spectrum is characteristic of the chemical structure of the dye; therefore, the spectrum is often used to differentiate fluorophores. As fluorescence lifetimes provide information about the kinetic interaction of the excited chromophore with its local environment, combining the spectrum with lifetime measurements can greatly assist the interpretation of FLI data, especially for samples with several fluorescence components. A spectrograph is compatible with most current FLI systems. Spectral-FLI has been demonstrated in various instrument setups, in both time- and frequency-domains (13–30, 32–40). As shown in Fig. 1, the emission light from the sample is dispersed by a spectrograph before the fluorescence passes to the image intensifier. Spectral-FLI acquires an additional dimension of data; therefore, the whole-field two-dimensional FLI image at different wavelengths is reconstructed by line-scanning the $y$-spatial dimension (the direction perpendicular to the slit height—this is called broom scanning) (Fig. 5). Conventional spectrographs can achieve high wavelength precision up to 0.1 nm (55), but often the signals from several pixels corresponding to neighboring wavelength channels can be averaged to achieve better signal-to-noise ratio. This limits the wavelength resolution, but the emission spectra of most fluorophores are usually broad. The width of the entrance slit of the spectrograph defines the spatial resolution in that direction. While a wider slit allows more photons to enter the spectrograph, the diffracted wavelengths from different image locations within the slit width overlap and decrease the spatial resolution in that direction. To acquire good spectra, a narrow entrance slit is preferable but the signal-to-noise ratio decreases. Fortunately, most fluorophores have broad spectrum ranges, and it does not require high wavelength resolution or many wavelength channels to differentiate fluorophores (see Subheading 2.6.3 for more details).

Figure 6 is an example of data acquired by Spectral-FLI. Intact avocado leaf samples were excited by a 440-nm laser at 100 MHz. The two samples, a mature dark green leaf and a young light green leaf from the same avocado tree, have different chlorophyll contents that result in different spectral shapes (56). They were measured at the same time point of the "Kautsky" induction

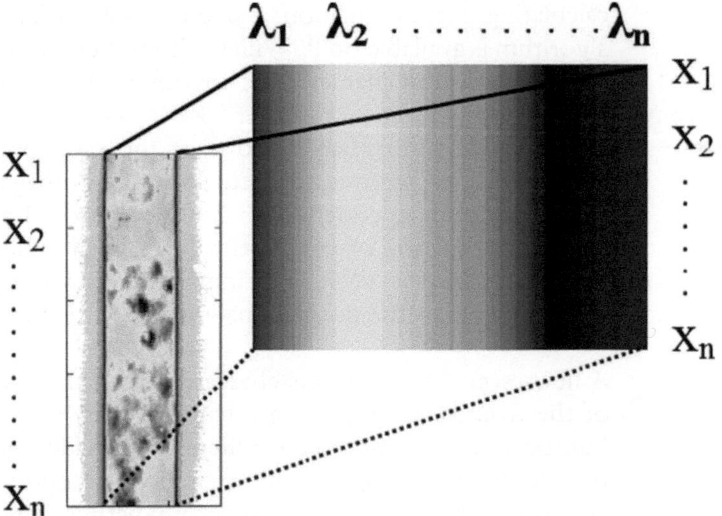


Fig. 5. Incorporating a spectrograph into the FLI can provide the spatial information in one axis, while the spectral information is dispersed along the other axis. See the text for a discussion of the spectrograph recording.

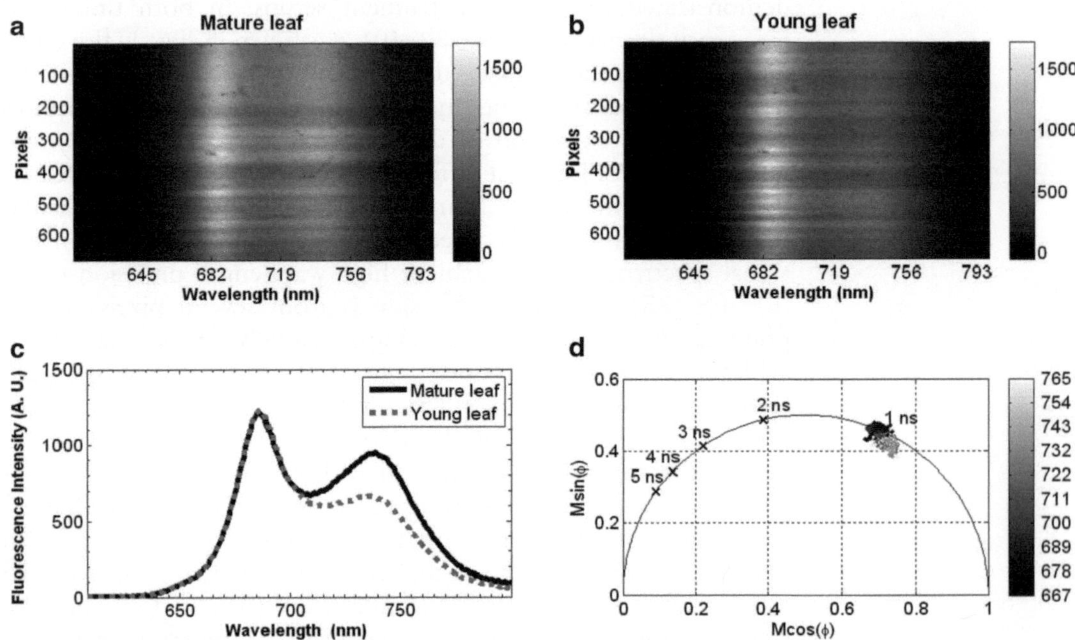


Fig. 6. The Spectral-FLIM measurement of a mature and a young avocado tree leaves. The excitation laser is 440 nm and the modulation frequency is 100 MHz. (**a**) and (**b**) are the fluorescence intensity images on the CCD acquired by a Spectral-FLIM setup. The *x*-axis is with the wavelength of the fluorescence spectrum, while the *y*-axis still preserves the spatial information of the sample within the limits of the spectrograph slit (see text). The gray-scale bar represents the fluorescence intensity in arbitrary units. (**c**) The emission spectra of the mature and young leaves, calculated by averaging along the *y*-axis of the fluorescence intensity image. They were measured at the same time point during the Kautsky transient (see text discussing the fluorescence response of a photosynthetic system). The spectra are from the raw data, which means there is no correction for the wavelength sensitivity of the instrument or any normalization. (**d**) The polar plot analysis of (**a**) and (**b**). The gray-scale bar represents the emission wavelength in nanometer. For both leaves, the longer lifetime pool belongs to the fluorescence peak at the shorter wavelength, and the faster lifetime pool comes from the fluorescence peak at the longer wavelength.

curve (53, 54) and both were acquired at room temperature. As shown in Fig. 1, the fluorescence signal passes through the spectrograph before it is demodulated by the image intensifier and then imaged on the CCD. Figure 6a, b shows the fluorescence signal after passing through a narrow entrance slit (0.5 mm) of the spectrograph. The spatial information along the slit is displayed on one axis of the image and the spectrum is dispersed along the other one. After calibrating the dispersion of the spectrograph, the wavelength information is indicated on the wavelength axis of the image. Figure 6c is the average calculated from Fig. 6a, b. The data have been acquired without correcting for the wavelength sensitivity of the instrument and before any normalization to certain wavelengths. The spectra from the two independent leaves show the same intensity at the peak ~685 nm, while the ~740 nm peak intensity varies. The difference between the spectra is an indication of the difference in molecular composition of the fluorophores and their environments.

Figure 6d is the polar plot analysis of both leaves, with the gray-scale bar representing the wavelength. It is shown that the ~685-nm peaks from both samples have the same longer lifetime location, and the ~740-nm lifetime pools also coincide. More detailed interpretations of the data are not discussed here; however, these data demonstrate the usefulness of Spectral-FLI in providing information on the different underlying fluorescence components. Some considerations regarding the spectrograph are provided in the following sections.

#### 2.6.1. Choosing a Spectrograph

Gratings are often used in spectrographs for their good resolving power in the visible, which is a measure of their ability to separate adjacent wavelengths. There are some concerns about using a grating as a dispersion device. One is the overlap of wavelengths from different diffraction orders, which introduces inaccuracy in the spectrum (57). But in fact, for the fluorescence emission measurement, the higher order signal usually does not appear within recording wavelength range. For example, when a fluorophore is excited at 488 nm, the emission spectra range is from about 500 to 800 nm. The higher order signal within this wavelength range starts to appear only at wavelengths approximately longer than 1,000 nm. Therefore, the superposition of different orders is seldom a problem. Another consideration is the imaging property of grating systems. Gratings have inherent artifacts such as astigmatism, background stray light, spectral aberration, etc. (57, 58). In order to form a flat-field image on the CCD, care must be taken to ensure that the design of the spectrograph meets the imaging criteria of the FLI setup. Czerny-Turner spectrographs are one of the most common setups used for full-field imaging (59, 60). This setup has two concave mirrors in addition to a plane grating, so that the dispersed spectrum is focused and linearly

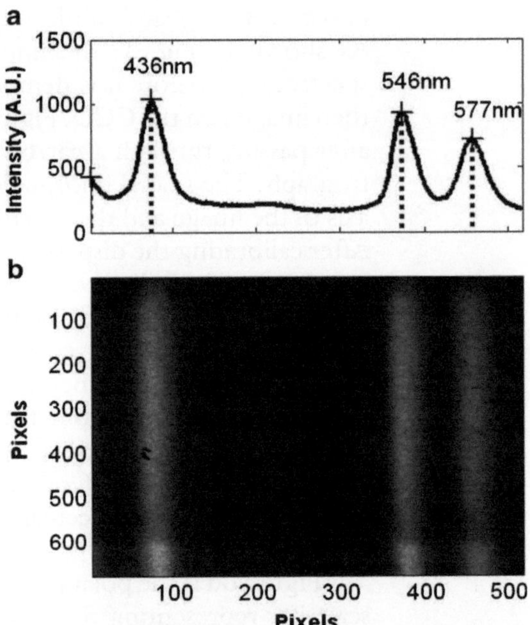


Fig. 7. Using the spectrum of mercury to calibrate the spectrograph. (**a**) is the partial spectrum of mercury, which is calculated from the Spectral-FLI image of a mercury lamp in (**b**).

dispersed. For different FLI setups (e.g., a scanning FLI that uses a multiple wavelength channels and PMTs), there might be different requirements for a spectrograph. It is important to be aware of the grating characteristics in relation to the particular FLI setup in order to interpret correctly FLI results.

*2.6.2. Characterizing a Spectrograph*

Conventional spectrographs are precalibrated and well aligned. After setting up the instrument, the first things to check are astigmatism and spectral aberration, which mean the focusing property of the tangential plane and capital plane, and the focusing property at different wavelengths (diffraction angle). A sample with specific known patterns or structures, such as a resolution mask, can be used to examine the focusing property. Another important consideration is the dispersion characteristic of the spectrograph. Light sources with known wavelength dispersions can be used to identify the peak wavelength location on the wavelength axis of the image. Figure 7 shows the Spectral-FLI image of the mercury spectrum. Mercury has hyperfine structures, and has several peaks over the visible light range. The linear dispersion relation and the reciprocal dispersion can be calculated by acquiring the Spectral-FLI image of a mercury lamp. Note that the intensity of a mercury lamp should be decreased greatly before it arrives at the FLI detector to avoid damage of the sensitive (and more expensive) electronics.

For the purpose of many FLI measurements, it is not required to acquire an absolute true spectrum in order to distinguish fluorescence lifetime components; the calibration of the instrument must simply be repeatable. But in some cases, the absolute spectrum is better for identifying the fluorescent molecules in the sample, such as data of auto fluorescence of tissue (33, 39). Several optics and electronics components in the instrument have wavelength-dependent sensitivities. For example, the dichroic mirror, grating, image intensifier, and the CCD are all wavelength dependent. The wavelength sensitivity of the instrument must be calculated if true spectra are desired; although, as mentioned above, often this is not necessary.

#### 2.6.3. Linear Unmixing to Separate Spectra

Linear unmixing is commonly used in spectral fluorescence microscopy (61–63). Fluorophore species can be distinguished by linearly unmixing the different spectral components from the recorded spectrum. It is especially useful when fluorophores of interest have overlapping spectra, which cannot be well separated by band pass filters. By applying linear unmixing in Spectral-FLI, the lifetime pool of each fluorophore can be distinguished with more reliability from other lifetime pools. As discussed in a previous section, the frequency-domain data recorded at each different phase shift constitute a steady-state fluorescence intensity image. Unmixing each separate phase-shifted FLI image is the same as unmixing the spectra in a conventional fluorescence spectrum. To be clear, we emphasize that the FLI intensity image recorded at each phase shift ($G = \sum_i A_i \tau_i E_0 G_0 (1 + M_E M_G M_{\tau_i} \cos(\phi_E + \phi_G + \phi_{\tau_i}))$) is not the overall steady-state fluorescence intensity ($G_{DC} = \sum A_i \tau_i E_0 G_0$). But the steady-state fluorescence intensity can be calculated correctly from the FLI data that has been unmixed at every separate phase shift.

Several considerations must be taken into account before performing the linear unmixing (61–63). First, the reference spectrum used to calibrate the system should be measured under the same conditions as the FLI data. In the previous section, it is mentioned that the final spectrum recorded on the CCD depends on several optical and electronic elements. It is more accurate to measure the reference dye directly than to correct the wavelength-sensitivity factors and then calculate the absolute spectrum value from literature. Second, the number of wavelength detection channels $n$ should be equal to or greater than the number of fluorophores $m$ to be determined, in order to solve the linear equation system as shown in (Eq. 12).

$$S(\lambda_n) = \sum_m a_m \cdot S_m(\lambda_n). \tag{12}$$

Here, $S(\lambda_n)$ is the spectrum measured from channel $n$, $S_m(\lambda_n)$ is the reference spectrum of dye molecule $m$, and $a_m$ accounts for the intensity contribution of fluoropohre *m*. Finally, since the signals from different channels are used to reconstruct the spectrum, the intensity variation between the individual detectors, that is, CCD pixels, should be calibrated before linear unmixing. And the background noise, such as the CCD read out noise, should be subtracted before the calculation.

## Acknowledgement

Yi-Chun Chen thanks the Taiwan Merit Scholarships (TMS-094-1-A-036) for financial support.

### References

1. Gadella T (2009) FRET and FLIM techniques, vol 33. Elsevier Science, Oxford
2. Periasamy A, Clegg RM (2009) FLIM microscopy in biology and medicine, 1st edn. Chapman & Hall, Boca Raton
3. Schneider PC, Clegg RM (1997) Rapid acquisition, analysis, and display of fluorescence lifetime-resolved images for real-time applications. Rev Sci Instrum 68:4107–4119
4. Buranachai C, Kamiyama D, Chiba A, Williams BD, Clegg R (2008) Rapid frequency-domain FLIM spinning disk confocal microscope: lifetime resolution, imageimprovement and wavelet analysis. J Fluoresc 18:929–942
5. Holub O, Seufferheld MJ, Gohlke C, Govindjee, Heiss GJ, Clegg RM (2007) Fluorescence lifetime imaging microscopy of Chlamydomonas reinhardtii: non-photochemical quenching mutants and the effect of photosynthetic inhibitors on the slow chlorophyll fluorescence transient. J Microsc 226:90–120
6. Hanley QS, Clayton AHA (2005) AB-plot assisted determination of fluorophore mixtures in a fluorescence lifetime microscope using spectra or quenchers. J Microsc 218:62–67
7. Redford G, Clegg R (2005) Polar plot representation for frequency-domain analysis of fluorescence lifetimes. J Fluoresc 15:805–815
8. Colyer R, Lee C, Gratton E (2008) A novel fluorescence lifetime imaging system that optimizes photon efficiency. Microsc Res Tech 71:201–213
9. Digman MA, Caiolfa VR, Zamai M, Gratton E (2008) The phasor approach to fluorescence lifetime imaging analysis. Biophys J 94: L14–L16
10. Willett RM, Nowak RD (2003) Platelets: a multiscale approach for recovering edges and surfaces in photon-limited medical imaging. IEEE Trans Med Imaging 22:332–350
11. Willett RM, Nowak RD (2004) Fast multiresolution photon-limited image reconstruction. In: IEEE international symposium on biomedical imaging: nano to macro. pp 1192–1195
12. Spring BQ, Clegg RM (2009) Image analysis for denoising full-field frequency-domain fluorescence lifetime images. J Microsc 235:221–237
13. Becker W, Bergmann A, Biskup C (2007) Multispectral fluorescence lifetime imaging by TCSPC. Microsc Res Tech 70:403–409
14. Volker U, Peter F, Iris R, Karsten K (2004) Compact multiphoton/single photon laser scanning microscope for spectral imaging and fluorescence lifetime imaging. Scanning 26:217–225
15. Vereb G, Jares-Erijman E, Selvin PR, Jovin TM (1998) Temporally and spectrally resolved imaging microscopy of lanthanide chelates. Biophys J 74:2210–2222
16. Tinnefeld P, Herten D-P, Sauer M (2001) Photophysical dynamics of single molecules studied by spectrally-resolved fluorescence lifetime imaging microscopy (SFLIM). J Phys Chem A 105:7989–8003
17. Thaler C, Koushik SV, Blank PS, Vogel SS (2005) Quantitative multiphoton spectral imaging and its use for measuring resonance energy transfer. Biophys J 89:2736–2749
18. Siegel J, Elson DS, Webb SED, Parsons-Karavassilis D, Lévêque-Fort S, Cole MJ, Lever MJ, French PMW, Neil MAA, Juskaitis R, Sucharov

LO, Wilson T (2001) Whole-field five-dimensional fluorescence microscopy combining lifetime and spectral resolution with optical sectioning. Opt Lett 26:1338–1340

19. Rück A, Hülshoff CH, Kinzler I, Becker W, Steiner R (2007) SLIM: a new method for molecular imaging. Microsc Res Tech 70:485–492
20. Rück A, Dolp F, Hülshoff C, Hauser C, Scalfi-Happ C (2005) Fluorescence lifetime imaging in PDT. An overview. Med Laser Appl 20:125–129
21. Riquelme BD, Dumas D, Valverde de Rasia J, Rasia RJ, Stoltz JF (2003) Analysis of the 3D structure of agglutinated erythrocyte using cell scan and confocal microscopy: characterization by FLIM-FRET. Proc SPIE 5139:190–198
22. Ramadass R, Becker D, Jendrach M, Bereiter-Hahn J (2007) Spectrally and spatially resolved fluorescence lifetime imaging in living cells: TRPV4-microfilament interactions. Arch Biochem Biophys 463:27–36
23. Provenzano P, Eliceiri K, Keely P (2009) Multiphoton microscopy and fluorescence lifetime imaging microscopy (FLIM) to monitor metastasis and the tumor microenvironment. Clin Exp Metastasis 26:357–370
24. De Pieter B, Dylan MO, Hugh BM, Clifford BT, Jose R-I, Christopher D, James M, Richard KPB, Daniel SE, Ian M, Lever MJ, Praveen A, Mark AAN, Paul MWF (2007) Rapid hyperspectral fluorescence lifetime imaging. Microsc Res Tech 70:481–484
25. Peter M, Ameer-Beg SM, Hughes MKY, Keppler MD, Prag S, Marsh M, Vojnovic B, Ng T (2005) Multiphoton-FLIM quantification of the EGFP-mRFP1 FRET pair for localization of membrane receptor-kinase interactions. Biophys J 88:1224–1237
26. Pan W, Qu J, Chen T, Sun L, Qi J (2009) FLIM and emission spectral analysis of caspase-3 activation inside single living cell during anticancer drug-induced cell death. Eur Biophys J 38:447–456
27. Nair DK, Jose M, Kuner T, Zuschratter W, Hartig R (2006) FRET-FLIM at nanometer spectral resolution from living cells. Opt Express 14:12217–12229
28. Hanley QS, Arndt-Jovin DJ, Jovin TM (2002) Spectrally resolved fluorescence lifetime imaging microscopy. Appl Spectrosc 56:155–166
29. Glanzmann T, Ballini J-P, van den Bergh H, Wagnieres G (1999) Time-resolved spectrofluorometer for clinical tissue characterization during endoscopy. Rev Sci Instrum 70:4067–4077
30. Bird DK, Eliceiri KW, Fan C-H, White JG (2004) Simultaneous two-photon spectral and lifetime fluorescence microscopy. Appl Opt 43:5173–5182
31. Biskup C, Zimmer T, Benndorf K (2004) FRET between cardiac Na+ channel subunits measured with a confocal microscope and a streak camera. Nat Biotechnol 22:220–224
32. Biskup C, Zimmer T, Kelbauskas L, Hoffmann B, Klöcker N, Becker W, Bergmann A, Benndorf K (2007) Multi-dimensional fluorescence lifetime and FRET measurements. Microsc Res Tech 70:442–451
33. De Beule PAA, Dunsby C, Galletly NP, Stamp GW, Chu AC, Anand U, Anand P, Benham CD, Naylor A, French PMW (2007) A hyperspectral fluorescence lifetime probe for skin cancer diagnosis. Rev Sci Instrum 78:123101–123107
34. Luong AK, Gradinaru CC, Chandler DW, Hayden CC (2005) Simultaneous time- and wavelength-resolved fluorescence microscopy of single molecules. J Phys Chem B 109:15691–15698
35. Pelet S, Previte MJR, Kim D, Kim KH, Su T-TJ, So PTC (2006) Frequency domain lifetime and spectral imaging microscopy. Microsc Res Tech 69:861–874
36. Spriet C, Trinel D, Laffray S, Landry M, Vandenbunder B, Heliot L, Barbillat J (2006) Setup of a fluorescence lifetime and spectral correlated acquisition system for two-photon microscopy. Rev Sci Instrum 77:123702–123706
37. Spriet C, Trinel D, Waharte F, Deslee D, Vandenbunder B, Barbillat J, Héliot L (2007) Correlated fluorescence lifetime and spectral measurements in living cells. Microsc Res Tech 70:85–94
38. Benninger RKP., McGinty J, Hofmann O, Requejo-Isidro J, Munro I, Elson DS, Dunsby C, Onfelt B, Davis DM, Neil MAA, deMello AJ, French PMW (2005) Application of multidimensional fluorescence imaging to microfluidic devices. In: conference on lasers and electro-optics Europe. p 620
39. Chorvat D, Chorvatova A (2006) Spectrally resolved time-correlated single photon counting: a novel approach for characterization of endogenous fluorescence in isolated cardiac myocytes. Eur Biophys J 36:73–83
40. Dumas D, Gaborit N, Grossin L, Riquelme B, Gigant-Huselstein C, Isla Nd, Gillet P, Netter P, Stoltz JF (2004) Spectral and lifetime fluorescence imaging microscopies: New modalities of multiphoton microscopy applied to tissue or cell engineering. Biorheology 41:459–467
41. Becker W, Bergmann A, Hink MA, König K, Benndorf K, Biskup C (2004) Fluorescence

lifetime imaging by time-correlated single-photon counting. Microsc Res Tech 63:58–66

42. Hanson KM, Behne MJ, Barry NP, Mauro TM, Gratton E, Clegg RM (2002) Two-photon fluorescence lifetime imaging of the skin stratum corneum pH gradient. Biophys J 83:1682–1690
43. Grant DM, Elson DS, Schimpf D, Dunsby C, Requejo-Isidro J, Auksorius E, Munro I, Neil MA, French PM, Nye E, Stamp G, Courtney P (2005) Optically sectioned fluorescence lifetime imaging using a Nipkow disk microscope and a tunable ultrafast continuum excitation source. Opt Lett 30:3353–3355
44. Holub O, Seufferheld MJ, Gohlke C, Govindjee, Clegg RM (2000) Fluorescence lifetime imaging (FLI) in real-time—a new technique in photosynthesis research. Photosynthetica 38:581–599
45. Mizeret J, Stepinac T, Hansroul M, Studzinski A, van den Bergh H, Wagnieres G (1999) Instrumentation for real-time fluorescence lifetime imaging in endoscopy. Rev Sci Instrum 70:4689–4701
46. Krishnan RV, Saitoh H, Terada H, Centonze VE, Herman B (2003) Development of a multiphoton fluorescence lifetime imaging microscopy system using a streak camera. Rev Sci Instrum 74:2714–2721
47. Chandler D, Majumdar Z, Heiss G, Clegg R (2006) Ruby crystal for demonstrating time- and frequency-domain methods of fluorescence lifetime measurements. J Fluoresc 16:793–807
48. vandeVen M, Ameloot M, Valeur B, Boens N (2005) Pitfalls and their remedies in time-resolved fluorescence spectroscopy and microscopy. J Fluoresc 15:377–413
49. Dartigalongue T, Hache F (2003) Precise alignment of a longitudinal pockels cell for time-resolved circular dichroism experiments. J Opt Soc Am B 20:1780–1787
50. Lakowicz JR (2006) Principles of fluorescence spectroscopy, 3rd edn. Springer, New York
51. Boens N, Qin W, Basaric N, Hofkens J, Ameloot M, Pouget J, Lefevre J-P, Valeur B, Gratton E, vandeVen M, Silva ND, Engelborghs Y, Willaert K, Sillen A, Rumbles G, Phillips D, Visser AJWG, van Hoek A, Lakowicz JR, Malak H, Gryczynski I, Szabo AG, Krajcarski DT, Tamai N, Miura A (2007) Fluorescence lifetime standards for time and frequency domain fluorescence spectroscopy. Anal Chem 79:2137–2149
52. Malachowski GC, Clegg RM, Redford GI (2007) Analytic solutions to modelling exponential and harmonic functions using chebyshev polynomials: fitting frequency-domain lifetime images with photobleaching. J Microsc 228:282–295
53. Papageorgiou GC, Govindjee (2004) Chlorophyll a fluorescence: a signature of photosynthesis, vol 19. Springer, Norwell
54. Lazár D (1999) Chlorophyll a fluorescence induction. BBA-Bioenergetics 1412:1–28
55. (2012) http://www.andor.com/
56. Lichtenthaler HK (1988) Applications of chlorophyll fluorescence: in photosynthesis research, stress physiology, hydrobiology and remote sensing, 1st edn. Springer, Norwell
57. (2012) http://gratings.newport.com/handbook/handbook.asp
58. Singh S (1999) Diffraction gratings: aberrations and applications. Optic Laser Tech 31:195–218
59. Simon JM, Gil MA, Fantino AN (1986) Czerny-turner monochromator: astigmatism in the classical and in the crossed beam dispositions. Appl Opt 25:3715–3720
60. Rosfjord KM, Villalaz RA, Gaylord TK (2000) Constant-bandwidth scanning of the Czerny-turner monochromator. Appl Opt 39:568–572
61. Zimmermann T, Rietdorf J, Pepperkok R (2003) Spectral imaging and its applications in live cell microscopy. FEBS Lett 546:87–92
62. Zimmermann T (2005) Spectral imaging and linear unmixing in light microscopy. In: Advances in biochemical engineering/biotechnology. Springer, Berlin, pp 245–265
63. Kraus B, Ziegler M, Wolff H (2007) Linear fluorescence unmixing in cell biological research. In: Mendez-Vilas A, Diaz J (eds) Modern research and educational topics in microscopy. Formatex, Badajoz, pp 863–872

# Chapter 2

# Förster Resonance Energy Transfer and Trapping in Selected Systems: Analysis by Monte-Carlo Simulation*

**P. Bojarski, A. Synak, L. Kułak, S. Rangelowa-Jankowska, A. Kubicki, and B. Grobelna**

## Abstract

Monte-Carlo simulation method is described and applied as an efficient tool to analyze experimental data in the presence of energy transfer in selected systems, where the use of analytical approaches is limited or even impossible. Several numerical and physical problems accompanying Monte-Carlo simulation are addressed. It is shown that the Monte-Carlo simulation enables to obtain orientation factor in partly ordered systems and other important energy transfer parameters unavailable directly from experiments. It is shown how Monte-Carlo simulation can predict some important features of energy transport like its directional character in ordered media.

**Key words:** Non-radiative energy transfer, Homotransfer, Trapping, Aggregates, Monte-Carlo simulations, Time-resolved emission spectra, Emission anisotropy, Emission anisotropy decay, Orientation factor, Mean square displacement of excitation energy, Directional energy transfer, Uniaxially stretched polymer films.

## 1. Introduction

The phenomenon of non-radiative energy transfer has been of continuous interest for numerous scientists active in different areas like solid-state physics, modern material sciences, physical chemistry, biochemistry, or medical sciences (1–6). This fact is both due to the common presence of this phenomenon in many natural systems and due to numerous applications of energy transfer anywhere the detailed knowledge on molecular systems and their interactions at nanoscale is required. Typical manifestations of energy transfer in molecular systems include concentrated dye solutions, polymers, glasses doped with luminescent species, some tunable lasers,

*In memory of Professor Czeslaw Bojarski.

Wlodek M. Bujalowski (ed.), *Spectroscopic Methods of Analysis: Methods and Protocols*, Methods in Molecular Biology, vol. 875, DOI 10.1007/978-1-61779-806-1_2, © Springer Science+Business Media New York 2012

monomer–aggregate systems, Langmuir-Blodgett films, membranes, molecular logic gates, solar concentrators, natural and artificial photoreception units, etc. Energy transfer is also one of the most important techniques to probe biological species. Förster resonance energy transfer (FRET) is often applied to determine the distance or distance distribution between two or more fluorophores that are separated by 1–10 nm on a biopolymer. More qualitatively, FRET is commonly used as a proximity indicator between a donor and an acceptor as a result of aggregation of proteins or oligonucleotides. These and many other fascinating applications of FRET in biological sciences have been described in detail in (6, 7).

Contrary to radiative energy transfer, which does not involve any interaction between fluorophores, the non-radiative energy transfer takes place due to the interaction between an excited and usually unexcited molecule separated by a distance shorter than 10 nm. Mention should be made that non-radiative transfer competes with any luminescence process. In liquid solutions or gases energy transfer may take place as a result of molecular collisions. However, for us more interesting are usually other mechanisms of energy transfer, which allow the excitation to be transferred over much longer distances characteristic for the dimensions of many nano-objects.

In this contribution we will focus on explanations of mechanisms of non-radiative energy transfer, and based on several examples some important problems and interesting results of energy transfer in partly ordered and disordered systems will be described.

## 2. Materials

Fluorophores: 3,3′-Diethylthiacarbocyanine Iodide (DTCI), 3,3′-Diethyl-9-Methylthiacarbocyanine Iodide (MDTCI), 3,3′-Diethylthiacyanine Iodide (DTTHCI), 3,3′-Diethyloxacarbocyanine Iodide (DOCI), Riboflavin-5′-phosphate sodium (FMN), Rhodamine 6G (R6G), Rhodamine 101 (R101), Acridone as well as polyvinyl alcohol (PVA) used in the experimental part of this study were spectroscopically pure and were purchased from Fluka. The dyes were dissolved in 5% ethanol–water solution of PVA at temperature $T = 323$ K to obtain homogeneous solution. Then samples were seasoned to allow water evaporation for about a week before stretching and measurements. Uniaxial stretching of the films was performed using homemade device at a temperature $T = 313$ K. The fluorescence signal should be recorded from the central small area of the sample to avoid effects connected with possible non-uniform film stretching. The optical density of the film should be quite low, possibly around 0.1 to neglect the inner filter effects (8).

## 3. Methods

### *3.1. Experimental Techniques*

1. Absorption spectra were measured using Shimadzu 1650 or Zeiss M40 spectrophotometer.
2. Fluorescence spectra and quantum yields were measured upon front-face excitation and observation of sample fluorescence (magic angle mode) using spectrofluorometer constructed earlier in our laboratory (9).
3. Steady-state emission anisotropy was recorded with two-channel single photon counting apparatus described at first in ref. (10) and then in ref. (11). The polarization plane of the exciting light was set vertical and was parallel to the direction of polymer stretching. In this unique equipment both components of fluorescence light intensity, i.e., the parallel and perpendicular one to the direction of the polarization plane of the excitation light are measured simultaneously, which makes it possible to eliminate the effect of fluctuations of incident light intensity and accelerates significantly the measurement. The measurements were performed upon front-face excitation and observation of sample fluorescence (magic angle mode).
4. Time-resolved emission anisotropy measurements were obtained with the Picoquant PQ 200 pulsed fluorometer using front-face configuration in the Center for Commercialization of Fluorescence Technology, University of North Texas. The polarization plane of the exciting light was set vertical and parallel to the direction of polymer stretching axis. Fluorescence light was recorded by the MCP-PMT (R3809) microchannel plate photomultiplier (Hamamatsu Photonics, Japan). Similar measurements have been also performed with a couple of pulsed spectrofluorometers assembled in our Institute (12, 13).
5. Streak camera-based system with solid-state laser and optical parametric generator as a source of light pulses was used for measurements of fluorescence spectra time evolution in ps and ns range. The laser system consists of: PL 2143A/SS laser and PG 401/SH optical parametric generator (Expla, Lithuania). The 2501S Spectrograph, Bruker Optics Inc. (USA) and the C4334-01 Streak Camera, Hamamatsu (Japan), constitute a central part of the detection system. The measurement is controlled by an original Hamamatsu HPDTA software which allows the real-time data analysis. This equipment enables any excitation and observation from 220 nm to 850 nm with 20–30 ps FWHM pulses. The apparatus has been described in detail in ref. (14).

### 3.2. Theoretical Methods

#### 3.2.1. Physical Mechanisms of Non-radiative Energy Transfer

Non-radiative energy transfer has been extensively studied theoretically, numerically, and experimentally (1, 2, 5, 15). It has been found that a variety of mechanisms is possible and that singlet and/or triplet states of interacting molecules can be involved in energy transfer. Non-radiative energy transfer is treated as a resonant process. By $D$ let us denote the donor molecule, whereas $A$ will stand for the acceptor. Selected mechanisms of energy transfer can occur according to the following schemes (1–4, 7, 15):

$$D_1^S + A_0^S = D_0^S + A_1^S, \tag{1}$$

$$D_1^S + A_1^T = D_0^S + A_2^T, \tag{2}$$

$$D_1^T + A_0^S = D_0^S + A_1^S, \tag{3}$$

$$D_1^T + A_0^S = D_0^S + A_1^T, \tag{4}$$

$$D_1^S + A_0^S = D_0^S + A_1^T, \tag{5}$$

$$D_1^T + D_1^T = D_1^S + D_0^S. \tag{6}$$

The superscript $S$ and $T$ corresponds to singlet and triplet state, respectively. The numbers in the subscript correspond to the ground state (0), first excited (1), or second excited (2) electronic state.

Besides the preservation of the total energy during energy transfer an equally important factor affecting strongly the transfer efficiency is the spin conservation of donor–acceptor system as a whole as well as individual spin of donor and acceptor. Scheme 1 represents singlet–singlet long-range energy transfer which is the most common process very broadly documented in literature (for details see reviews and the references therein (1–7, 15)). Scheme 2 represents in turn a less common process though it is also allowed by spin conservation rules. This process is much less common because acceptor molecules must stay initially also in the excited triplet state (at sufficiently high concentration), which requires specific experimental conditions and/or setup. Perylene (donor)–phenantrene (acceptor) in a rigid matrix is a typical example (16).

Long-range process represented by Scheme 3 can be observed in the case of strong spin-orbit coupling as it weakens the spin selection rule due to mixing the $S$ and $T$ excited states (17). This kind of energy transfer can be sometimes detected both in rigid matrices and liquid solutions (18, 19, 99). Contrary to processes 1–3, scheme 4 allows for energy transfer despite the change in the acceptor spin. This triplet–triplet energy transfer takes place due to the exchange resonant mechanism (20) and it is often diffusion controlled or diffusion affected (21), though first observations of this type of transfer have been recorded in rigid cooled solutions (22). Process 5 called singlet–triplet energy transfer is rather uncommon and usually ineffective, however it has been observed for such donors and acceptors which contain heavy atoms (23). Scheme 6 represents a different class of energy transfer processes, where a great number of donors, i.e., chemically identical molecules remain excited at the same time. As a result of energy homotransfer the excitation energy is collected by a single molecule in a different excited state of higher energy, while the second molecule becomes deactivated. To observe this kind of process it is necessary to populate significantly excited triplet states of concentrated rigid solutions, crystals or some polymers. In this way we can obtain blue shifted light compared to the excitation wavelength of the system (24). Many other interesting energy transfer mechanisms involving two or more excited molecules have also been discovered, however due to space limitations they cannot be herein reviewed.

Generally, non-radiative energy transfer can take place as a result of coulombic or exchange interaction. Coulombic interaction plays a major role if the transition between the ground and excited state of acceptor is allowed as far as spin selection rule is concerned. On the other hand, acceptor transitions forbidden by spin selection rule occur mostly as a result of exchange interaction.

Further, we will focus on the case of the most common long-range singlet–singlet energy transfer. Modern theories of energy transfer originate directly from the concepts developed by Förster (25). In his model it is assumed that donor and acceptor molecules are separated by a distance $R$ and are statistically distributed in space. Next, donor–acceptor pair can remain only in two states in which one of the molecules is electronically excited, while the other one must be unexcited. According to the quantum mechanics the rate constant (probability per time unit) is in the first approximation of the perturbation theory expressed as:

$$k_{DA} = \frac{4\pi^2}{h}\left|\left\langle \Psi_I \middle| \hat{H} \middle| \Psi_F \right\rangle\right|^2 \rho(E), \tag{7}$$

where $\Psi_I$ stands for the initial and $\Psi_F$ for the final state of the donor–acceptor system, $\rho(E)$ denotes the density of the final states with energy $E$ and the matrix element $U_{DA} = \left\langle \Psi_I \middle| \hat{H} \middle| \Psi_F \right\rangle$ represents the coulombic interaction between virtual charge distributions, which can usually be limited to dipole–dipole interaction. In this case:

$$U_{DA} = \frac{1}{n^2 R^3}\left[\vec{m}_D \cdot \vec{m}_A - \frac{3}{R^2}(\vec{m}_D \cdot \vec{R})(\vec{m}_A \cdot \vec{R})\right], \tag{8}$$

where $\vec{m}_D$ and $\vec{m}_A$ are transition dipole moments of donor and acceptor, respectively.

Interaction energy in the case of Eq. 8 takes the following form:

$$U_{DA}^2 = \frac{\kappa^2 \left|\vec{m}_D\right|^2 \cdot \left|\vec{m}_A\right|^2}{n^4 R^6}, \tag{9}$$

where $\kappa^2$ is so-called real orientation factor describing the mutual spatial orientation of donor transition moment in emission and acceptor transition moment in absorption (for a single donor–acceptor pair). It follows from Eq. 9 that the rate constant for energy transfer is proportional to $R^{-6}$.

The mechanism of non-radiative resonance energy transfer depends also on the relation between the interaction energy $U$, Franck–Condon energy, $\Delta\varepsilon$, resulting from the change in the equilibrium nuclei configuration in the excited and ground state and the energetic width of the single vibrational level $\Delta\varepsilon'$. Usually, three rather well-defined types of coupling are introduced: strong coupling ($\Delta\varepsilon' \ll \Delta\varepsilon \ll |U|$), weak coupling ($\Delta\varepsilon' \ll |U| \ll \Delta\varepsilon$) and a very weak coupling ($|U| \ll \Delta\varepsilon' \ll \Delta\varepsilon$).

Herein, we will focus on the case of a very weak coupling leading to the well-known Förster case. In this case the full relaxation over vibrational levels in the excited state occurs before the act of energy transfer. The total rate constant for energy transfer for continuous spectra of polyatomic molecules in condensed phase can be expressed by the following practical formula (25, 26):

$$k_{DA}(dd) = \frac{9 < \kappa^2 > (\ln 10)}{128\pi^5 n^4 N' \tau_{0D}} \frac{1}{R^6} \int f_D(v)\varepsilon_A(v)\frac{dv}{v^4} = \frac{1}{\tau_{0D}}\left(\frac{R_0}{R}\right)^6, \quad (10)$$

where $n$ is the refractive index of the medium, $N'$denotes the number of fluorophores in 1 mmol, $\tau_{0D}$ is the mean donor fluorescence lifetime (in the absence of acceptor), $v$ is the wavenumber, $\varepsilon_A(v)$ is the molar decimal extinction coefficient of acceptor, $f_D(v)$ is the normalized fluorescence spectral distribution $\int f_D(v)dv = 1$ and $\langle\kappa^2\rangle$ is the orientation factor averaged over all molecular configurations (averaged orientation factor), where $\kappa^2$ is a real orientation factor defined for a given donor–acceptor pair as:

$$\kappa^2 = (\cos\phi_{DA} - 3\cos\phi_D \cos\phi_A)^2, \quad (11)$$

where $\phi_{DA}$ denotes the angle between the directions of donor transition moment in emission and acceptor transition moment in absorption, $\phi_D$ — the angle between donor transition moment in emission and the axis connecting the interacting molecules, $\phi_A$ — the angle between acceptor transition moment in absorption and the axis connecting the interacting molecules.

The values of $\kappa^2$ may change from 0 to 4. The knowledge of $\kappa^2$ or its averaged value $\langle\kappa^2\rangle$ may be crucial to understand the properties of energy transfer especially in systems exhibiting some spatial order. Usually, the averaging of $\kappa^2$ leads to strong narrowing of the possible values range of $\langle\kappa^2\rangle$. In the case of disordered system (randomly distributed fluorophores) $\langle\kappa^2\rangle$ may change from 0.476

(the case of statistically distributed immobile dipoles) to 2/3 (fast rotating dipoles) (27–29). The intermediate values characteristic for many realistic systems can be obtained from polarization data and various interpolation methods (6, 30, 31) as well as from Monte-Carlo simulations (32, 33). In the case of partly ordered systems $\langle\kappa^2\rangle$ may change in a much wider range and it is generally hard to calculate analytically $\langle\kappa^2\rangle$. However, the problem can be solved by Monte-Carlo method provided there is some knowledge on the geometry of the system (for example, spatial distribution functions of donors and acceptors are known) (34). This issue will be addressed in more detail in subsequent sections.

$R_0$ appearing in Eq. 10 is so-called critical distance or critical radius for energy transfer and it has a physical meaning of such a distance at which the transfer probability is equal to ½. In other words it is such a distance at which the probability of energy transfer is equal to the sum of probabilities of all other processes deactivating the excited state of the donor. $R_0$ is a very important characteristics of a given donor–acceptor system as it enables quantitative evaluation and prediction of energy transfer effect, is indispensible to determine distance or distance distribution in macromolecules, and allows for comparisons between different donor–acceptor systems. Determination of $R_0$ is rather easy from the measurements of donor emission and acceptor absorption spectra as indicated by the following formula:

$$R_0^6 = \frac{9\langle\kappa^2\rangle(\ln 10)\eta_{0D}}{128\pi^5 n^4 N'} I_\nu(D, A), \quad (12)$$

where:

$$I_\nu(D, A) = \int_0^\infty f_D(\nu)\varepsilon_A(\nu)\frac{\mathrm{d}\nu}{\nu^4} \quad (13)$$

is the overlap integral and $\eta_{0D}$ is the absolute donor fluorescence quantum yield in the absence of acceptor.

In some cases critical distance can be also quite easily determined as a best fit parameter from fluorescence intensity decays of donors in the presence of acceptors (35) or from concentration dependence of polarization data (36). However, generally one should be cautious with such a determination in view of possible artifacts. Our preference is to determine critical distance from spectral data if they are only available.

#### 3.2.2. Multistep Energy Transfer and Trapping

Original Förster model is a commonly accepted milestone on the way to the most recent developments. It has evolved in many different ways and important progress concerned, for example, multistep energy transfer followed by its trapping by aggregates, systems of different dimensionality, donor–acceptor systems of particular spatial distribution of chromophores, diffusion affected systems, energy transfer in different nanostructures, energy transfer enhanced by interactions with plasmons on metallic nanofilms, metallic islands, and many others (2–6, 12, 37, 38). It is impossible to describe all these contributions in this particular paper due to space limitations, therefore this time we will focus on multistep energy transfer and trapping in disordered and some partly ordered systems.

In Förster model it is assumed that the energy transfer takes place directly from an excited donor to unexcited acceptor. This situation takes place, for example, for donor–acceptor pairs linked to biologically active macromolecules or in such donor–acceptor solutions for which the concentration of donor is much smaller than that of acceptors. However, solutions, glasses, polymers, gel matrices, photosynthetic units, and many other media doped with high concentration of fluorescent molecules do not fulfill this requirement of the Förster model. In such a case of concentrated donor–acceptor solution excitation energy can migrate within a set of chemically identical chromophores an arbitrary number of times, before it is trapped by a chemically different molecule. This process can symbolically be represented as follows:

$$D^* + D + \cdots + A \rightarrow D + D^* + \cdots + A \rightarrow \cdots \rightarrow D + D + \cdots + A^* . \tag{14}$$

If homotransfer (energy migration) occurs as an exclusive transfer process (no chemically different acceptors), then only concentration depolarization of fluorescence or emission anisotropy decay measured at several fluorophore concentrations can confirm its presence. Non-radiative energy migration does not affect fluorescence lifetime, fluorescence quantum yield or fluorescence intensity decay (2, 32), but it affects strongly the emission anisotropy since only molecules primarily excited by light absorption contribute to emission anisotropy (2, 4, 32, 39, 40). From the previous sentence its follows also that the homotransfer (energy migration) and the self-quenching are not the same but two different phenomena which is often a source of misunderstanding in the literature. Self-quenching leading to the decrease of fluorescence quantum yield and lifetime shortening is caused by the presence of aggregates at sufficiently high fluorophore concentrations.

Usually, we deal with multicomponent systems, which allow also other than polarization experimental techniques to be effectively used to study homotransfer and trapping (2, 4). The homotransfer process has been originally described within the framework of so-called hopping model which is based on a set of specific kinetic equations (41). Then, a number of improvements and some different formalisms have been applied to develop the description of systems with multistep energy migration and trapping in disordered systems (42–58). In particular, the process of reverse energy transfer from acceptor or imperfect trap (fluorescent aggregate) to the donor ensemble has been taken into account (48–52, 56–58). The models mentioned (42–58) have been successfully verified in many one- and two-component disordered systems and may be considered standards to which experimental data obtained for more complex systems are often referred (4, 57).

Nowadays, a lot of attention is paid to such media in which fluorophores and their aggregates are not uniformly distributed, but are organized in a specific way. In such a case, the description of multistep energy transfer and trapping is even far more complex and it cannot be directly expressed in terms of analytical theories. However, if we have some knowledge on fluorophores distribution in a medium or local concentrations and organization of the dyes, Monte-Carlo method can successfully be used to obtain information on the mechanism of energy transfer in practically any medium. On the other hand, often, local concentrations of the dyes or their organization can be determined based on the results of Monte-Carlo simulation of energy transport. This is, for example, the case of gel matrices with nanopores, in which dye molecules are incorporated up to extremely high concentrations despite the low average fluorophore bulk concentration. One of the simplest case of effective use of Monte-Carlo simulation is also the uniaxially stretched polymer film, which allows for high-precision control over the elongated fluorophores distribution.

Below we present in more detail some procedures and examples of Monte-Carlo analysis in this medium.

### 3.3. Monte-Carlo Simulation Method

Basic parameters used in the Monte-Carlo simulation to investigate excitation energy transfer in the two-component systems are: concentrations of acceptors and donors, $C_A$, $C_D$, respectively, critical radii for the energy transfer for homotransfer between donors, donor–acceptor transfer, and homotransfer between acceptors: $R_0^{DD}, R_0^{DA}, R_0^{AA}$, respectively, stretching coefficient of polymer film $R_S$, linear dichroism ratio $R_D$ and the number of fluorophores ($N$ donors and $M$ acceptors). Using a uniform pseudorandom numbers generator the molecules are randomly distributed in a cube representing a given simulated system and the spatial orientations of transition dipole moments in the absorption and emission are generated. Each molecular configuration $\Re = (x_1, x_2, ..., x_{N+M})$, where $x_i = \left(\vec{r}_i, \Omega_i, E_i\right)$, of the system is characterized by the location of molecules $\vec{r_i}$ and orientation $\Omega_i$ representing the set of angular coordinates, $E_i$ is the energy of excited electronic state.

Polymer matrix is stretched along the selected axis, for example the $z$ axis of the coordinate system. Let us denote by $\phi_j$ the angle between the direction of transition dipole moment in the

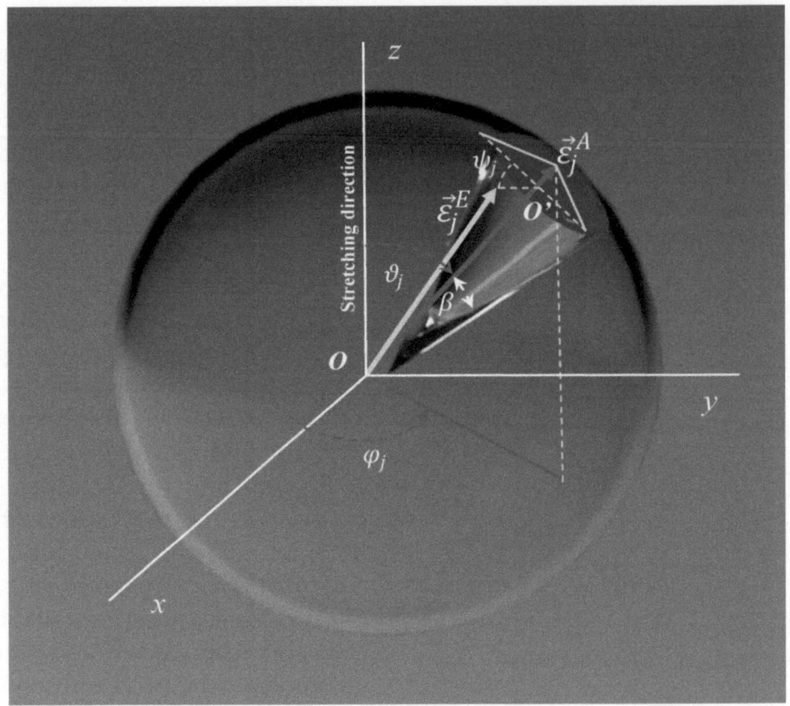


Fig. 1. The concept of the orientation factor for a molecular donor–acceptor pair participating in energy transfer.

absorption $\vec{\varepsilon}_j^A$ projected on the surface $Oxy$ and the axis $x$. $\vartheta_j$ is the angle between transition moment $\vec{\varepsilon}_j^A$ and the chosen axis $z$, and $\beta$ denotes the angle between transition dipole moments in the absorption and the emission. In the case of unstretched polymer matrix both angles are described by the uniform distribution of transition moments orientations on a unit sphere, i.e., the variable $\cos\vartheta_j$ has uniform distribution on the interval $[-1, 1]$ and $\phi_j$ on the interval $[0, 2\pi]$, respectively. The angle $\beta$ for different fluorophores can vary from 0 to 90°. The described geometry of the system is visualized in Fig. 1.

In the case of stretched polymer films we assume that doped dye molecules are located inside the spherical cavities (with the random distribution) present in the film. During the stretching of the film along the $z$ axis spherical cavities are deformed into ellipsoids.

According to the definition of the stretching coefficient $R_s$ of a polymer film which is equal to the ratio of the length of ellipsoid semi-axes formed by the deformation from the spherical cavity:

$$R_S = b/a \tag{15}$$

and assuming that the volume of the cavity with radius $r$ does not change after stretching of the polymer film, one obtains the following relation between the length of the ellipsoid semi-axes and the coefficient of stretching (34):

$$a = \frac{r}{\sqrt[3]{R_S}}, \tag{16}$$

$$b = r\sqrt[3]{R_S^2}. \tag{17}$$

The Monte-Carlo simulation is performed in the unit cube, which is stretched according to the above equations. Within a very good approximation we can tie distances among molecules before and after stretching as follows:

$$x'_{ij} = \frac{x_{ij}}{\sqrt[3]{R_S}}, \tag{18}$$

$$y'_{ij} = \frac{y_{ij}}{\sqrt[3]{R_S}}, \tag{19}$$

$$z'_{ij} = z_{ij}\sqrt[3]{R_S^2}. \tag{20}$$

In the Monte-Carlo simulation the value of the angle $\vartheta_j$ for the $j$-th molecule in uniaxially stretched polymer film, is obtained by a method of inverting probability distribution, i.e. the random number $r_j$ is generated from uniform probability distribution on the interval $[-1, 1]$. Taking into account

$$r_j = \int_0^{\vartheta_j} f_\vartheta(\omega)\mathrm{d}\omega, \tag{21}$$

and using the explicit form of the Tanizaki distribution function (59)

$$f_\vartheta(\vartheta_j) = R_S^2 \sin\vartheta_j \left[1 + (R_S^2 - 1)\sin^2\vartheta_j\right]^{-3/2}, \tag{22}$$

we obtain (60):

$$\cos\vartheta_j = \frac{R_S r_j}{\sqrt{1 + (R_S^2 - 1)r_j^2}}. \tag{23}$$

The distance and dipole orientation-dependent transfer rate from the $j$-th $X$ molecule to the $i$-th Y molecule ($X, \mathrm{Y} \in \{D, A\}$) is denoted by $w_{x_i x_j}^{X\mathrm{Y}}$. For the Förster mechanism (resonance energy transfer) and transfer rate independent of energy $E_i$, the quantity $w_{x_i x_j}^{X\mathrm{Y}}$ is given by:

$$w_{x_i x_j}^{X\mathrm{Y}} = \frac{\kappa_{ij}^2}{\tau_{0X}}\left(\frac{R_0^{X\mathrm{Y}}}{r_{ij}}\right), \tag{24}$$

where $\kappa_{ij}^2$ is the real orientation factor for a given molecular pair, $r_{ij}$ is the distance between the $i$-th and $j$-th molecule, and $R_0^{X\mathrm{Y}}$ is the Förster radius for $X$–Y excitation transfer. The constants $\tau_{0D}$ and

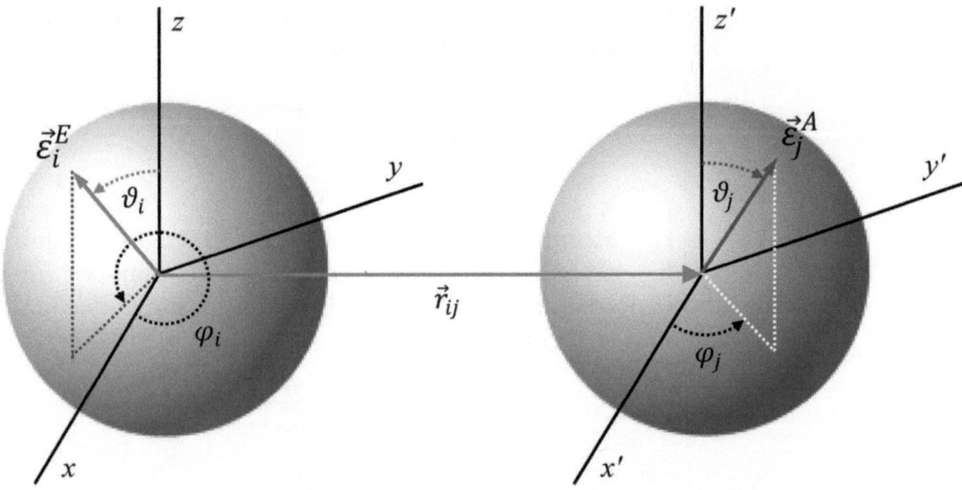


Fig. 2. Geometrical relations between the directions of transition dipole moments for a given fluorophore in disordered and partly ordered systems.

$\tau_{0A}$ are the lifetimes of donor and acceptor molecules, respectively, measured in the absence of the intermolecular transfer.

For the purpose of simulation, the orientation factor, $\kappa_{ij}^2$, is defined as:

$$\kappa_{ij}^2 = \left[\vec{\varepsilon}_i^E \cdot \vec{\varepsilon}_j^A - 3\left(\vec{\varepsilon}_i^E \cdot \vec{r}_{ij}\right)\left(\vec{\varepsilon}_j^A \cdot \vec{r}_{ij}\right)\right]^2 \tag{25}$$

(equivalent to the previously given definition (Eq. 11)), where $\vec{r}_{ij}$ denotes the distance unitary vector between the interacting molecules, $\vec{\varepsilon}_i^E$and $\vec{\varepsilon}_j^A$ are transition moments in the emission and absorption, respectively. Including a real value of the orientation factor $\kappa_{ij}^2$ into the transfer rate allows for investigations of the influence of a mutual orientation of transition moments on excitation energy transfer. The concept of $\kappa_{ij}^2$ is given in Fig. 2.

The dynamics of the simulated system, i.e. conditional probability that an excitation is found on $i$-th molecule at time $t$, $P_{x_ix_j}(t) \equiv P\left(\vec{r}_i, E_i; t|\vec{r}_j, E_j; t=0\right)$, provided it was at time $t=0$ on $j$-th molecule, for a given fixed molecular configuration, obeys the following *master equation* (58, 61):

$$\frac{\mathrm{d}P_{x_ix_j}(t)}{\mathrm{d}t} = -\frac{1}{\tau_{0D}}P_{x_ix_j}(t) + \sum_{k=1,k\neq i}^{N} w_{x_ix_k}^{DD}P_{x_kx_j}(t)$$

$$- \sum_{k=1,k\neq i}^{N} w_{x_kx_i}^{DD}P_{x_ix_j}(t) - \sum_{k=N+1}^{N+M} w_{x_kx_i}^{DA}P_{x_ix_j}(t), 1 \leq i \leq N,$$

$$\frac{\mathrm{d}P_{x_ix_j}(t)}{\mathrm{d}t} = -\frac{1}{\tau_{0A}}P_{x_ix_j}(t) + \sum_{k=1}^{N} w_{x_ix_k}^{DA}P_{x_kx_j}(t) + \sum_{k=N+1,k\neq i}^{N+M} w_{x_ix_k}^{AA}P_{x_kx_j}(t)$$
$$- \sum_{k=N+1,k\neq i}^{N+M} w_{x_kx_i}^{AA}P_{x_ix_j}(t),\ N+1 \leq i \leq N+M, \qquad (26)$$

with the initial condition $P_{x_ix_j}(0) \equiv \delta_{ij}$, where $\delta_{ij}$ is the Kronecker delta.

The step-by-step Monte-Carlo simulation method (Metropolis algorithm) (62) consists in the use of the random-number generator for the cyclic formulation of answers to the following questions: (1) When does any of the assumed processes occur? (2) What a process it is?

To identify originally excited molecule the photoselection procedure is applied (60). Molecules originally excited are selected in accordance to the $\cos^2\vartheta_j$ probability distribution, what means, that with the largest probability those molecules become excited which have the transition moments parallel to the direction of the electric vector of the excitation light. Hence, a random number $r_j$ from the uniform distribution on [0,1] interval is generated. If the following inequality is fulfilled:

$$\cos^2\vartheta_j \geq r_j \qquad (27)$$

the $j$-th molecule becomes excited. The molecules breaking this inequality are rejected from the set of originally excited molecules. The idea of photoselection is presented in Fig. 3. Linearly polarized light excites, more probably, those molecules whose transition moments (in absorption) form smaller angles with the direction of the electric vector $\vec{E}$.

In the Monte-Carlo simulation the deactivation of $j$-th molecule may occur through the following processes (34) in the donor ensemble:

($P_1$) process 1: $D^* \to D$, photon emission or non-radiative energy conversion, with the rate $1/\tau_{0D}$;

($P_2$) process 2: $D^* + D \to D + D^*$, energy migration (energy transfer to the molecules of the same kind), with the transfer rate $w_{x_ix_j}^{DD}$;

($P_3$) process 3: $D^* + A \to D + A^*$ non-radiative energy transfer from the excited donor to an acceptor, with the transfer rate $w_{x_ix_j}^{DA}$;

In the acceptor ensemble:

($P_1'$) process 1′: $A^* \to A$, photon emission or non-radiative energy conversion, with the rate $1/\tau_{0A}$;

($P_2'$) process 2′: $A^* + A \to A + A^*$, energy migration (energy transfer to the molecules of the same kind), with the transfer rate $w_{x_ix_j}^{AA}$;

Next, for the $j$-th excited donor molecule the values of the following total transfer rates are calculated:

$$c_{1j} = 1/\tau_{0D}, \qquad (28)$$

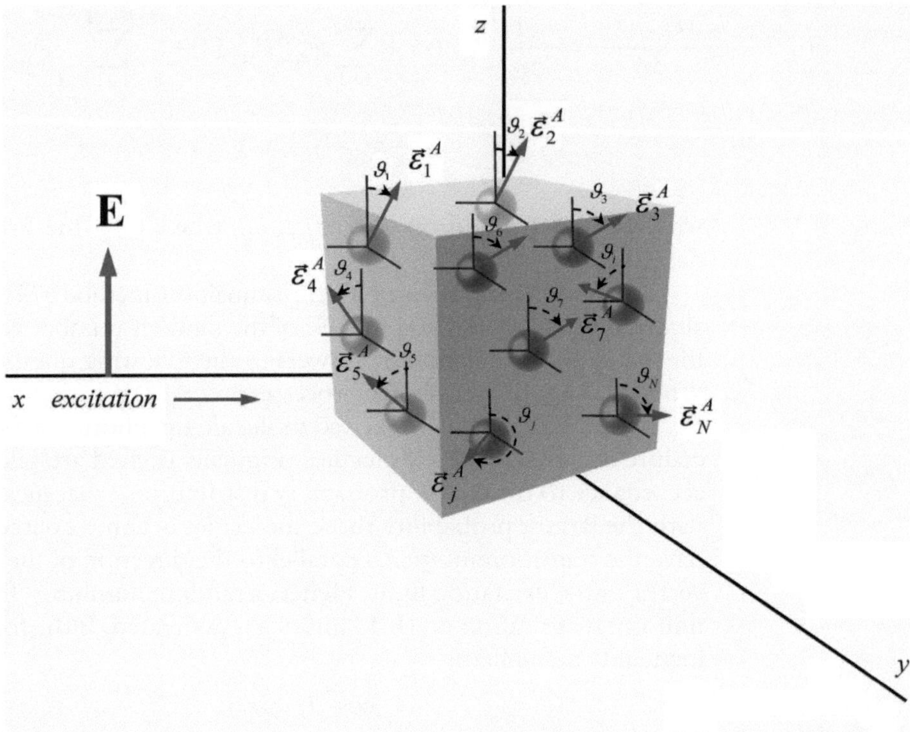


Fig. 3. Illustration of the photoselection phenomenon. Excitation with linearly polarized light is more effective for molecules transition moments of which form smaller angles with the direction of the electric vector of the exciting light.

$$c_{2j} = \sum_{i=1, i \neq j}^{N} w_{x_i x_j}^{DD}, \tag{29}$$

$$c_{3j} = \sum_{i=N+1}^{N+M} w_{x_i x_j}^{DA}, \tag{30}$$

$$c_j = c_{1j} + c_{2j} + c_{3j}. \tag{31}$$

Otherwise, when $i$-th acceptor is excited, the values of:

$$c'_{1i} = 1/\tau_{0A}, \tag{32}$$

$$c'_{2i} = \sum_{j=N+1, i \neq j}^{N+M} w_{x_j x_i}^{AA}, \tag{33}$$

$$c'_i = c'_{1i} + c'_{2i}, \tag{34}$$

are calculated.

The time at which any of the investigated processes occurs is calculated by inverting the distribution function of the probability, $p_j(t, P_k)dt$, that if at time $t$ the $j$-th molecule is excited, then the process $P_k$ appears in the time interval $(t, t + \mathrm{d}t)$:

$$p_j(t) = \sum_{k=1}^{3} p_j(t, P_k) = c_j \exp(-c_j t). \tag{35}$$

Generating a random number $r_{1j}$ from the interval [0, 1] the time at which any process takes place is obtained by inverting the distribution function $p_j(t)$:

$$\int_0^{t_j} \mathrm{p_j}(t)\mathrm{d}t = r_{1j}, \quad \text{i.e. } t_j = -(1/c_j)\ln(1 - r_{1j}). \tag{36}$$

The same procedure can be applied to the excited acceptor. In this case all constants $c_{ij}$ should be replaced by $c'_{ji}, j = 1, 2$.

Next, the process that took place at time $t_j$ is identified. By generating next random number, $r_{2j}$, such a value of index $k$ can be found for which the following inequality is satisfied:

$$\sum_{i=1}^{k-1} c_{ij} < r_{2j}c_j \leq \sum_{j=1}^{k} c_{ij}, \quad k = 1, 2, 3. \tag{37}$$

If $k = 1$, then the activated molecule is quenched by a photon emission or non-radiative excitation energy conversion and it means that this pass of simulation is finished. If $k = 2$ or $k = 3$, the energy migration or energy transfer process takes place, and it is necessary to determine next activated molecule. For this reason third random number, $r_{3j}$, is generated and the value of index $n$ is found which fulfills one of the inequalities:

$$\sum_{i=1}^{n-1} w_{x_i x_j}^{DD} < r_{3j}c_{2j} \leq \sum_{i=1}^{n} w_{x_i x_j}^{DD}, \text{ for } \quad k = 2, n \leq N, \tag{38}$$

Or

$$\sum_{i=N+1}^{n-1} w_{x_i x_j}^{DA} < r_{3j}c_{3j} \leq \sum_{i=N+1}^{n} w_{x_i x_j}^{DA}, \text{ for } \quad k = 3, n > N \tag{39}$$

where $n$ is the number of next activated donor or acceptor. Then, after inserting the value of $n$ for the index $i$, the simulation goes to the next step. The simulation run is finished when, after several migration or energy transfer acts, the process with $k = 1$ occurs (photon emission or non-radiative energy conversion). After that simulation starts again (i.e., for a new donor and acceptor spatial and angular configuration). If an acceptor is excited, the number of the next excited molecule, $n$, is determined in a similar way. This algorithm can be easily modified depending on the number of fluorescent species, their distribution in a medium and additional processes which can take place in the system.

To eliminate inaccuracies connected with the effect of: (1) finite size of the generated system, (2) finite number of molecules in the simulation, and (3) boundary effects, we introduce periodical boundary condition (the simulation box is surrounded by replicas of itself) with the minimum image convention (the molecule interacts with another molecule or its periodic image, Fig. 4).

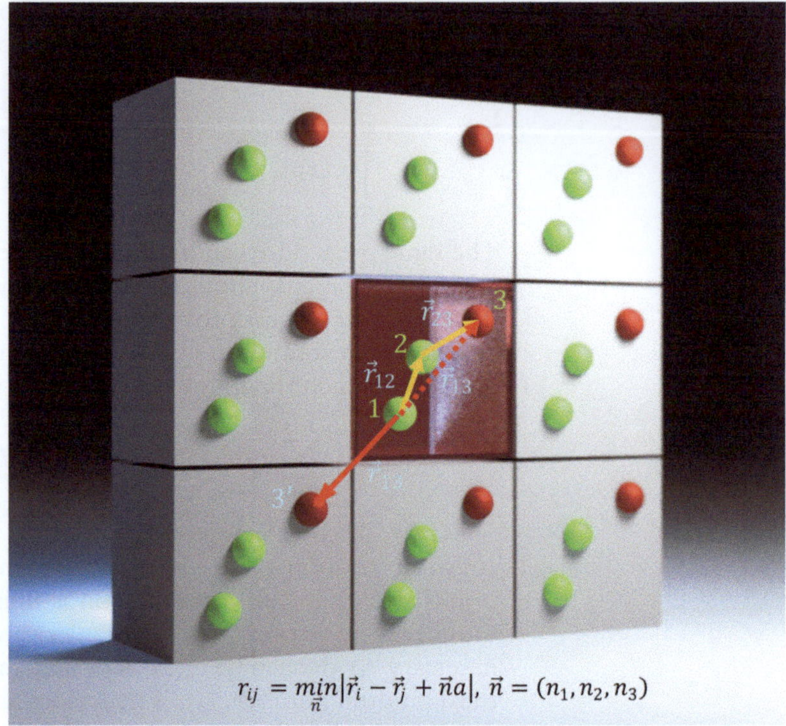


Fig. 4. Illustration of periodic boundary conditions. A given molecule can interact with another molecule from the same cube or with its periodic image from neighboring cube. This boundary condition is very important for such molecules which are located close to a cube wall.

The relative fluorescence quantum yield is determined by dividing the number of simulation runs completed with the donor fluorescence by the number of all runs.

The fluorescence intensity decay curve is obtained similarly to the real experiment, i.e., the time scale is divided into appropriate number of intervals (e.g. 4,096 channels) and, if photon emission at the time $t_i$ is "observed," then the number of photons is increased in the respective "channel." Finally, the normalized decay curve (histogram) is obtained as:

$$\Phi_D(\Delta t_k) = 1 - \left( \sum_{j=1}^{k} n_j / \sum_{j=1}^{k_{\max}} n_j \right), \tag{40}$$

$$\Delta t_k = (k/k_{\max})t, \tag{41}$$

where $n_k$ denotes the number of photons in the $k$-th channel, $k_{\max}$ is the total number of all channels.

The components of the emission intensity along the axes, $I_x(t)$, $I_y(t)$ and $I_z(t)$ are calculated as follows (13, 63):

$$I_z(t) = I_0 \cos^2 \vartheta_j(t), \tag{42}$$

$$I_x(t) = I_0 \sin^2\vartheta_j(t)\cos^2\phi_j(t), \tag{43}$$

$$I_y(t) = I_0 \sin^2\vartheta_j(t)\sin^2\phi_j(t), \tag{44}$$

where $\vartheta_j(t)$ and $\phi_j(t)$ are the respective angles of the $j$-th fluorophore emission transition moment.

The emission anisotropy as a function of time is obtained from the following definition:

$$r(t) = \left(I_z - I_y\right)/\left(I_x + I_y + I_z\right). \tag{45}$$

The final results of emission anisotropy and other observables of interest are calculated by averaging them over suitably big number of molecular configurations (simulation runs). During such one full simulation cycle, all characteristics can be calculated simultaneously. It is worth of note that not only typical fluorescence observables such as emission anisotropy, emission anisotropy decay, fluorescence quantum yield, mean fluorescence lifetime, or fluorescence intensity decay can be calculated but also other important characteristics which are not directly available during the measurements of any type such as averaged orientation factor, the number of excitation jumps, mean localization time of excitation on molecules, range of energy transfer in any direction, etc.

### 3.4. Examples and Discussion

For numerous partly ordered systems our knowledge on energy transport processes is highly insufficient. Uniaxially stretched polymer film is an example, for which the non-uniform angular distribution of transition dipole moment directions can be obtained in an easily controllable way. The films oriented in such a way have been among others successfully used to determine the directions of transition moments of many elongated fluorophores (64–67, 100, 101) as well as to study molecular conformations and distance distributions of donor–acceptor pairs linked to biologically important macromolecules (68–70).

Figure 5 shows typical concentration-dependent course of emission anisotropy for stretched and unstretched polyvinyl alcohol films with DTCI fluorophore (60). It can be seen from the figure that for unstretched films emission anisotropy strongly decreases with the increase in concentration as a result of homotransfer between randomly distributed fluorophores. Experimental data are well described by the Monte-Carlo simulation. The reason for such a course of emission anisotropy is that only primarily excited molecules contribute to emission anisotropy, and with the concentration increase the number of these molecules decreases rapidly as a result of excitation energy escape to other fluorophores emitting almost depolarized light (39, 40). A completely different behavior of emission anisotropy can be, however, observed for stretched films. This time much weaker concentration depolarization is visible for strongly ordered system. This fact can be explained as

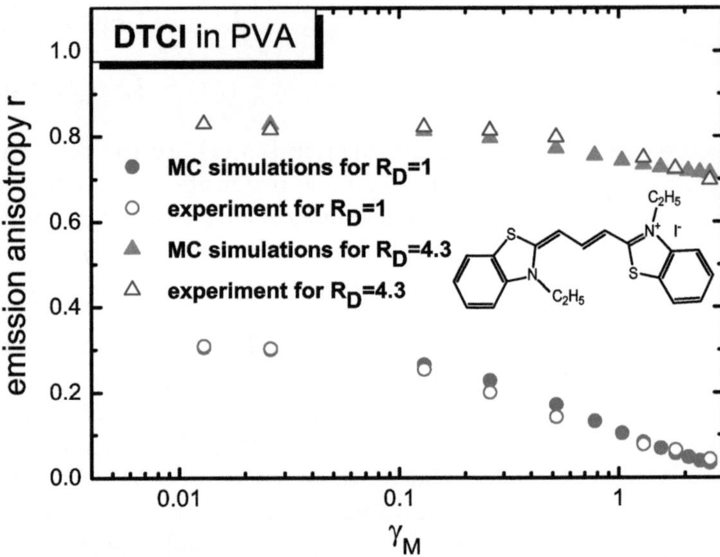


Fig. 5. Concentration depolarization of fluorescence for DTCI in PVA for disordered ($R_D = 1$) and ordered polymer film ($R_D = 4.3$). Triangles correspond to the ordered system, whereas circles to disordered system. Full triangles/circles correspond to the results of Monte-Carlo simulation and empty triangles/circles correspond to the results of experiment. Chemical structure of elongated DTCI molecule is inserted into the figure. $\gamma_M$ is the so-called reduced monomer concentration which is related to molar concentration as: $\gamma_M = (\sqrt{\pi}/2) C_M/C^0_{MM}$, where $C_M$ is the molar monomer concentration and $C^0_{MM}$ is the critical concentration for homotransfer corresponding to the respective critical distance.

follows: with the increasing concentration, more and more light is emitted by molecules others than those primarily excited due to homotransfer. However, as a result of preferential mutual orientation of transition dipole moments of molecules participating in homotransfer in stretched polymer films, these later excited molecules (via homotransfer) still strongly contribute to the emission anisotropy. Thus emission anisotropy decreases weakly in ordered system compared to the disordered one. A conclusion can be drawn that the assumption present in all modern theories of homotransfer in disordered systems that only primarily excited molecules contribute to emission anisotropy is absolutely invalid for systems exhibiting macroscopic partial order like uniaxially stretched films. Simultaneously, it can be seen that the absolute values of emission anisotropy in partly ordered stretched polymer matrix are much higher. This is a result of transition moments reorientation towards the direction of film stretching and their highly anisotropic angular distribution (given, for example, by Eq. 22).

Similar regularities have been observed for emission anisotropy decay for many elongated molecules in a polymer film (13, 63, 71). An example is given in Fig. 6a–c (71), which shows emission anisotropy decay of MDTCI in disordered and ordered polymer matrix. Extremely slow fluorescence anisotropy decay in stretched

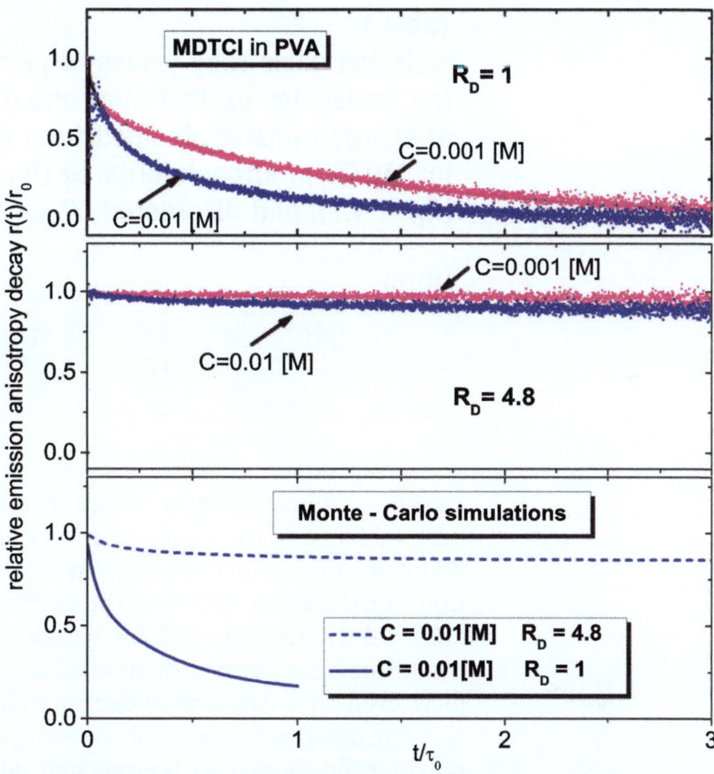


Fig. 6. Relative emission anisotropy decay of MDTCI in unstretched (**a**) and stretched (**b**) polymer films at high ($C = 0.01$ M) and moderate concentration ($C = 0.001$ M). Figures (**a**) and (**b**) show experimental data, whereas (**c**) shows selected Monte-Carlo results (for $C = 0.01$ M).

polymer matrix results again from the fact that energy homotransfer takes place between strongly directionally correlated transition moments. It means also that unlike disordered system in stretched films molecules excited via energy transfer still emit strongly polarized light. Figure 6c shows corresponding results of Monte-Carlo simulations of emission anisotropy decay for $C = 0.01$ M. The same regularities can be seen as in the case of experiment evidencing the correctness of our analysis. Very simple, but powerful and direct proof for that is another result obtained from Monte-Carlo simulations concerning the emission anisotropy emitted by primarily excited molecules, molecules excited after one and two steps of homotransfer. Table 1 shows several results of such Monte-Carlo calculations for MDTCI for three systems differing in linear dichroism ratio $R_D$. It can be seen that for disordered polymer film ($R_D = 1$) the light emitted even after one act of non-radiative energy transfer is virtually depolarized. The obtained values of emission anisotropy are consistent with experimental data and calculations performed by Jablonski (40). Further acts of transfer lead to practically total depolarization of fluorescence light.

**Table 1**
**Emission anisotropy values of primarily excited molecules ($r_0$), molecules excited after one ($r_1$) and two ($r_2$) acts of energy transfer obtained from Monte Carlo simulations for MDTCI in strongly ordered ($R_D = 4.8$), weakly ordered ($R_D = 1.3$), and disordered ($R_D = 1$) PVA films**

| MDTCI | | $r_0$ | $r_1$ | $r_2$ |
|---|---|---|---|---|
| $C = 0.01$ [M] | $R_D = 4.8$ | 0.81683 | 0.69511 | 0.69117 |
| | $R_D = 1.3$ | 0.57719 | 0.29147 | 0.28218 |
| | $R_D = 1$ | 0.38584 | 0.0084 | 0.0009 |

However, quite different situation can be observed for strongly ordered films. In this case excitation transfer does not lead to strong depolarization, which means that information on the spatial direction of the electric vector of the exciting light is kept in the system ($R_D = 4.8$). As expected for weakly organized films ($R_D = 1.3$) the fluorescence depolarization effect is significant, but not complete even after two acts of energy transfer.

The described case involves strong mechanical deformation of a polymer film doped with elongated fluorophores which easily orientate in a polymer film upon stretching. However, it occurs that even fluorophores, which weakly orientate in a polymer matrix (as a result of their nonlinear shape), exhibit quite detectable effect on emission anisotropy in the ordered matrix. Similarly, even small stretching degrees of polymers (10–20%) doped with elongated fluorophores lead to detectable emission anisotropy changes.

Figure 7 shows the concentration course of emission anisotropy for FMN (72), a member of flavins, which plays as a coenzyme and photoreceptor an important role in numerous biological processes (73–75). From FMN structure shown in Fig. 7, it can be seen that it is a nonlinear molecule which does not orientate effectively upon uniaxial stretching of the polymer film. Indeed, linear dichroism studies reveal that $R_D = 1.25$, despite extremely strong stretching of the film ($R_S = 7$). Formally good agreement is obtained between the experimental data and Monte-Carlo simulations for $R_D = 2.25$. However, for disordered systems excellent agreement between experimental data and Monte-Carlo is obtained for the same value of $R_D = 1$. This is an example that one should be very cautious during such comparisons for nonlinear molecules as the formally fitted value of $R_D$ and that experimentally determined in the stretched film are significantly different. This inconsistency arises in our opinion from the choice of a specific distribution function in Monte-Carlo simulations as for nonlinear molecules the mechanism of molecular reorientation upon film stretching is different than in the case of linear elongated molecules (64).

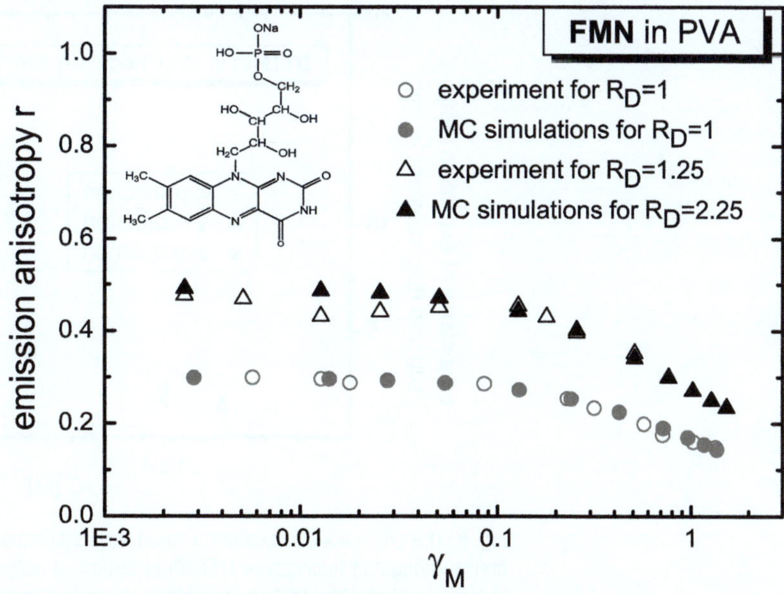


Fig. 7. The dependence of flavinmononucleotide emission anisotropy on the reduced monomer concentration of the fluorophore. Description the same as in Fig. 5.

Monte-Carlo simulations can provide also other valuable information on energy transfer properties and/or on the objects or systems studied. For example, in many systems information on the preferable spatial direction of the excitation energy expansion through its non-radiative transfer is important. Though such information cannot be derived directly from the measurement, it can be obtained from Monte-Carlo simulations through the values of the mean square displacement of the excitation energy and its projections to *x*, *y*, and *z* axes of the coordinate system. Figure 8 shows concentration dependence of these characteristics for an elongated DTTHCI molecule in a strongly ordered PVA matrix ($R_D = 3.5$). It can be seen from the figure that the values of $\sqrt{\langle z^2 \rangle}$are significantly higher at increased fluorophore concentrations, where homotransfer is effective, than those of $\sqrt{\langle x^2 \rangle}$ and $\sqrt{\langle y^2 \rangle}$ indicating that the range of excitation energy is longer along the *z* axis.

One should be aware that one-component system in which only homotransfer takes place does not reflect the properties of a real system at sufficiently high concentrations as a result of intermolecular aggregation. If the aggregates are formed in the ground state, their presence is first of all demonstrated by the change of absorption spectra profiles and formation of new bands with increasing concentration. In viscous solutions, polymers, glasses, Langmuir–Blodgett films, or hybrid materials such aggregates can sometimes emit their own fluorescence (76–94). In such a case also aggregate fluorescence characteristics can deliver important information on the photophysical properties of monomer-aggregate

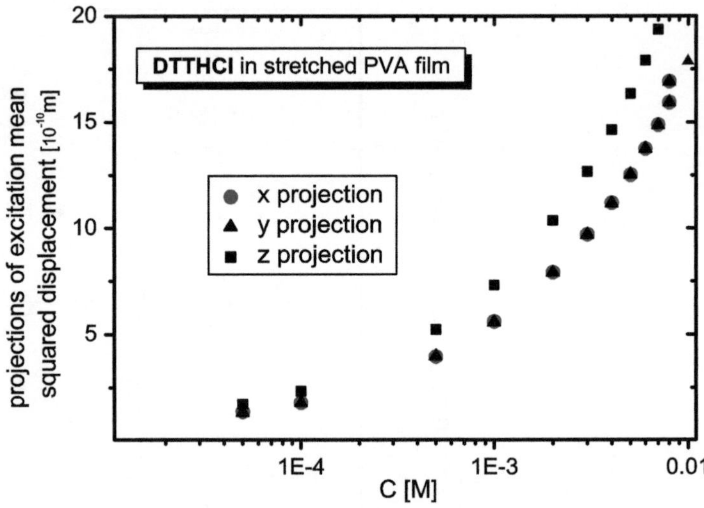


Fig. 8. The projections of excitation mean squared displacement versus molar concentration of elongated fluorophore DTTHCl in stretched polymer film ($R_D = 3.5$). The results obtained with the Monte-Carlo technique. Excitation energy is transmitted more efficiently towards the *z* axis, which evidences directional energy transfer. Mean squared displacements in *x* and *y* directions remain the same as a result of axial symmetry of the problem.

system and energy transfer. In particular, if donor absorption and acceptor fluorescence spectra partly overlap energy transfer can take place also in the reverse direction, i.e., from the aggregate to the monomer. Figure 9a, b shows an example of absorption, fluorescence and fluorescence anisotropy spectra of an elongated molecule MDTCI at its low $C = 0.0005$ M and high concentration $C = 0.02$ M in disordered and strongly ordered PVA film ($R_D = 4.8$) (87). It can be seen from the figure that at high concentration the maximum of absorption spectra is shifted to shorter wavelength (for disordered system from 562 nm at low concentration to 550 nm at high concentration and for the ordered system from 558 nm to 552 nm, respectively). Simultaneously, at high concentration a new weak absorption band appears at the red part of absorption spectra (with the maximum at around 620 nm) evidencing the formation of aggregates in the ground state. It occurs that these aggregates are able to emit fluorescence as evidenced by fluorescence excitation spectra (not shown), fluorescence spectra, emission anisotropy spectra and time-resolved emission spectra (87). Fluorescence spectra shown in the figure exhibit long wavelength shift accompanied by a formation of a new aggregate band with the maximum at around 650 nm. It is also interesting to take a look at the emission anisotropy spectra visible in the figure. Rapid drop of emission anisotropy can be observed when passing from the monomer to the aggregate band (triangles), suggesting that these fluorescent aggregates emit almost depolarized light. In uniaxially stretched PVA films this emission anisotropy also decreases strongly despite weaker

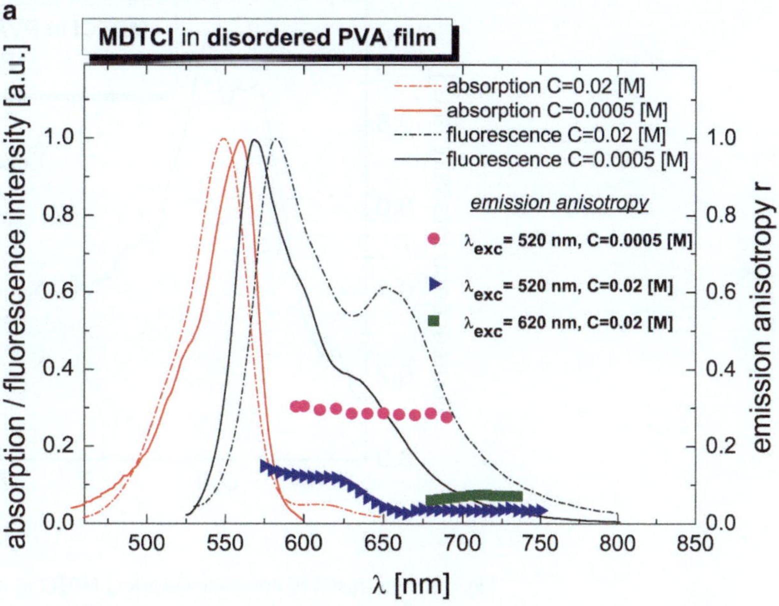


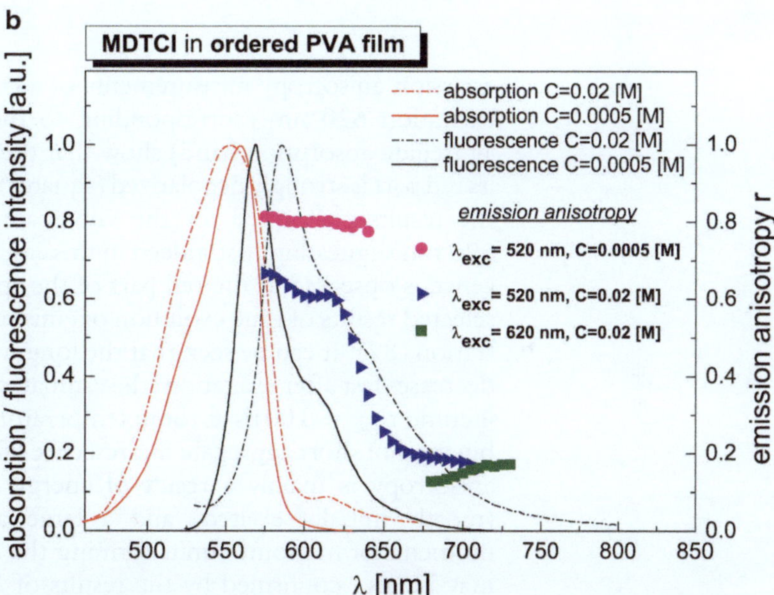


Fig. 9. Absorption, fluorescence, and fluorescence anisotropy spectra of MDTCI in disordered (**a**) and ordered polymer film (**b**) at low and high fluorophore concentration.

aggregation suggesting that the detection of fluorescent aggregates in polymers may be easier after their stretching. Simultaneously, excitation of less concentrated MDTCI sample ($C = 0.0005$ M) yields much higher emission anisotropies in the monomer band (circles) with no fluorescence signal at $\lambda > 650$ nm, as in this case practically only monomers exist in the sample. On the other hand, the

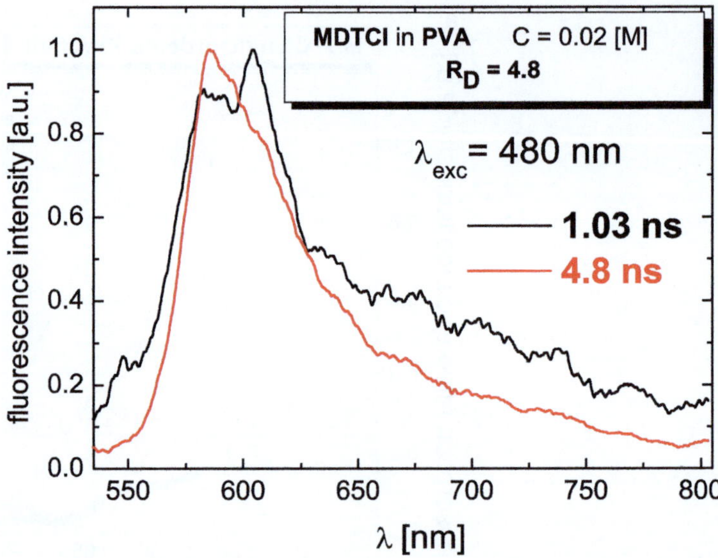


Fig. 10. Time-resolved emission spectra of MDTCI at high concentration $C = 0.02$ M in stretched polymer film. *Red part* of the spectra corresponding to aggregate fluorescence decay fast with time compared to the monomer fluorescence.

emission anisotropy measurements of a concentrated sample at the excitation 620 nm (corresponding to the maximum of the weak aggregate absorption band) show that the fluorescence spectrum in its red part is strongly depolarized (squares). This is in agreement with the results performed for the same sample but at the excitation 520 nm suggesting that indeed in this case mostly aggregate fluorescence is observed at the red part of the spectrum. Figure 10 shows selected results of time evolution of emission spectra at high concentration (87). It can be seen that the long wavelength aggregate band decreases fast after excitation (the estimated mean MDTCI aggregate lifetime $\tau_{aggr} \approx 100$ ps at room temperature). The uncommon combination of short aggregate fluorescence lifetime and its low emission anisotropy is mainly a result of energy transfer from monomers (mostly initially excited) and a large angle between transition moments of monomer units forming the aggregate. This latter fact may be also confirmed by the results of Monte-Carlo simulations. Monte-Carlo simulations can deliver especially important information on aggregate properties and energy transfer in rigid media, if the fluorescent aggregates can be treated as dimers. Then, we deal with a two component, monomer–dimer system. Generally, in such a system, energy transfer can take place as homotransfer in the donor and acceptor ensemble as well as heterotransfer from donor to acceptor and the other way round (reverse transfer). Such a case can be considered, for example, for some rhodamines or FMN (78–86).

From the Monte-Carlo analysis some important parameters like dimer fluorescence quantum yield or its emission anisotropy can be retrieved, which are not available using other methods. Especially, the case of FMN is interesting and important due to the possible role of FMN dimers in the photoreception process. As the fluorescence spectrum of the FMN dimer overlaps with that of the monomer absorption, not only forward but also reverse energy transfer takes place in this case. Dimers never exist alone in the solution, but they coexist with monomers. Energy transfer processes strongly affect monomer and dimer quantum yield, therefore the dimer fluorescence quantum yield cannot be determined only from the experiment. However, Monte-Carlo analysis (and sometimes the use of analytical approaches) together with experimental data enables to obtain these characteristics (86). For example, it has been determined that the FMN dimer quantum yield in PVA at room temperature amounts to 0.02, whereas its emission anisotropy is 0.17.

Monomer–dimer system represents a special, natural case of a more general donor–acceptor system. In a homogeneous solution donors and acceptors or donor–acceptor pairs are usually randomly distributed in space. Sometimes, however, donors and acceptors can be organized in a special medium, and moreover different organization for donors and acceptors in the same medium can be encountered. Such a case leads to very interesting physical effects which seems practically useful. As mentioned, energy transfer from donor to acceptor leads always to almost total acceptor fluorescence depolarization in disordered systems. However, upon certain experimental conditions it is possible to observe quite an opposite effect (i.e., polarized acceptor emission) in a multicomponent ordered system (34, 95–98).

Contrary to one-component systems, it is possible to create differently oriented transition moments of donors and acceptors in the same stretched PVA matrix depending on the physical and chemical properties of molecules. If, for example, transition dipole moments of both energy donor and acceptor are directed along their long molecular axes, then they both tend towards the axis of film stretching (95). If, however, transition dipole moment of the donor is perpendicular to its long molecular axis, but acceptor transition moment is parallel to its long molecular axis, then both orientations of donor and acceptor transition moments are completely different leading to repolarization effect in the acceptor band after energy transfer (97). It is worthy of notice that energy-transfer parameters like the values of orientation factor and critical distances are also strongly affected by specific donor–acceptor distribution. The number of such manipulations on donor–acceptor transition moment orientations is high and it increases strongly with the number of fluorescent components. This problem is interesting having in mind molecular devices for storage, coding and decoding and transmittance of information in space and time. It can help also to understand the behavior of complex organized

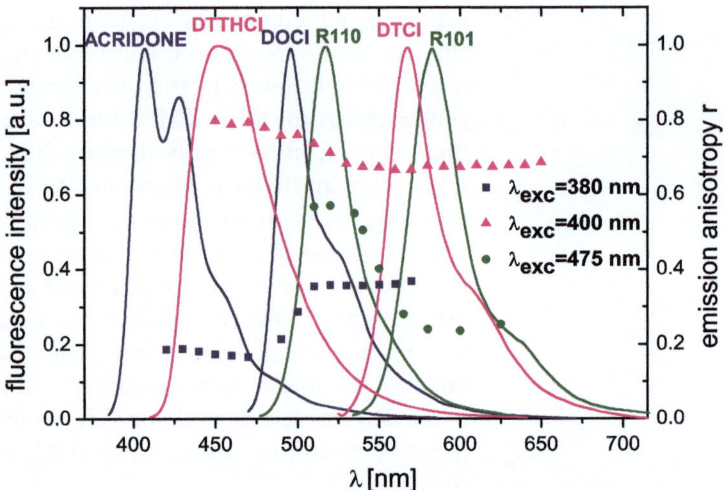


Fig. 11. Fluorescence spectra and fluorescence anisotropy spectra for three donor–acceptor systems differing in the orientation of transition moments in the PVA film.

fluorescent molecular assemblies. Figure 11 shows an example of emission anisotropy spectra together with fluorescence spectra recorded for different ordered donor–acceptor systems: (1) DTTHCI – DTCI (96); (2) acridone – DOCI (98); (3) R6G – R101 (98). For the system (1) the transition moments of donors and acceptors are located parallel to the long molecular axes, for the system (2) transition moment of acridone (donor) is perpendicular to its long molecular axis, whereas for DOCI (acceptor) it is parallel to its long axis, and finally for system (3) donors and acceptors are nonlinear molecules and weakly orientate upon strong film stretching. It can be seen from the figure that in the case of system (1) emission anisotropy is strongly preserved when shifting the observation towards acceptor fluorescence band (triangles); however for system (2) strong repolarization effect can be seen in the acceptor fluorescence band (squares). In the case of system (3) the emission anisotropy is significantly diminished in the acceptor fluorescence band despite strong film stretching (circles). All the experiments are performed at such an excitation wavelength that mostly donors are excited and the emission of acceptors results almost only from energy transfer. The preservation of emission anisotropy in the acceptor fluorescence band means that the information on the orientation of the electric vector of the exciting linearly polarized light is strongly kept in the system (system 1). The repolarization effect can be interpreted as a possibility of regaining such an information despite its initial partial loss (system 2) and finally the fluorescence depolarization denotes the loss of information (system 3). Further confirmation of these results is the emission anisotropy decay. Just one example of donor and acceptor emission anisotropy decay is given below in Fig. 12a, b for DOCI–DTCI system. It can be seen in the

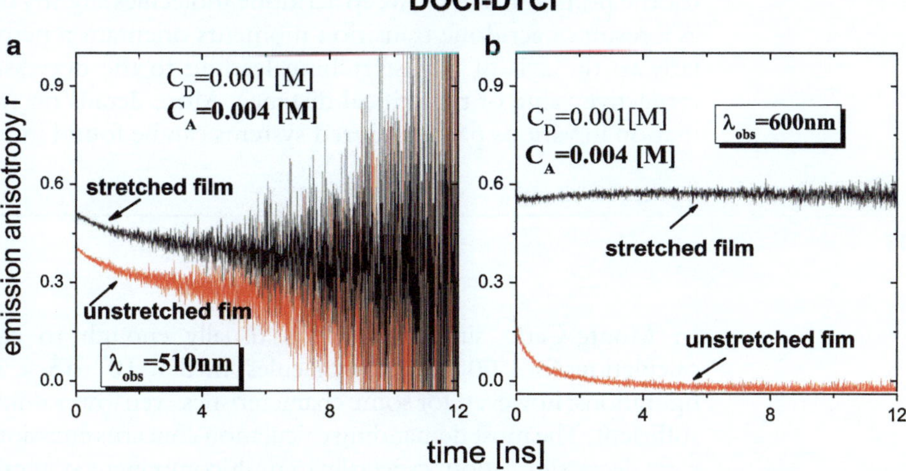


Fig. 12. Emission anisotropy decays of DOCI–DTCI system in stretched and unstretched film. Panel (**a**) shows the donor emission anisotropy, whereas panel (**b**) shows emission anisotropy measured in the acceptor fluorescence band but, of course, upon donor excitation.

figure that for disordered system acceptor emission anisotropy decays very fast compared to that of the donor (red curves in Fig. 12a, b, respectively). However, after film stretching, acceptor emission anisotropy is preserved over the whole timescale of the experiment indicating the memory of the initial polarization of the exciting light (black curve in Fig. 12b).

It occurs that different orientations of donors and acceptors (on a macroscopic scale) affect strongly the values of the averaged orientation factor and leads in this way to the change of critical distance for energy transfer. Besides, contrary to disordered system, where we introduce only one value of the averaged orientation factor (for example, 0.476 for immobile dipoles in the viscous solution), we have to deal with several generally different averaged orientation factors corresponding to energy homotransfer in the donor ensemble, energy heterotransfer or energy homotransfer in the acceptor ensemble if donor and acceptor transition moments are differently distributed in the same ordered film.

However, it occurs that the most spectacular change of the averaged orientation factor between disordered and ordered films occurs for elongated, strongly and similarly oriented donors and acceptors. For example, for the DTTHCI-DTCI system or DOCI-DTCI system the value of $<\kappa^2>$ obtained from Monte-Carlo calculations increases from 0.476 (disordered system) to 1.27 for energy migration and transfer ($R_D = 6$) (34). Such a change leads to around 20% increase of the critical distance, which may also help to investigate distance distributions of more separated donor–acceptor pairs linked to macromolecules. On the other hand, for acridone–DOCI system averaged orientation factor for the homotransfer between acridone molecules slightly decreases as a result of acridone transition moments orientation perpendicularly to the axis of film stretching leading to the decrease in the respective value of the critical distance. More details on $\langle \kappa^2 \rangle$ estimation in various partly ordered systems can be found in ref. (34).

## 4. Notes

In Monte-Carlo simulations it is usually enough to carry out calculations for 1,000–3,000 molecules and $5 \times 10^4$ to $5 \times 10^6$ configurations; however, for some characteristics even lower numbers are sufficient. The most demanding calculation concerns emission anisotropy decay simulation, especially in multicomponent systems of very different concentrations of donors and acceptors.

Imagine that the donor concentration is 100 times higher than that of acceptor. If a typical number of 1,000 donors is used in such a simulation then only ten acceptors participate in the simulation which is much too less to attain proper statistics and numerical stability.

From the last passage also another question arises: how to determine in practice sufficient number of simulation runs? The answer is quite simple: we stop simulation at a point, where further changes of a given observable are insignificant (usually when the variance is less than 0.1%). There are also several other tricky steps one has to pay attention to, for example, (1) the determination of a suitable number of molecules (2) operating on the physically sensible results of pseudorandom number generation. For obvious statistical reasons the number of simulated molecules should not be too small. Nevertheless, it should also be not too high as the calculation time rises approximately with $N^3$. Therefore, there is a question of a compromise and the determination of the sufficient number of molecules is usually made in such a way that we analyze, whether the results are affected by the increased number of molecules. If the results do not change above certain small margin, the number of molecules is usually considered as sufficient. Certain help in the limitation of this number and the finite size of the system is given by the employment of periodic boundary conditions (the cube is surrounded by replicas of itself) with the minimum image convention meaning that the molecule interacts either with another real molecule or its periodic image. The decision whether the real molecule or its periodic image is chosen is based on a calculation of the shortest distance from the excited molecule. The idea of periodic boundary conditions is given in Fig. 4 (please, note that for legibility only several molecules are drawn in the figure. In real simulations this number usually exceeds 1,000 in a single box). There are several issues concerning point 2; however, we will say a few words just about one of them. Sometimes, generation of the locations of all molecules in a cube or cuboid may lead to such a case that two molecules are separated by a very small distance. However, real molecules have finite dimensions which prevent quite arbitrary mutual locations of them. Therefore, a cutoff radius should be introduced into the simulation process which excludes all unphysical locations of molecular pairs. Negligence or a bad choice of a cutoff radius can lead to a simulation run infinite in time as a result of extremely high rate constant for energy transfer for such a closely localized molecular pair. It is usually sensible to assume the value of the cutoff radius similar to the doubled molecular radius. The idea of the cutoff radius may be used also to evaluate the effects connected with real dimensions of macromolecules on energy transfer (so-called excluded volume effect) (102).

## Acknowledgment

This paper has been supported by the grant: NR 15 0029/2009. S.R.J. has been supported by the European Social Fund and Foundation for Development of the University of Gdańsk.

### References

1. Förster T (1967) Mechanisms of energy transfer. In: Florkin M, Stolz EH (eds) Comprehensive biochemistry. Elsevier, Amsterdam, 22:61–80
2. Bojarski C, Sienicki K (1990) Energy transfer and migration in fluorescent solution. In: Rabek JA (ed) Photophysics and photochemistry. CRC, Boca Raton, pp 1–57
3. Van der Meer BW, Coker Ill G, Chen SY (1994) Resonance energy transfer: theory and data. VCH Publishers (Now Wiley-VCH), Inc., New York, pp 1–49
4. Bojarski P, Kułak L, Kamińska A (2002) Nonradiative excitation energy transport and its analysis in concentrated systems. Asian J Spectrosc 5:145–163
5. Clegg RM (1996) Fluorescence resonance energy transfer. In: Wang XF, Herman B (eds) Fluorescence imaging spectroscopy and microscopy. Wiley, New York, pp 179–252
6. Lakowicz JR (2006) Principles of fluorescence spectroscopy. Kluwer/Plenum, New York
7. Valeur B (2001) Molecular fluorescence: principles and applications. Wiley-VCHVerlag GmbH, pp 247–271
8. Ketskemety I, Dombi J, Horvai R, Hevesi J, Kozma L (1961) Experimentelle Prüfung des Wawilowschen Gesetzes im Falle Fluoresziender Lösungen. Acta Physica Chem 7:17–21
9. Kawski A, Piszczek G, Kukliński B, Nowosielski T (1994) Isomerization of diphenyl polyenes. Part VIII. Absorption and fluorescence properties of 1-phenyl-4-diphenylthiophosphinyl butadiene in poly(vinyl alcohol). Z Naturforsch 49a:824–828
10. Kubicki A (1989) A universal photon-counting measuring system for polarized spectroscopy. Exper Techn Phys 37:329–333
11. Bojarski P, Kawski A (1992) The influence of reverse energy transfer on emission anisotropy in two-component viscous solutions. J Fluoresc 2:133–139
12. Jankowski D, Bojarski P, Kwiek P, Rangełowa-Jankowska S (2010) Chem Phys 373:238–242
13. Synak A, Gondek G, Bojarski P, Kułak L, Kubicki A, Szabelski M, Kwiek P (2004) Fluorescence depolarization in the presence of excitation energy migration in partly ordered polymer films. Chem Phys Lett 399:114–119

14. Kubicki A, Bojarski P, Grinberg M, Sadownik M, Kukliński B (2006) Time-resolved streak camera system with solid state laser and optical parametric generator in different spectroscopic applications. Optic Comm 263:275–280
15. Bojarski C (1974) Nonradiative excitation energy transport and some concentration effects in fluorescent solution. Scientific Dissertations Technical University of Gdansk, 13
16. Benett RG (1964) Radiationless intermolecular energy transfer. V. Singlet-triplet transfer. J Chem Phys 41:3048–3050
17. Hameka HF (1967) The triplet state. Cambridge University Press
18. Ermolaev WL, Bodunov EN, Schveschnikova EB, Szachverdov TA (1977) Nonradiative electronic excitation energy transfer. Nauka (in Russian)
19. Vassilev RF (1962) Secondary processes in chemiluminescent solutions. Nature 196:668–669
20. Dexter DL (1953) A theory of sensitized luminescence in solids. J Chem Phys 21:836–851
21. Bäckström HL, Sandros K (1958) The quenching of the long-lived fluorescence of biacetyl in solution. Acta Chem Scand 12:823–832
22. Terenin AN, Ermolaev VL (1952) Sensibilized phosphorescence of organic molecules at low temperature. Intermolecular energy transfer from the excited triplet state. DAN SSSR (in Russian) 85: 547–550
23. Ermolaev WL, Antipenko BM, Schveschnikova JB, Tachin WS, Schaverdov TA (1970) Molecular photonics. Izd Lenigrad (in Russian)
24. Parker CA, Hatchard CG (1962) Delayed fluorescence from solutions of anthracene and phenanthrene. Proc R Soc Lond 269:574
25. Förster T (1948) Intermolecular energy migration and fluorescence. Ann Physik 2:55–75
26. Förster T (1966) Modern quantum chemistry. vol 3. Academic, New York
27. Steinberg IZ (1968) Nonradiative energy transfer in systems in which rotatory brownian motion is frozen. J Chem Phys 48:2411–2414
28. Maksimow MZ, Rozman I (1962) On the energy transfer in rigid solutions. Opt Spectr 12:606–609
29. Dale RE, Eisinger J (1976) Intramolecular energy transfer and molecular conformation. Proc Natl Acad Sci USA 73:271–273
30. Bojarski C, Dudkiewicz J (1979) Orientation factor in concentration effects due to nonradiative energy transfer in luminescent systems. Chem Phys Lett 67:450–454
31. Knoester J, Van Himbergen JE (1986) Theory of concentration depolarization in the presence of orientational correlations. J Chem Phys 84:2990–2998
32. Bojarski P, Kułak L, Bojarski C, Kawski A (1995) Nonradiative excitation energy transport in one- component disordered systems. J Fluoresc 5:307–319
33. Kułak L (2009) Hybrid Monte-Carlo simulations of fluorescence anisotropy decay in three-component donor–mediator–acceptor systems in the presence of energy transfer. Chem Phys Lett 467:435–438
34. Bojarski P, Sadownik M, Rangełowa-Jankowska S, Kułak L, Dasiak K (2008) Unusual fluorescence anisotropy spectra of three-component donor–mediator–acceptor systems in uniaxially stretched polymer films in the presence of energy transfer. Chem Phys Lett 456:166–169
35. Scully AD, Matsumoto A, Hirayama S (1991) A time-resolved fluorescence study of electronic excitation energy transport in concentrated dye solutions. Chem Phys 157:253–269
36. Bojarski C, Kawski A (1959) Über die Bestimmung des kritischen Molekülabstandes bei der Konzentrationsdepolarisation der Fluoreszenz. Ann Phys 5:31–34
37. Kusba J, Lakowicz JR (1994) Diffusion modulated energy transfer and quenching: analysis by numerical integration of the diffusion equation in the laplace space. Methods Enzymol 224:216–262
38. Gryczynski ZI, Lakowicz JR (2005) Basics of fluorescence and FRET. In: Periasamy A, Day R (eds) Molecular imaging: fret microscopy and spectroscopy. Oxford, pp 21–55
39. Eriksen EL, Ore A (1967) On mathematico-physical models for self-depolarization of fluorescence. Phys Norv 2:159–171
40. Jablonski A (1970) Anisotropy of fluorescence of molecules excited by excitation transfer. Acta Phys Pol A38:453–458, Errata (1971) A39, 87
41. Bojarski C, Domsta J (1971) Theory of the influence of concentration on the luminescence of solid solutions. Acta Phys Acad Sci Hung 30:145–166
42. Huber DL (1979) Fluorescence in the presence of traps. Phys Rev B 20:2307–2314
43. Huber DL, Hamilton DS, Barnett D (1977) Time-dependent effects in fluorescent line narrowing. Phys Rev B 16:4642–4650

44. Huber DL (1987) Transfer and trapping of optical excitation. In: Grassano UM, Terzi N (eds) Excited-state spectroscopy in solids. North-Holland, Amsterdam
45. Twardowski R, Kusba J, Bojarski C (1982) Donor fluorescence decay in solid solution. Chem Phys 64:239–248
46. Twardowski R, Bojarski C (1985) Remarks on the theory of concentration depolarization of fluorescence. J Lumin 33:79–85
47. Burstein AI (1985) Quantum yields of selective and non-selective luminescence in solid solutions. J Lumin 34:201–209
48. Bojarski C (1984) Influence of the reversible energy transfer on the donor fluorescence quantum yield in donor-acceptor systems. Z Naturforsch 39:948–951
49. Twardowski R, Kusba J (1988) Reversible energy transfer and fluorescence decay in solid solutions. Z Naturforsch 43:627–632
50. Sienicki K, Winnik MA (1988) Donor-acceptor kinetics in the presence of energy migration. Forward and reverse energy transfer. Chem Phys 121:163–174
51. Sienicki K, Mattice WL (1989) Forward and reverse energy transfer in the presence of energy migration and correlations. J Chem Phys 90:6187–6193
52. Kulak L, Bojarski C (1992) Direct and reverse energy transport in systems of monomers and imperfect traps: Monte Carlo simulations. J Fluoresc 2:123–131
53. Haan SW, Zwanzig R (1978) Forster migration of electronic excitation between randomly distributed molecules. J Chem Phys 68:1879–1884
54. Gochanour CR, Andersen HC, Fayer MD (1979) Electronic excited state transport in solution. J Chem Phys 70:4254–4271
55. Loring RF, Andersen HC, Fayer MD (1982) Electronic excited state transport and trapping in solution. J Chem Phys 76:2015–2027
56. Kulak L, Bojarski C (1995) Forward and reverse electronic energy transport and trapping in solution. I. Theory; Forward and reverse electronic energy transport and trapping in solution. II. Numerical results and Monte Carlo simulations. Chem Phys 191:43–66, (1995) 191: 67–86
57. Bojarski P, Kulak L (1997) Forward and reverse excitation energy transport in concentrated two-component systems. Chem Phys 220:323–336
58. Rangelowa S, Kulak L, Gryczynski I, Sakar P, Bojarski P (2008) Fluorescence anisotropy decay in the presence of multistep energy migration and back transfer in disordered two-component systems. Chem Phys Lett 452:105–109
59. Tanizaki Y (1959) Dichroism of dyes in stretched PVA sheet. Bull Chem Soc Jap 32:1362–1363, (1965) 38: 1798–1799
60. Bojarski P, Synak A, Kulak L, Sadownik M (2003) Excitation energy migration in uniaxially oriented PVA films. Chem Phys Lett 375:547–552
61. Kulak L (2008) Hybrid Monte-Carlo simulations of fluorescence anisotropy decay in disordered two-component systems in the presence of forward and back energy transfer. Chem Phys Lett 457:259–262
62. Metropolis N, Rozenbluth A, Rozenbluth M, Teller M, Teller E (1953) Equation of state calculations by fast computing machines. J Chem Phys 21:1087–1093
63. Bojarski P, Gryczynski I, Kulak L, Synak A, Barnett A (2006) Excitation energy migration between elongated fluorophores in uniaxially oriented polyvinyl alcohol films. Chem Phys Lett 431:94–99
64. Michl J, Thulstrup EW (1986) Spectroscopy with polarized light. VCH, New York
65. Kawski A, Gryczynski Z (1986) On the determination of transition-moment directions from emission anisotropy measurements. Z Naturforsch 41a:1195–1199
66. Gryczynski Z, Kawski A (1987) Relation between the emission anisotropy and the dichroic ratio for solute alignment in stretched polimer films. Z Naturforsch 42a:1396–1398
67. Synak A, Bojarski P (2005) Transition moment directions of selected carbocyanines from emission anisotropy and linear dichroism measurements in uniaxially stretched polimer films. Chem Phys Lett 416:300–304
68. Gryczynski I, Gryczynski Z, Wiczk W, Kusba J, Lakowicz R (1992) Effect of molecular ordering on distance distributions of flexible donor-acceptor pairs. SPIE 1640:622–631
69. Hasegawa M, Enomoto S, Hoshi T, Igarashi K, Yamazaki T, Nishimura Y, Speiser Y, Yamazaki I (2002) Intramolecular excitation energy transfer in bichromophoric compounds in stretched polymer films. J Phys Chem B 106:4925–4931
70. Szabelski M, Bojarski P, Wiczk W, Gryczynski I (2007) Fluorescence resonance energy transfer in short linear peptides carrying 3-[2-(2-benzofuranyl)benzoxazol-5-yl]-alanine and 3-nitro-l-tyrosine molecules in poly(vinyl alcohol) film. Chem Phys Lett 442:418–423

71. Bojarski P, Gryczynski I, Kulak L, Synak A, Bharill S, Rangelowa S, Szabelski M (2007) Multistep energy migration between 3,3'-diethyl-9-methylthiacarbocyanine iodide monomers in uniaxially oriented polymer films. Chem Phys Lett 439:332–336
72. Bojarski P, Synak A, Kulak L, Baszanowska E, Kubicki A, Grajek H, Szabelski M (2006) Excitation energy migration in uniaxially oriented polymer films: a comparison between strongly and weakly organized systems. Chem Phys Lett 421:91–95
73. Heelis PF (1982) The photophysical and photochemical properties of flavins (isoalloxazines). Chem Soc Rev 11:15–39
74. Ninnemann H (1980) Blue light photoreceptors. Bio Sci 30:166–170
75. Gabrys H (1985) Chloroplast movement in Mougeotia induced by blue light pulses. Planta 166:134–140
76. Tian Ch-H, Liu D-J, Gronheid R, Van der Auweraer M, De Schryver FC (2004) Mesoscopic organization of two-dimensional j-aggregates of thiacyanine in langmuir-schaefer films. Langmuir 20:11569–11576
77. Takahashi D, Oda H, Izumi T, Hirohashi R (2005) Substituent effects on aggregation phenomena in aqueous solution of thiacarbocyanine dyes. Dyes Pigments 66:1–6
78. Bojarski P, Matczuk A, Bojarski C, Kawski A, Kuklinski B, Zurkowska G, Diehl H (1996) Fluorescent dimers of rhodamine 6G in concentrated ethylene glycol solution. Chem Phys Lett 210:485–499
79. Bojarski P (1997) Concentration quenching and depolarization of rhodamine 6G in the presence of fluorescent dimers in polyvinyl alcohol films. Chem Phys Lett 278:225–232
80. Bojarski P, Kulak L (1997) Nonradiative excitation energy transport between monomers and fluorescent dimers of rhodamine 6G in ethylene glycol. Asian J Spectros 1:107–119
81. Grajek H, Zurkowska G, Bojarski P, Kuklinski B, Smyk B, Drabant R, Bojarski C (1998) Spectroscopic manifestations of flavomononucleotide dimers in polyvinyl alcohol films. Biochim Biophys Acta 1384:253–267
82. Bojarski P, Kulak L (1998) Forward and reverse excitation energy transport between monomers and fluorescent dimers of rhodamine 6 G in polyvinyl alcohol films. Asian J Spectros 2:91–102
83. Bojarski P, Matczuk A, Kulak L, Bojarski C (1999) Quantitative analysis of concentration fluorescence quenching in condensed systems. Asian J Spectros 5:1–21
84. Bojarski P, Grajek H, Zurkowska G, Smyk B, Kuklinski B, Drabent R (1999) Concentration quenching of flavomononucleotide in polyvinyl alcohol films. J Fluoresc 3(4):391–396
85. Bojarski P (2000) Temperature effect on dimerization constant of dye molecules in polyvinyl alcohol films. Asian J Spectrosc 4:57–66
86. Bojarski P, Kulak L, Grajek H, Zurkowska G, Kaminska A, Kuklinski B, Bojarski C (2003) Excitation energy transport and trapping in concentrated solid solutions of flavomononucleotide. Biochim Biophys Acta 1619:201–208
87. Synak A, Bojarski P, Gryczynski I, Gryczynski Z, Rangelowa-Jankowska S, Kulak L, Sadownik M, Kubicki A, Koprowska E (2008) Aggregation and excitation trapping of 3,3'-diethyl-9-methylthiacarbocyanine iodide in disordered and uniaxially oriented polymer films. Chem Phys Lett 461:222–225
88. Drobizhev M, Sigel Ch, Rebane A (2000) Picosecond fluorescence decay and exciton dynamics in a new far-red molecular J-aggregate system. J Luminesc 86:107–116
89. Janssens G, Touhari F, Gerritsen JW, Van Kempen H, Callant P, Deroover G, Vanderbroucke D (2001) Chemical structure, aggregate structure and optical properties of adsorbed dye molecules investigated by scanning tunneling microscopy. Chem Phys Lett 344:1–6
90. Zakharova GV, Kombaev AR, Chibisov AK (2004) J Aggregation of meso-ethylsubstituted carbocyanine dyes in polymer films. High Energy Chem 38:180–183
91. Del Monte F, Levy D (1999) Identification of oblique and coplanar inclined fluorescent J-dimers in rhodamine 110 doped sol-gel-glasses. J Phys Chem 103:8080–8086
92. Lopez-Arbeloa F, Martinez-Martinez V, Banuelos-Prieto J, Lopez-Arbeloa I (2002) Adsorption of rhodamine 3B dye in saponite colloidal particles in aqueous suspensions. Langmuir 18:2658–2664
93. Ferrer ML, del Monte F, Levy D (2003) Rhodamine 19 fluorescent dimers resulting from dye aggregation on the porous surface of sol-gel silica glasses. Langmuir 19:2782–2786
94. Ferrer ML, del Monte F (2005) Enhanced emission of nile red fluorescent nanoparticles embedded in hybrid sol-gel glasses. J Phys Chem 109:80–86
95. Sadownik M, Bojarski P (2004) The effect of intermolecular donor–acceptor energy transfer on emission anisotropy in uniaxially oriented polymer films. Chem Phys Lett 396:293–297

96. Sadownik M, Bojarski P, Kwiek P, Rangelowa S (2008) Energy transfer between unlike fluorophores in uniaxially oriented polymer films monitored by time – resolved and steady – state emission anisotropy. Opt Mater 30:810–813
97. Bojarski P, Sadownik M, Kułak L, Rangełowa-Jankowska S, Synak A, Jankowski D, Gryczynski I, Grobelna P, Kubicki A, Directional energy transfer and acceptor fluorescence repolarization in two-component anisotropic polymer films, Chem Phys (submitted)
98. Sadownik M (2009) Nonradiative excitation energy transfer in two and three component systems of controllable degree. PhD thesis, University of Gdansk (in Polish)
99. Vassilev RF (1963) Spin-orbit coupling and intermolecular energy transfer. Nature 200:773–774
100. Kawski A, Gryczynski Z (1987) Determination of transition-moment directions from photoselection in partially oriented systems. Z Naturforsch 42a:808–812
101. Gryczynski Z, Kawski A (1988) Directions of the electronic transition moments in dioxide-p-terphenyl. Z Naturforsch 43a:193–195
102. Rangełowa-Jankowska S, Kułak L, Bojarski P (2008) Nonradiative long range energy transfer in donor–acceptor systems with excluded volume. Chem Phys Lett 460:306–310

# Chapter 3

# Molecular Organization of Polyene Antibiotic Amphotericin B Studied by Means of Fluorescence Technique

**Wieslaw I. Gruszecki, Rafal Luchowski, Piotr Wasko, Zygmunt Gryczynski, and Ignacy Gryczynski**

## Abstract

Amphotericin B (AmB) is a polyene antibiotic used to treat deep-seated mycoses. Both the pharmaceutical and toxic activities of AmB depend on the molecular organization of the drug. The fluorescence of AmB has proven to be a powerful technique of studying the drug's association state. In particular, fluorescence lifetime appeared to be sensitive to the formation of AmB dimers and aggregated structures. This paper addresses the application of the fluorescence technique in the study of the molecular organization of AmB, and perspectives on future application of this approach are addressed briefly.

**Key words:** Amphotericin B, Polyene antibiotic, Fluorescence, Fluorescence lifetime, FLIM

## 1. Introduction

Amphotericin B (AmB, see Fig. 1) is a polyene antibiotic used to treat deep-seated mycotic infections (1). Despite severe toxic side effects to patients, the drug has been in use several decades, owing to its effectiveness (2). Both the pharmaceutical activity and toxic side effects of AmB are directly linked to molecular organization of the drug in the environment of biomembranes (3). According to general conviction, AmB forms barrel-shaped aggregated structures that can act as transmembrane pores (channels) that disturb physiological ion transport (4, 5) severely. The fact that such structures are formed preferentially in the lipid membranes containing in their composition ergosterol, a sterol of fungi, seems to be a central paradigm of selectivity of the drug (6). On the other hand, similar structures may be also formed in the environment of ergosterol-free membranes, e.g. containing cholesterol, which is most probably related to the toxic side effects of this popular antibiotic. The amphipatic

Wlodek M. Bujalowski (ed.), *Spectroscopic Methods of Analysis: Methods and Protocols*, Methods in Molecular Biology, vol. 875, DOI 10.1007/978-1-61779-806-1_3, © Springer Science+Business Media New York 2012

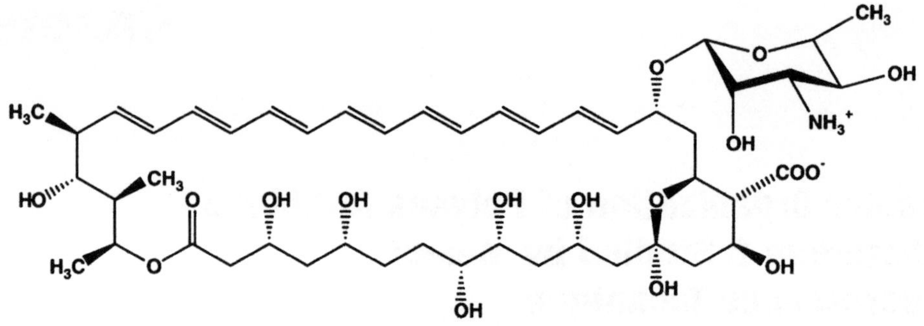

Fig. 1. Chemical structure of amphotericin B.

structure of AmB implies formation of molecular structures of the drug, in which hydrophobic polyene portions of the molecules are exposed to an unpolar environment (e.g. hydrophobic lipid membrane core) (6). One can expect formation of inversed structures in a polar environment (7), such as headgroup region of the lipid membrane or water phase. AmB dimer is the simplest associated molecular form of the drug (8) most probably involved in the formation of larger aggregated structures (9). Owing to the excitonic interactions of polyene chromophores in the molecular structures of AmB, the electronic absorption spectra of aggregated forms are distinctly different as compared to monomeric forms (9–11). The most pronounced spectral effect is a hypsochromic shift in the spectrum and partial loss of the vibrational substructure. Unfortunately, the spectral effects accompanying the formation of AmB dimers, relevant to the physiological standpoint, are usually covered by the spectral effects accompanying the formation of larger molecular structures (12). Such a fact limits application of the electronic absorption spectroscopy as a technique of monitoring the molecular organization of AmB. Interestingly, AmB dimers have been shown to present fluorescence emission spectra (12–15) and emission lifetimes (16) which depend on the molecular organization of the drug. This dependency opens perspectives for a fluorescence technique as a diagnostic method, which can be applied in the analysis of binding AmB to biomembranes from the water phase and its organization in the lipid membrane environment.

## 2. Materials

### *2.1. Purification of AmB*

1. AmB can be purchased from pharmaceutical or chemical companies such as Sigma-Aldrich (from *Streptomyces* ~80% pure).
2. In order to conduct precise spectroscopic measurements, the antibiotic has to be purified by means of HPLC. A possible

HPLC system of choice is a Supelco PKB-100 phase reversed column (length 250 nm, internal diameter 4.6 mm) with 40% 2-propanol in $H_2O$ as a mobile phase.

3. Final concentration of AmB can be calculated from electronic absorption spectra on the basis of the molar extinction coefficient $1.3 \times 10^5$ $M^{-1}$ $cm^{-1}$.

## 3. Methods

### 3.1. Preparation of AmB Samples

1. Purified and vacuum dried AmB may be transferred to any organic solvents. Owing to the extremely low solubility of the drug, solution has to be centrifugated before use (typically 10 min 15,000 × *g*) in order to eliminate large particles present in the solution.
2. In order to incorporate AmB to liposomes, a mixture of lipid (e.g. synthetic dipalmitoylphosphatidylcholine, DPPC, from Sigma-Aldrich) and AmB (1 mol% or less with respect to lipid) is deposited by evaporation to the surface of a glass test tube from the chloroform solution. Possible residuals of the solvent have to be eliminated by vacuum incubation (typically 1 h or longer). A thin film of DPPC-AmB is hydrated with buffer (typically 10-mM Tricine, pH 7.6) to a final lipid concentration lower than 1 mg/ml. Sample is vigorously vortexed for 30 min at temperature exceeding a main phase transition temperature (in the case of DPPC ~41°C) to yield large multilamellar liposomes. In order to obtain small unilamellar liposomes the suspension has to be subjected to sonication (e.g. with a 20 kHz ultrasonic disintegrator, for 1 min, at 5°C) or passed through a polycarbonate filter (e.g. pore size 100 nm) more than 50 times with use of liposome extruder (e.g. Avestin Inc.). Stability and quality of the AmB-containing liposomes have been analyzed in our group by several techniques, including small-angle neutron scattering, FTIR, $^1$H-NMR among others (13, 17–19).

### 3.2. Light Absorption and Fluorescence Measurements

3. Electronic absorption spectra were recorded from the samples placed in rectangular quartz cells (optical path 1 cm) on a Shimadzu UV-160A-PC spectrophotometer (Japan).
4. Fluorescence excitation and fluorescence emission spectra were recorded from the samples placed in rectangular quartz cells (1 × 1 cm or with lower optical pathlength) with Cary Eclipse fluorescence spectrophotometer from Varian (Australia). Spectra were corrected for the Xenon lamp characteristics and for the sensitivity of photomultiplier. Excitation and emission slits were set to 3 nm. Fluorescence emission spectra of AmB have to be corrected for contribution of Raman scattering components (see Note 1).

5. Time-resolved imaging and single molecule measurements were performed on a confocal MicroTime 200 (Picoquant, Germany) system coupled with OLYMPUS IX71 microscope. Photons were collected from different points on the chosen area with 60× water-immersed infinity-corrected objective (NA 1.2) (OLYMPUS). Samples were placed on nonfluorescent Menzel-Glaser #1 cover slips and mounted on a stage piezo-scanner. The fluorescence was excited by a solid-state pulsed laser (405 nm) with repetition rate of 20 MHz. The excitation light was spectrally cleaned by a z430/25x-HT bandpass filter (Chroma Technology Corp.) and delivered to the main optical unit by a single mode fiber. Fluorescence photons were collected with the Perkin-Elmer SPCM-AQR-14 single-photon-sensitive avalanche photodiode (APD) and processed by the PicoHarp300 time-correlated single-photon counting (TCSPC) module based on detection of photons from a periodical light signal. Decay data analysis was performed using a SymPhoTime (v. 5.0) software package, which also controlled the data acquisition.

### 3.3. AmB Fluorescence Measurements and Interpretation

A chromophore of AmB, responsible for electronic absorption and fluorescence properties of the drug, is formally a heptaene: conjugated double-bond system composed of seven C=C bonds. Figure 2a presents the electronic absorption spectrum of monomeric AmB, consisting of a single band corresponding to the $S_0 \rightarrow S_2$ transition. This particular transition is strongly allowed owing to the different symmetries and parities of the ground and excited energy states (formally $^1A_g$ and $^1B_u$ respectively, see Fig. 3). On the other hand, the transition from the ground to the first excited singlet energy level ($S_1$) is not allowed because of the symmetry (Fig. 3) and therefore is not represented in the absorption spectrum (Fig. 2a). Energy of the $S_1$ state of AmB can be deduced from the fluorescence emission spectrum (Fig. 2b), owing to the fact that, in the case of heptaenes, this state can be populated via the internal conversion from the excited $S_2$ state and therefore radiative de-excitation originates from both the $S_2$ and $S_1$ energy levels. These two emission bands can be very clearly resolved in the fluorescence emission spectra (Fig. 2b). Figure 4 presents the electronic absorption and fluorescence spectra of AmB in hexane. Owing to very low solubility of AmB in most solvents, molecules of the drug appear in different organization states in such a system: monomeric, dimeric, and aggregated (11). As can be deduced from the energy diagram presented in Fig. 3, there is an overlap in practically the entire spectral region, making it very difficult to assess the precise contribution of the different molecular organization forms of AmB to the absorption and fluorescence spectra. Interestingly, according to a recent report (16), molecular organization forms of AmB emit fluorescence with distinctly different lifetimes: in monomeric form

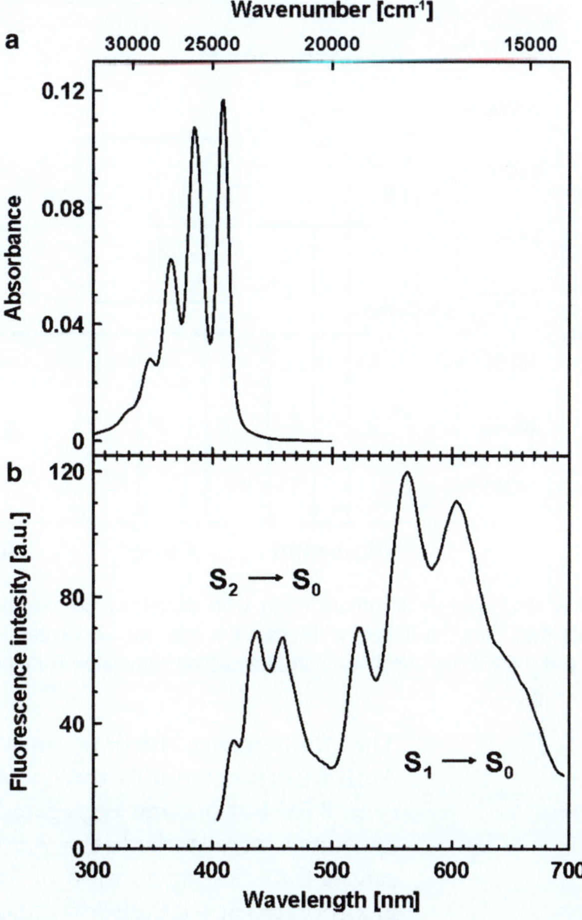


Fig. 2. (**a**) Absorption spectrum of monomeric amphotericin B sample dissolved in 40% 2-propanol in $H_2O$ and (**b**) fluorescence emission spectrum of the same sample. The fluorescence emission spectrum was corrected according to the procedure described in Note 1.

from the $S_1$ state $\tau_{M1} = 0.36$ ns, from the $S_2$ state $\tau_{M2} < 10$ ps, in a dimeric state $\tau_D = 14$ ns, and in an associated dimmers (aggregated) state $\tau_A = 3.5$ ns (see also Fig. 3). Such a pronounced difference enables accurate assignment of molecular organization forms of AmB in any experimental system. Based on the fluorescence lifetime analysis, Gruszecki et al. (16) concluded recently that molecular organization of AmB in a lipid monolayer formed with DPPC containing cholesterol (the sterol present in human cells) was different than in the monolayer containing ergosterol (the sterol present in the cells of fungi). In the case of the cholesterol-containing system, AmB appears mostly in the aggregated state, while in the case of the ergosterol-modified membranes the drug also appears in the monomeric and dimeric states (16). Figure 5 presents FLIM analysis of the $7 \times 7$-μm area of the glass support covered partially with the single layer and in part with the stacked layer of small unilamellar liposomes formed with DPPC and containing 0.1 mol% AmB.

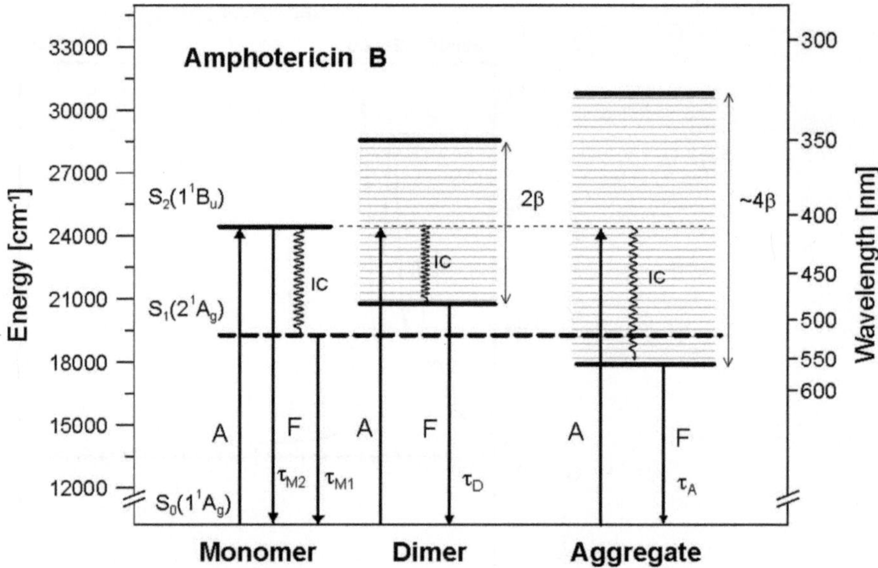


Fig. 3. Energy level diagram of amphotericin B in different molecular organization forms. Energy levels based on spectroscopic data from the literature discussed in the text. *A* denotes light absorption, *F* fluorescence, *IC* internal conversion, and $\beta$ the dipole–dipole excitonic interaction parameter. For more explanation see the text.

The fluorescence lifetime analysis indicates that molecules of AmB incorporated into the liposomes remained in a dimeric form ($\tau > 8$ ns) but became aggregated (formed associated dimers) while liposomes were stacked ($\tau = 3.5$ ns). Such a finding not only indicates a dimeric organization of AmB in lipid membranes (the drug incorporated at relatively low concentration) but also suggests localization of the drug in the surface membrane zone. We propose applying fluorescence to the analysis of the molecular organization and localization of AmB in lipid membranes and other systems relevant to the biological standpoint.

## 4. Note

1. Fluorescence emission spectra of AmB consist of two bands: one corresponding to the $S_2 \rightarrow S_0$ transition (between 400 and 500 nm) and the second corresponding to the $S_1 \rightarrow S_0$ transition (between 500 and 700 nm). Excitation of AmB samples in wavelengths below 350 nm carries the possible risks the excitation of traces of shorter polyenes (degradation products) (20) or aggregated forms of the drug (13). On the other hand, excitation of the AmB sample at longer wavelengths (e.g. at 405 nm, 0-0 vibrational transition) may result in distortion of the emission spectra from Raman scattering, particularly in the case of solvents with molecular structures which give rise to

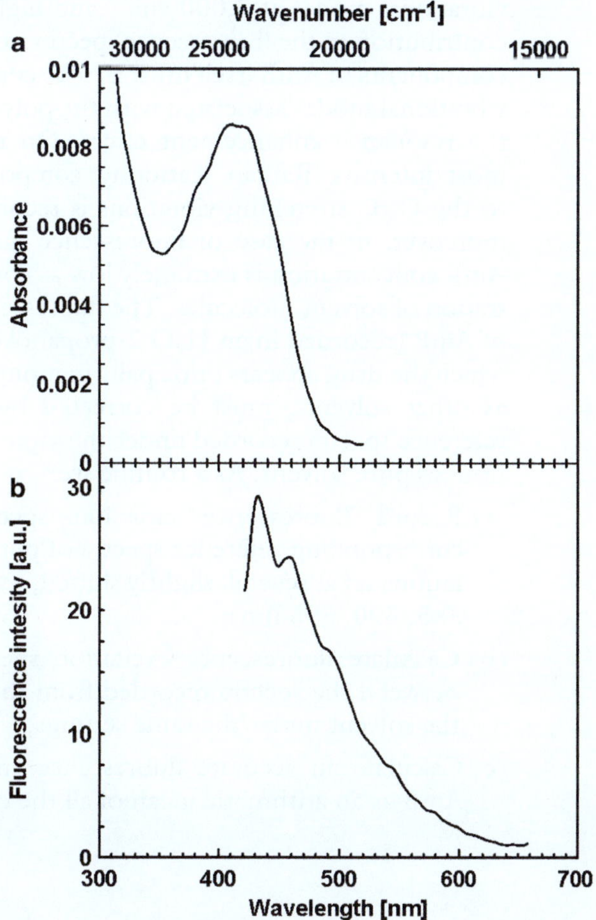


Fig. 4. (**a**) Absorption spectrum of amphotericin B in hexane and (**b**) fluorescence emission spectrum of the same sample. Purified sample was dissolved in hexane, and large particles of AmB were eliminated from the solution by centrifugation for 10 min at 15,000 × *g*.

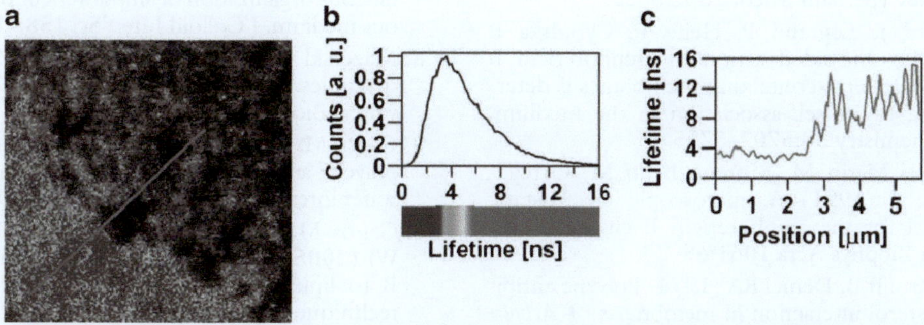


Fig. 5. (**a**) FLIM image of the 7 × 7-μm area of the glass support covered by spin-coating with the suspension of small unilamellar liposomes formed with DPPC and containing 0.1 mol% amphotericin B. The sample is composed of a single liposome layer (*top right hand corner*) and stacked liposome region (*bottom left hand corner*). (**b**) AmB fluorescence lifetime histogram analysis of the entire sample presented in *panel* (**a**). (**c**) AmB fluorescence lifetime analysis along the section line presented in *panel* (**a**).

vibrational modes at 3,000 $cm^{-1}$ and higher frequencies. The contribution to the fluorescence spectra from Raman scattering components of AmB itself must also be considered in particular vibrational modes associated with the polyene chains, owing to the resonance enhancement effect. On the other hand, the most intensive Raman scattering component corresponding to the C=C stretching vibrations is recorded at 1,560 $cm^{-1}$; moreover, in the case of fluorescence measurements, typical AmB concentration is extremely low as compared to a concentration of solvent molecules. The fluorescence emission spectra of AmB (recorded in an $H_2O$:2-propanol mixture (6:4, v:v) in which the drug appears principally in monomeric form), as well as other solvents, must be corrected by subtraction of the reference spectra recorded under the same experimental conditions as pure solvent. As a routine we:

(a) Record fluorescence emission spectra of AmB and corresponding reference spectra of pure solvents with excitations set at several, slightly shifted, wavelengths (e.g. 380, 385, 390, 395 nm);

(b) Calculate fluorescence excitation spectra as a difference between the spectra recorded from the AmB solution and the solvent under the same settings;

(c) Calculate an accurate fluorescence emission spectrum of AmB as an arithmetic mean of all the difference spectra.

## References

1. Torrado JJ, Espada R, Ballesteros MP, Torrado-Santiago S (2008) Amphotericin B formulations and drug targeting. J Pharm Sci 97:2405–2425
2. Laniado-Laborin R, Cabrales-Vargas MN (2009) Amphotericin B: side effects and toxicity. Rev Iberoam Micol 26:223–227
3. Bolard J, Legrand P, Heitz F, Cybulska B (1991) One-sided action of amphotericin B on cholesterol-containing membranes is determined by its self-association in the medium. Biochemistry 30:5707–5715
4. Bonilla-Marin M, Moreno-Bello M, Ortega-Blake I (1991) A microscopic electrostatic model for the amphotericin B channel. Biochim Biophys Acta 1061:65–77
5. De Kruijff B, Demel RA (1974) Polyene antibiotic-sterol interaction in membranes of *Acholeplasma laidlawii* cells and lecithin liposomes. III. Molecular structure of the polyene antibiotic-cholesterol complex. Biochim Biophys Acta 339:57–70
6. Baginski M, Resat H, Borowski E (2002) Comparative molecular dynamics simulations of amphotericin B-cholesterol/ergosterol membrane channels. Biochim Biophys Acta 1567:63–78
7. Barwicz J, Gruszecki WI, Gruda I (1993) Spontaneous organization of amphotericin B in aqueous medium. J Colloid Interf Sci 158:71–76
8. Mazerski J, Borowski E (1996) Molecular dynamics of amphotericin B. II. Dimer in water. Biophys Chem 57:205–217
9. Gagos M, Gruszecki WI (2008) Organization of polyene antibiotic amphotericin B at the argon-water interface. Biophys Chem 137:110–115
10. Gagos M, Gabrielska J, Dalla SM, Gruszecki WI (2005) Binding of antibiotic amphotericin B to lipid membranes: monomolecular layer technique and linear dichroism-FTIR studies. Mol Membr Biol 22:433–442
11. Gagos M, Herec M, Arczewska M, Czernel G, Dalla SM, Gruszecki WI (2008) Anomalously high aggregation level of the polyene antibiotic

amphotericin B in acidic medium: implications for the biological action. Biophys Chem 136:44–49

12. Gruszecki WI, Gagos M, Herec M, Kernen P (2003) Organization of antibiotic amphotericin B in model lipid membranes. A mini review. Cell Mol Biol Lett 8:161–170
13. Gruszecki WI, Gagos M, Herec M (2003) Dimers of polyene antibiotic amphotericin B detected by means of fluorescence spectroscopy: molecular organization in solution and in lipid membranes. J Photochem Photobiol B 69:49–57
14. Gruszecki WI, Herec M (2003) Dimers of polyene antibiotic amphotericin B. J Photochem Photobiol B 72:103–105
15. Stoodley R, Wasan KM, Bizzotto D (2007) Fluorescence of amphotericin B-deoxycholate (fungizone) monomers and aggregates and the effect of heat-treatment. Langmuir 23:8718–8725
16. Gruszecki WI, Luchowski R, Gagos M et al (2009) Molecular organization of antifungal antibiotic amphotericin B in lipid monolayers studied by means of Fluorescence Lifetime Imaging Microscopy. Biophys Chem 143:95–101
17. Herec M, Dziubinska H, Trebacz K, Morzycki JW, Gruszecki WI (2005) An effect of antibiotic amphotericin B on ion transport across model lipid membranes and tonoplast membranes. Biochem Pharmacol 70:668–675
18. Gabrielska J, Gagos M, Gubernator J, Gruszecki WI (2006) Binding of antibiotic amphotericin B to lipid membranes: a 1H NMR study. FEBS Lett 580:2677–2685
19. Herec M, Islamov A, Kuklin A, Gagos M, Gruszecki WI (2007) Effect of antibiotic amphotericin B on structural and dynamic properties of lipid membranes formed with egg yolk phosphatidylcholine. Chem Phys Lipids 147:78–86
20. Bolard J, Cleary JD, Kramer RE (2009) Evidence that impurities contribute to the fluorescence of the polyene antibiotic amphotericin B. J Antimicrob Chemother 63:921–927

# Chapter 4

# Spectroscopic Probes of RNA Structure and Dynamics

**Kathleen B. Hall**

## Abstract

A single technique is insufficient to characterize the properties of an RNA molecule so this Guide provides advice and suggestions for use of several spectroscopic methods applied to RNA. It begins with a discussion of design features to synthesize a suitable molecule for study, assuming that the reader is familiar with in vitro enzymatic and chemical synthesis of RNA. With the RNA in hand, the application of UV melting studies to characterize its folding/unfolding transitions is presented, followed by instructions for fluorescence experiments to augment the UV data. Since RNAs are flexible molecules, it is important to characterize their mobility, and so time-resolved fluorescence data are analyzed. A brief exposition of the power of NMR spectroscopy to identify ion-binding sites is provided. RNA examples are described for each method to give a perspective on what can and cannot be learned.

**Key words:** RNA hairpin, UV melting, 2-aminopurine fluorescence, Ion binding, Imino protons

## 1. Introduction

As an example of solution methods available to study an RNA molecule, this paper focuses on a small RNA hairpin. RNA hairpins are found in all long RNAs, whether it be a 2,000-nucleotide ribosomal RNA or a 53-nucleotide riboswitch, and as such are intrinsic structural elements that confer specific properties onto the larger RNA molecule. While there are exceptions to every rule, it is generally true that an RNA hairpin can be removed from a larger RNA for detailed characterization. Its properties as an isolated component are most likely to represent its properties in its natural context, unless it is part of a tertiary interaction. It is this property of RNAs that in general makes the study of their component pieces both tractable and biologically relevant. There are some outstanding examples of studies of these RNA fragments that have made major contributions to the greater field of RNA biology, and they will be used here as guides and models.

Wlodek M. Bujalowski (ed.), *Spectroscopic Methods of Analysis: Methods and Protocols*, Methods in Molecular Biology, vol. 875, DOI 10.1007/978-1-61779-806-1_4, 

The goal here is to illustrate the use of various spectroscopic methods to describe the properties of a novel RNA hairpin. The format focuses on interpretation of the data and the caveats rather than step-by-step procedures, in part because detailed protocols of some methods (thermal denaturation) are in the literature but some methods (fluorescence and NMR) would require a book. My intent is to describe what a given spectroscopic method will reveal about an RNA, and to provide an example for illustration, a hairpin from Drosophila U2 snRNA will be used. In vivo, this RNA is bound by a protein, which motivates interest in its intrinsic properties. With only that information as a starting point, what would happen next?

## 2. Materials

All experiments are performed in the buffers optimized for the system stability over the time period necessary to execute the experiments.

## 3. Methods

### 3.1. Construction of the Hairpin RNA

In the context of the entire U2 snRNA, this hairpin contains an 11-nucleotide loop (A1UU3GCA7GUA9CC11) on a six-base pair stem (Fig. 1). The first experiment to do is a computational one: predict the

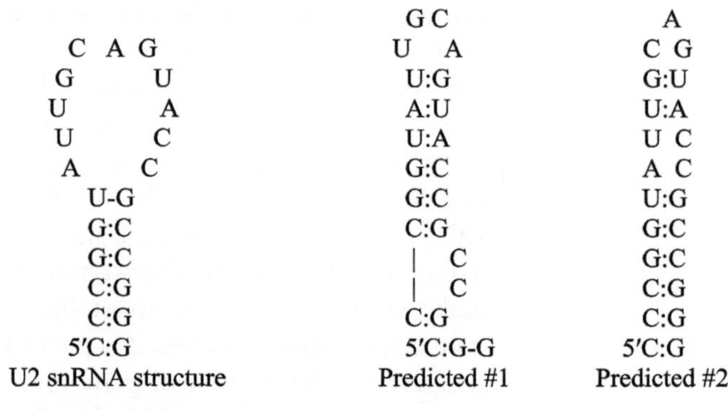


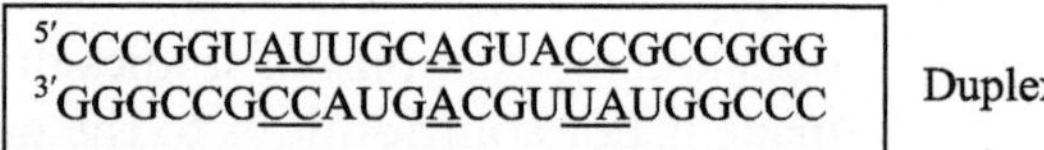


Fig. 1. The secondary structures of the U2 snRNA stemloop as found in the full length U2 snRNA and alternative structures predicted by *mfold* (1). Predicted alternative structures #1 and #2 have folding free energies within 1 kcal/mol of the most stable U2 stemloop. The hairpin could also form a duplex; the unpaired bases are *underlined*.

folding free energy and the alternative structures that the RNA can adopt. Two of the most popular platforms for predicting the secondary structures of RNA are *mfold* (mfold.bioinfo.rpi.edu/cgi-bin/rna-form1.cgi) and Vienna RNA (http://www.tbi.univie.ac.at/~ivo/RNA/). Type in the RNA sequence and submit it directly to the servers, and if it is a short sequence (such as the hairpin) the results will be returned within minutes. Follow the directions at each site, which in *mfold* includes a query for the number of alternative folds to be presented (percent suboptimality). The default value is 5, but prudence dictates at least 25 should be entered. For the U2 hairpin, the most favorable free energy structure calculated by *mfold* (1, 2) is the hairpin found in the snRNA, with $\Delta G_{(\text{folding, } 37^\circ\text{C})} = -8.4$ kcal/mol.

Interpretation of these results requires knowledge of the origins of the thermodynamic parameters upon which they are based. These calculated free energies come from experimental data of short duplex RNA melting in 1 M NaCl with an ad hoc addition of a penalty for adding a loop of N-nucleotides (3). While the duplex data are very robust, the loop penalties are not: for example, a tetraloop is a loop of $N = 4$ and as such would incur a penalty of +4.9 kcal/mol at 37°C in calculations of folding free energy. Of course, if the tetraloop is UUCG with a C:G loop-closing base pair, then it has a free energy contribution of −2.00 kcal/mol at 37°C, making a more stable hairpin. By analogy, if an $N = 11$ loop had an unusual structure that stabilized the hairpin (such as the sarcin-ricin loop in 23 S rRNA), then its contribution to the folding free energy would not be present in the algorithm, where an 11-nucleotide loop is assigned a "penalty" of +5.40 kcal/mol at 37°C. The other point to note about the predicted folding free energies is that the experimental data were collected in 1 M NaCl where effects of divalent ions are effectively eliminated; melting is equivalent to solution conditions of 0.1 M $Na^+$ or $K^+$ plus millimolar $Mg^{2+}$ (4). These are important considerations for characterization of an RNA, since many loops associate with divalent ions (especially $Mg^{2+}$) as a part of their folding process, and since 1 M NaCl is not a solution condition where typical biochemical processes are studied (such as ligand binding). So what do these calculations mean for subsequent hairpin studies?

The relative folding free energies calculated for the hairpin show that the natural structure is the most probable conformation. However, there are two alternative conformations within 1 kcal/mol of the hairpin (Fig. 1) that might be favored under other solution conditions (such as lower monovalent salt concentration with/without addition of $MgCl_2$). Of the two predicted alternative structures, one only affects the loop, which is allowed to form additional base pairs. The other structure involves a rearrangement of the loop and stem, and that structure must be discouraged through modification of the stem sequence to increase its stability. Any new sequence must be folded again to assess alternative structures.

Since the hairpins will be synthesized in vitro using T7 RNA polymerase, a change in the natural stem sequence will be required to optimize the polymerase efficiency. The preferred transcription start sites for T7 RNA polymerase were extensively investigated by the Uhlenbeck lab (5) for the purpose of synthesizing biochemical quantities of RNAs. The efficiency depends on two parameters: the sequence of the first seven nucleotides and the length of the synthesized strand. The enzyme prefers a 5′GGG start, followed by ANN. Sequences as short as ten nucleotides produce very small amounts of product; 25-nucleotide sequences can give great yields. 5′gggcccgcuAUUGCAGUACCggcgggucc is an adjusted U2 hairpin sequence that meets the criteria of no significant alternative structures, an efficient T7 start site, and a suitable length for high yield. An in vitro transcription reaction must be done to verify that the practice and theory agree, but assuming success, the RNA hairpin is ready for study.

### *3.2. Thermal Denaturation*

Solutions to make: 1 M sodium cacodylate, pH 7.0 (100 mL); 2 M KCl (100 mL), 1 M $MgCl_2$ (optical grade) (50 mL), in RNase-free water (remember that autoclaving doesn't inactivate RNases). All solutions are filtered through 0.45-μm Nalgene nitrocellulose filters to remove all protein (nucleases) and stored in plastic. Quartz cuvettes are acid-washed and rinsed extensively with RNase-free water. Beyond the obvious need for RNase-free solutions, these methods require careful sample handling to protect the RNA from the nucleases we all carry on our fingers.

Unfolding thermodynamics will be measured using UV absorption at 260 and 280 nm as a function of temperature. (The absorption of nucleic acid bases increases when they are not stacked—the famous hyperchromic effect.) These measurements are designed to determine the contribution of the RNA loop to the energetics of unfolding, and thus obtain the "loop penalty" (2), and also to examine the transition to determine whether it is two-state. If bases/riboses in a loop make no hydrogen-bonding interactions, then loop structure will be controlled by base-stacking interactions, which will propagate from the loop-closing base pair. Base stacking is not sensitive to salt concentration (and it is not a hydrophobic interaction), but it does decrease with increasing temperature, so the absorbance change from loop unstacking will give a sloping baseline at the start of the melt. If there is inter-nucleotide hydrogen bonding within the loop, then there could be two separable unfolding transitions, with the expectation that loop structure melts out at lower temperature.

To prepare the RNA samples, heat the RNA to at least 65°C in RNase-free distilled and deionized water, place on ice to snap cool (to favor intramolecular folding), then add buffer to the final concentration to provide salt to fold the RNA. Dilute the RNA into degassed buffer + salt to make samples at various RNA concentrations. As a general protocol, the following steps are recommended:

1. Heat all empty cuvettes to 90°C (in steps of 10°C to prevent cracking), then cool to room temperature. This will clean off walls and prevent bubble formation during the run.
2. Add samples and cool to the starting temperature (typically 2–8°C); one cuvette contains only buffer for background subtraction. Allow 30–60 min for equilibration.
3. Heat the RNA at the selected ramp rate, recording absorbance and temperature. For melting, the temperature should be increased at a rate of 0.2 to 1°C/min from 2 to 98°C. If the ramp rate is too rapid, the incremental unfolding of the RNA will lag behind the measured temperature, and the calculated melting temperature will be too high.
4. After the melting experiments, cool down the samples either in steps of 10°C or with a slow temperature ramp, and remeasure the absorbance to confirm that it has not changed.
5. Vary the RNA concentration over at least a 10× range. If melting temperatures are identical, then the RNA undergoes unimolecular unfolding. If the melting temperature increases with RNA concentration, then a dimer has formed and solution conditions need to be revised.
6. Repeat the melting experiments using the same samples to assess reversibility of the transition. An irreproducible melt could have several causes: either hydrolysis of the RNA has occurred at high temperature (particularly if divalent ions are present), or the state of the RNA has altered from a hairpin to a dimer upon cooling.

When selecting the concentration of your RNA, remember Beer's law $A = \varepsilon c \cdot l$ where $A$ is absorbance, $c$ the molar concentration, $\varepsilon$ the molar extinction coefficient, and $l$ the pathlength of the cuvette; work in the linear range of the spectrophotometer. Using *very* short pathlength cuvettes, 1 mM RNA can be measured to anticipate NMR concentrations. If dimer formation becomes a problem, then reducing the salt concentration to a minimum will favor the hairpin; this minimum is typically 10 mM sodium cacodylate, pH 7.0, with 10 μM EDTA to scavenge divalent ions. (Cacodylate is our preferred buffer, since its pH is relatively insensitive to temperature but use with handling precautions appropriate to an arsenic compound. Phosphate should not be used since it precipitates at high temperature together with the RNA.)

Particulars of data analysis depend on the instrument used and the software available. Parameter of interest is the melting temperature, obtained by taking the first derivative of the absorbance with respect to the temperature ($\partial A/\partial T$). By measuring absorbance

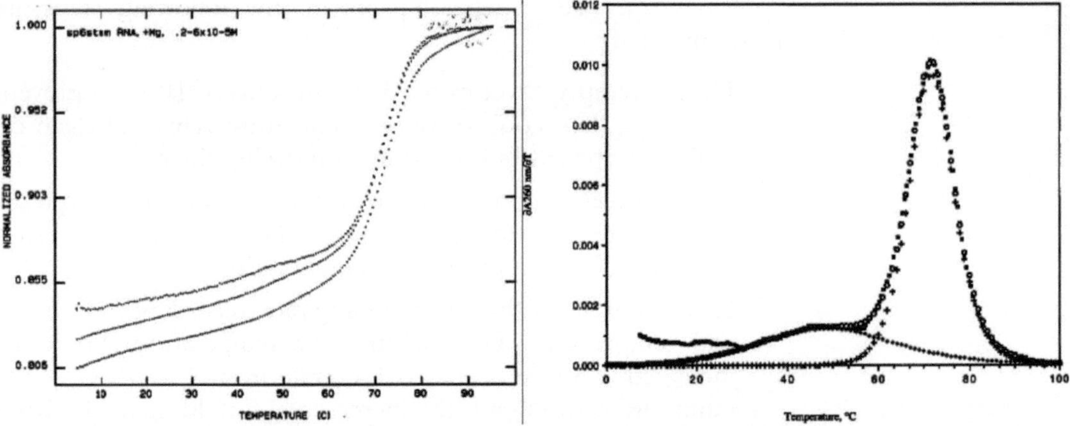


Fig. 2. Predicted thermal denaturation curves (*melting*) of the U2 hairpin over a 30× range of concentration shows the sloping baseline at lower temperatures followed by a sharp cooperative transition. Taking the first derivative of the melting curve reveals two transitions: the broad one centered near 45°C could be the loop structure unstacking (even single-stranded RNAs exhibit hyperchromicity since the bases prefer to stack), while the higher temperature transition is the stem.

at 260 nm, where all the bases absorb, and at 280 nm where guanosine absorbs, it is possible to identify transitions from different structures in the RNA when there is an uneven distribution of guanosines. More information comes from fitting the melting curves, assuming a two-state unfolding transition (6). A recommended method is to use nonlinear least-squares parameter estimation of $T_{m,i}$ (melting temperature of transition $i$) and $\Delta H_i$ (van't Hoff enthalpy) (4). Free energies of unfolding are extrapolated to 25°C using the equation $\Delta G^\circ = \Delta H^\circ(1 - 298/T_M)$.

Theoretical melting curves of the U2 RNA hairpin over a 10× range of concentration are shown in Fig. 2. The curves are offset by their relative absorbance values, but it is apparent that the melting temperatures in this solution of 100 mM KCl, 10 mM sodium cacodylate, pH 7.0, 10 μM EDTA are nearly identical. A first derivative of all melting curves shows a broad transition centered near 50°C and a sharp transition at 72°C (Fig. 2). The high temperature transition certainly is the stem melting; but what is the lower transition? Since it is present at all concentrations of RNA, it must come from the loop as the bases unstack. Perhaps there is at least one base pair within the loop (for example, a U3:A9 pair) and its disruption leads to further unstacking of the loop bases.

The next logical probe of the RNA structure is introduction of $Mg^{2+}$ ion. The addition of 1 mM $MgCl_2$ to 100 mM KCl will raise the ionic strength only slightly, but because the divalent ion is able to associate preferentially with the RNA phosphates (7), it will raise the RNA melting temperature. However, if there is a specific interaction of the loop structure with $Mg^{2+}$, then its melting profile will be altered, probably to become sharper and more defined. $Mg^{2+}$ can be titrated into the RNA to measure its effect on both stem and loop melting, with the caution that heating to 90–95°C to melt the RNA will also lead to cleavage of its phosphodiester bonds by $Mg^{2+}$, making these melting experiments irreversible. Of course, any ion of interest can be added to the RNA, provided it doesn't interfere with the absorbance measurements.

UV melting data can often be fit with two transitions, typically assuming a two-state cooperative model for the folding ↔ unfolding pathway. This assumption is convenient but not necessarily accurate, as suggested by several studies that show multistate transitions of hairpins (8, 9). Certainly it is easy to imagine that the unfolded RNA is flexible and floppy, and that it transiently adopts many conformations: it is an ensemble of structures. The same description should be applied to the folded state: typically, the first two base pairs at the free end of a hairpin stem fray—their hydrogen bonds break and reform very rapidly. The loop-closing base pairs can also be unstable when the loop is large with an unconstrained structure. The bases in such a loop will stack but be free to move, which they do with some temperature-dependent frequency. However, in a typical melting experiment that measures the solution average conformation, intermediate states are unlikely to be detected and the two-state approximation will be appropriate.

### 3.3. Fluorescence and RNA Structure

Fluorescent probes can be used to monitor RNA folding and local dynamics. Although there are some fluorescent base analogs that are available, most are modified by addition of bulky substituents; while these bases are useful for DNA duplex studies, they are likely to alter a folded RNA structure. 2-Aminopurine is a safe substitution for most RNAs and 8-azaguanosine can also work (10) in addition to fluorophores that are attached to the 5′ and 3′ ends of a strand.

Fluorescently modified RNAs can be purchased from Dharmacon (Colorado), TriLink (SanDiego, CA) or IBA (Gottingen, Germany). Only IBA has the popular Atto fluorophores as well as standard Alexa dyes. RNAs containing 2-aminopurine (2AP) must be chemically synthesized. Not all fluorophores are commercially available as phosphoramidites for in-house synthesis, but some can be obtained from Glen Research (Sterling VA). Addition of selected fluorophores at the 3′ end of any T7 transcript can be accomplished through hydrazide chemistry.

#### 3.3.1. Oxidation Step

1. Make fresh 2× Oxidizing solution in a microtube: 10 mM $KIO_4$ in 200 mM sodium acetate (pH 5.3). Use 406 µl water and 27 µl 3 M sodium acetate (pH 5.3) per mg. Do not filter—avoid RNAses. This is near the solubility limit of $KIO_4$ so mix well by pipetting with a 200-µl pipet tip. Precool the solution on ice.
2. In the dark, add 40 µl of the cooled 2× oxidizing solution to 40 µl of 50–200-µM RNA (in water).
3. Place on ice, in the dark, for 30 min.

4. Add 8.8 μl (1/10 dilution) of 100-mM ethylene glycol stock (in water), on ice and in the dark another 5 min. This quenches the $KIO_4$.
5. Increase the sodium acetate concentration to 0.3 M by adding 6.5 μl of 3 M NaOAc (pH 5.3).
6. Add 290 μl ethanol and precipitate on dry ice as usual, including washing the pellet with 70% cold ethanol.
7. The pellet from this last step should be dissolved in a volume to assure at least a 3:1 label:RNA ratio. A 10:1 ratio is preferred. Another question is whether the RNA should be melted/quench-cooled before the next step to free up possible occluded 3′-ends.

*3.3.2. Labeling Step*

Texas Red hydrazide has much lower solubility (about 12 μM in aqueous buffer) than Alexa488 hydrazide (at least 10 mM). However, the Texas Red hydrazide is completely soluble in DMSO up to at least 2.7 mM so the stock solution is in DMSO.

1. For Alexa488 [Texas Red] labeling reconstitute (in the dark) the pellet in 46 μl of 0.1-M sodium acetate, then add 4 μl [10 μl] of 10-mM Alexa488 hydrazide [2.7 mM Texas Red hydrazide] (this is a 10:1 [7:1] ratio if final [RNA] was 50 μM in step 2 above).
2. Let sit at room temperature in the dark for 5 h (or overnight). Tap the tube periodically.
3. For Alexa488, a lot of free dye can be extracted with an equal volume of phenol solution (without indicator dye). Do not phenol-extract Texas Red reactions as the RNA dye prefers the solvent interface.
4. Add 3 M sodium acetate (pH 5.3) to bring to 0.3 M (3.5 μl for a 50 μl reaction) and precipitate with 3 volumes ethanol on dry ice (or overnight at –20°C).
5. Reconstitute final pellet in 50 μl of 10 mM Tris (pH 8.0)/100 mM NaCl.

Free dye can be removed by FPLC using a small (100 × 6 mm) P2 column in 10 mM Tris (pH 8.0)/100 mM NaCl. Run at ~0.2 mL/min. The RNA dyes appear to be somewhat labile after several months. Repeated freeze/thawing cycles may affect this. It may be best to buffer the final reconstitution solution. Available hydrazide fluorophores (as of 2009) from Molecular Probes: Alexa 350, 488, 555, 568, 594, 633, and 647 dyes, Cascade Blue, Bodipy-FL, Fluorescein (2 derivatives), Lucifer Yellow, FMOC, Dansyl, diethylamino-coumarin-3 carboxylic acid, QSY9 (a quencher), plus some biotin derivatives. Amersham (GE): Cy3, Cy3.5, Cy5, and Cy5.5.

### 3.4. Fluorescence and Hairpin Melting

For illustration, an RNA hairpin has been chemically synthesized with a Dabcyl quencher on its 5′ end and TAMRA (tetramethylrhodamine) on its 3′ end (DT-RNA). When the hairpin stem is intact, TAMRA fluorescence is almost completely quenched by Dabcyl, but as the hairpin stem opens, TAMRA fluorescence increases until upon complete melting its fluorescence is maximized. Constructs like this one have been used in fine detailed studies of DNA hairpin unfolding/folding by the Van Orden group (11), to look for intermediates on the folding pathway, and aficionados will appreciate that the two-state model is being revisited both experimentally and computationally (9, 12, 13).

These fluorescence melting experiments are complicated by the photophysics of TAMRA, specifically its temperature-dependent fluorescence intensity. Experiments done in 10-mM sodium cacodylate pH 7 from 0 to 100-mM KCl show that the salt concentration does not affect TAMRA fluorescence. However, as the temperature increases, TAMRA fluorescence decreases, and that loss must be taken into account in the interpretation of the RNA melting experiment. Not all fluorophores have such a dramatic temperature dependence, but their independent characterization must be done in the solution conditions of the experiment. A control DNA strand of $dT_{25}$-3′-TAMRA was chemically synthesized, and used in parallel temperature measurements, since poly (dT) has no structure to complicate interpretation of fluorescence. Shown in Fig. 3 are the steady-state fluorescence intensity data for $dT_{25}$-3′-TAMRA and DT-RNA, and the processed melting curve. $dT_{25}$-TAMRA fluorescence was expressed as a fraction of its initial value: for example, at 78°C, the endpoint of the experiment, it had lost ~75% of its fluorescence (its $f_{\mathrm{TAMRA}}$ fractional intensity = 0.23). DT-RNA fluorescence increased as expected as the RNA stem melted but then began to decrease; however, when its fluorescence intensity ($I_{\mathrm{DT\text{-}RNA}}$) is divided by the TAMRA fractional intensity at each point, the melting curve shown is obtained (melt = $(I_{\mathrm{DT\text{-}RNA}})/f_{\mathrm{TAMRA}}$).

Measuring melting by fluorescence has some advantage over melting by UV absorbance because placement of the fluorophore can reveal properties of a local environment. The fluorophore must be carefully placed to ensure that it does not disrupt structure, which restricts its position. Spectacular use of fluorescent probes in the study of RNA dynamics has been at the level of single molecule measurements of folding, which is beyond this chapter but must be noted (14).

### 3.5. Hairpin Structure/Dynamics

Another use of fluorescence probes is to provide detailed information on both global and local RNA structure and mobility. For characterization of the U2 RNA hairpin, 2-aminopurine is the fluorophore of choice since it should not introduce any new intramolecular structure (Fig. 4). 2AP fluorescence properties are

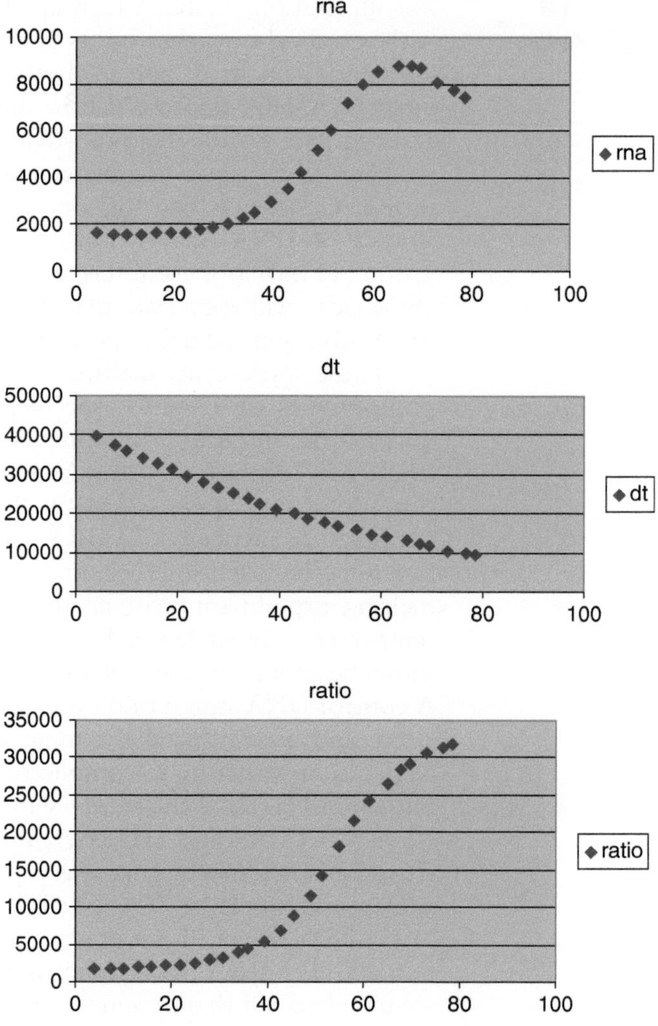


Fig. 3. Another method to monitor melting of the RNA takes advantage of fluorescent labeling of the 5′ and 3′ ends of the hairpin. In this example, TAMRA fluorescence is quenched by the proximal dabcyl molecule when the hairpin is intact, but as it melts and the base pairs break (first fraying, then the stem unzippers), TAMRA fluorescence increases (*top curve*). TAMRA fluorescence is temperature dependent, however, (*middle curve*) decreasing significantly as the temperature increases. Correcting for this intrinsic property of the fluorophore yields the melting curve (*lower*).

determined by its context. It is not a bright fluorophore (quantum yield of 0.68), and when it is stacked with another nucleobase, its fluorescence is quenched (15, 16). Hydrogen bonding has little effect on its fluorescence properties, so a loss of fluorescence intensity is typically interpreted as evidence that it is stacked.

In the context of the U2 hairpin, 2-aminopurine introduced at each of two positions in the loop would be particularly informative: at the apex of the loop (A7) and at the (putative) base pair site A9.

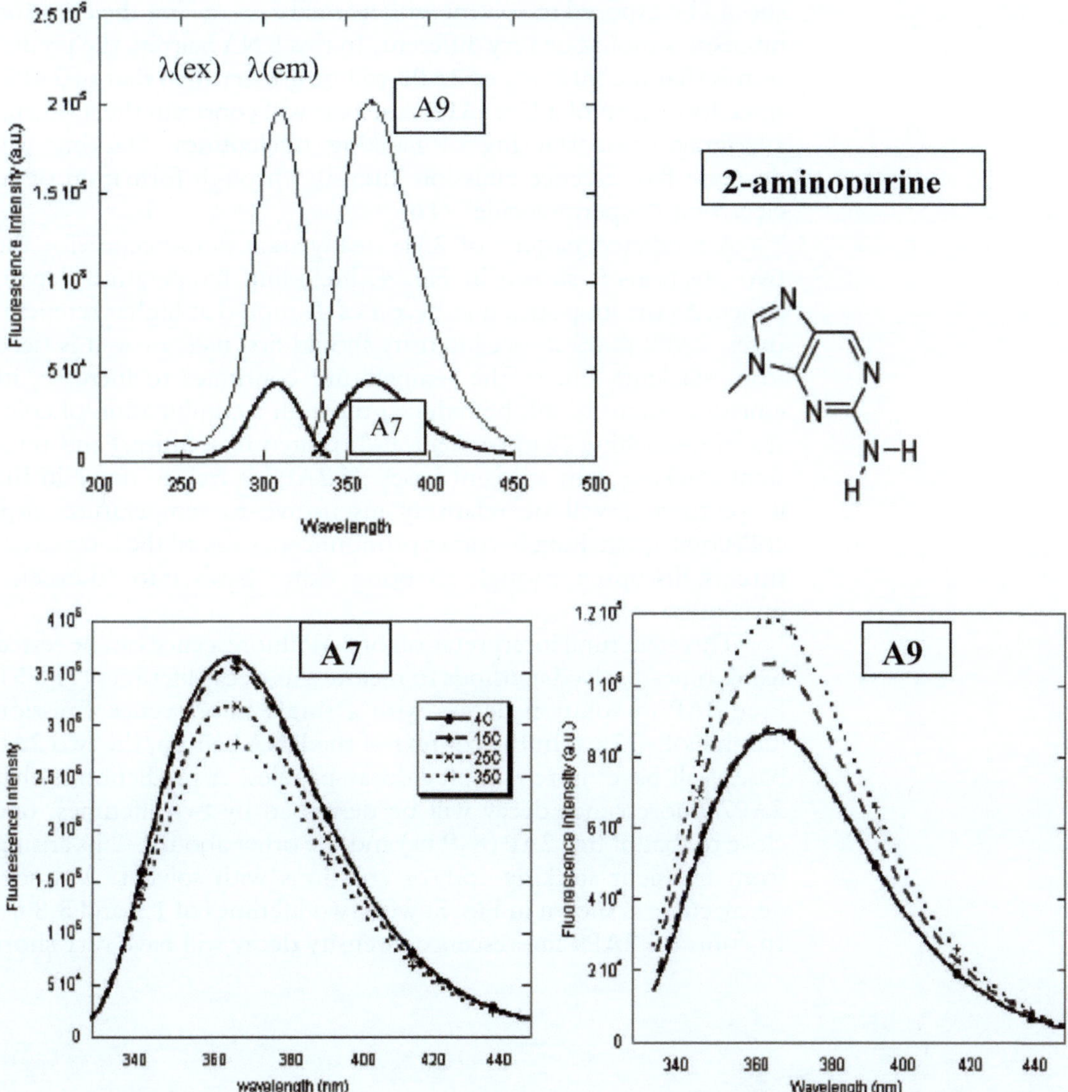


Fig. 4. 2-aminopurine fluorescence is another excellent probe of nucleobase environment and flexibility, since it can base-pair with uridine in a duplex and has no extraneous moieties that disrupt RNA folding. The predicted fluorescence of 2AP at positions 7 and 9 of the U2 loop is shown: 2AP7 fluorescence would be quenched if it stacks with flanking bases. The temperature dependence of 2AP fluorescence in different structural environments also provides information on the loop structure. Here, the temperature dependence of 2AP at position 7 is predicted to decrease with increasing temperature, in the same way that free 2AP fluorescence would be quenched by more frequent collisions with solvent and oxygen as the temperature increases. In contrast, the fluorescence intensity of 2AP at position A9 is quenched due to its stacking interactions (*in our model*) so its fluorescence intensity increases with temperature as the stacking interactions are disrupted. This pattern of 2AP fluorescence was observed with the six nucleotide iron responsive element RNA hairpin (20) with the striking result that two adjacent nucleotides exhibited two very different properties.

The prediction is that 2AP7 should be flexible, whereas 2AP9 would be hydrogen bonded to U3 and stacked. For solvent-accessible 2AP, the excitation maximum $\lambda_{\mathrm{Ex(max)}} = 305$ nm; any deviation from this indicates an unusual environment. In the U2 RNA, both 2AP sites should be exposed to solvent with normal $\lambda_{Ex(max)}$ but the emission intensities should be very different. In this RNA hairpin, the prediction is that the intensity of 2AP7 will be greater than that of 2AP9, since formation of a U3:2AP9 base pair will constrain the loop and encourage base stacking of flanking nucleotides. Stacking will decrease fluorescence emission intensity through formation of an electronic "supermolecule" (16).

A predicted pattern of 2AP steady-state fluorescence for the two positions is shown in Fig. 4, including temperature dependence. As the loop structure becomes disrupted at higher temperatures, 2AP9 fluorescence intensity should first increase as it is freed from stacking, but as the temperature continues to increase, its emission intensity will be reduced through a combination of collisional quenching (with solvent and oxygen in solution) and transient stacking with adjacent bases. If 2AP7 is free to move in the loop, then it will be relatively insensitive to temperature until collisional quenching becomes pronounced, or until the loop structure is disrupted enough to bring other bases into (transient) proximity.

This structural interpretation of 2AP fluorescence can be tested using time-resolved methods to measure its decay lifetimes (Fig. 5). Free 2AP in solution decays with a single fluorescence emission lifetime of ~12 ns. In the context of the RNA hairpin, the two 2AP bases will have more complex decay profiles. A prediction is that 2AP7 fluorescence decay will be described by two lifetimes, one close to that of free 2AP (8–9 ns) and the other about 1–2 ns arising from transient stacking and/or collisions with solvent. A model decay curve is shown in Fig. 5, with two lifetimes of 1.2 and 5.8 ns. In contrast, 2AP9 fluorescence intensity decay will have very short

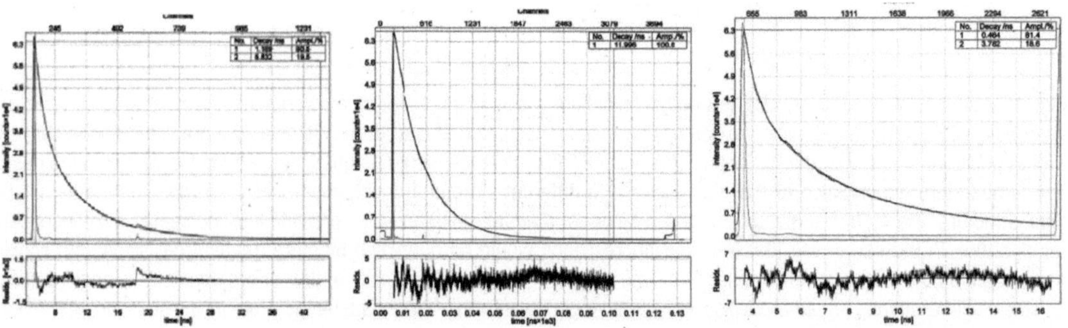

Fig. 5. Time-resolved fluorescence can provide information about the timescale of motions and the environment of bases. Three fluorescence emission decay curves are shown: *left*: 2AP alone, which is fit by $\tau = 12$ ns. *Mid*: 2AP9; $\tau_1 = 0.46$ ns with 81% amplitude and $\tau_2 = 3.8$ ns with 19% amplitude. *Right*: 2AP7; $\tau_1 = 1.2$ ns (80% amplitude) and $\tau_2 = 5.8$ ns with 20% amplitude. Fitting of the data is not trivial, and the solution is not necessarily unique, so great care must be exercised in interpretation. The shorter lifetime would come from stacking (and is certain to be an average of lifetimes), while the longer one reflects excursions of the base away from a stacked conformation. 2AP in both positions clearly spends time stacked, but the stacking (with its much shorter lifetime) is more pronounced at the 2AP9 position.

components due to stacking with the bases. If 2AP9 does have a long lifetime component, it will be present at very low amplitudes at room temperature, and come from its excursions out of its stacked structure. A sample decay curve with two lifetimes of 0.46 and 3.8 ns is shown in Fig. 5. Compare the time axes for the three samples: for 2AP the data were collected over 130 ns, but for 2AP9 the timescale is only 13 ns. Clearly 2AP9 decay is very rapid, indicative of stable stacking with other bases.

Time-resolved fluorescence anisotropy provides another measure of base structure and dynamics. The anisotropy of fluorescence decay (or the depolarization of the emission) is dependent on the overall tumbling of the molecule. 2AP alone is small and tumbles very rapidly; its fluorescence depolarization occurs in much less than 1 ns. When 2AP is attached to an RNA, its anisotropy will reflect its ability to move independently of the overall tumbling of the larger RNA molecule. If its motion is restricted by its structure in the RNA, then its depolarization (anisotropy) will be determined by the reorientation of the RNA. If it can move independently, for example if it is free to rotate about its glycosidic bond, then there will be two components, one short and one longer. Conventional interpretation of the anisotropy decay is that one value will report on the tumbling time of the entire molecule while the other (shorter) one reports on local motion.

The combination of steady-state and time-resolved 2AP fluorescence will reveal the environments of the loop body (2AP9) and its apex (2AP7), particularly indicating how flexible the bases are in these two regions. Base flexibility is an important part of the description of any RNA, for example to understand how a ligand is able to bind to the hairpin: a stable stacked hairpin loop looks very different than an overall floppy loop and the energetics of ligand binding will reflect its conformational states. Flexibility of even a single nucleotide could impact how an RNA molecule interacts with ligands, but the current state of the art does not allow accurate prediction of these properties.

### 3.6. NMR Spectroscopy of RNA

Nuclear magnetic resonance is the most powerful method for looking at the solution state of an RNA, especially for smaller constructs, and an RNA hairpin is an ideal candidate for a comprehensive study by NMR. Many spectroscopic methods require low concentrations (nM to μM) of RNA, but nuclear magnetic resonance is an exception. For NMR experiments, an ideal RNA concentration is $\geq$1 mM (hairpin), and that creates a different problem for characterization of structure and dynamics. Le Chatelier is not your friend here, for any hairpin can form a duplex at sufficiently high concentration: A + A $\leftrightarrow$ A:A. In the case of the U2 hairpin, duplex formation results in quite a stable structure (Fig. 1). It can be tricky to detect the hairpin-to-dimer transition, unless one form has a unique NMR spectroscopic signature.

This U2 RNA dimer has two A:U base pairs, two C:G base pairs, and two G:U base pairs that are not present in the hairpin, and their imino proton resonances would be indicative of the dimer.

One notable study of a larger and more structurally complex RNA (17), a 30-kDa GAAA:receptor RNA, will be used here to illustrate some valuable methods that can be applied to any RNA that can be synthesized in quantity and does not aggregate. The construct used for this work was very cleverly engineered to form a symmetric dimer, which has the effect of giving a spectrum equivalent to a monomer but with a twofold increase in the signal intensity. This simplification is necessary because the resolution of an NMR spectrum is limited by the redundancy in the chemical shift of bases, riboses, and phosphates. The chemical shift is the resonance frequency of a nucleus, expressed as parts per million of the static magnetic field. Ribose illustrates the problem: all nucleotides have a ribose, and all riboses in a duplex have the same sugar pucker ($C_{3'}$*-endo*). In a duplex, the ribose $^{13}C$ and $^{1}H$ at each position will have virtually the same magnetic environment, and so will have virtually the same chemical shift. Noncanonical environments, such as an internal bulge, alter the regular environment and lead to different chemical shifts, and riboses in a loop can have other puckers. Nucleobases have similar problems, since there are only four (to be compared to the 21 amino acid sidechains in a protein) and their $^{13}C$, $^{1}H$, and $^{15}N$ all resonate at similar frequencies (they appear at nearly the same chemical shifts). This level of chemical shift overlap can be overcome somewhat by selective labeling, to make RNAs with only $^{13}C$ or $^{15}N$ or $^{13}C/^{15}N$ labeled guanosine, or uridine, etc., and by selective pulse sequences. Thus the structure of the 30-kDa GAAA:receptor RNA was an exceptional undertaking.

The structure of the GAAA:receptor was solved using an RNA that contained both the tetraloop and the receptor in the same strand. A tetraloop in one strand docked with its receptor in another strand to create a stable symmetric dimer. Figure 6 illustrates how the imino protons of the RNA respond to ions. What are these imino protons? These are exchangeable $^{1}H$-N protons from guanosine and uridine that are base-paired in duplexes and can be hydrogen-bonded in noncanonical structures (Fig. 6). Formation of a stable hydrogen bond protects them from rapid exchange with water protons and allows them to be observed, although it is important to remember that imino protons are in constant exchange with water protons even in duplex structures. [As an aside, the mechanism of their exchange is not clear, but must require breakage of the base pairing and movement of the base into solution. The frequency of this motion (and the base pair lifetime), its dependence on sequence, and the geometry of the displacement remain subjects of research.] Counting the number of imino protons provides a quick assessment of the RNA structure, since there will be one

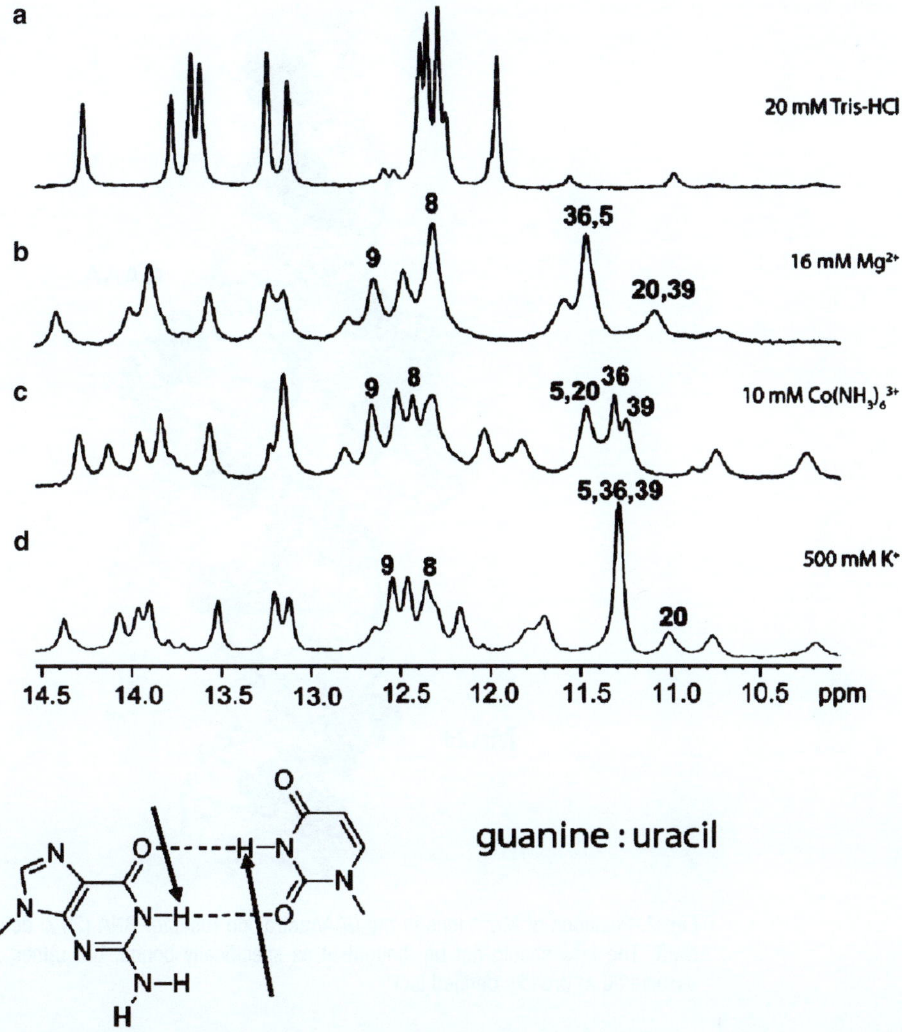


Fig. 6. Imino proton spectra of the GAAA:tetraloop receptor RNA as a function of ions. In the Tris–HCl solution, the tetraloop does not dock with its receptor, and the imino protons observed are from the duplexes. In the presence of $Mg^{2+}$, cobalt hexammine, or 500 mM KCl, new resonances appear that come from the tetraloop G (20) and the U's (5,36) and G's (8,9,39) in the receptor. Their chemical shifts are distinct to the ions in solution. The imino protons of U and G are noted at the bottom as they would appear in a wobble G:U pair.

imino proton resonance from each G:C and A:U pair and two from a G:U pair. Assignment of imino protons was confirmed using NOESY (through-space) experiments that essentially walk up the duplex using dipole–dipole energy transfer from proximal nuclei. As the figure illustrates, in Tris–HCl, the spectrum contains only imino protons from the duplex regions. When $Mg^{2+}$ is added, the imino protons from the GAAA/receptor interaction appear.

NMR is uniquely suited to study interactions of RNA with ions. This is a powerful application, because many RNAs require ions to form a specific structure, yet there are few spectroscopic methods to measure the interactions, much less identify the binding sites.

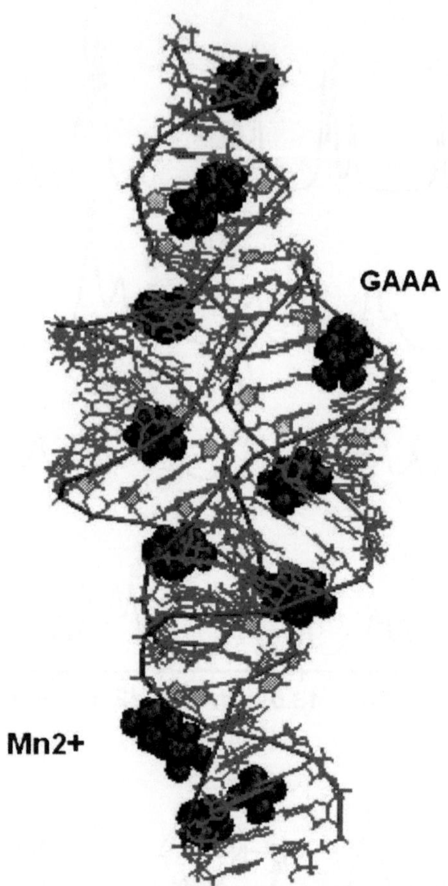


Fig. 7. Positions of $Mn^{2+}$ ions in the GAAA:tetraloop receptor RNA (2I7Z) determined by NMR. The ions should not be thought of as specifically bound, but rather on average associated at broadly defined loci.

The exquisite sensitivity of RNA to ions is readily seen by a comparison of the imino proton chemical shifts of the GAAA-receptor in $Mg^{2+}$, cobalt hexammine, and $K^+$ (Fig. 6) (18). Other NMR experiments showed that the structure was not altered as a function of the specific ion, so the changes in chemical shift were ascribed to ion associations. Using paramagnetic $Mn^{2+}$ broadening to detect sites of contact, the ions were localized on the edges of the bases and in the major groove of the RNA (Fig. 7). $Mn^{2+}$ is a hexahydrate as is $Mg^{2+}$ so these ions could use their waters as hydrogen bond donors for the hydrogen bond acceptors on the bases. These ions are not specific site binders, but rather diffusely associated with positions dictated by the electrostatic surface of the RNA. In this structure, there are no ions closely associated with the GAAA tetraloop.

Applying these methods to the U2 RNA hairpin, what ions might be found to interact with its hairpin loop? Assume that the first three and last three bases of the loop are stacked in a nearly

A-form structure that is anchored by the U3:A9 base pair. Then, the juxtaposed A1:C11 and U2:U10 bases form noncanonical base pairs that will attract ions to a deformed helix. The five nucleotides at the apex of the loop could form a G4CAGC8 pentaloop structure that uses hydrated ions to hydrogen bond to exposed G and A moieties. Alternatively, a G4:C8 pair could form, resulting in a triloop (19), which in turn would attract ions to hydrogen bond with its exposed bases. If the entire loop is unstructured, however, then ions would be diffusely associated.

UV melting studies of the U2 loop in the presence and absence of $Mg^{2+}$ might have suggested some contribution of the ion to the structure, to provide the impetus for subsequent analysis by NMR. Assuming that the NMR assignments of the U2 RNA have been completed, then a study analogous to that of the GAAA:receptor would be carried out. The first experiment is to simply add $Mg^{2+}$ to look for chemical shift changes of any resonance, beginning with the imino protons. If this looks promising, then more extensive studies using paramagnetic broadening by $Mn^{2+}$ or NOEs from cobalt hexammine to the bases would be done. An excellent study of ion binding to a small hairpin is that from the Legault lab (20) who probed a U-turn RNA with $Mg^{2+}$, $Mn^{2+}$, and $Co(NH_3)_6{}^{3+}$ using NMR. In that RNA, the divalent ions directly stabilize a compacted conformation of the U-turn. The atomic detail of ion binding provided by NMR must not be interpreted to mean that the ions are stuck to the RNA, only that as a solution average structure (averaged in time and space), there is a concentration of ions associated with a specific site in the structure.

## 4. Conclusions

Even a small RNA has complex behavior, and to describe that RNA requires experimental methods that measure its thermodynamic, dynamic, and structural parameters. Each RNA will be slightly different in its ion requirements, temperature dependence, the $pK_a$ of its bases, and its intrinsic flexibility. This narrative hopefully encourages thoughtful and creative application of the concepts to any novel RNA molecule.

## Acknowledgments

I thank W. Tom Stump for instrument maintenance, Artem Melynkov for fluorescence melting data, and Professor Sam Butcher for Fig. 6. Funding in part comes from NIH GM77231 and GM46318.

## References


1. Zuker M (2003) Mfold web server for nucleic acid folding and hybridization prediction. Nucleic Acids Res 31:3406–3415
2. Mathews DH, Sabina J, Zuker M, Turner DH (1999) Expanded sequence dependence of thermodynamics parameters improves prediction of RNA secondary structure. J Mol Biol 288:911–940
3. Mathews DH, Disney MD, Childs JL, Schroeder SJ, Zuker M, Turner DH (2004) Incorporating chemical modification constraints into a dynamic programming algorithm for prediction of RNA secondary structure. Proc Natl Acad Sci USA 101:7287–7292
4. Xia T, SantaLucia J, Burkard ME, Kierzek R, Schroeder SJ, Jiao X, Cox C, Turner DH (1998) Thermodynamic parameters for an expanded nearest-neighbor model for formation of RNA duplexes with Watson-Crick base pairs. Biochemistry 37:14719–14735
5. Milligan JF, Groebe DR, Witherall GW, Uhlenbeck OC (1987) Oligoribonucleotide synthesis using T7 RNA polymerase and synthetic DNA templates. Nucleic Acids Res 15:8783–8799
6. Laing LG, Draper DE (1994) Thermodynamics of RNA folding in a conserved ribosomal RNA domain. J Mol Biol 237:560–576
7. Soto AM, Misra V, Draper DE (2007) Tertiary structure of an RNA pseudoknot is stabilized by "diffuse" Mg2+ ions. Biochemistry 46:2973–2983
8. Chen SJ, Dill KA (2000) RNA folding landscapes. Proc Natl Acad Sci USA 97:646–651
9. Ma H, Proctor DJ, Kierzek E, Kierzek R, Bevilacqua PC, Gruebele M (2006) Exploring the energy landscape of a small RNA hairpin. J Am Chem Soc 128:1523–1530
10. Liu L, Cottrell JW, Scott LG, Fedor MJ (2009) Direct measurement of the ionization state of an essential guanine in the hairpin ribozyme. Nat Chem Biol 5:351–357
11. Jung J, Ihly R, Scott E, Yu M, Van Orden A (2008) Probing the complete folding trajectory of a DNA hairpin using dual beam fluorescence fluctuation spectroscopy. J Phys Chem B 112:127–133
12. Lin MM, Meinhold L, Shorokhov D, Zewail AH (2008) Unfolding and melting of DNA (RNA) hairpins: the concept of structure-specific 2D dynamic landscapes. Phys Chem Chem Phys 10:4227–4239
13. Garcia AE, Paschek D (2008) Simulation of the pressure and temperature folding/unfolding equilibrium of a small RNA hairpin. J Am Chem Soc 130:815–817
14. Lemay J-F, Penedo JC, Muhlbacher J, Lafontaine DA (2009) Molecular basis of RNA-mediated gene regulation on the adenine riboswitch by single-molecule approaches. Methods Mol Biol 540:1064–3745
15. Rachofsky EL, Osman R, Ross JBA (2001) Probing structure and dynamics of DNA with 2-aminopurine: effects of local environment on fluorescence. Biochemistry 40:946–956
16. Jean JM, Hall KB (2001) 2-aminopurine fluorescence quenching and lifetimes: role of base stacking. Proc Natl Acad Sci USA 98:37–41
17. Davis JH, Tonelli M, Scott LG, Jaeger L, Williamson JR, Butcher SE (2005) RNA helical packing in solution: NMR structure of a 30 kDa GAAA tetraloop-receptor complex. J Mol Biol 351:371–382
18. Davis JH, Foster TR, Tonelli M, Butcher SE (2007) Role of metal ions in the tetraloop-receptor complex as analyzed by NMR. RNA 13:76–86
19. Lisi V, Major F (2007) A comparative analysis of the triloops in all high-resolution RNA structures reveals sequence structure relationships. RNA 13:1537–1545
20. Campbell DO, Bouchard P, Desjardins G, Legault P (2006) NMR structure of Varkud satellite ribozyme stem-loop V in the presence of magnesium ions and localization of metal-binding sites. Biochemistry 45:10591–10605
21. Hall KB, Williams DJ (2004) Dynamics of the IRE RNA hairpin probed by 2-aminopurine fluorescence and stochastic dynamics simulations. RNA 10:34–47

# Chapter 5

# Fluorescence Methods to Study DNA Translocation and Unwinding Kinetics by Nucleic Acid Motors

Christopher J. Fischer, Eric J. Tomko, Colin G. Wu, and Timothy M. Lohman

## Abstract

Translocation of nucleic acid motor proteins (translocases) along linear nucleic acids can be studied by monitoring either the time course of the arrival of the motor protein at one end of the nucleic acid or the kinetics of ATP hydrolysis by the motor protein during translocation using pre-steady state ensemble kinetic methods in a stopped-flow instrument. Similarly, the unwinding of double-stranded DNA or RNA by helicases can be studied in ensemble experiments by monitoring either the kinetics of the conversion of the double-stranded nucleic acid into its complementary single strands by the helicase or the kinetics of ATP hydrolysis by the helicase during unwinding. Such experiments monitor translocation of the enzyme along or unwinding of a series of nucleic acids labeled at one position (usually the end) with a fluorophore or a pair of fluorophores that undergo changes in fluorescence intensity or efficiency of fluorescence resonance energy transfer (FRET). We discuss how the pre-steady state kinetic data collected in these ensemble experiments can be analyzed by simultaneous global nonlinear least squares (NLLS) analysis using simple sequential "*n-step*" mechanisms to obtain estimates of the macroscopic rates and processivities of translocation and/or unwinding, the rate-limiting step(s) in these mechanisms, the average "kinetic step-size," and the stoichiometry of coupling ATP binding and hydrolysis to movement along the nucleic acid.

**Key words:** Translocase, Helicase, ATPase, Motor protein, Stopped-flow kinetics

## 1. Introduction

The ability to translocate processively and with biased directionality along a nucleic acid (NA) filament is central to the biological function of many enzymes involved in nucleic acid metabolism including DNA and RNA polymerases (1), helicases (2–6), chromatin remodelers (7–10), some nucleases (11, 12), and some restriction enzymes (13–15). These "molecular motors" all use the chemical potential energy obtained through the binding and hydrolysis of nucleoside triphosphates (NTP or dNTP) to perform the mechanical work of directional translocation along the NA filament.

Wlodek M. Bujalowski (ed.), *Spectroscopic Methods of Analysis: Methods and Protocols*, Methods in Molecular Biology, vol. 875, DOI 10.1007/978-1-61779-806-1_5, © Springer Science+Business Media New York 2012

Helicases are also capable of the NTP-dependent unwinding of double-stranded nucleic acids (dsNA) (2, 3, 16, 17). An understanding of the translocation and unwinding mechanisms of these motor proteins requires quantitative kinetic information to obtain the rate constants, processivities, kinetic step-sizes, and ATP coupling stoichiometries associated with these processes.

Here, we describe the design and analysis of pre-steady state ensemble stopped-flow kinetic experiments (3, 6, 8, 14, 15, 18–23) to probe either the mechanism of single-stranded nucleic acid (ssNA) translocation or of double-stranded nucleic acid (dsNA) unwinding by processive nucleic acid motor proteins using a simple sequential "*n*-step" kinetic model. The application of this methodology provides an accurate determination of macroscopic kinetic parameters such as the rate of net forward motion of the motor protein along the NA and the net efficiency at which the hydrolysis of ATP is coupled to this net forward motion. However, the estimates of microscopic kinetic parameters, such as the kinetic step size of translocation, can be inflated under some circumstances if nonuniform motion or persistent heterogeneity (static disorder) (24–27) occurs during translocation. We also compare the approach of analyzing the full time course of the kinetic reaction to a simpler "time to peak" analysis and show that the simpler method can grossly overestimate the rate of translocation under a number of circumstances.

## 2. Materials

The experiments discussed below are applicable to any NA motor that can translocate along ssNA or unwind dsNA substrates. Generally, the solution conditions (i.e., buffer components, pH, temperature, etc.) that are used are those in which the motor enzyme is stable, well behaved, can interact with the NA substrate, and can function. Special consideration is needed when designing the NA substrates and selecting a trap for free motor enzyme.

### 2.1. ssNA and dsNA Substrates

Three important features need to be considered when designing the fluorescent ssNA and dsNA substrates: length, base composition, and fluorophore. In order to determine the kinetic parameters for translocation along ssNA or unwinding of dsNA, experiments should be carried out with a series of NA substrates of different lengths. In our studies we generally start by testing 4–5 different lengths spanning the range from 25 to 124 nucleotides or base pairs. Since these nucleic acids are generally synthesized, the lengths are limited to less than 150 nucleotides or base pairs.

When monitoring translocation along single-stranded DNA (ssDNA), we have used nucleic acids composed of a single base type, generally oligodeoxythymidylates (oligo(dT)) because these do not form internal base pairs and are the easiest oligodeoxynucleotides to synthesize. Similarly, oligouridylates (oligo(U)) could be used if one is studying single-stranded RNA (ssRNA) translocation (6).

When selecting a fluorophore for monitoring translocation along ssNA the fluorophore must yield a detectable fluorescent change upon interaction with the motor enzyme. A number of fluorescent dyes are commercially available as phosphoramadite derivatives and thus can be incorporated directly into the ssDNA using an automated oligodeoxynucleotide synthesizer. For translocation substrates the dye is usually placed at either the 3′-or 5′-end of the ssNA. We have found that Cy3 and Fluorescein have generally yielded good signal changes for the translocases that we have studied (20, 28–30). For DNA or RNA unwinding substrates the complementary strands are labeled with a FRET pair such as Cy3 and Cy5. Upon separation of the complementary strands, a change in FRET is detected.

### *2.2. Traps for Free Enzyme*

All the experiments discussed below are performed as single-round kinetic experiments in that they monitor a single round of NA translocation or NA unwinding by an enzyme initially bound to the NA substrate before the initiation of the translocation or unwinding reaction. That is, any enzyme that dissociates during translocation or unwinding is prevented from reinitiating translocation or unwinding on the NA and any enzyme that is initially free in solution is prevented from binding the NA. This condition is essential if one is to apply the analysis in Subheading 3.5 to extract the kinetic parameters for ssNA translocation or dsNA unwinding from the data. Single-round conditions can be maintained experimentally by including a trap for free enzyme in the reaction that prevents rebinding of enzyme to the NA substrate.

In principle any nucleic acid that binds to the motor enzyme can be employed as a trap. In addition, we have often used the polyanion heparin which binds non-specifically to most nucleic acid-binding proteins. For the enzymes we have studied we use heparin, obtained from porcine mucosa (available commercially), as a trap (6, 20, 21, 28, 30, 31). Heparin is relatively inexpensive, has high solubility (~50 mg/ml) in aqueous buffers, and spectroscopic assays are available for readily determining the concentration (32) (see Note 1). The advantage of heparin is that the ATPase activity of the enzyme is not generally stimulated by heparin (30). This will allow for more straightforward and simple analysis of the ATPase activity of the motor protein that is associated with translocation or unwinding. The concentration of trap used for a given experiment needs to be empirically determined for a particular trap, enzyme, and solution condition.

## 3. Methods

The kinetic models and associated equations used to analyze the ssNA translocation and dsNA unwinding experiments assume that no more than one molecular motor is bound to each NA substrate. Thus, the application of the kinetic models to the analysis of the data requires the experiments to be performed under these conditions. Generally, for experiments monitoring translocation along ssNA this is achieved by performing the experiments under conditions where the ssNA is in molar excess over the motor protein. For experiments monitoring dsNA unwinding, where the helicase generally initiates at a particular site, the relative concentration of dsNA to the concentration of motor protein will vary depending upon the system. As mentioned in Subheading 2.2, analysis of the data collected in these experiments also assumes single round or single turnover conditions, i.e., that any motor enzyme that is initially free in solution at the start of the reaction or that dissociates from the DNA during translocation or unwinding does not rebind to the DNA. This is accomplished experimentally by including a trap for free enzyme (e.g., heparin).

### 3.1. Monitoring the Kinetics of the Arrival of the Translocase at a Specific Site on ssNA

The first experimental method we discuss to study the kinetics of translocation of an enzyme along ssNA is a stopped-flow fluorescence approach first introduced by Dillingham et al. (33), and subsequently modified by Fischer et al. (19, 20, 30), is depicted in Fig. 1a. For studies of ssDNA translocation, the method utilizes a series of oligodeoxythymidylates of varying lengths, $L$ ($(\mathrm{d}T)_L$), that have a fluorophore attached covalently to either the 3′ or 5′ end of the DNA. When the translocase reaches and interacts with the fluorophore a change in fluorescence signal occurs. In this way, one can monitor the time-dependent concentration of the translocase at the DNA end resulting from arrival of the translocase due to translocation from other sites on the DNA and dissociation of translocases from the end.

These experiments are performed by pre-incubating the translocase with ssNA in one syringe of the stopped-flow and initiating translocation by rapidly mixing the enzyme:ssNA complex with ATP, $MgCl_2$, and trap. When the experiments are performed as a function of ssNA length the resulting time courses can be analyzed using a sequential $n$-step model discussed in Subheading 3.5 to determine the microscopic kinetic parameters associated with translocation of the enzyme along the NA. An example of time courses for translocation along ssDNA by the monomeric UvrD translocase is shown in Fig. 2. Similar experiments could be performed for an RNA translocase with a series of fluorescently end-labeled ssRNA, such as oligouridylate (oligo(U)) (6).

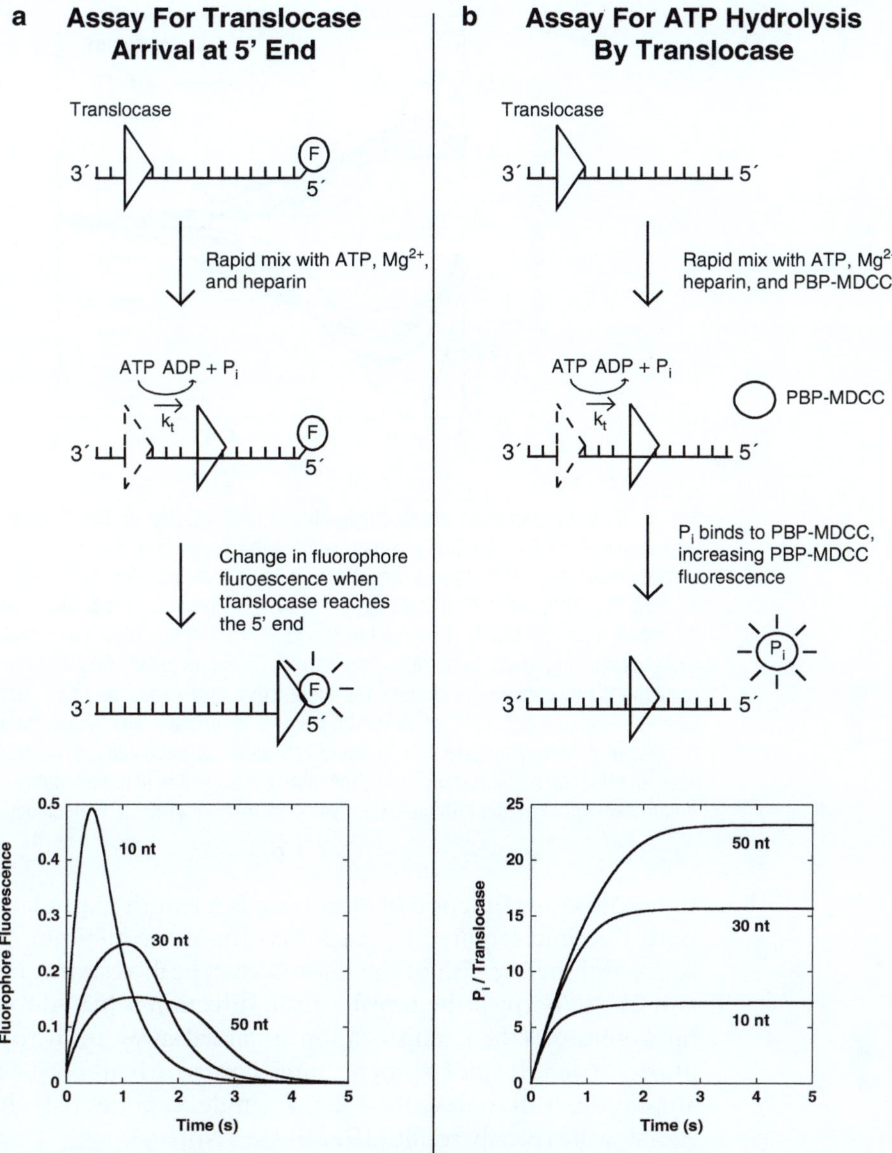


Fig. 1. Stopped-flow assays for monitoring the pre-steady state kinetics of enzyme translocation along ssNA. Panel (**a**): A translocase is pre-bound to single-stranded ssNA labeled at the 5′-end with a fluorescent dye, then rapidly mixed with ATP, $Mg^{2+}$, and heparin (protein trap) to initiate translocation. When the translocase nears the 5′-end of the ssNA, the fluorescence of the dye is either quenched or enhanced. Example time courses are shown for three different lengths of DNA. Panel (**b**): A translocase is pre-bound to ssNA and then rapidly mixed with ATP, $Mg^{2+}$, heparin, and an excess concentration of fluorescently labeled phosphate-binding protein (PBP-MDCC) to initiate translocation. As the translocase moves along the filament, ATP is hydrolyzed into ADP and inorganic phosphate ($P_i$). PBP-MDCC rapidly binds the $P_i$ resulting in an increase in the PBP-MDCC fluorescence. Example time courses are shown for three different lengths of ssNA.

The directionality bias of translocation along the ssNA can be determined by comparing the time courses observed when the fluorophore is attached to the 3′ versus the 5′ end of the ssDNA (19, 20, 33). Specifically, characteristic changes in the fluorescence

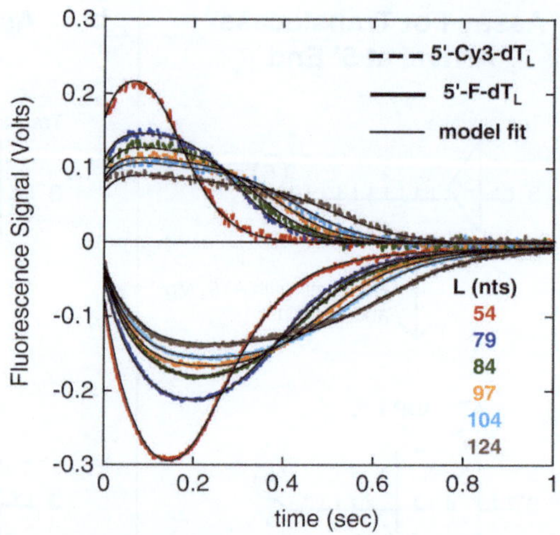


Fig. 2. UvrD translocation along single-strand DNA labeled at the 5′-end with either fluorescein (F) or Cy3. UvrD is pre-incubated with excess DNA then rapidly mixed in the stopped-flow with ATP, $MgCl_2$, and heparin to initiate translocation (final conditions: 10 mM Tris–HCl, pH 8.3, 20 mM NaCl, 20% (v/v) glycerol, 25 nM UvrD, 50 nM DNA, 0.5 mM ATP, 2 mM $MgCl_2$, 4 mg/ml heparin at 25°C). Heparin serves as a protein trap for UvrD, preventing UvrD from rebinding to the DNA and reinitiating translocation. The resulting fluorescence time courses for fluorescein- (*solid lines*) and Cy3- (*dashed lines*) labeled single-strand DNA of differing lengths is shown. The single-strand DNA is composed of deoxythymidylates to avoid formation of secondary structures that may have an effect on translocation. The *black curves* are a global fit to both data sets using an *n*-step sequential model (30) yielding a value of $m^*k_t = (193 \pm 1)$ nucleotides/s.

time course as a function of increasing NA length (e.g. an increase in both the time required to reach maximum (or minimum) fluorescence and the breadth of the fluorescence peak as shown in Figs. 1a and 2) will occur if the translocation direction is biased toward the fluorophore. When translocation is biased away from the fluorophore, a length-independent time course when normalized for amplitude, often described by a single exponential change in signal, will typically result (19, 20) (see Note 6).

### 3.2. Monitoring the Kinetics of ATP Hydrolysis by the Translocase During Translocation

Enzyme translocation along ssNA can also be monitored by measuring the amount of ATP hydrolyzed by the enzyme during translocation. This approach also requires transient pre-steady state kinetic experiments rather than steady-state ATPase experiments since steady-state rates of ATP hydrolysis will generally be limited by other kinetic processes that are slower than protein translocation (e.g. dissociation and/or rebinding of protein to another NA molecule). The pre-steady state rate and extent of ATP hydrolysis by the translocase can be monitored, for example, by directly measuring the conversion of ATP to ADP using a radioactive assay (34, 35) or by monitoring the release of inorganic phosphate using a fluorescently labeled phosphate-binding protein (30, 36) as depicted in Fig. 1b.

Analysis of a series of time courses of ATP hydrolysis during translocation performed as a function of ssNA length can be analyzed using a sequential $n$-step model (Subheading 3.5) to estimate the ATP coupling stoichiometry during translocation. This analysis requires knowledge of the kinetic parameters obtained from independent analysis of translocation time courses using methods in Subheading 3.1 to be used as constraints, due to parameter correlation in the $n$-step model (30).

### 3.3. Monitoring the Kinetics of dsNA Unwinding

A generalized stopped-flow fluorescence-based technique for monitoring the helicase catalyzed unwinding of dsDNA is depicted in Fig. 3a (22, 37). This method employs a series of dsNA substrates, of varying lengths, $L$, that have donor and acceptor fluorophores attached covalently to either strand of the dsNA. In Fig. 3a, the two fluorophores are shown adjacent to each other in order to improve the fluorescence resonance energy transfer (FRET) between the two fluorophores; however, alternate orientations are also possible. The FRET efficiency between the two fluorophores will decrease significantly upon unwinding of the duplex and subsequent separation of the two single strands. This decrease in FRET efficiency will result in an increase in the fluorescence intensity of the donor fluorophore and a decrease in the fluorescence intensity of the acceptor fluorophore, assuming only FRET changes occur. Quantitative analysis of a series of these time courses performed as a function of $L$ using Eq. 10 allows one to estimate the microscopic kinetic parameters associated with translocation of the enzyme along the DNA. A subsequent estimate of the kinetic step size, $m$, can then be obtained from the analysis of the dependence of $n$ on $L$ through Eq. 12. An example of time courses for unwinding of double-stranded DNA by the RecBC helicase (38) is shown in Fig. 4.

### 3.4. Monitoring the Kinetics of ATP Hydrolysis by the Helicase During Double-Stranded DNA Unwinding

The helicase catalyzed unwinding of double-stranded DNA can also be monitored by measuring the amount of ATP hydrolyzed by the helicase during the unwinding reaction. As with measurements of ssNA translocation, the pre-steady state rate and extent of ATP hydrolysis by the helicase as it unwinds dsNA can be monitored using either a radioactive assay (34, 35) or a fluorescence-based assay (30, 36) as depicted in Fig. 3b.

Analysis of a series of time courses of ATP hydrolysis during unwinding performed as a function of dsNA length, $L$, can be analyzed using Eq. 11 to determine estimates of the microscopic parameters $c$, $k_a$, $k_{\mathrm{end}}$, and the macroscopic ATP coupling stoichiometry $c/m$. In this analysis, the values of the microscopic parameters obtained from the analysis of unwinding time courses using methods in Subheading 3.4 ($k_u$, $k_d$, $k_{\mathrm{end}}$,) are used as fixed constraints in the application of Eq. 10.

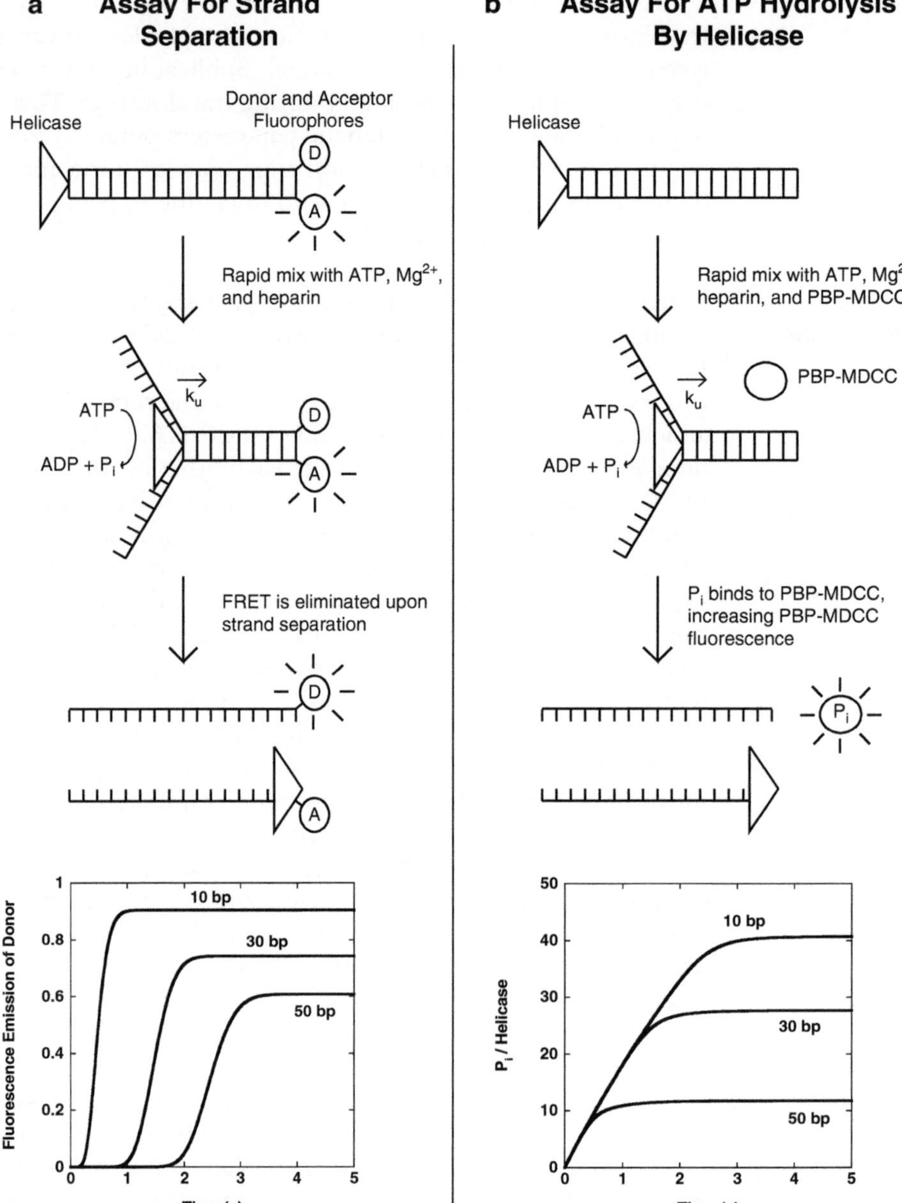


Fig. 3. Stopped-flow assays for monitoring the pre-steady state kinetics of helicase catalyzed dsNA unwinding. Panel (**a**): A helicase is pre-bound to dsNA labeled at the opposite end with donor and acceptor fluorophores. The close proximity of the two fluorophores results in large FRET between them. The helicase–dsNA complexes are then rapidly mixed with ATP, $Mg^{2+}$, and heparin (protein trap) to initiate dsNA unwinding. When the helicase completely unwinds the dsNA, the two single strands separate and the FRET efficiency is dramatically decreased; this decrease in FRET efficiency will result in an increase in the fluorescence emission of the donor fluorophore and a decrease in the fluorescence emission of the acceptor fluorophore. Example time courses are shown for the unwinding of three different lengths of dsNA. Panel (**b**): A helicase is pre-bound to dsNA and then rapidly mixed with ATP, $Mg^{2+}$, heparin, and an excess concentration of fluorescently labeled phosphate-binding protein (PBP-MDCC) to initiate translocation. As the helicase unwinds the dsNA, ATP is hydrolyzed into ADP and inorganic phosphate ($P_i$). PBP-MDCC rapidly binds the $P_i$ resulting in an increase in the PBP-MDCC fluorescence. Example time courses are shown for the unwinding of three different lengths of dsDNA.

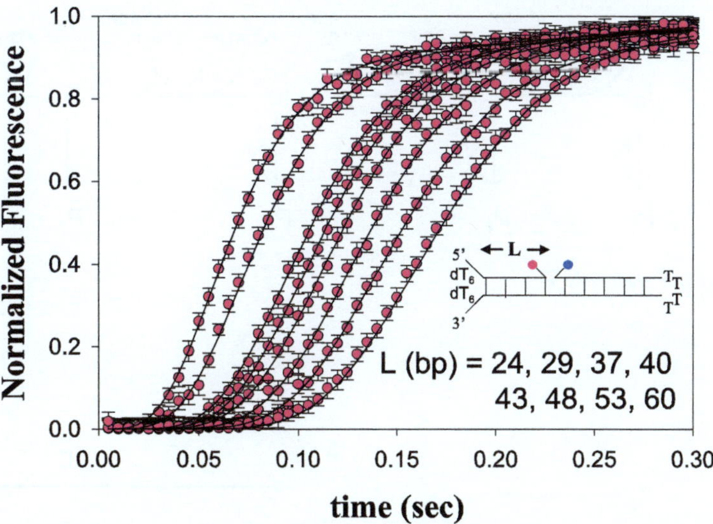


Fig. 4. Stopped-flow fluorescence time courses of RecBC catalyzed DNA unwinding. 40 nM of DNA duplexes of varying length *L* are pre-incubated with 200 nM RecBC. DNA unwinding is initiated by rapid mixing with 10 mM ATP and 15 mg/ml heparin trap (20 mM Mops-KOH (pH 7.0 at 25°C), 30 mM NaCl, 10 mM $MgCl_2$, 1 mM 2-ME, 5% (v/v) glycerol). Each DNA substrate is labeled with Cy3 (*pink*) and Cy5 (*blue*) as indicated, and DNA unwinding is observed by monitoring both Cy3 and Cy5 fluorescence simultaneously. Cy3 time courses are shown and the data is described well by Scheme 2 using Eq. 10 with an associated value of $m{*}k_u = (372 \pm 15)$ base pairs/s.

### 3.5. Sequential "n-step" Models for Analyzing Translocation and DNA Unwinding Time Courses

#### 3.5.1. Translocation Along ssNA

The sequential "*n*-step" kinetic mechanism shown in Scheme 1 has been used to model enzyme translocation and its coupling to ATP hydrolysis (19, 30). In this mechanism (19), depicted in Fig. 5, a translocase with an occluded site size of *p* nucleotides and a contact size of *d* nucleotides binds with polarity to an ssNA molecule, *L* nucleotides long. The contact size, *d*, represents the number of consecutive nucleotides required to satisfy all contacts with the translocase and is thus less than or equal to the occluded site size. In this discussion we will assume that translocation along the ssNA is directionally biased from 3′ to 5′, but the results are equally applicable to a translocase which exhibits the opposite directional bias (see Note 9).

The translocase is initially bound *i* translocation steps away from the 5′-end, with concentration, $I_i$. The number of translocation steps, *i*, is constrained ($1 \leq i \leq n$), where *n* is the maximum number of translocation steps needed for a translocase bound initially at the 3′ end to move to the 5′ end of the NA. In the discussion here we consider two initial binding states for the translocase: one in which all proteins initiate translocation from the same site on the ssNA (in this case at the 3′ end of the ssNA, which is *n* steps away from the 5′ end) and one in which the proteins initiate translocation from random binding sites on the ssNA.

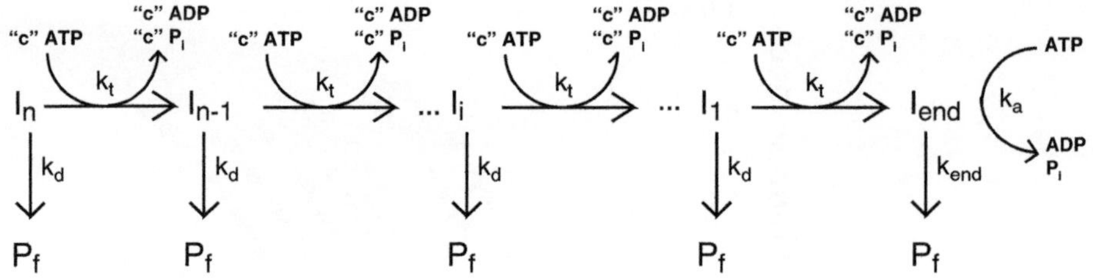

Scheme 1. (see Note 7)

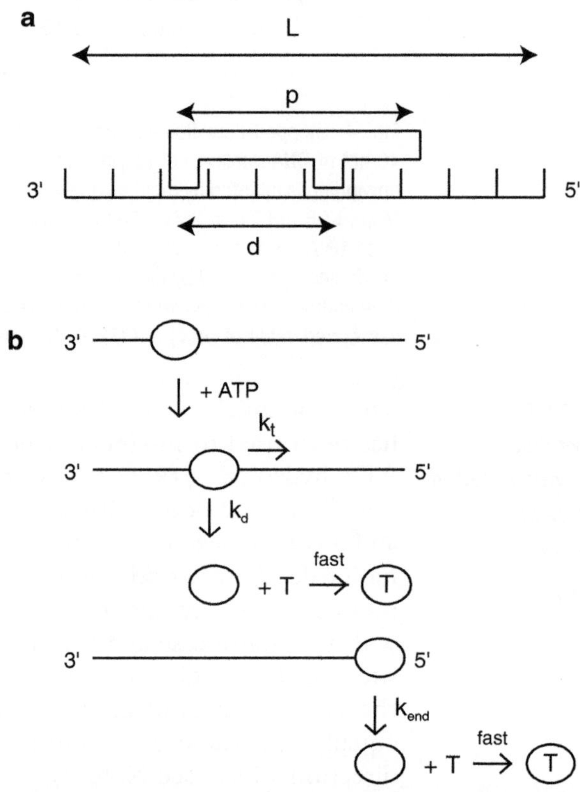


Fig. 5. Kinetic model for ATP-dependent protein translocation along an ssNA filament. Panel (**a**): A cartoon depicting the binding of a translocase with a contact size *d* and occluded site size *p* to an ssNA filament of length *L*. As shown in this cartoon, the contact size, *d*, is always less than or equal to the occluded site size, *b*. Panel (**b**): Cartoon showing the model used to describe enzyme translocation along an ssNA filament. The line segments represent the ssNA and the triangles represent the translocase. The translocase binds randomly, but with polarity, to the ssNA and upon binding and hydrolysis of ATP proceeds to translocate toward the 5′ end of the filament in discrete steps with rate constant $k_t$. The rate constant of dissociation during translocation is $k_d$. Upon reaching the 5′ end of the filament, the translocase dissociates with a rate constant $k_{end}$. Dissociated translocases bind to a protein trap, *T*, and are thereby prevented from rebinding the ssNA.

Upon addition of ATP, the translocase moves with directional bias along the ssNA via a series of repeated rate-limiting translocation steps each associated with the same rate constant, $k_t$. The rate constant for protein dissociation during translocation is $k_d$. The processivity of translocation along the NA can be defined as $P = \frac{k_t}{k_d + k_t}$. Between two successive rate-limiting translocation steps the enzyme moves $m$ nucleotides, while hydrolyzing $c$ ATP molecules. Therefore, $c/m$ is defined as the macroscopic ATP coupling stoichiometry and corresponds to the average number of ATP molecules hydrolyzed per nucleotide translocated along the ssNA. Similarly the product $m^*k_t$ is the macroscopic translocation rate in units of nucleotides/second. When the translocase reaches the 5′-end of the NA it continues to hydrolyze ATP with rate constant $k_a$ and dissociates from the NA with rate constant $k_{\text{end}}$. This hydrolysis of ATP at the end of the ssNA is not coupled to the physical movement of the enzyme along the ssNA and thus is referred to as futile hydrolysis (30, 36).

We note that, in general, $k_t$ represents the rate constant for the rate-limiting step that occurs within each repeated translocation cycle and does not necessarily correspond to the rate constant for physical movement of the translocase along the ssNA (19). Similarly, the average number of nucleotides translocated between two successive rate-limiting steps, defined as the translocation "kinetic step-size" ($m$), can be larger than the length of ssNA traversed during hydrolysis of a single ATP.

Based on Scheme 1, the expressions in Equations and can be derived (19, 22) for the time-dependent accumulation of protein at the 5′ end of the ssNA. In these equations, $\mathcal{L}^{-1}$ is the inverse Laplace transform operator, $s$ is the Laplace variable (22) and the parameters $k_t$, $k_d$, $k_{\text{end}}$, $c$, $k_a$, and $n$ are as defined above and $r$ is the initial (at time, $t = 0$) ratio of the probability of the translocase binding to any one binding position on the ssNA other than the 5′ end to the probability of the translocase binding to the 5′ end (19, 20) (see Notes 3, 4, and 8). For the case where all the proteins are initially bound at the same position (in this case taken to be the 3′ end of the ssNA), the equation for the time-dependent accumulation of protein at the 5′ end of the ssNA is given by Eq. 1.

$$f_{5'}(t) = A^*\mathcal{L}^{-1}\left[\frac{1}{k_{\text{end}} + s}\left(\frac{k_t}{k_t + k_d + s}\right)^n\right] \tag{1}$$

For the case where all the proteins are initially bound at random positions along the NA, the equation for the time-dependent accumulation of protein at the 5′ end of the NA is given by Eq. 2.

$$f_{5'}(t) = \frac{A}{1 + n^*r}\mathcal{L}^{-1}\left[\frac{1}{s + k_{\text{end}}} * \left(1 + \frac{k_t * r}{s + k_d}\left(1 - \left(\frac{k_t}{s + k_t + k_d}\right)^n\right)\right)\right] \tag{2}$$

The scalar $A$ in Eqs. 1 and 2 allows for conversion of the concentration of protein bound at the 5′ end of the NA into a signal that can be measured experimentally (e.g., a spectroscopic change) (19, 20, 30).

Similarly, Eqs. 3 and 4 are expressions for the time-dependent production of ADP or $P_i$, due to ATP hydrolysis by the translocase (19). In Eqs. 3 and 4, $I(0)$ is the concentration of translocase initially bound to the NA (at time, $t = 0$) and the scalar A allows for the conversion of the concentration of ADP (or $P_i$) produced by the translocase into a signal that can be measured experimentally. Equation 3 describes the ATP production occurring when all proteins are initially bound at the same position (e.g., the 3′ end of the NA) and Eq. 4 describes the ATP production occurring when all the proteins are initially bound at random positions along the NA.

$$\mathrm{ADP}(t) = A^{*}I(0)^{*}\mathcal{L}^{-1}\left[\frac{1}{s}\left(\frac{c^{*}k_t^{*}\left(1-\left(\frac{k_t}{k_t+k_d+s}\right)^{n}\right)}{k_d+s}+\frac{k_a}{k_{\mathrm{end}}+s}\left(\frac{k_t}{k_t+k_d+s}\right)^{n}\right)\right] \tag{3}$$

$$\mathrm{ADP}(t) = \frac{A^{*}I(0)}{1+n^{*}r}\mathcal{L}^{-1}\left[\frac{1}{s}\left(\frac{c^{*}k_t^{*}r^{*}\left(n(k_d+s)+k_t\left(\left(\frac{k_t}{k_t+k_d+s}\right)^{n}-1\right)\right)}{(k_d+s)^2}+\frac{k_a\left(1+\frac{k_t * r\left(1-\left(\frac{k_t}{k_t+k_d+s}\right)^{n}\right)}{k_d+s}\right)}{k_{\mathrm{end}}+s}\right)\right] \tag{4}$$

The maximum number of translocation steps, $n$, for a NA of a length, $L$, is related to the translocation kinetic step-size, $m$, and the translocase contact size by Eq. 5.

$$n = \frac{L-d}{m} \tag{5}$$

Equation 5 can be re-expressed as Eq. 6 which allows for the determination of $m$ from the experimentally determined dependence of $L$ on $n$.

$$L = mn + d \tag{6}$$

*3.5.2. Unwinding of dsNA*

The sequential "$n$-step" kinetic mechanism shown in Scheme 2 can be used to model helicase catalyzed dsNA unwinding and the coupling of this process to ATP hydrolysis (22). In this mechanism, a helicase binds to a unique site at one end of region of dsNA $L$ base pairs long. Upon addition of ATP, the helicase unwinds the dsNA through a series of repeated rate-limiting steps each associated with the same rate constant, $k_u$. The rate constant for protein dissociation during unwinding is $k_d$. An expression for the processivity of unwinding, identical to the expression for the processivity of translocation, can be defined as $P = \frac{k_t}{k_d+k_t}$. Between two successive rate-limiting

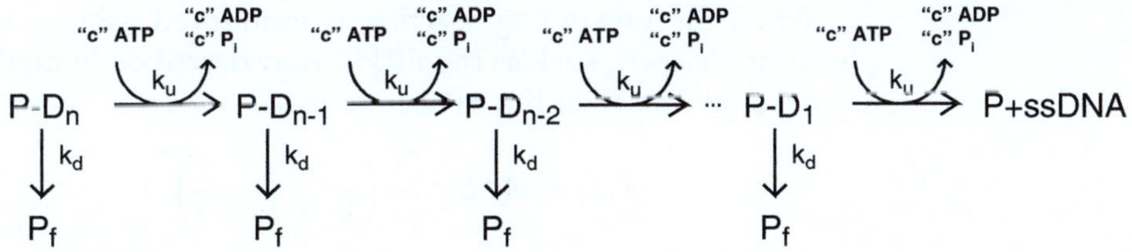

"c" ATP / "c" ADP "c" $P_i$

$$P\text{-}D_n \xrightarrow{k_u} P\text{-}D_{n-1} \xrightarrow{k_u} P\text{-}D_{n-2} \xrightarrow{k_u} \cdots\ P\text{-}D_1 \xrightarrow{k_u} P\text{+ssDNA}$$

$$P\text{-}D_n \xrightarrow{k_d} P_f, \quad P\text{-}D_{n-1} \xrightarrow{k_d} P_f, \quad P\text{-}D_{n-2} \xrightarrow{k_d} P_f, \quad P\text{-}D_1 \xrightarrow{k_d} P_f$$

Scheme 2. (see Note 7)

unwinding steps the enzyme unwinds $m$ base pairs of the dsNA, while hydrolyzing $c$ ATP molecules. As for the case of ssNA translocation, the ratio $c/m$ is defined as the macroscopic ATP coupling stoichiometry and, in this case, corresponds to the average number of ATP molecules hydrolyzed per base pair of dsNA unwound. Similarly the product $m^*k_u$ is the macroscopic unwinding rate in units of base pairs/second. When the helicase completes the unwinding of all $L$ base pairs of the dsNA, it may remain bound to one of the resulting ssNA strands and continue to hydrolyze ATP (futile hydrolysis) with rate constant $k_a$ and dissociates from the DNA with rate constant $k_{\text{end}}$.

We note that, in general, $k_u$ represents the microscopic rate constant for the rate-limiting step that occurs within each repeated unwinding cycle and does not necessarily correspond to the rate constant for physical movement of the helicase through the dsNA (19). Similarly, the average number of base pairs unwound between two successive rate-limiting steps, defined as the translocation "kinetic step-size" ($m$), can be larger than the length of dsNA traversed during hydrolysis of a single ATP.

Based on Scheme 2, the expression in Eq. 7 can be derived (22) for the time-dependent formation of ssNA (or the corresponding decrease in the concentration of dsNA) resulting from the activity of the helicase.

$$\text{ssDNA}(t) = P^n\left(1 - \frac{\Gamma(n,(k_u + k_d)t)}{\Gamma(n)}\right) \tag{7}$$

In this equation, the parameters P, $k_u$, $k_d$, and $n$ are defined as above. $\Gamma(n)$ is the Gamma function, defined in Eq. 8, and $\Gamma(n, (k_u + k_d)t)$ is the incomplete Gamma function, defined in Eq. 9.

$$\Gamma(n) = \int_0^\infty r^{n-1}e^{-r}\mathrm{d}r = (n-1)! \tag{8}$$

$$\Gamma(n,(k_u + k_d)t) = \int_{(k_u+k_d)t}^\infty r^{n-1}e^{-r}\mathrm{d}r \tag{9}$$

When the unwinding reaction is monitored using a spectrophotometric assay, such as the FRET assay described in methods in Subheading 3.3, then Eq. 10 can be used.

$$f(t) = A^{*}\mathcal{L}^{-1}\left[\frac{1}{s}\left(\frac{k_u}{k_u + k_d + s}\right)^{n}\right] \tag{10}$$

In Eq. 10, $\mathcal{L}^{-1}$ is the inverse Laplace transform operator, $s$ is the Laplace variable (22), and the parameters $k_u$, $k_d$, and $n$ are as defined above. The scalar $A$ allows for the conversion of the formation of ssNA into a signal that can be measured spectrophotometrically (see Note 5).

Similarly, Eq. 11 is the expression for the time-dependent production of ADP or $P_i$, due to ATP hydrolysis by the helicase during the unwinding of the dsNA.

$$\mathrm{ADP}(t) = [P - D_n]_{t=0} * \mathcal{L}^{-1}\left[\frac{1}{s}\left(\frac{c^{*}k_u{}^{*}\left(1 - \left(\frac{k_u}{k_u+k_d+s}\right)^{n}\right)}{k_d + s} + \frac{k_a}{k_{\mathrm{end}} + s}\left(\frac{k_u}{k_u + k_d + s}\right)^{n}\right)\right] \tag{11}$$

In Eq. 11, $\mathcal{L}^{-1}$ is the inverse Laplace transform operator, $s$ is the Laplace variable (22) and the parameters $k_u$, $k_d$, $k_{\mathrm{end}}$, $k_a$, and $n$ are as defined above. $[PD_n]_{t=0}$ is the concentration of helicase initially bound to the dsNA (at time, $t = 0$). The maximum number of unwinding steps, $n$, for a dsNA of length $L$ is related to the kinetic step size, $m$, by Eq. 12.

$$n = \frac{L}{m} \tag{12}$$

## 4. Notes

1. In preparing the trap for the motor enzyme, it is important to determine the concentration of trap that will be effective in trapping any free enzyme. This is especially true if heparin is used as the trap, since heparin preparations obtained commercially are not homogeneous and in our experience each new lot of heparin varies in its effectiveness. Trap effectiveness can be assayed most directly by monitoring the kinetics of enzyme binding to the ssNA or dsNA used in the translocation or unwinding experiments, respectively, in the presence of varying concentrations of the trap. The minimum concentration of the trap that can effectively compete with the NA for enzyme binding is the concentration that should be used in all experiments as the presence of trap may affect the kinetics of ssNA translocation or dsNA unwinding (see Note 2).

2. It has been observed that the enzyme trap can often facilitate the dissociation of the motor enzyme from the NA substrate. We have observed this in our studies of *E. coli* UvrD using heparin as a trap (20). This can be determined by performing a series of experiments as a function of the trap concentration to see if the time courses change. In the case of UvrD, the only kinetic parameters affected by the heparin concentration are the rate constants for dissociation from internal sites and for dissociation from the 5′-end of the ssDNA (20), which increase with heparin concentration. If this situation occurs the processivity will also be dependent on trap concentration due to the effect on the dissociation rate constant.

3. When monitoring motor enzyme translocation along ssNA, a single set of fluorescent labeled ssNA substrates is sufficient in obtaining an initial estimate of the macroscopic translocation rate ($m{*}k_t$). However, the microscopic parameters ($k_d$, $k_t$, $r$, $m$, $n$, $d$, and $k_{end}$) can be fluorophore dependent. This can result due to the different mechanisms by which the fluorophore intensities can be affected by the presence of the enzyme. In addition, the intensities of different fluorophores can be affected by the enzyme over different distances. In order to obtain fluorophore independent estimates of the macroscopic and microscopic kinetic parameters data should be collected from at least two sets of experiments using ssNA substrates labeled with different fluorescent probe. Simultaneous, global analysis of the resulting data together using the appropriate sequential $n$-step model yields fluorophore independent estimates of $k_t$, $m$, $n$, $d$, and $r$ (20, 30).

4. Significant correlations can occur among the parameters in Eqs. 1 through 4. For this reason, it is always best to determine independently as many parameters as possible so that they can be constrained in the NLLS analysis to determine the remaining parameters. For example, the rate of dissociation during translocation ($k_d$ in Scheme 1) can often be determined independently by monitoring the dissociation of enzyme during translocation along an infinitely long DNA (19, 20, 30).

5. As seen in Eqs. 7 and 10, it is the sum of $k_u$ and $k_d$ that dominates the kinetics of dsNA unwinding (22). This sum is often termed $k_{obs} = k_u + k_d$, and can be determined from the dependence of the kinetics of double-stranded DNA unwinding on DNA length, regardless of the processivity of the helicase (21–23).

6. We note that the presence of the fluorophore on the DNA can potentially influence the rate of translocation, unwinding, or the rate of dissociation near the fluorophore (20, 30). It is also possible that variations in the electrostatics of the NA molecule

near its ends may contribute to differences in $k_t$ and $k_d$ at binding positions near the ends. Thus, for an enzyme that translocates in a 3′ to 5′ direction, the values for $k_t$ and $k_d$ obtained from fitting experimental time courses obtained with NA labeled with a fluorophore at the 3′ end may not be the same as the values of $k_t$ and $k_d$ that apply in the absence of the fluorophore or to interior regions of the NA. Similarly, the value of $k_{\text{end}}$ obtained from fitting experimental time courses obtained with DNA labeled with a fluorophore at the 5′ end may not equal the value of $k_{\text{end}}$ that applies in the absence of the fluorophore (19, 30).

7. The models (Schemes 1 and 2) presented here assume that translocation and unwinding occur via a uniform repetition of irreversible rate-limiting steps and ignore any nonuniformity in these processes. Our analysis of simulated kinetic data that includes nonuniformity has demonstrated that the values of the macroscopic translocation rate ($mk_t$), macroscopic unwinding rate ($mk_u$), and the coupling stoichiometry ($c/m$) obtained using these equations reliably reflect the actual input values used for the simulations *regardless* of the presence of any nonuniform motion of the motor protein. The estimates of the microscopic kinetic parameters, especially $k_u$, $k_t$, and $m$, will be affected by the presence of the nonuniformity, however. Generally, the estimate of m is increased by the presence of nonuniformity; because of this, the estimate of $m$ should be considered an upper limit.
8. Use of the uniform sequential "$n$-step" model to analyze time courses of ssNA translocation obtained from experiments in which the translocase initiates at random positions along the NA requires inclusion of the $r$ parameter in Eqs. 2 and 4. If the translocase has equal affinity for all potential binding sites on the DNA then $r$ must have a value between 1 and $m$ depending upon the specific details of the translocation mechanism near the end of the DNA (see ref. 19 for more details). Thus, estimated values of $r$ for which $r > m$ may indicate a failure of the simple model to correctly describe the translocation process, *regardless* of the quality of the fits. In other words, $r$ can serve as an indicator of potential nonuniformity in the translocation mechanism.
9. We note that an approximate analysis, which we refer to as the "time to peak fluorescence" analysis has been used to estimate the macroscopic translocation rate from data collected using methods in Subheading 3.2. This approximate analysis is based upon the dependence on the length of the NA of the time required to reach the maximum (or minimum) fluorescence signal during a translocation time course (33, 39). We have examined the utility of this procedure by using it to analyze computer-simulated

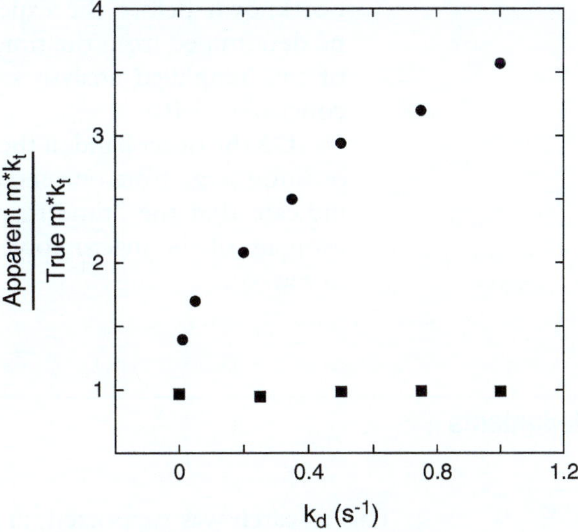


Fig. 6. The dependence of the ratio of the apparent value of the macroscopic translocation rate ($m^*k_t$) to the true value of $m^*k_t$ used in the simulation on the value of $k_d$ used in the simulation. The apparent value of the $m^*k_t$ was determined from the dependence on the length of the DNA of the maximum fluorescence signal (i.e., population of protein bound at the end of the DNA) that occurs during the time course of protein translocation along the DNA (33). In the simulations used to generate these time courses, we used constant values of $k_t = 30$ steps/s, $m = 1$ nt/step, and $k_{\text{end}} = 3\ \text{s}^{-1}$. Qualitatively similar results are obtained with other sets of parameters (data not shown).

translocation time courses and have determined that it does not generally provide a reliable estimate of even the macroscopic translocation rate; in fact, the procedure results consistently in an overestimate of the rate for systems in which the enzyme can initiate translocation randomly along the ssNA (40). The exact magnitude of the overestimation is a function of the microscopic translocation rate ($k_t$ in Scheme 1), the microscopic rate of dissociation during translocation ($k_d$ in Scheme 1), and the microscopic rate of dissociation from the end of the DNA ($k_{end}$ in Scheme 1). For example, as shown in Fig. 6 (filled symbols), the ratio of the apparent macroscopic translocation rate ($m^*k_t$) determined from the "time to peak" analysis to the true value used in the simulation is strongly dependent on the value of $k_d$ in the simulation, with the overestimation of the rate increasing with increasing $k_d$. Similar results were obtained when either $k_t$ or $k_{\text{end}}$ was varied independently, while other simulation parameters were held constant (data not shown). We found that this method of determining the value of $m^*k_t$ was most accurate when the magnitude of $k_t$ was much larger than the magnitude of $k_{\text{end}}$ and when the processivity of translocation was sufficient for nearly all the translocating enzymes to reach the end of the NA. However, since such microscopic kinetic information is not known before the experiment is performed, and cannot be determined from the time to peak fluorescence analysis, use of this simplified analysis can result in incorrect quantitative conclusions (40).

On the other hand, if the translocase initiates from a unique position (e.g., from one end of the DNA), then our simulations indicate that the "time to peak" analysis returns an accurate estimate of the macroscopic translocation rate (open symbols in Fig. 6).

## Acknowledgments

This research was supported, in part, by startup funding from the University of Kansas (to C.J.F.) and by NIH grants GM045948 and GM030498 (to T.M.L) and P20 RR17708 from the Institutional Development Award (IDeA) Program of the National Center for Research Resources (to C.J.F.).

### References

1. Kornberg RD (2007) The molecular basis of eukaryotic transcription. Proc Natl Acad Sci USA 104:12955–12961
2. Lohman TM, Bjornson KP (1996) Mechanisms of helicase-catalyzed DNA unwinding. Annu Rev Biochem 65:169–214
3. Lohman TM, Hsieh J, Maluf NK, Cheng W, Lucius AL, Fischer CJ, Brendza KM, Korolev S, Waksman G (2003) DNA helicases, motors that move along nucleic acids: lessons from the SF1 helicase superfamily. In: Hackney DD, Tamanoi F (eds) The Enzymes ATP and Molecular Motors, vol XXIII, pp 303–369 (Academic Press)
4. Matson SW, Kaiser-Rogers KA (1990) DNA helicases. Annu Rev Biochem 59:289–329
5. Lohman TM, Tomko EJ, Wu CG (2008) Non-hexameric DNA helicases and translocases: mechanisms and regulation. Nat Rev Mol Cell Biol 9:391–401
6. Khaki AR, Field C, Malik S, Niedziela-Majka A, Leavitt SA, Wang R, Hung M, Sakowicz R, Brendza KM, Fischer CJ (2010) The macroscopic rate of nucleic acid translocation by Hepatitis C virus helicase NS3h is dependent on both sugar and base moieties. J Mol Biol 400:354–378
7. Becker PB (2005) Nucleosome remodelers on track. Nat Struct Mol Biol 12:732–733
8. Fischer CJ, Saha A, Cairns BR (2007) Kinetic model for the ATP-dependent translocation of *Saccharomyces cerevisiae* RSC along double-stranded DNA. Biochemistry 46:12416–12426
9. Fischer CJ, Yamada K, Fitzgerald DJ (2009) Kinetic mechanism for single stranded DNA binding and translocation by *S. cerevisiae* Isw2. Biochemistry 48:2960–2968
10. Sirinakis G, Clapier CR, Gao Y, Viswanathan R, Cairns BR, Zhang Y (2011) The RSC chromatin remodelling ATPase translocates DNA with high force and small step size. EMBO J 30:2364–2372
11. Kovall RA, Matthews BW (1998) Structural, functional, and evolutionary relationships between lambda-exonuclease and the type II restriction endonucleases. Proc Natl Acad Sci USA 95:7893–7897
12. Kovall RA, Matthews BW (1999) Type II restriction endonucleases: structural, functional and evolutionary relationships. Curr Opin Chem Biol 3:578–583
13. Szczelkun MD (2002) Kinetic models of translocation, head-on collision, and DNA cleavage by type I restriction endonucleases. Biochemistry 41:2067–2074
14. Firman K, Szczelkun MD (2000) Measuring motion on DNA by the type I restriction

endonuclease EcoR124I using triplex displacement. EMBO J 19:2094–2102
15. McClelland SE, Dryden DT, Szczelkun MD (2005) Continuous assays for DNA translocation using fluorescent triplex dissociation: application to type I restriction endonucleases. J Mol Biol 348:895–915
16. Delagoutte E, von Hippel PH (2003) Helicase mechanisms and the coupling of helicases within macromolecular machines. Part II: integration of helicases into cellular processes. Q Rev Biophys 36:1–69
17. Patel SS, Donmez I (2006) Mechanisms of helicases. J Biol Chem 281:18265–18268
18. Ali JA, Lohman TM (1997) Kinetic measurement of the step size of DNA unwinding by *Escherichia coli* UvrD helicase. Science 275: 377–380
19. Fischer CJ, Lohman TM (2004) ATP-dependent translocation of proteins along single-stranded DNA: models and methods of analysis of pre-steady state kinetics. J Mol Biol 344:1265–1286
20. Fischer CJ, Maluf NK, Lohman TM (2004) Mechanism of ATP-dependent translocation of *E. coli* UvrD monomers along single-stranded DNA. J Mol Biol 344:1287–1309
21. Lucius AL, Jason Wong C, Lohman TM (2004) Fluorescence stopped-flow studies of single turnover kinetics of *E. coli* RecBCD helicase-catalyzed DNA unwinding. J Mol Biol 339:731–750
22. Lucius AL, Maluf NK, Fischer CJ, Lohman TM (2003) General methods for analysis of sequential "n-step" kinetic mechanisms: application to single turnover kinetics of helicase-catalyzed DNA unwinding. Biophys J 85:2224–2239
23. Lucius AL, Vindigni A, Gregorian R, Ali JA, Taylor AF, Smith GR, Lohman TM (2002) DNA unwinding step-size of *E. coli* RecBCD helicase determined from single turnover chemical quenched-flow kinetic studies. J Mol Biol 324:409–428
24. Fuller DN, Raymer DM, Kottadiel VI, Rao VB, Smith DE (2007) Single phage T4 DNA packaging motors exhibit large force generation, high velocity, and dynamic variability. Proc Natl Acad Sci USA 104:16868–16873
25. Bianco PR, Brewer LR, Corzett M, Balhorn R, Yeh Y, Kowalczykowski SC, Baskin RJ (2001) Processive translocation and DNA unwinding by individual RecBCD enzyme molecules. Nature 409:374–378
26. Lionnet T, Dawid A, Bigot S, Barre FX, Saleh OA, Heslot F, Allemand JF, Bensimon D, Croquette V (2006) DNA mechanics as a tool to probe helicase and translocase activity. Nucleic Acids Res 34:4232–4244
27. Sisakova E, Weiserova M, Dekker C, Seidel R, Szczelkun MD (2008) The interrelationship of helicase and nuclease domains during DNA translocation by the molecular motor EcoR124I. J Mol Biol 384:1273–1286
28. Brendza KM, Cheng W, Fischer CJ, Chesnik MA, Niedziela-Majka A, Lohman TM (2005) Autoinhibition of *Escherichia coli* Rep monomer helicase activity by its 2B subdomain. Proc Natl Acad Sci USA 102: 10076–10081
29. Niedziela-Majka A, Chesnik MA, Tomko EJ, Lohman TM (2007) Bacillus stearothermophilus PcrA monomer is a single-stranded DNA translocase but not a processive helicase in vitro. J Biol Chem 282:27076–27085
30. Tomko EJ, Fischer CJ, Niedziela-Majka A, Lohman TM (2007) A nonuniform stepping mechanism for *E. coli* UvrD monomer translocation along single-stranded DNA. Mol Cell 26:335–347
31. Lucius AL, Lohman TM (2004) Effects of temperature and ATP on the kinetic mechanism and kinetic step-size for *E. coli* RecBCD helicase-catalyzed DNA unwinding. J Mol Biol 339:751–771
32. Jaques LB (1977) Determination of heparin and related sulfated mucopolysaccharides. Methods Biochem Anal 24:203–312
33. Dillingham MS, Wigley DB, Webb MR (2002) Direct measurement of single-stranded DNA translocation by PcrA helicase using the fluorescent base analogue 2-aminopurine. Biochemistry 41:643–651
34. Hsieh J, Moore KJ, Lohman TM (1999) A two-site kinetic mechanism for ATP binding and hydrolysis by *E. coli* Rep helicase dimer bound to a single-stranded oligodeoxynucleotide. J Mol Biol 288:255–274
35. Wong I, Moore KJ, Bjornson KP, Hsieh J, Lohman TM (1996) ATPase activity of *Escherichia coli* Rep helicase is dramatically dependent on DNA ligation and protein oligomeric states. Biochemistry 35:5726–5734
36. Dillingham MS, Wigley DB, Webb MR (2000) Demonstration of unidirectional single-stranded DNA translocation by PcrA helicase: measurement of step size and translocation speed. Biochemistry 39:205–212
37. Bjornson KP, Amaratunga M, Moore KJ, Lohman TM (1994) Single-turnover kinetics of helicase-catalyzed DNA unwinding monitored continuously by fluorescence energy transfer. Biochemistry 33:14306–14316

38. Wu CG, Lohman TM (2008) Influence of DNA end structure on the mechanism of initiation of DNA unwinding by the *Escherichia coli* RecBCD and RecBC helicases. J Mol Biol 382:312–326
39. Saikrishnan K, Powell B, Cook NJ, Webb MR, Wigley DB (2009) Mechanistic basis of 5′-3′ translocation in SF1B helicases. Cell 137: 849–859
40. Tomko EJ, Fischer CJ, Lohman TM (2010). Ensemble methods for monitoring enzyme translocation along single stranded nucleic acids. Methods 51:269–276

Chapter 6

# Fluorescence Intensity, Anisotropy, and Transient Dynamic Quenching Stopped-Flow Kinetics

Wlodek M. Bujalowski and Maria J. Jezewska

## Abstract

The kinetic mechanisms of biological reactions are predominantly addressed by spectroscopic stopped-flow or temperature-jump methods. Both the stopped-flow and the temperature-jump methods are relaxation kinetic techniques, i.e., they rely on examining the effect of perturbation on the reaction system under study. The relaxation kinetic measurements of the approach to equilibrium of the ligand–macromolecule reactions provide two independent sets of data, relaxation times and amplitudes. Although the traditional matrix method is a powerful approach, the matrix projection operator technique is an exceptionally convenient approach to analyze stopped-flow kinetics. The numerical analysis of a complex multistep reaction is reduced to finding only the eigenvalues of the original coefficient matrix. The method is illustrated by examination of the kinetics of a fluorescent nucleotide analog binding to the *E. coli* replicative helicase, the DnaB protein.

Fluorescence intensity is one of the most often used spectroscopic signals to monitor the progress of biochemical reactions. Its properties also give an opportunity to address various structural aspects of the intermediates unavailable by any other method. The relative molar fluorescence intensities of different intermediates provide information about the physical environment surrounding the fluorophore during the course of the reaction. On the other hand, time-dependence of the fluorescence anisotropy in stopped-flow experiments provides information about the mobility of the fluorescing species in each intermediate of the observed kinetic process. Moreover, transient anisotropy data may also put additional light on the mechanism of the reaction, not obvious in studies using the emission intensity alone. Finally, collisional dynamic quenching of the fluorescence emission allows the experimenter to assess the solvent accessibility of the fluorophore. The method is mostly applied to steady-state fluorescence intensity in equilibrium. However, the same approach can be applied to address the solvent accessibility of the different intermediates, during the time course of the reaction monitored in the stopped-flow experiment.

**Key words:** Ligand–macromolecule interactions, Fluorescence anisotropy, Fluorescence stopped-flow kinetics, Fluorescence quenching, Relaxation kinetics, Matrix method

## 1. Introduction

Spectroscopic relaxation kinetic methods, like the stopped-flow and temperature-jump techniques, are the most powerful methods in examining the kinetic mechanisms of ligand–macromolecule

Wlodek M. Bujalowski (ed.), *Spectroscopic Methods of Analysis: Methods and Protocols*, Methods in Molecular Biology, vol. 875, DOI 10.1007/978-1-61779-806-1_6, © Springer Science+Business Media New York 2012

interactions (1–5). While the temperature-jump method is limited to the systems approaching the equilibrium state, the stopped-flow method can also be used to study irreversible reactions. In our discussion, we concentrate on the stopped-flow method and on the interacting systems, which approach the equilibrium state. This is because the stopped-flow instruments are more accessible to the experimenter than the temperature-jump equipments and the reaction systems approaching the equilibrium state are most often encountered in biological kinetic studies of the ligand–macromolecular recognition processes. Moreover, we focus on the quantitative method of extracting kinetic and spectroscopic parameters, characterizing the intermediates of the reaction, not their molecular interpretations. While the methods of quantitative extraction of interaction and spectroscopic parameters are shared by most of the examined interacting systems, the molecular interpretations of the obtained results are profoundly dependent upon the physical characteristics of a particular system under study. Both the stopped-flow and temperature-jump methods rely on examining the effect of a perturbation on the reaction system under study.

The relaxation kinetic measurements of the approach to equilibrium of the ligand–macromolecule reactions provide two independent sets of data, relaxation times and amplitudes, characterizing the normal modes of the observed relaxation processes (3, 6–9). In general, the characteristic behaviors of the relaxation times and the amplitudes, as functions of the ligand or macromolecule concentration, serve as the major criterion as to what the mechanism of the observed reaction is. It is unfortunate that in the literature the conclusions about the behavior of the interacting system is often exclusively based on the analysis of the relaxation times and the analysis of amplitudes of the observed relaxation processes is not considered. As we discuss it below, quantitative studies of both the relaxation times and the amplitudes of the examined relaxation process provide information not only on the mechanistic details, but also on the nature of the formed intermediates unavailable by any other method (3, 6–8, 10–20). Applying dynamic fluorescence quenching and fluorescence anisotropy to the experimental repertoire further reinforces the resolution of such kinetic analyses. Over last two decades, the described methods and analyses have been used in our laboratory in examining various ligand–macromolecular systems, including protein–nucleic acid and proteins–nucleotide complexes (7, 10–18).

## 2. Materials

All kinetic experiments are performed in the same buffers as the thermodynamic binding studies on the corresponding interacting systems. The buffers and, in general, solution conditions are optimized for the system stability over the time period necessary to execute the experiments.

## 3. Methods

### 3.1. General Analysis of Relaxation Stopped-Flow Kinetic Data Using the Matrix Projection Operator Technique

Analyses of the stopped-flow kinetics of the ligand–macromolecule interactions can be conveniently accomplished using the matrix projection operator technique (10–18, 21, 22). It is an extremely useful approach for the analysis of complex stopped-flow kinetics, particularly, because it provides closed-form expressions for the relaxation amplitudes. In general, the return to the equilibrium state by any system of the first-order reactions, which is subjected to perturbation, i.e., the relaxation process of the system after perturbation always occurs as a sum of normal modes, mathematically described as exponential functions (1–5). Therefore, the observed spectroscopic signal accompanying the relaxation process, e.g., fluorescence intensity, is a sum of the exponential functions, defined as

$$F(t) = F(\infty) + \sum_{i=1}^{n} A_i \exp(-\lambda_i t), \tag{1}$$

where $F(t)$ is the fluorescence intensity at time $t$, $F(\infty)$ is the fluorescence intensity at $t = \infty$, $A_i$ is the amplitude corresponding to $i$th relaxation process, $\lambda_i$ is the time constant (reciprocal relaxation time), characterizing $i$th relaxation process, and $n$ is the number of relaxation processes (normal modes of the reaction). Because of the high sensitivity of the fluorescence and the fact that most of the proteins are fluorescing, fluorescence emission changes are the most often used spectroscopic signal in biological kinetics. In experimental practice, a reaction system is rendered as a first-order system by performing the kinetic measurement under pseudo-first-order conditions, i.e., in a large excess of one of the participating components in the process (1–5, 10–18). Nevertheless, it should be stressed that relaxation times are not rate constants, characterizing a particular kinetic step in the multiple-step, kinetic mechanism underlying the relaxation process, but a complex function of all rate constants of the interacting system (see below) (3, 4, 7–9). By the same token, amplitudes of the multiple-step relaxation process are not molar fluorescence intensities of particular intermediates, but complex functions of rate constants, relaxation times, and molar fluorescence intensities of all the species participating in the relaxation process (3, 4, 7–9).

The stopped-flow experiments can be conventionally performed by directly measuring the fluorescence intensity as a function of time, as defined by (1). A less conventional measurement relies on monitoring the total fluorescence intensity, which is defined as

$$F = I_{\mathrm{VV}} + 2GI_{\mathrm{VH}}, \tag{2}$$

where, $I_{\mathrm{VV}}$ and $I_{\mathrm{VH}}$, are the fluorescence intensities and the first and second subscripts refer to vertical (V) polarization of the excitation and vertical (V) or horizontal (H) polarization of the emitted light. The factor, $G = I_{\mathrm{HV}}/I_{\mathrm{HH}}$, corrects for the different sensitivities of the emission monochromator for vertically and horizontally polarized light (15, 16, 23). The use of the total fluorescence intensity is particularly useful when there is a large change in the fluorescence anisotropy of the sample, as a result of the examined reaction, which may introduce artifacts in direct intensity measurements (6, 15, 24). Time-dependence of the total fluorescence intensity is then

$$F(t) = \sum_{i=1}^{n} I_{VV}(t) + 2G\sum_{i=1}^{n} I_{VH}(t). \tag{3}$$

Since both, $I_{VV}$ and $I_{VH}$, are the same function of time, the time-dependence of the total emission is then

$$F(t) = \sum_{i=1}^{n} (I_{VV} + 2GI_{VH})_i \exp(-\lambda_i t). \tag{4}$$

The expression in parentheses in (4) symbolizes the $i$th relaxation amplitude measured using the total emission of the sample. Thus, the mathematical form of the expression defining the time-dependence of the total emission is the same as that of the direct fluorescence emission measurement (1).

To illustrate the matrix projection operator approach, as an example, we consider a complex sequential reaction between a macromolecule, D, and a ligand, L, of the type

$$D_0 + L \underset{k_{-1}}{\overset{k_1}{\rightleftarrows}} D_1 \underset{k_{-2}}{\overset{k_2}{\rightleftarrows}} D_2 \underset{k_{-3}}{\overset{k_3}{\rightleftarrows}} D_3 \underset{k_{-4}}{\overset{k_4}{\rightleftarrows}} D_4, \tag{5}$$

where the initial bimolecular process is followed by three isomerization reactions of the formed complex, monitored by the changes of the fluorescence emission, $F(t)$, of the macromolecule. Four relaxation times and four amplitudes characterize the considered relaxation process, i.e., there are four normal modes of the reaction (3, 4). The process is examined under pseudo-first-order conditions with respect to the ligand, i.e., the total ligand concentration, $[L]_T$, is much larger than the total concentration of the macromolecule, $[D]_T$. In matrix notation, the differential equations describing the time course of reaction (5) is then defined as

$$\begin{pmatrix} \frac{dD_0}{dt} \\ \frac{dD_1}{dt} \\ \frac{dD_2}{dt} \\ \frac{dD_3}{dt} \\ \frac{dD_4}{dt} \end{pmatrix} = \begin{pmatrix} -k_1[L]_T & k_{-1} & 0 & 0 & 0 \\ k_1[L]_T & -(k_{-1}+k_2) & k_{-2} & 0 & 0 \\ 0 & k_2 & -(k_{-2}+k_3) & k_3 & 0 \\ 0 & 0 & k_3 & -(k_3+k_4) & -k_{-4} \\ 0 & 0 & 0 & k_4 & k_{-4} \end{pmatrix} \begin{pmatrix} D_0 \\ D_1 \\ D_2 \\ D_3 \\ D_4 \end{pmatrix} \tag{6}$$

or

$$\dot{\mathbf{D}} = \exp(\mathbf{M}t)\mathbf{D} \tag{7}$$

where $\dot{\mathbf{D}}$ is a vector of time derivatives, $\mathbf{M}$ is the coefficient matrix, and $\mathbf{D}$ is the vector of concentrations of different macromolecule species in solution (10–18). In the standard matrix approach, the solution of the system 6 is

$$\mathbf{D} = \exp(\mathbf{M}t) = \mathbf{V}\begin{pmatrix} \exp(\lambda_0 t) & 0 & 0 & 0 & 0 \\ 0 & \exp(\lambda_1 t) & 0 & 0 & 0 \\ 0 & 0 & \exp(\lambda_2 t) & 0 & 0 \\ 0 & 0 & 0 & \exp(\lambda_3 t) & 0 \\ 0 & 0 & 0 & 0 & \exp(\lambda_4 t) \end{pmatrix}\mathbf{V}^{-1}\mathbf{D}_0, \tag{8}$$

where, $\lambda_0$, $\lambda_1$, $\lambda_2$, $\lambda_3$, and $\lambda_4$ are eigenvalues of matrix $\mathbf{M}$, $\mathbf{V}$ is a matrix whose columns are the eigenvectors of matrix $\mathbf{M}$, and $\mathbf{D}_0$ is the vector of the initial concentrations of the different macromolecule species. In the considered sequential reaction (5), $\mathbf{D}_0$ is a column vector ($[\mathrm{D}]_\mathrm{T}$, 0, 0, 0, 0) where $[\mathrm{D}]_\mathrm{T}$ is the total concentration of the macromolecule. The form of the vector $\mathbf{D}_0$ reflects the fact that at $t = 0$, in a stopped-flow experiment, the concentration of the free macromolecule is equal to its total concentration, while the concentrations of all other macromolecular species are zero. In the standard matrix approach, system 6 is solved by first obtaining the eigenvalues of matrix $\mathbf{M}$, and then the corresponding eigenvectors. For a multistep mechanism, like (5), this can be achieved only through cumbersome numerical analyses, particularly for the eigenvectors.

However, instead of finding eigenvectors corresponding to each eigenvalue, $\lambda_i$, of matrix $\mathbf{M}$, one can expand the matrix, exp ($\mathbf{M}t$), using its eigenvalues, $\exp(\lambda_i t)$, and corresponding projection operators, $\mathbf{Q}_i$, as (10–18, 21, 22)

$$\exp(\mathbf{M}t) = \sum_{i=0}^{4} \mathbf{Q}_i \exp(\lambda_i t). \tag{9}$$

The projection operators, $\mathbf{Q}_i$, can easily be found by Sylvester's theorem, using the original coefficient matrix $\mathbf{M}$ and its eigenvalues, $\lambda_i$ (21, 22). The general algebraic formula for a projection operator, $\mathbf{Q}_i$, corresponding to an eigenvalue, $\lambda_i$, is then (21, 22)

$$\mathbf{Q}_i = \frac{\prod_{j\neq i}^{n} (\mathbf{M} - \lambda_j \mathbf{I})}{\prod_{j\neq i}^{n} (\lambda_i - \lambda_j)}, \tag{10}$$

where $n$ is the number of eigenvalues and $\mathbf{I}$ is the identity matrix of the same size as $\mathbf{M}$. In the considered reaction, there are five eigenvalues, $\lambda_0$, $\lambda_1$, $\lambda_2$, $\lambda_3$, and $\lambda_4$. However, one of the eigenvalues,

$\lambda_0 = 0$, because of the mass conservation in the reaction system. In other words, in the equilibrium state, the sum of the macromolecular species must be equal to the total concentration of the macromolecule, at the beginning of the reaction. For the reaction system considered here, using (10), one obtains

$$\mathbf{Q}_0 = \frac{(\mathbf{M} - \lambda_1\mathbf{I})(\mathbf{M} - \lambda_2\mathbf{I})(\mathbf{M} - \lambda_3\mathbf{I})(\mathbf{M} - \lambda_4\mathbf{I})}{\lambda_1\lambda_2\lambda_3\lambda_4}, \quad (11a)$$

$$\mathbf{Q}_1 = \frac{\mathbf{M}(\mathbf{M} - \lambda_2\mathbf{I})(\mathbf{M} - \lambda_3\mathbf{I})(\mathbf{M} - \lambda_4\mathbf{I})}{\lambda_1(\lambda_1 - \lambda_2)(\lambda_1 - \lambda_3)(\lambda_1 - \lambda_4)}, \quad (11b)$$

$$\mathbf{Q}_2 = \frac{\mathbf{M}(\mathbf{M} - \lambda_1\mathbf{I})(\mathbf{M} - \lambda_3\mathbf{I})(M - \lambda_4\mathbf{I})}{\lambda_2(\lambda_2 - \lambda_1)(\lambda_2 - \lambda_3)(\lambda_2 - \lambda_4)}, \quad (11c)$$

$$\mathbf{Q}_3 = \frac{\mathbf{M}(\mathbf{M} - \lambda_1\mathbf{I})(\mathbf{M} - \lambda_2\mathbf{I})(\mathbf{M} - \lambda_4\mathbf{I})}{\lambda_3(\lambda_3 - \lambda_1)(\lambda_3 - \lambda_2)(\lambda_3 - \lambda_4)}, \quad (11d)$$

$$\mathbf{Q}_4 = \frac{\mathbf{M}(\mathbf{M} - \lambda_1\mathbf{I})(\mathbf{M} - \lambda_2\mathbf{I})(\mathbf{M} - \lambda_3\mathbf{I})}{\lambda_4(\lambda_4 - \lambda_1)(\lambda_4 - \lambda_2)(\lambda_4 - \lambda_3)}. \quad (11e)$$

The solution of the system of the differential equations (6), expressed in terms of the matrix projection operators, is then

$$\mathbf{D} = \mathbf{Q}_0\mathbf{D}_0 + \mathbf{Q}_1\mathbf{D}_0 \exp(\lambda_1 t) + \mathbf{Q}_2\mathbf{D}_0 \exp(\lambda_2 t) + \mathbf{Q}_3\mathbf{D}_0 \exp(\lambda_3 t) + \mathbf{Q}_4\mathbf{D}_0 \exp(\lambda_4 t), \quad (12)$$

where $\mathbf{Q}_i$ is defined by (11a)–(11e). The products $\mathbf{Q}_i\mathbf{D}_0$ are column vectors, $\mathbf{P}_i$, which are the projections of $\mathbf{D}_0$ on each eigenvector of matrix $\mathbf{M}$, i.e.,

$$\begin{pmatrix} \mathbf{D}_0 \\ \mathbf{D}_1 \\ \mathbf{D}_2 \\ \mathbf{D}_3 \\ \mathbf{D}_4 \end{pmatrix} = \begin{pmatrix} \mathbf{P}_{01} \\ \mathbf{P}_{02} \\ \mathbf{P}_{03} \\ \mathbf{P}_{04} \\ \mathbf{P}_{05} \end{pmatrix} + \begin{pmatrix} \mathbf{P}_{11} \\ \mathbf{P}_{12} \\ \mathbf{P}_{13} \\ \mathbf{P}_{14} \\ \mathbf{P}_{15} \end{pmatrix} \exp(\lambda_1 t) + \begin{pmatrix} \mathbf{P}_{21} \\ \mathbf{P}_{22} \\ \mathbf{P}_{23} \\ \mathbf{P}_{24} \\ \mathbf{P}_{25} \end{pmatrix} \exp(\lambda_2 t) + \begin{pmatrix} \mathbf{P}_{31} \\ \mathbf{P}_{32} \\ \mathbf{P}_{33} \\ \mathbf{P}_{34} \\ \mathbf{P}_{35} \end{pmatrix} \exp(\lambda_3 t) + \begin{pmatrix} \mathbf{P}_{41} \\ \mathbf{P}_{42} \\ \mathbf{P}_{43} \\ \mathbf{P}_{44} \\ \mathbf{P}_{45} \end{pmatrix} \exp(\lambda_4 t), \quad (13)$$

where $P_{ij}$ is the *j*th element of the projection of the vector of the initial concentrations $\mathbf{D}_0$ on the eigenvector, corresponding to the *i*th eigenvalue of matrix $\mathbf{M}$. In stopped-flow experiments, the concentrations of all protein species change from the concentrations at $t = 0$ to the equilibrium concentrations at $t = \infty$, defined by the elements of vector $\mathbf{P}_0$. It should be pointed out that each element of $P_{ij}$ in (13) is an algebraic expression, in terms of eigenvalues, $\lambda_i$, rate constants of the system, and total ligand and macromolecule concentrations, defined by the products, $\mathbf{Q}_i\mathbf{D}_0$.

Therefore, using projection operators the numerical analysis of a complex multistep reaction is reduced to finding only the eigenvalues of the original coefficient matrix **M** (10–18).

There are four normal modes of the reaction and four amplitudes, $A_1$, $A_2$, $A_3$, and $A_4$, corresponding to relaxation times $\tau_1 = -1/\lambda_1$, $\tau_2 = -1/\lambda_2$, $\tau_3 = -1/\lambda_3$, and $\tau_4 = -1/\lambda_4$. In spectroscopic stopped-flow experiments, concentrations of the reactants and products are indirectly monitored through some spectroscopic parameter, in our case, fluorescence, characterizing interacting species. In general, each intermediate will have different fluorescence properties. Thus, there are five molar fluorescence intensities, $F_0$, $F_1$, $F_2$, $F_3$, and $F_4$, characterizing $D_0$, $D_1$, $D_2$, $D_3$, and $D_4$ states of the macromolecule, free and in different intermediates of the complex with the ligand. Notice, the quantities, $D_0$, $D_1$, $D_2$, $D_3$, and $D_4$ indicate the states of the macromolecules in different intermediates. For simplicity, we also use the same symbols to indicate the concentrations of the same states.

The fluorescence emission of the examined interacting system, at any time of the reaction, $F(t)$, is then defined, by a "signal conservation equation," as

$$F(t) = F_0 D_0 + F_1 D_1 + F_2 D_2 + F_3 D_3 + F_4 D_4, \tag{14}$$

where $D_0$, $D_1$, $D_2$, $D_3$, and $D_4$ are now the concentrations of different states of the macromolecule in different intermediates (see above). Introducing (13) into (14), one obtains the value of $F(t)$, in terms of the normal modes of the reaction, as

$$F(t) = \begin{pmatrix} F_0 & F_1 & F_2 & F_3 & F_4 \end{pmatrix} \times \begin{pmatrix} P_{01} & P_{11} & P_{21} & P_{31} & P_{41} \\ P_{02} & P_{12} & P_{22} & P_{32} & P_{42} \\ P_{03} & P_{13} & P_{23} & P_{33} & P_{43} \\ P_{04} & P_{14} & P_{24} & P_{34} & P_{44} \\ P_{05} & P_{15} & P_{25} & P_{35} & P_{45} \end{pmatrix} \begin{pmatrix} 1 \\ \exp(\lambda_1 t) \\ \exp(\lambda_2 t) \\ \exp(\lambda_3 t) \\ \exp(\lambda_4 t) \end{pmatrix}. \tag{15}$$

The total amplitude, $A_T$, of the fluorescence stopped-flow trace, at any concentration of the ligand, is the sum of individual amplitudes of all normal modes of the relaxation process, as

$$A_T = A_1 + A_2 + A_3 + A_4. \tag{16}$$

The experimentally observed total amplitude, $A_T$, of the fluorescence stopped-flow trace, is described by

$$A_T = F(0) - F(\infty). \tag{17}$$

where $F(0)$ and $F(\infty)$ are the observed fluorescence intensities, $F(t)$, of the system at $t = 0$ and $t = \infty$, respectively. Thus, introducing $t = 0$ for $F(0)$ and $t = \infty$ for $F(\infty)$ in (15) provides the values of $F(t)$ at the beginning and at the end of the reaction, at a given total ligand concentration, as

$$F(0) = \begin{pmatrix} F_0 & F_1 & F_2 & F_3 & F_4 \end{pmatrix} \begin{pmatrix} P_{01} & P_{11} & P_{21} & P_{31} & P_{41} \\ P_{02} & P_{12} & P_{22} & P_{32} & P_{42} \\ P_{03} & P_{13} & P_{23} & P_{33} & P_{43} \\ P_{04} & P_{14} & P_{24} & P_{34} & P_{44} \\ P_{05} & P_{15} & P_{25} & P_{35} & P_{45} \end{pmatrix} \begin{pmatrix} 1 \\ 1 \\ 1 \\ 1 \\ 1 \end{pmatrix} \tag{18}$$

and

$$F(\infty) = \begin{pmatrix} F_0 & F_1 & F_2 & F_3 & F_4 \end{pmatrix} \begin{pmatrix} P_{01} & P_{11} & P_{21} & P_{31} & P_{41} \\ P_{02} & P_{12} & P_{22} & P_{32} & P_{42} \\ P_{03} & P_{13} & P_{23} & P_{33} & P_{43} \\ P_{04} & P_{14} & P_{24} & P_{34} & P_{44} \\ P_{05} & P_{15} & P_{25} & P_{35} & P_{45} \end{pmatrix} \begin{pmatrix} 1 \\ 0 \\ 0 \\ 0 \\ 0 \end{pmatrix} \tag{19}$$

Introducing (18) and (19) into (17), provides the total amplitude, $A_T$, as

$$A_T = \begin{pmatrix} F_0 & F_1 & F_2 & F_3 & F_4 \end{pmatrix} \begin{pmatrix} P_{11} & P_{21} & P_{31} & P_{41} \\ P_{12} & P_{22} & P_{32} & P_{42} \\ P_{13} & P_{23} & P_{33} & P_{43} \\ P_{14} & P_{24} & P_{34} & P_{44} \\ P_{15} & P_{25} & P_{35} & P_{45} \end{pmatrix} \begin{pmatrix} 1 \\ 1 \\ 1 \\ 1 \end{pmatrix} \tag{20}$$

where each individual amplitude, $A_1$, $A_2$, $A_3$, and $A_4$, is a product of the row vector, ($F_0$, $F_1$, $F_2$, $F_3$, $F_4$), with the corresponding column in (20). The final vector multiplication in (20) returns the scalar value for the total amplitude. Thus, the individual amplitudes, $A_1$, $A_2$, $A_3$, and $A_4$, for each normal mode are then

$$A_1 = \begin{pmatrix} P_{11} & P_{12} & P_{13} & P_{14} & P_{15} \end{pmatrix} \begin{pmatrix} F_0 \\ F_1 \\ F_2 \\ F_3 \\ F_4 \end{pmatrix} \tag{21a}$$

$$A_2 = \begin{pmatrix} P_{21} & P_{22} & P_{23} & P_{24} & P_{25} \end{pmatrix} \begin{pmatrix} F_0 \\ F_1 \\ F_2 \\ F_3 \\ F_4 \end{pmatrix} \tag{21b}$$

$$A_3 = \begin{pmatrix} P_{31} & P_{32} & P_{33} & P_{34} & P_{35} \end{pmatrix} \begin{pmatrix} F_0 \\ F_1 \\ F_2 \\ F_3 \\ F_4 \end{pmatrix} \tag{21c}$$

and

$$A_4 = \begin{pmatrix} P_{41} & P_{42} & P_{43} & P_{44} & P_{45} \end{pmatrix} \begin{pmatrix} F_0 \\ F_1 \\ F_2 \\ F_3 \\ F_4 \end{pmatrix} \quad (21d)$$

Expressions (20) and (21a–d) are closed-form, explicit relationships for the total and individual amplitudes for the four-step reaction mechanism described by (5). Thus, once the matrix operators are formulated in terms of the original matrix of coefficients, **M**, and its eigenvalues, the total and individual amplitudes of the reaction system can be easily defined (10–18). It should be obvious that analogous relationships apply when the pseudo-first-order conditions refer to the macromolecule concentration, i.e., the macromolecule is in large excess of the ligand and, $[L]_T$, is replaced by, $[D]_T$. In such a situation, the fluorescence of the ligand is usually used to monitor the reaction (12, 14, 16, 25, 26).

The relationships derived above also provide an important intuitive insight into the effect of different values of spectroscopic properties of the intermediates of the reaction on the observed amplitudes. For instance, even if all intermediates have the same fluorescence properties ($F_2 = F_3 = F_4$), the amplitudes of all normal modes will be observed, in spite of the fact that there are no additional fluorescence changes in all transitions following $D_1$. The fluorescence changes in subsequent transitions depend on the difference between the fluorescence intensity of the free macromolecule and the fluorescence intensities of all remaining intermediates, even if it would concern only one intermediate. Thus, the fact that in some kinetic systems not all normal modes of the reaction are detectable is more a result of the combined effect of the rate constants, relaxation times, and spectroscopic changes than the similarity in the spectroscopic properties of the intermediates. Extraction of the spectroscopic properties of each intermediate of the reaction, which provide information about the nature and structure of the intermediate, is achieved by setting the fluorescence of the free macromolecule, $F_0 = 1$. Then, all remaining molar fluorescence intensities, $F_1$, $F_2$, $F_3$, and $F_4$, are uniquely determined, relative to $F_0$. Nevertheless, the quantum yield for the free macromolecule can be independently obtained. Therefore, if needed, the true quantum yields for all other intermediates can also be obtained.

#### 3.1.1. Relaxation Times

The first fundamental step in establishing the mechanism of a complex reaction and determining the rate constants of particular elementary processes relies on examining the determined reciprocal relaxation times as a function of the total ligand concentration (3, 4). The reciprocal relaxation times for the sequential four-step reaction (5), as a function of the free ligand concentration, are

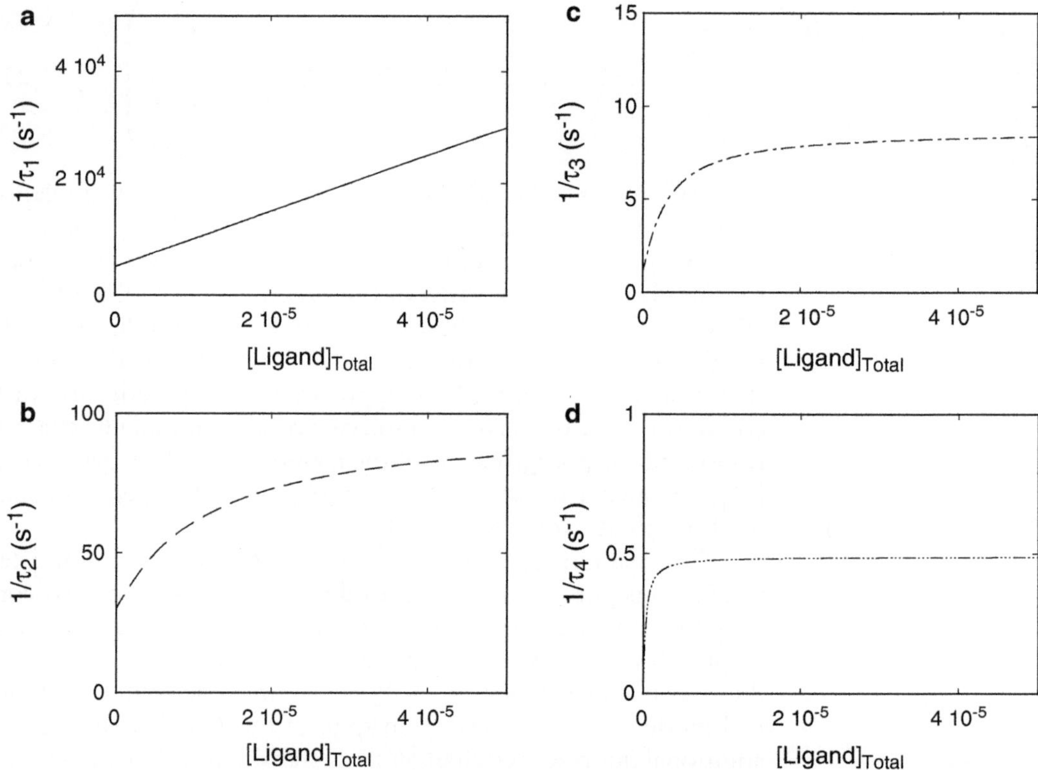


Fig. 1. Computer simulation of the dependence of reciprocal relaxation times for the four-step sequential mechanism of ligand binding to a single site on a macromolecule, defined by (5), upon total ligand concentration. Relaxation times have been obtained by numerically determining the eigenvalues of the coefficient matrix **M** ($\lambda_1$, $\lambda_2$, $\lambda_3$, $\lambda_4$) then using identities: $1/\tau_1 = -\lambda_1$, $1/\tau_2 = -\lambda_2$, $1/\tau_3 = -\lambda_3$, and $1/\tau_4 = -\lambda_4$. The simulations have been performed using rate constants: $k_1 = 5 \times 10^8\ \mathrm{M^{-1}\,s^{-1}}$, $k_{-1} = 5{,}000\ \mathrm{s^{-1}}$, $k_2 = 75\ \mathrm{s^{-1}}$, $k_{-2} = 20\ \mathrm{s^{-1}}$, $k_3 = 10\ \mathrm{s^{-1}}$, $k_{-3} = 1\ \mathrm{s^{-1}}$, $k_4 = 0.5\ \mathrm{s^{-1}}$, and $k_{-4} = 0.05\ \mathrm{s^{-1}}$. The selected total macromolecule concentration, $[\mathrm{D}]_\mathrm{T}$, is $1 \times 10^{-8}$ M; **(a)** $1/\tau_1$, **(b)** $1/\tau_2$, **(c)** $1/\tau_3$, **(d)** $1/\tau_4$.

shown in Fig. 1a–d. The reciprocal of relaxation times have been obtained by numerical determination of the eigenvalues, $\lambda_1$, $\lambda_2$, $\lambda_3$, and $\lambda_4$, of the matrix **M**, at a given total ligand concentration, $[\mathrm{L}]_\mathrm{T}$, using the identities of $1/\tau_1 = -\lambda_1$, $1/\tau_2 = -\lambda_2$, $1/\tau_3 = -\lambda_3$, and $1/\tau_4 = -\lambda_4$. The largest reciprocal relaxation time, $1/\tau_1$, increases linearly with the ligand concentration, which is typical behavior for the bimolecular binding process (3, 4, 10–18). However, for the selected values of the rate constants, the values of $1/\tau_1$ in Fig. 1a would be well beyond the resolution of the stopped-flow experiment, which is a typical situation in most studies (10–18). Also, what is not detectable in Fig. 1a is the short nonlinear part of the plot at a very low ligand concentration, which may appear in experimental studies (see below). All remaining reciprocal relaxation times, $1/\tau_2$, $1/\tau_3$, and $1/\tau_4$, for the considered sequential four-step reaction (5), show hyperbolic dependence upon, $[\mathrm{L}]_\mathrm{T}$, and reach plateaus at high $[\mathrm{L}]_\mathrm{T}$. In other words, $1/\tau_2$, $1/\tau_3$, and $1/\tau_4$, become independent of the ligand concentration at high, $[\mathrm{L}]_\mathrm{T}$.

Such behavior indicates that the considered relaxation times characterize intramolecular transitions (1, 3, 4). Experimentally, the hyperbolic dependence of $1/\tau_2$, $1/\tau_3$, and $1/\tau_4$, upon the ligand concentration, may often be undetectable when the experimental data are already placed in the plateaus under the applied pseudo-first-order conditions (Fig. 1) (10–18). In other words, independence of the values of $1/\tau_2$, $1/\tau_3$, and $1/\tau_4$ on the ligand concentration will be observed (10–18).

#### 3.1.2. Relaxation Amplitudes

The dependence of normalized individual amplitudes, $A_1$, $A_2$, $A_3$, and $A_4$, expressed as fractions, $A_i/\sum A_i$, of the total amplitude ($A_T = \sum A_i$) upon the logarithm of the ligand concentration is shown in Fig. 2. The plots have been performed using (21a)–(21d). The selected molar fluorescence intensities are $F_0 = 1.0$, $F_1 = 0.30$, $F_2 = 0.50$, $F_3 = 0.40$, and $F_4 = 0.30$. The reaction produces a quenching of the macromolecular fluorescence with $\Delta F_{max} \approx 0.70$ at saturation of the macromolecule with the ligand. The rate constants are the same as in Fig. 1. For the selected values of the rate constants, at low ligand concentrations, only the amplitudes of the third, $A_3$, and fourth, $A_4$, normal modes of the reaction contribute significantly to the observed $A_T$. This is in spite of the fact that the major fluorescence change, as compared to the fluorescence of the free nucleic acid, accompanies the formation of $D_1$. Such behavior is

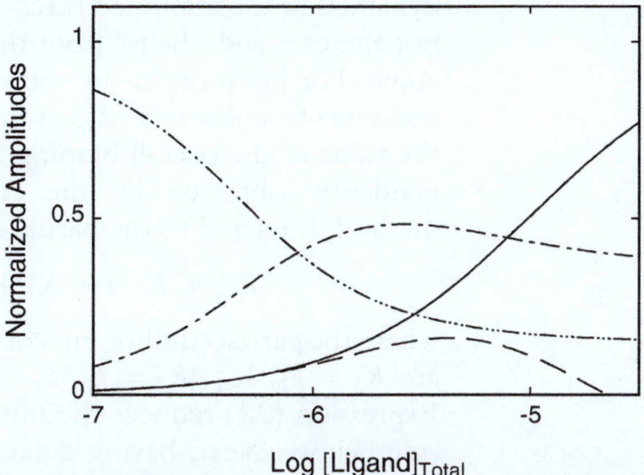


Fig. 2. Computer simulation of the dependence of normalized individual relaxation amplitudes, $A_1$, $A_2$, $A_3$, and $A_4$ for the four-step sequential mechanism of ligand binding to a single site on a macromolecule, defined by (5), upon the logarithm of the total ligand concentration. The relative fluorescence intensities, $F_1$, $F_2$, $F_3$, and $F_4$, characterizing corresponding intermediates, $D_1$, $D_2$, $D_3$, and $D_4$ are 0.3, 0.5, 0.4, and 0.3, respectively. The fluorescence of the free macromolecule, $D_0$, is taken as $F_0 = 1$. The individual amplitudes are normalized with respect to the total amplitude. The simulations have been performed using closed-form expressions defined by (21a)–(21d), with the rate constants: $k_1 = 5 \times 10^8$ M$^{-1}$ s$^{-1}$, $k_{-1} = 5{,}000$ s$^{-1}$, $k_2 = 75$ s$^{-1}$, $k_{-2} = 20$ s$^{-1}$, $k_3 = 10$ s$^{-1}$, $k_{-3} = 1$ s$^{-1}$, $k_4 = 0.5$ s$^{-1}$, and $k_{-4} = 0.05$ s$^{-1}$; $A_1$ (—), $A_2$ (— — —), $A_3$ (— · —), $A_4$ (— · · · · —).

the result of the low efficiency of the $C_1$ complex formation at low $[L]_T$, while the formed complex still relaxes in the third and fourth normal mode. At higher ligand concentrations, the amplitude of the third normal mode, $A_3$, strongly contributes to the relaxation process, although the intermediate, $D_3$, does not have the lowest values of fluorescence quenching. At the highest ligand concentrations the amplitude of the first normal mode, $A_1$ is little visible in the low ligand concentration range and begins to dominate the relaxation process. The computer simulations in Fig. 2 show that all four amplitudes of the relaxation modes of the system are detectable in the examined ligand concentration range. However, if the experiment was performed at $[L]_T > 10^{-5}$ M, the amplitude, $A_2$, strongly decreases in this total ligand concentration range and it would be hardly detectable, becoming even negative at higher $[L]_T$. A comment is necessary here. The computer simulations shown in Fig. 2 were performed with given values of the relative fluorescence intensities for all participating intermediates. In experimental studies of a kinetic system, this process is reversed, i.e., from the dependence of the amplitudes of the system upon ligand concentrations one determines the spectroscopic parameters characterizing all intermediates (3, 4, 10–18).

#### 3.1.3. Help from Thermodynamics

If all relaxation modes are easily accessible from the experiment, kinetic measurements can completely address the energetics and dynamics of the examined reaction system. In practice, this is often not the case and "help" from thermodynamic studies is very welcome. For instance, in the numerical analyses, to extract the rate constants from the relaxation time data, one can utilize the fact that the value of the overall binding constant, $K_{ov}$, which can be independently obtained by the equilibrium fluorescence titration method, is related to the partial equilibrium steps in (5), as

$$K_{ov} = K_1(1 + K_2 + K_2K_3 + K_2K_3K_4), \quad (22)$$

where the partial equilibrium constants for each step of the reaction are $K_1 = k_1/k_{-1}$, $K_2 = k_2/k_{-2}$, $K_3 = k_3/k_{-3}$, and $K_4 = k_4/k_{-4}$. Expression (22) reduces the number of fitting parameters by one (see below). Next, having determined individual amplitudes, one can address the molar fluorescence intensities characterizing each intermediate of the reaction, using the maximum, fractional quenching (increase) of the macromolecule fluorescence, $\Delta F_{max}$, once again, obtained in independent equilibrium titrations studies. The values of $\Delta F_{max}$ is analytically expressed as

$$\Delta F_{max} = \frac{\Delta F_1 + K_2\Delta F_2 + K_2K_3\Delta F_3 + K_2K_3K_4\Delta F_4}{1 + K_2 + K_2K_3 + K_2K_3K_4}, \quad (23)$$

where $\Delta F_1$, $\Delta F_2$, $\Delta F_3$, and $\Delta F_4$ are fractional fluorescence intensities of the corresponding intermediates, relative to the fluorescence of the free macromolecule, $F_0$, i.e., $\Delta F_i = (F_i - F_0)/F_0$. The value of $F_0$ can be taken as 1. Expression (23) furnishes an additional relationship among the fluorescence parameters, with the value of $\Delta F_{max}$ playing the role of a scaling factor against which the entire set of molar fluorescence intensities can be examined (10–18).

#### 3.1.4. Mechanism of the E. coli DnaB Helicase–Nucleotide Interactions

As pointed out above, the same approach in analyzing the relaxation kinetic data applies when the signal is generated from the ligand. In such a case, the kinetic experiments are performed under pseudo-first-order conditions, with respect to the macromolecule concentration. The difference in the analysis is that the total macromolecule, $[M]_T$, concentration replaces the total concentration of the ligand, $[L]_T$, in all the kinetic equations discussed above. An example of the analogous analysis, where the spectroscopic signal originates from the ligand, is the kinetics of binding of the fluorescent nucleotide analog, MANT-AMP-PNP, to the *E. coli* primary replicative helicase the DnaB protein (14). The DnaB helicase forms a stable hexamer built of six chemically identical subunits (10, 11, 14). Binding of the fluorescence analogs of nucleotides, MANT-AMP-PNP or MANT-ADP, to the enzyme is accompanied by a dramatic increase of the nucleotide fluorescence, providing an excellent signal to monitor the interactions. The DnaB hexamer binds six molecules of the nucleotide cofactor. Initially all six binding sites are equivalent, although there is a negative cooperativity among the nucleotide binding sites. In order to examine the intrinsic kinetic mechanism of the cofactor binding to a single noninteracting site of the enzyme, the concentration of the nucleotide has to be low enough to make the population of the complexes with cooperative interactions negligible (see below).

#### 3.1.5. Experimental Approach

1. Prepare several samples of the *E. Coli* DnaB helicase solution, ~1.5–2 ml each, in the binding buffer, which is 50 mM Tris/HCl, pH 8.1, 100 mM NaCl, 5 mM $MgCl_2$, 10% glycerol, differing by the enzyme concentration. The volumes of the samples are not specific for the DnaB protein, but dictated by the requirement of the currently available, conventional stopped-flow instruments. The concentrations of the samples should differ by at least ~30% and the difference between the lowest and the highest concentration should be at least approximately one order of magnitude. However, the total lowest concentration of the protein must be ~5 times higher than the total concentration of the nucleotide cofactor, after mixing, to assure the pseudo-first-order conditions (see below). On the other hand, the highest protein concentration should correspond to the concentration, where at least ~80% of the cofactor is saturated with the enzyme. The required total concentration is calculated using the binding constant determined in independent equilibrium experiments. Also, remember, after

mixing in the stopped-flow apparatus, the concentration of the enzyme will be half of the concentration of the prepared sample. Usually 7–12 samples of the protein are enough to cover the required protein concentration range. The exact DnaB concentration is measured using the extinction coefficient $\varepsilon_{280}$ = 185,000 $M^{-1}$ $cm^{-1}$ (Hexamer) (10, 11, 14). All samples should be kept on ice *prior* to the kinetic experiment.

2. Prepare approximately ~20–25 ml of the stock solution of the selected nucleotide cofactor, e.g., MANT-AMP-PNP or MANT-ADP in the binding buffer. This total volume of the cofactor solution should be enough for all planned experimental points, i.e., it should correspond to the number of the DnaB samples, with ~2 ml of the cofactor solution for each protein sample. The nucleotide concentration in the stock sample is $6 \times 10^{-7}$ M. Thus, after mixing in the stopped-flow instrument, the final concentration will be $3 \times 10^{-7}$ M. The stock solution of the cofactor should be kept on ice *prior* to the kinetic experiment.
3. Set the excitation wavelength at 356 nm on the stopped-flow instrument (absorption maximum of MANT-AMP-PNP or MANT-ADP). Use the 400-nm cut-off filter on the emission channel to extract the cofactor fluorescence and to eliminate any artifact resulting from the light scattering of the sample. Set up the temperature of the measurement. To collect the total emission of the sample, the stopped-flow instrument must be equipped with polarizers and two emission channels. Set up the excitation polarizer at the vertical position (V), one emission channel at the vertical position, and the other emission channel at the horizontal position (H). Load one syringe of the stopped-flow instrument with the nucleotide solution and the other syringe with the binding buffer. Perform at least six to seven shots to fill the observation chamber of the instrument with the nucleotide. Set the voltage on the photomultipliers to obtain a possible maximum signal, which gives you the reference point, i.e., the emission intensity of the free nucleotide. Next, because the fluorescence of the cofactor strongly increases upon binding to the helicase, decrease the voltage to a value of ~60% of the maximum signal. This will allow you to perform measurements of all samples at a single value of the voltage, which, in turn, will allow you to obtain individual and total amplitudes, without recalibrating the instrument at different voltage values.
4. Record the fluorescence stopped-flow kinetic trace for just the nucleotide at the selected time range. This is your initial signal with respect to which you will be determining the total changes of the fluorescence intensity for the nucleotide–DnaB protein complexes at different protein concentrations. Replace the

buffer solution in the syringe with the sample containing the lowest protein concentration. Wait approximately 10–15 min to equilibrate the temperature of the protein sample with the temperature of the kinetic measurement. Perform at least six to seven shots to fill the observation chamber of the instrument with the nucleotide and the protein. Select the time range of the kinetic measurement by observing where the system reaches the equilibrium for the last couple of traces. If the instrument allows and usually it does these days, it is prudent to use two time bases to obtain a high resolution at the short- and long-time ranges of the observed kinetic trace.

5. Depending on the quality of the kinetic sample (noise), collect 5–15 kinetic traces and average them, using the averaging software provided with the instrument. Repeat the same kinetic measurements for all protein samples. Perform nonlinear least squares fitting of the average kinetic traces for each sample, using a function, which is the sum of the exponential terms. This is a pseudo-first-order system and it decays to equilibrium as an exponential function or as a sum of exponential functions (see above). Observe both the value of the variance and, particularly, the residuals of the fit. Select the fitting function, which provides the best residuals, evenly distributed over the signal. Also, your eye is a very good discriminator in selecting what is a "good fit." A change of variance by ~10% does not justify adding an extra exponent. Remember the Ockham's razor, i.e., do not add more exponents to the fitting function than are needed to obtain a good fit. Extract the reciprocal relaxation times and the individual amplitudes as functions of the total protein concentration. Extract the total amplitude for each sample, as the difference between the end point of the kinetic trace recorded in the presence of the protein and the signal obtained for the nucleotide alone. Normalize the individual amplitudes with respect to the total amplitude. Plot the reciprocal relaxation times as a function of the total concentration of the DnaB helicase and the normalized amplitudes as a function of the logarithm of the total concentration of the protein. Although the above procedure has been described as applied to the *E. coli* DnaB–nucleotide system, it applies to any interacting system under pseudo-first-order conditions, where the fluorescence intensity is used to monitor the formation of the complex. Only excitation and emission wavelengths will be different and specifically selected for the examined system.

6. Experimental kinetic measurements of the fluorescence anisotropy are identical to that described above for the total emission. The difference is that one collects a transformed signal using the instrumental setup, specific for the particular stopped-flow instrument equipped with polarizers and two emission photomultipliers (see below).

7. Kinetic experiments in the presence of a dynamic quencher, e.g., acrylamide, are performed analogously as described above for the total emission. In this case, one collects the total emission of the sample, but the buffer, containing the protein and the fluorescent nucleotide analog, also contains a selected quencher concentration. In order to analyze data in terms of the quenching of the individual intermediates of the reaction, one must determine the Stern–Volmer quenching constant for the examined fluorophore, free in the same buffer and using the same excitation and emission wavelengths, as applied in the kinetic measurements (see below).

#### *3.1.6. Stopped-Flow Kinetics of the E. coli DnaB Helicase–Nucleotide Interactions*

The stopped-flow kinetic trace of the MANT-AMP-PNP fluorescence, after mixing the analog with the large excess of the DnaB helicase, is shown in Fig. 3. The curve has been recorded in two time bases, 1 and 200 s. Inspection of the plot shows that the observed kinetic process is complex, clearly showing the presence of multiple steps. The solid line in Fig. 3 is a nonlinear least-squares fit of the experimental curve using the four-exponential fit. The three-exponential function does not provide an adequate description of the experimentally observed kinetics (data not shown). On the other hand, a higher number of exponents does not significantly improve the statistics of the fit. Therefore, the association of MANT-AMP-PNP with the DnaB helicase is at least a four-step process.

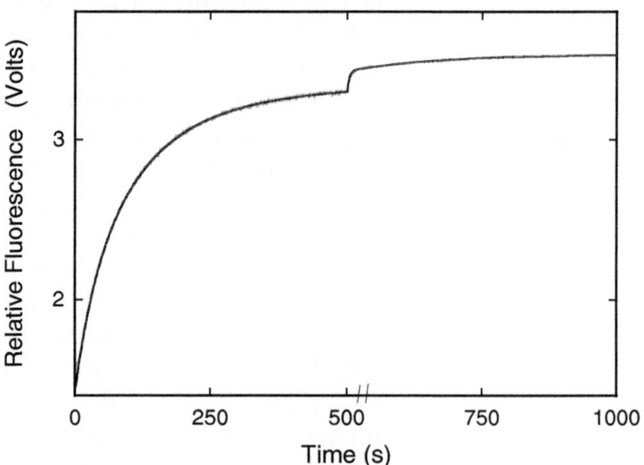


Fig. 3. The fluorescence stopped-flow kinetic trace, recorded in two time bases, 1 and 200 s, after mixing the DnaB helicase with MANT-AMP-PNP (pH 8.1, 100 NaCl, 20°C) ($\lambda_{ex} = 295$ nm, $\lambda_{em} > 400$ nm) (14). The final concentrations of the DnaB helicase and the nucleotide are $1.2 \times 10^{-5}$ M (Protomer) and $3 \times 10^{-7}$ M, respectively. The *solid line* is the four-exponential, nonlinear least-square fit of the experimental curve using (1) (Reproduced from *Biochemistry* 2000 (ref. 14) with permission from American Chemical Society).

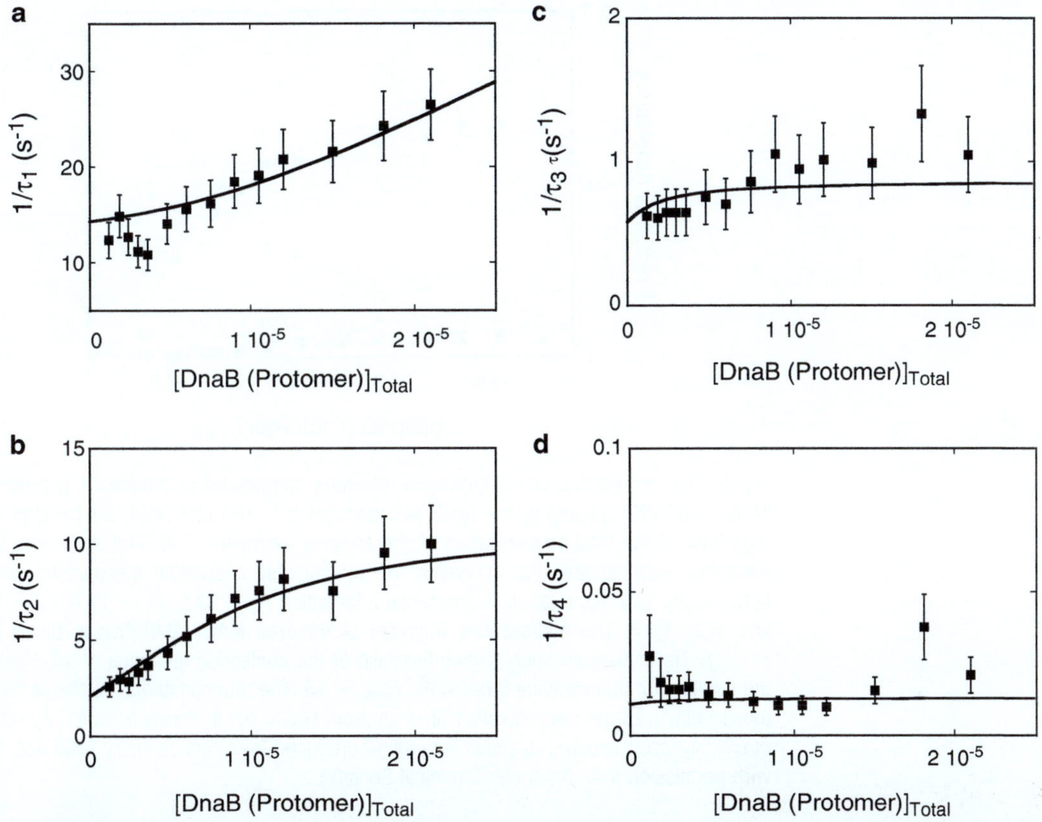


Fig. 4. The dependence of the reciprocal of the relaxation times for the binding of MANT-AMP-PNP to the DnaB helicase (pH 8.1, 100 mM NaCl, 20°C) upon the total concentration of the enzyme (Protomer) (14). The *solid lines* are nonlinear least squares fits of the experimental data to the four-step sequential mechanism, defined by (5), using the rate constants: $k_1 = 9 \times 10^5\ M^{-1}\ s^{-1}$, $k_{-1} = 5\ s^{-1}$, $k_2 = 6\ s^{-1}$, $k_{-2} = 5\ s^{-1}$, $k_3 = 0.3\ s^{-1}$, $k_{-3} = 0.7\ s^{-1}$, $k_4 = 0.011\ s^{-1}$, and $k_{-4} = 0.011\ s^{-1}$; (**a**) $1/\tau_1$, (**b**) $1/\tau_2$, (**c**) $1/\tau_3$, (**d**) $1/\tau_4$. The error bars are standard deviations obtained from three to four independent experiments (Reproduced from *Biochemistry* 2000 (ref. 14) with permission from American Chemical Society).

The reciprocal relaxation times, $1/\tau_1$, $1/\tau_2$ $1/\tau_3$, and $1/\tau_4$, characterizing the four normal modes of the reaction, as a function of the total DnaB helicase concentration, are shown in Fig. 4a–d (14). The largest reciprocal relaxation time, $1/\tau_1$, increases with the increasing concentration of the enzyme, and the dependence becomes linear at high enzyme concentrations, although there is a nonlinear phase at the very low DnaB concentration. Such behavior is typical for the relaxation time characterizing the bimolecular binding step (see above). The values of $1/\tau_2$ and $1/\tau_3$ show hyperbolic dependence upon the enzyme concentration, while $1/\tau_4$ is, within experimental accuracy, independent of $[DnaB]_T$. The simplest minimum mechanism that can account for the observed dependence of the relaxation times upon the enzyme concentration

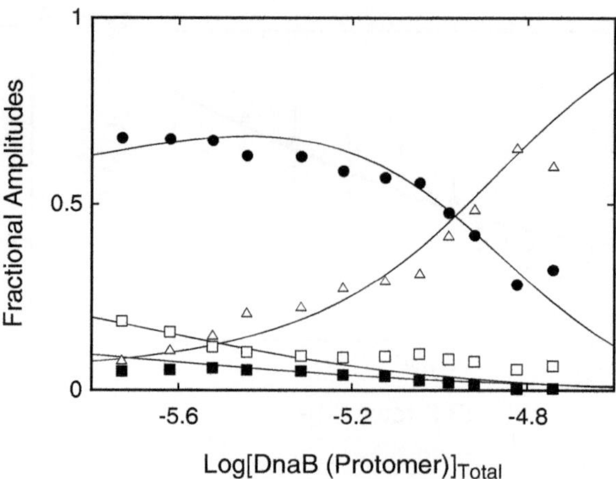


Fig. 5. The dependence of the individual relaxation amplitudes of the kinetic process of MANT-AMP-PNP binding to the DnaB helicase (pH 8.1, 100 mM NaCl, 20°C) upon the logarithm of the total concentration of the enzyme (protomer) (14). The *solid lines* are nonlinear least squares fits according to the four-step sequential mechanism, using (21a)–(21d), with the relative fluorescence intensities: $F_2 = 13.6$, $F_3 = 11.9$, $F_4 = 12$, and $F_5 = 12.7$. The fluorescence intensity of the free MANT-AMP-PNP is taken as, $F_1 = 1$. The maximum fluorescence increase of the nucleotide has been determined in the equilibrium fluorescence titration as, $F_{max} = 13$. The rate constants are the same as those obtained from the relaxation time analysis (Table 2); $A_1$ (*open triangle*), $A_2$ (*filled circle*), $A_3$ (*open square*), $A_4$ (*filled square*) (Reproduced from *Biochemistry* 2000 (ref. 14) with permission from American Chemical Society).

is a four-step, sequential binding process in which the bimolecular association is followed by three isomerization steps, as described by (5). The solid lines in Fig. 4a–d are nonlinear least squares fits of the relaxation times according to the above mechanism, which also utilize the fact that the overall binding constant, $K$, is known from the independent equilibrium titration studies (14).

The dependence of the individual amplitudes, $A_1$, $A_2$, $A_3$, and $A_4$, of all four normal modes of the reaction upon the logarithm of the DnaB protein concentration, is shown in Fig. 5. The individual amplitudes are normalized, i.e., expressed as a fraction of the total amplitude, $A_i/\sum A_i$. At low helicase concentrations, the amplitude of the first bimolecular process does not significantly contribute to the observed kinetics. The amplitude of the second relaxation process, $A_2$, dominates the total amplitude. The amplitudes, $A_3$ and $A_4$, of the third and fourth relaxation steps contribute to the observed kinetics mainly at low enzyme concentrations. As the concentration of the DnaB helicase increases, the amplitude of the bimolecular step, $A_1$, increases and at a high enzyme concentration range becomes a dominant relaxation effect (Fig. 4). Such behavior of the individual amplitudes is in full agreement with the proposed sequential mechanism (5). The molar fluorescence intensities, characterizing each intermediate of the reaction relative to the fluorescence of the free nucleotide, have been determined using the matrix projection operator technique. Moreover, the analysis utilized the fact that the maximum fluorescence increase, $\Delta F_{max}$, accompanying the complex formation, is known from independent equilibrium fluorescence titrations (14).

## 4. Stopped-Flow Fluorescence Anisotropy Studies of the Kinetic Mechanisms of the Ligand–Macromolecule Interactions

### *4.1. Steady-State Anisotropies of the Reaction Intermediates*

In a typical, fluorescence stopped-flow kinetic experiment, the dynamics of the ligand–macromolecule association is followed by monitoring the emission or total emission of the sample, $F(t)$. Because at the concentration ranges used in biological kinetic studies, the observed fluorescence emission of the examined sample is proportional to the molar concentrations of the fluorescing species, the change of the emission of the sample is described by the sum of the exponential functions as defined by (1). In other words, the fluorescence emission changes the same way as the normal modes of the reaction. Fluorescence intensity of a given macromolecular or ligand species is usually dependent upon the nature of the environment surrounding the fluorophore (27). Thus, relative molar fluorescence intensities of different intermediates, obtained from the relaxation amplitude analysis, provide information about the physical environment surrounding the fluorophore during the course of the reaction (see below).

On the other hand, fluorescence anisotropy reflects the rotational mobility of the fluorescing species during the fluorescence lifetime of the fluorophore (28). Notice, the mechanism of the reaction can be obtained using the change of the fluorescence intensity during the course of the reaction. Once the mechanism is known, together with the rate constants and the relative molar intensities of all intermediates, time-dependence of the fluorescence anisotropy in stopped-flow experiments provides information about the mobility of the fluorescing species in each intermediate of the observed kinetic process (16, 24). The lifetime of a typical fluorophore used in kinetic studies is ~4–10 ns, i.e., it is orders of magnitude shorter than the lifetime of any intermediate identified in stopped-flow studies. Thus, stopped-flow fluorescence anisotropy studies provide the values of the steady-state anisotropy of the ligand or macromolecule in each intermediate. Moreover, transient anisotropy data may also shed additional light on the mechanism of the reaction, not obvious in the studies using the emission intensity (see below) (24). First, we discuss the extraction of the steady-state anisotropies of the intermediates of the reaction, using the sequential, four-step reaction mechanism discussed above. Subsequently, we address the mechanism of the reaction.

In general, unlike the emission intensity, stopped-flow kinetics traces recorded using the fluorescence anisotropy are not a simple sum of exponential functions, even under pseudo-first-order conditions (16, 24). This is an often forgotten difference between these two spectroscopic signals. The time-course of the fluorescence anisotropy of the sample, containing "$n$" intermediates, is defined as

$$r(t) = \frac{\sum_{i=1}^{n} I_{VV}(t) - \sum_{i=1}^{n} I_{VH}(t)}{\sum_{i=1}^{n} I_{VV}(t) + 2G\sum_{i=1}^{n} I_{VH}(t)}, \quad (24)$$

where the denominator is the total emission of the sample, $F(t)$ (3) or (4). The above expression can be written as

$$r(t) = \sum_{i=1}^{n} f_i(t) r_i, \quad (25)$$

where $f_i(t)$ is the time-dependent *fractional* contribution of $i$th fluorescing intermediate of the reaction to the emission of the sample and $r_i$ is its corresponding steady-state fluorescence anisotropy. For instance, for the sequential four-step mechanism discussed above, relationship 25 takes the form of

$$r(t) = f_0(t) r_{D0} + f_1(t) r_{D1} + f_2(t) r_{D2} + f_3(t) r_{D3} + f_4(t) r_{D4}, \quad (26)$$

where $r_{D1}$, $r_{D2}$, $r_{D3}$, and $r_{D4}$ are steady-state anisotropies of the free macromolecule $D_0$ and the macromolecule in $D_1$, $D_2$, $D_3$, and $D_4$ intermediates. Using molar fluorescence intensities of the fluorescing intermediates, $F_0$, $F_1$, $F_2$, $F_3$, and $F_4$, and their corresponding molar concentrations, $D_0$, $D_1$, $D_2$, $D_3$, and $D_4$, obtained using the matrix projection operator method (13), the time course of the fluorescence anisotropy in (26) is then defined as

$$r(t) = \frac{\begin{pmatrix} F_0 r_{D0} & F_1 r_{D1} & F_2 r_{D2} & F_3 r_{D3} & F_4 r_{D4} \end{pmatrix} \begin{pmatrix} P_{01} & P_{11} & P_{21} & P_{31} & P_{41} \\ P_{02} & P_{12} & P_{22} & P_{32} & P_{42} \\ P_{03} & P_{13} & P_{23} & P_{33} & P_{43} \\ P_{04} & P_{14} & P_{24} & P_{34} & P_{44} \\ P_{05} & P_{15} & P_{25} & P_{35} & P_{45} \end{pmatrix} \begin{pmatrix} 1 \\ \exp(\lambda_1 t) \\ \exp(\lambda_2 t) \\ \exp(\lambda_3 t) \\ \exp(\lambda_4 t) \end{pmatrix}}{F(t) = \begin{pmatrix} F_0 & F_1 & F_2 & F_3 & F_4 \end{pmatrix} \begin{pmatrix} P_{01} & P_{11} & P_{21} & P_{31} & P_{41} \\ P_{02} & P_{12} & P_{22} & P_{32} & P_{42} \\ P_{03} & P_{13} & P_{23} & P_{33} & P_{43} \\ P_{04} & P_{14} & P_{24} & P_{34} & P_{44} \\ P_{05} & P_{15} & P_{25} & P_{35} & P_{45} \end{pmatrix} \begin{pmatrix} 1 \\ \exp(\lambda_1 t) \\ \exp(\lambda_2 t) \\ \exp(\lambda_3 t) \\ \exp(\lambda_4 t) \end{pmatrix}} \quad (27)$$

The generalization of the above expression to any kinetic mechanism is straightforward (16).

There are 18 parameters, $k_1$, $k_{-1}$, $k_2$, $k_{-2}$, $k_3$, $k_{-3}$, $k_4$, $k_{-4}$, $F_0$, $F_1$, $F_2$, $F_3$, $F_4$, $r_{D0}$, $r_{D1}$, $r_{D2}$, $r_{D3}$, and $r_{D4}$, in (25). Accurate fitting of any kinetic curve with this number of parameters is a rather hopeless task. However, as pointed out above, kinetic measurements, using the fluorescence intensity as observable, provide 13 of them, $k_1$, $k_{-1}$, $k_2$, $k_{-2}$, $k_3$, $k_{-3}$, $k_4$, $k_{-4}$, $F_0$, $F_1$, $F_2$, $F_3$, and $F_4$. The value of $r_{D0}$, i.e., the anisotropy of the macromolecule *prior* to the ligand binding is also known. Therefore, there are only four parameters, $r_{D1}$, $r_{D2}$, $r_{D3}$, and $r_{D4}$, which remain to be determined, using any efficient, nonlinear least squares fitting routine.

Notice, if there is not any change of the fluorescence intensity accompanying the reaction then, $F(t) = F_0 = F_1 = F_2 = F_3 = F_4$, and the time-dependence of the anisotropy of the sample is

$$r(t) = \begin{pmatrix} r_{D0} & r_{D1} & r_{D2} & r_{D3} & r_{D4} \end{pmatrix} \times \begin{pmatrix} P_{01} & P_{11} & P_{21} & P_{31} & P_{41} \\ P_{02} & P_{12} & P_{22} & P_{32} & P_{42} \\ P_{03} & P_{13} & P_{23} & P_{33} & P_{43} \\ P_{04} & P_{14} & P_{24} & P_{34} & P_{44} \\ P_{05} & P_{15} & P_{25} & P_{35} & P_{45} \end{pmatrix} \begin{pmatrix} 1 \\ \exp(\lambda_1 t) \\ \exp(\lambda_2 t) \\ \exp(\lambda_3 t) \\ \exp(\lambda_4 t) \end{pmatrix} \quad (28)$$

This expression is completely analogous to (15) for the time-dependence of the fluorescence intensity, $F(t)$. Therefore, in the absence of the emission change in the course of the reaction and only in this case, the fluorescence anisotropy can be used to directly address the kinetic mechanism of the studied process, in the same way as the fluorescence intensity.

### 4.2. Stopped-Flow Fluorescence Anisotropy Analysis of the Reaction Mechanisms

The discussion above concentrated on extracting steady-state fluorescence anisotropies of all intermediates of the reaction, once the kinetic mechanism and the molar fluorescence intensities, characterizing each intermediate, are known from the fluorescence intensity studies of the relaxation process. However, it often happens that the intensity measurements will provide limited knowledge about the kinetic mechanism. For instance, this is the case when one or two steps of the reaction are so fast that they are beyond, or hardly in the range of an accurate resolution in the conventional stopped-flow measurement. The simulated stopped-flow trace of a sequential reaction, described by (5), is shown in Fig. 6. The reaction is followed by monitoring the quenching of the macromolecule fluorescence. The solid line in Fig. 6 is the numerical solution for the four-step mechanism. The corresponding reciprocal relaxation times and amplitudes, as functions of the ligand concentration, are shown in Figs. 7 and 8. It is evident that, with the selected rate constants, the largest reciprocal relaxation time, $1/\tau_1$, is beyond the resolution of the stopped flow experiment, while the values of $1/\tau_2$ are high and barely in the measuring range of the stopped-flow method. The experimenter would be prone to fit a

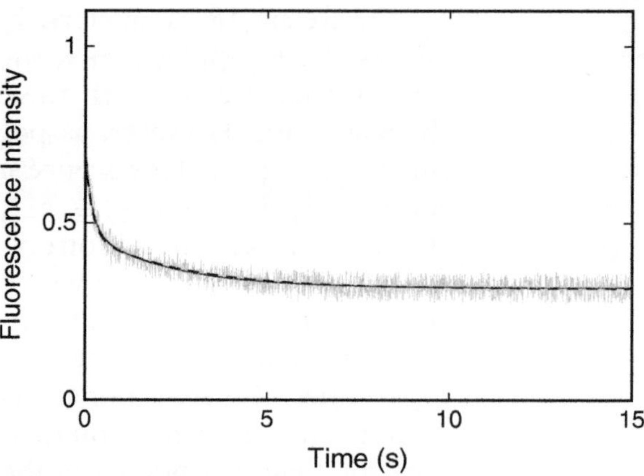


Fig. 6. Computer simulation of the fluorescence stopped-flow kinetic trace, after mixing a fluorescing macromolecule with the ligand, with the final concentrations of the macromolecule and the ligand $1 \times 10^{-8}$ and $2.5 \times 10^{-6}$ M, respectively. The simulations have been performed using (27), with the rate constants: $k_1 = 5 \times 10^8$ $M^{-1}$ $s^{-1}$, $k_{-1} = 5{,}000$ $s^{-1}$, $k_2 = 3{,}000$ $s^{-1}$, $k_{-2} = 800$ $s^{-1}$, $k_3 = 10$ $s^{-1}$, $k_{-3} = 1$ $s^{-1}$, $k_4 = 0.5$ $s^{-1}$, and $k_{-4} = 0.05$ $s^{-1}$, with added Gaussian random noise amounting to 15% of the signal. The selected relative fluorescence intensities are $F_1 = 0.30$, $F_2 = 0.50$, $F_3 = 0.40$, and $F_4 = 0.30$. The *solid line* is the analytical representation of the trace using the four-step sequential reaction mechanism (1). The *dashed line* is the analogous representation, using three-step sequential reaction mechanism, with the rate constants: $k_1 = 1.5 \times 10^9$ $M^{-1}$ $s^{-1}$, $k_{-1} = 5{,}000$ $s^{-1}$, $k_2 = 11$ $s^{-1}$, $k_{-2} = 0.80$ $s^{-1}$, $k_3 = 0.40$ $s^{-1}$, and $k_{-3} = 0.0443$ $s^{-1}$. The relative fluorescence intensities are $F_1 = 0.300$, $F_2 = 0.410$, and $F_3 = 0.297$.

trace like the one in Fig. 6 using the three-exponential function and such "a fit" is included in Fig. 6. Within accuracy of the measurement, both four-exponential and three-exponential functions would be acceptable representations of the observed relaxation process. The thermodynamic data do not help in this case, for both analytical representations the same overall binding constant and the maximum fluorescence change are used. On the basis of the intensity measurement alone, the correct mechanism and, by the same token, rate and spectroscopic parameters characterizing the intermediates of the relaxation process, could not be accurately determined.

The situation where the fluorescence anisotropy is examined is different and if applied, helps in establishing the correct mechanism. The computer simulation of the stopped-flow trace of the fluorescence anisotropy for the same system is shown in Fig. 9. The fluorescence of the anisotropy of the macromolecule alone in the absence of the ligand is also included. The solid line in Fig. 9 is the numerical representation of the time-dependence of the correct, four-step mechanism, using (27). The dashed line is the best fit of the same data using the kinetic and spectroscopic parameters of the three-step mechanism, which can represent fluorescence intensity

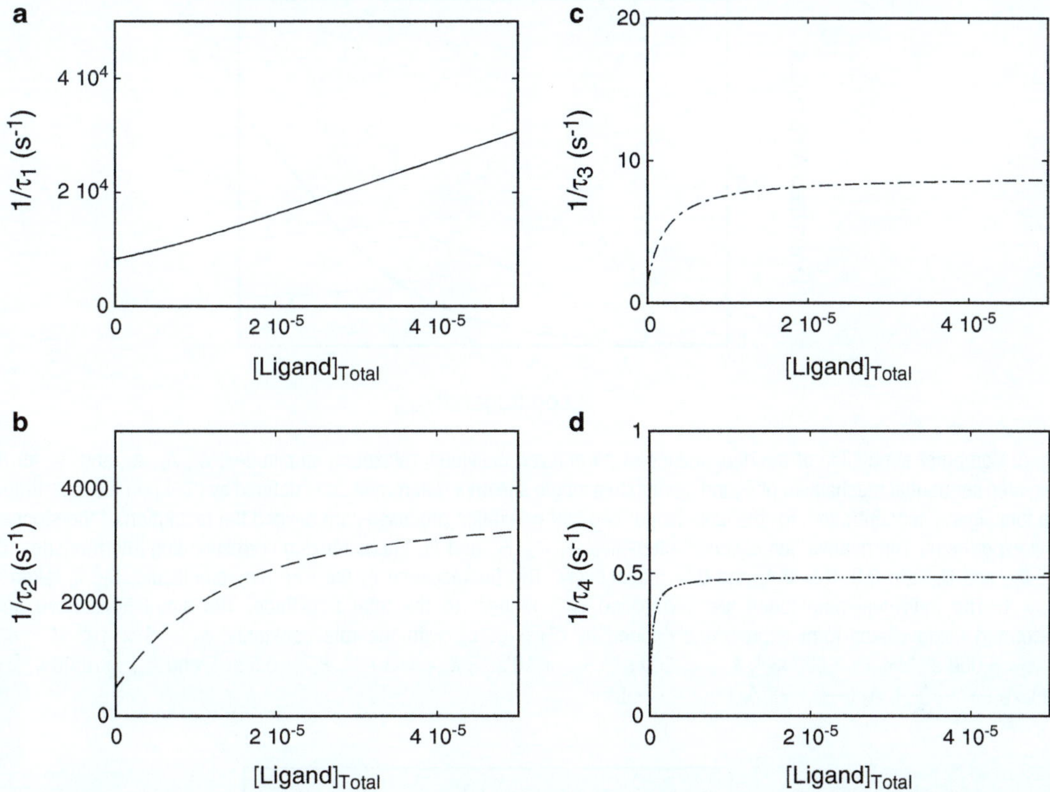


Fig. 7. Computer simulation of the dependence of the reciprocal relaxation times for the four-step sequential mechanism of ligand binding to a single site on a macromolecule, defined by (5), upon the total ligand concentration, for the case where two fast relaxation processes are beyond the resolution of the stopped-flow experiment. Relaxation times have been obtained by numerically determining the eigenvalues of the coefficient matrix **M**, ($\lambda_1$, $\lambda_2$, $\lambda_3$, $\lambda_4$), then using identities $1/\tau_1 = -\lambda_1$, $1/\tau_2 = -\lambda_2$, $1/\tau_3 = -\lambda_3$, and $1/\tau_4 = -\lambda_4$. The simulations have been performed using rate constants: $k_1 = 5 \times 10^8\ \text{M}^{-1}\ \text{s}^{-1}$, $k_{-1} = 5{,}000\ \text{s}^{-1}$, $k_2 = 3{,}000\ \text{s}^{-1}$, $k_{-2} = 800\ \text{s}^{-1}$, $k_3 = 10\ \text{s}^{-1}$, $k_{-3} = 1\ \text{s}^{-1}$, $k_4 = 0.5\ \text{s}^{-1}$, and $k_{-4} = 0.05\ \text{s}^{-1}$. The selected total macromolecule concentration, $[\text{D}]_\text{T}$, is $1 \times 10^{-8}$ M; (**a**) $1/\tau_1$, (**b**) $1/\tau_2$, (**c**) $1/\tau_3$, (**d**) $1/\tau_4$.

data. It is clear that the three-step model does not adequately describe the observed time-dependence of the fluorescence anisotropy, particularly at the initial part of the curve. In order to adequately represent the data, one would have to assume the value of the anisotropy of the free macromolecule to be significantly higher than experimentally observed. Thus, anisotropy data provide the experimenter a clear evidence that the mechanism includes an additional step. Moreover, the step is faster than the identified transitions in the three-step mechanism. The next step in the analysis would be to redefine the mechanism as the correct four-step process, with the first two steps undergoing fast equilibration *prior* to the observed relaxation processes.

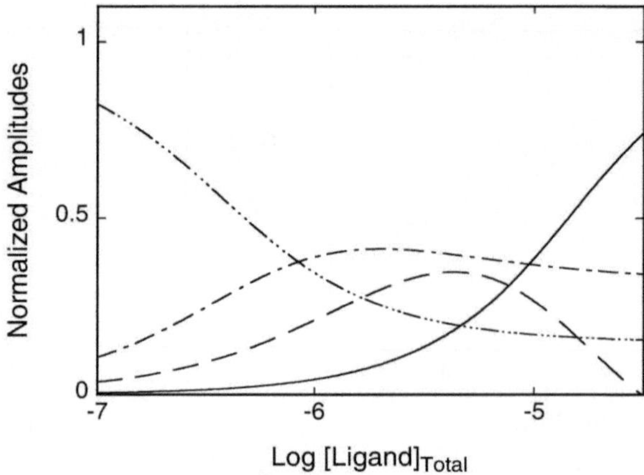


Fig. 8. Computer simulation of the dependence of normalized individual relaxation amplitudes, $A_1$, $A_2$, $A_3$, and $A_4$, for the four-step sequential mechanism of ligand binding to a single site on a macromolecule, defined by (5), upon the logarithm of the total ligand concentration, for the case where two fast relaxation processes are beyond the resolution of the stopped-flow experiment. The relative fluorescence intensities, $F_1$, $F_2$, $F_3$, and $F_4$, characterizing corresponding intermediates, $D_1$, $D_2$, $D_3$, and $D_4$, are 0.3, 0.5, 0.4, and 0.3, respectively. The fluorescence of the free macromolecule, $D_0$, is taken as $F_0 = 1$. The individual amplitudes are normalized with respect to the total amplitude. The simulations have been performed using closed-form expressions defined by (21a)–(21d), with the rate constants: $k_1 = 5 \times 10^8$ $M^{-1}$ $s^{-1}$, $k_{-1} = 5{,}000$ $s^{-1}$, $k_2 = 3{,}000$ $s^{-1}$, $k_{-2} = 800$ $s^{-1}$, $k_3 = 10$ $s^{-1}$, $k_{-3} = 1$ $s^{-1}$, $k_4 = 0.5$ $s^{-1}$, and $k_{-4} = 0.05$ $s^{-1}$; $A_1$ (—), $A_2$ (— — —), $A_3$ (— - —), $A_4$ (— - - - —).

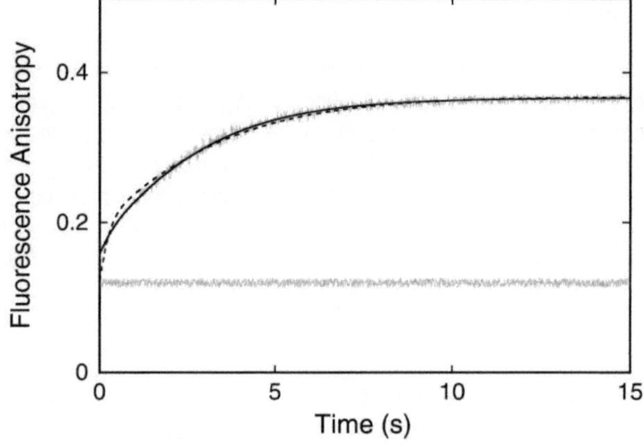


Fig. 9. Computer simulation of the fluorescence anisotropy, stopped-flow kinetic trace, after mixing a fluorescing macromolecule with the ligand. The final concentrations of the macromolecule and the ligand are $1 \times 10^{-8}$ and $2.5 \times 10^{-6}$ M, respectively. The simulations have been performed using (27), with the rate constants: $k_1 = 5 \times 10^8$ $M^{-1}$ $s^{-1}$, $k_{-1} = 5{,}000$ $s^{-1}$, $k_2 = 3{,}000$ $s^{-1}$, $k_{-2} = 800$ $s^{-1}$, $k_3 = 10$ $s^{-1}$, $k_{-3} = 1$ $s^{-1}$, $k_4 = 0.5$ $s^{-1}$, and $k_{-4} = 0.05$ $s^{-1}$, with added Gaussian random noise amounting to 15% of the signal. The selected relative fluorescence intensities are $F_1 = 0.30$, $F_2 = 0.50$, $F_3 = 0.40$, and $F_4 = 0.30$, and the steady-state fluorescence anisotropies of the intermediates are $r_{D0} = 0.120$, $r_{D1} = 0.400$, and $r_{D2} = 0.200$, $r_{D3} = 0.200$, and $r_{D4} = 0.400$. The *solid line* is the analytical representation of the trace using four-step sequential reaction mechanism (1). The *dashed line* is the analogous representation using three-step sequential reaction mechanism, with the rate constants: $k_1 = 1.5 \times 10^9$ $M^{-1}$ $s^{-1}$, $k_{-1} = 5{,}000$ $s^{-1}$, $k_2 = 11$ $s^{-1}$, $k_{-2} = 0.80$ $s^{-1}$, $k_3 = 0.40$ $s^{-1}$, and $k_{-3} = 0.0443$ $s^{-1}$. The relative fluorescence intensities are $F_1 = 0.300$, $F_2 = 0.410$, and $F_3 = 0.297$, and the steady-state fluorescence anisotropies of the intermediates are $r_{D0} = 0.120$, $r_{D1} = 0.120$, and $r_{D2} = 0.233$, and $r_{D3} = 0.400$. The horizontal trace is the analogous computer simulation of the stopped-flow trace of the macromolecule in the absence of the ligand.

## 5. Transient, Dynamic Fluorescence Quenching Studies of Individual Intermediates. Stern–Volmer Quenching Constant for Individual Intermediates of the Reaction

Collisional dynamic quenching of the fluorescence emission is one of the most often used experimental methods to assess the solvent accessibility of the fluorophore (29–31). The method is based on introducing an additional collisional quenching of the intrinsic fluorescence of the macromolecule or the ligand by the presence of an efficient, preferably nonionic, collisional quencher. Acrylamide fulfills these conditions and is usually the collisional quencher of choice in solvent accessibility studies (29–31). Dynamic or collisional quenching is a unique property of fluorescence and results from the fact that an *extra radiationless process* can be introduced to the experimental system without affecting, to any significant degree, the thermodynamic equilibrium and kinetic properties of the system, provided that the concentration of the quencher is low. The method is mostly applied to steady-state fluorescence intensity of the systems in equilibrium. If the collisional quenching efficiency is different for the different states of the fluorophore, the presence of these different states will be manifested by the change of the fluorescence of the macromolecule or of the ligand (16). Nevertheless, the same methodology can be applied to address the solvent accessibility of the different intermediates, during the time course of the reaction monitored in the stopped-flow experiment (16). Because the time of the dynamic collision is in the range of $10^{-12}$–$10^{-9}$ s, i.e., it is orders of magnitude shorter than the lifetimes of the identified intermediates in the stopped-flow experiment, the observed quenching of the fluorescence emission of the transient intermediates corresponds to the dynamic quenching of the steady-state fluorescence of the equilibrium system (16).

The quenching of the fluorescence emission of a fluorophore by an external collisional quencher, *Q*, is conveniently described by the Stern–Volmer equation, as (16, 29, 30)

$$\frac{F_{\mathrm{L}}^{\mathrm{o}}}{F_{\mathrm{L}}^{\mathrm{q}}} = 1 + k_{\mathrm{q}}\tau_{\mathrm{o}}[\mathrm{Q}] = 1 + K_{\mathrm{SV}}^{\mathrm{o}}[\mathrm{Q}], \tag{29}$$

where $F_{\mathrm{L}}{}^{\mathrm{o}}$ and $F_{\mathrm{L}}{}^{\mathrm{q}}$, are the steady-state fluorescence intensities of the of the fluorescent ligand in the absence and presence of the quencher, $k_{\mathrm{q}}$, is the bimolecular quenching constant, $\tau_{\mathrm{o}}$, is the fluorescence lifetime of the fluorophore, and [Q] is the total concentration of the quencher. The quantity, $K_{\mathrm{SV}}{}^{\mathrm{o}} = k_{\mathrm{q}}\tau_{\mathrm{o}}$, is the Stern–Volmer quenching constant, which is the parameter used to describe the quenching process (16, 29, 30).

Expression (29) describes the effect of the collisional quencher on the fluorescence of the free fluorophore in solution. An analogous equation can be derived for any intermediate of the reaction, in which the same fluorophore participates in terms of the relative molar fluorescence intensities, obtained from the amplitude analyses of the relaxation process. As an example, we consider an association reaction between a fluorescing ligand and a macromolecule. Nevertheless, the derived relationships are equally applicable to the cases where the fluorescing ligand, or vice versa macromolecule is concerned. The relative or normalized molar fluorescence intensity of an *i*th intermediate, $A_i$, with respect to the emission of the free fluorescing ligand, in the absence of the collisional quencher, is

$$F_{\mathrm{i}} = \frac{I_{\mathrm{i}}}{F_{\mathrm{L}}^{\mathrm{o}}}, \tag{30}$$

where, $I_{\mathrm{i}}$ is the fluorescence intensity of the intermediate, not normalized with respect to the fluorescence of the free ligand. This emission ratio is directly provided by the amplitude analysis of the relaxation process (16). In the presence of the collisional quencher, the absolute molar fluorescence intensity of the *i*th intermediate, $I_i^{\mathrm{q}}$, is defined as

$$I_i^{\mathrm{q}} = \frac{I_{\mathrm{i}}}{1 + K_{\mathrm{SV}i}[Q]}, \tag{31}$$

where, $K_{\mathrm{SV}i}$, is the Stern–Volmer quenching constant of an *i*th intermediate. Therefore, the relative molar fluorescence intensity of the intermediate, $F_{\mathrm{q}}$, directly determined from the amplitude analysis of the kinetic data and obtained in the presence of the quencher, is

$$F_{\mathrm{q}} = \frac{I_i^{\mathrm{q}}}{F_{\mathrm{L}}^{\mathrm{q}}} = \frac{I_{\mathrm{i}}(1 + K_{\mathrm{SV}}^{\mathrm{o}}[Q])}{F_{\mathrm{L}}^{\mathrm{o}}(1 + K_{\mathrm{SV}i}[Q])} \tag{32}$$

and

$$F_{\mathrm{q}} = \frac{F_{\mathrm{i}}(1 + K_{\mathrm{SV}}^{\mathrm{o}}[Q])}{(1 + K_{\mathrm{SV}i}[Q])}. \tag{33}$$

Rearranging (33) provides the required relationship for any intermediate of the reaction, analogous to the Stern–Volmer equation (29) for the steady-state fluorescence of the system at equilibrium, in terms of the experimentally determined relative molar fluorescence of the given intermediate, obtained in the absence and presence of the quencher, $F_{\mathrm{i}}$ and $F_{\mathrm{q}}$, respectively, as

$$\left(\frac{F_{\mathrm{i}}}{F_{\mathrm{q}}}\right)(1 + K_{\mathrm{SV}}^{\mathrm{o}}[Q]) = 1 + K_{\mathrm{SV}i}[Q]. \tag{34}$$

With the exception of $K_{\mathrm{SV}i}$, this parametric expression above contains all known quantities. The plot of the left side of (34), as a function of [Q], is a straight line with the intercept equal to 1 and a slope that is equal to the Stern–Volmer quenching constant, $K_{\mathrm{SV}i}$, specific for a given *i*th intermediate (16).

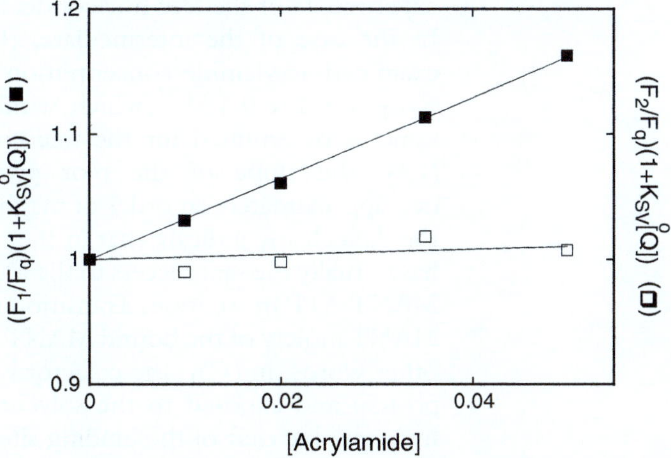


Fig. 10. Stern–Volmer plots of fluorescence quenching ($\lambda_{ex}$ = 356 nm, $\lambda_{em}$ > 400 nm), using acrylamide as a collisional quencher, for the first $(C)_1$ (*filled square*) and the second $(C)_2$ (*open square*) intermediate in the intrinsic binding of MANT-ATP to the DnaC protein (pH 8.1, 100 mM NaCl, 20°C). The relative molar fluorescence intensities of the intermediates, at different acrylamide concentrations, were obtained from the amplitude analysis of the relaxation process. The *solid lines* are linear least squares fits of the intermediate Stern–Volmer equation (34) to the data points, using $K_{SV1} = 3.2\ M^{-1}$ (*filled square*) and $K_{SV2} = 0.21\ M^{-1}$ (*open square*), respectively (16) (Reproduced from *Biochemistry* 2002 (ref. 16) with permission from American Chemical Society).

The analysis described above has been applied in kinetic studies of the nucleotide binding to the *E. coli* DnaC protein (16). The protein is an essential replication factor in the *E. coli* cell, involved in initiation and elongation stages of the DNA replication. It possesses a single nucleotide-binding site specific for the adenosine cofactors (15, 16). Stopped-flow kinetic data indicate that binding of fluorescent analogs of adenosine nucleotide cofactors, MANT-ATP and MANT-ADP, to the DnaC protein proceeds through the mechanism

$$\begin{gathered}(\mathrm{DnaC})_1 \underset{k_{-1}}{\overset{k_1}{\leftrightarrow}} (\mathrm{DnaC})_2 \\ (\mathrm{DnaC})_2 + \mathrm{N} \underset{k_{-2}}{\overset{k_2}{\leftrightarrow}} (\mathrm{C})_1 \underset{k_{-3}}{\overset{k_3}{\leftrightarrow}} (\mathrm{C})_2.\end{gathered} \tag{35}$$

Therefore, the protein exists in two conformational states *prior* to the binding of the nucleotide cofactor, N, and the cofactor associates with only one conformational state of the protein in a two-step, sequential reaction. Molar relative fluorescence intensities, $F_1$ and $F_2$, characterizing the two intermediates, $(C)_1$ and $(C)_2$, have been obtained as a function of the concentration of the collisional quencher, acrylamide. The Stern–Volmer plots (34) for the two intermediates, $(C)_1$ and $(C)_2$, in the binding of MANT-ATP to the DnaC protein, are shown in Fig. 10. For the free MANT-ATP, the Stern–Volmer quenching constant is, $K_{SV}{}^{o} \approx 3.1 \pm 0.1\ M^{-1}$ and has been determined in independent quenching experiments of the free nucleotide, in the same buffer conditions. In the case of the intermediate, $(C)_1$, the plot is linear in the examined acrylamide concentration range and its slope provides, $K_{SV1} = 3.2 \pm 0.4\ M^{-1}$, which, within experimental accuracy, is the same as determined for the free cofactor. For the intermediate, $(C)_2$, the slope of the plot gives $K_{SV2} = 0.21 \pm 0.1\ M^{-1}$, i.e., approximately an order of magnitude lower than, $K_{SV}{}^{o}$. Thus, the data clearly indicate that in the intermediate, $(C)_1$, acrylamide has virtually the same access to the MANT group as it has to the free MANT-ATP in solution. Transition to $(C)_2$ completely shields the MANT moiety of the bound MANT-ATP from the solvent (16). In other words, in $(C)_1$, the cofactor is located on the surface of the protein and exposed to the solvent, while in $(C)_2$, it enters the hydrophobic cleft of the binding site (16).

## Acknowledgments

This work was supported by NIH Grants GM46679 and GM58565 (to W.B.). We wish to thank Gloria Drennan Bellard for a careful reading of the manuscript.

### References

1. Hammes GG, Schimmel PR (1970) The Enzymes. Kinetics and Mechanism. Vol. II. Academic, New York, pp 67–114
2. Fierke CA, Hammes GG (1995) Transient kinetic approaches to enzyme mechanisms. Methods Enzymol 249:3–37
3. Bujalowski W (2006) Thermodynamic and kinetic methods of analyses of protein–nucleic acid interactions. From Simpler to more complex systems. Chem Rev 106:556–606
4. Bernasconi CJ (1976) Relaxation kinetics. Academic, New York
5. Font J, Torrent J, Ribó M, Laurents DV, Balny C, Vilanova M, Lange R (2006) Pressure-jump-induced kinetics reveals a hydration dependent folding/unfolding mechanism of ribonuclease A. Biophys J 15:2264–2274
6. Eigen M, De Maeyer L (1963) Relaxation methods. In: Friess SL, Lewis S, Weissberger A (eds) Technique of organic chemistry VIII part II. Wiley, New York, pp 895–1054
7. Galletto R, Jezewska MJ, Bujalowski W (2004) Multi-step sequential mechanism of *E. coli* Helicase PriA protein–ssDNA interactions kinetics and energetics of the active ssDNA-searching site of the enzyme. Biochemistry 43:11002–11016
8. Bujalowski W, Greaser E, McLaughlin LW, Porschke D (1986) Anticodon loop of $tRNA^{Phe}$: structure, dynamics, and $Mg^{2+}$ binding. Biochemistry 25:6365–6371
9. Porschke D (1998) Time-resolved analysis of macromolecular structures during reactions by stopped-flow electro-optics. J Biophys 75: 528–537
10. Bujalowski W, Jezewska MJ (2000) Kinetic mechanism of the single-stranded DNA recognition by *Escherichia coli* replicative helicase DnaB protein. Application of the matrix projection operator technique to analyze stopped-flow kinetics. J Mol Biol 295:831–852
11. Rajendran S, Jezewska MJ, Bujalowski W (2000) Multiple-step kinetic mechanism of DNA-independent ATP binding and hydrolysis by *Escherichia coli* replicative helicase DnaB protein: quantitative analysis using the rapid quench-flow method. J Mol Biol 303: 773–795
12. Galletto R, Jezewska MJ, Bujalowski W (2004) Multi-step sequential mechanism of *E. coli*

helicase PriA protein–ssDNA interactions. Kinetics and energetics of the active ssDNA-searching site of the enzyme. Biochemistry 43: 11002–11016

13. Bujalowski W, Jezewska MJ, Galletto R (2002) Dynamics of gapped DNA recognition by human polymerase β. J Biol Chem 277: 20316–20327
14. Bujalowski W, Jezewska MJ (2000) Kinetic mechanism of nucleotide cofactor binding to *Escherichia coli* replicative helicase DnaB protein. Stopped-flow kinetic studies using fluorescent, ribose-, and base-modified nucleotide analog. Biochemistry 39:2106–2122
15. Galletto R, Bujalowski W (2002) The *E. coli* replication factor DnaC protein exists in two conformations with different nucleotide binding capabilities. I. Determination of the binding mechanism using ATP and ADP fluorescent analogues. Biochemistry 41:8907–8920
16. Galletto R, Bujalowski W (2002) Kinetics of the *E. coli* replication factor DnaC protein–nucleotide interactions. II. fluorescence anisotropy and transient, dynamic quenching stopped-flow studies of the reaction intermediates. Biochemistry 41:8921–8934
17. Jezewska MJ, Galletto R, Bujalowski W (2003) Rat polymerase β–gapped DNA interactions: antagonistic effects of the 5′ terminal $PO^{4-}$ group and magnesium on the enzyme binding to the gapped DNAs with different ssDNA gaps. Cell Biochem Biophys 38:125–160
18. Galletto R, Jezewska MJ, Bujalowski W (2005) Kinetics of allosteric conformational transition of a macromolecule prior to ligand binding. Analysis of stopped-flow kinetic experiments. Cell Biochem Biophys 42:121–144
19. Quast U, Schimerlik MI, Raftery MA (1979) Ligand-induced changes in membrane-bound acetylcholine receptor observed by ethidium fluorescence. 2. Stopped-flow studies with agonists and antagonist. Biochemistry 18: 1891–1901
20. Galletto R, Jezewska MJ, Bujalowski W (2004) Multi-step sequential mechanism of *E. coli* helicase PriA protein–ssDNA interactions. Kinetics and energetics of the active ssDNA-searching site of the enzyme. Biochemistry 43: 11002–11016
21. Pilar FL (1968) Elementary quantum chemistry. McGraw-Hill, New York, pp 196–233
22. Fraser RA, Duncan WJ, Collar AR (1965) Elementary matrices and some applications to dynamics and differential equations. Cambridge University Press, Cambridge, pp 57–96
23. Lakowicz JR (1999) Principles of fluorescence spectroscopy, 2nd edn. Kluwer Academic, New York, pp 291–347
24. Otto R, Lillo MP, Beechem JM (1994) Resolution of multi-phasic reactions by the combination of fluorescence total-intensity and anisotropy stopped-flow kinetic experiments. Biophysical J 67:2511–2521
25. Moore KJ, Lohman TM (1994) Kinetic mechanism of adenine nucleotide binding to and hydrolysis by the *Escherichia coli* Rep monomer. 1. Use of fluorescent nucleotide analogues. Biochemistry 33:14550–14564
26. Moore KJ, Lohman TM (1994) Kinetic mechanism of adenine nucleotide binding to and hydrolysis by the *Escherichia coli* Rep monomer. 2. Application of a kinetic competition approach. Biochemistry 33:14565–14578
27. Valeur B (2002) Molecular fluorescence. Principles and applications. Wiley, Weinheim, NY, pp 200–247
28. Steiner RF (1991) Fluorescence anisotropy: theory and applications. In: Lakowicz J (ed) Topics in fluorescence spectroscopy, vol 2. Plenum Press, New York, pp 127–176
29. Eftink MR (1991) Fluorescence quenching: theory and applications. In: Lakowicz J (ed) Topics in fluorescence spectroscopy, vol 2. Plenum, New York, pp 53–128
30. Eftink MR, Ghiron CA (1981) Fluorescence quenching studies with proteins. Anal Biochem 114:199–227
31. Jezewska MJ, Bujalowski W (1997) Quantitative analysis of ligand–macromolecule interactions using differential quenching of the ligand fluorescence to monitor the binding. Biophys Chem 64:253–269

# Chapter 7

# Using Structure–Function Constraints in FRET Studies of Large Macromolecular Complexes

**Wlodek M. Bujalowski and Maria J. Jezewska**

## Abstract

The structural aspects of large macromolecular systems in solution can be conveniently addressed using the fluorescence resonance energy transfer (FRET) approach. FRET efficiency is the major parameter examined in such studies. However, its quantitative determination in associating macromolecular systems requires careful incorporation of thermodynamic quantities into specific expressions defining the FRET efficiencies. There are two widely used methods of obtaining FRET efficiencies, examination of both the donor quenching and of the sensitized emission of the FRET acceptor. Both approaches provide only apparent FRET efficiencies, not the true Förster FRET efficiency, which should be independent of the means to measure the efficiency.

The accuracy of the determined distances in macromolecular systems depends on the accuracy of the determination of the FRET efficiency and the estimate of the parameter, $\kappa^2$, which depends on the mutual orientation of the donor and the acceptor. Known procedures, based on limiting anisotropy measurements, to estimate $\kappa^2$ are of limited use to deducing the functional conclusions about the studied systems. On the other hand, using multiple donor–acceptor pairs and/or donors and acceptors placed in interchanged locations in the macromolecular system is an equally rigorous procedure to empirically evaluate the possible effect of $\kappa^2$ on the measured distance.

Protein–nucleic acid systems are particularly suited for FRET methodology. There is a plethora of commercial fluorescent markers, which can serve as donor–acceptor pairs. In the case of the nucleic acid, the markers can specifically be introduced in practically any location of the molecule. Application of the FRET measurements to examine structures of the large protein–nucleic acid complexes is particularly fruitful in cases where the presence of known structural constraints allows the experimenter to address the fundamental topology of the complexes. The discussed methodology can be applied to any associating macromolecular system.

**Key words:** FRET, Helicases, DNA replication, Protein–nucleic acid interactions, Motor proteins

## 1. Introduction

Fluorescence resonance energy transfer (FRET) measurements are one of the most common methods used to estimate distances in macromolecular systems in solution (1–16). The approach is applicable to both the structural studies of the systems in equilibrium, using

Wlodek M. Bujalowski (ed.), *Spectroscopic Methods of Analysis: Methods and Protocols*, Methods in Molecular Biology, vol. 875, DOI 10.1007/978-1-61779-806-1_7, © Springer Science+Business Media New York 2012

steady-state fluorescence intensity and to the dynamic studies of the conformational transitions and complex formations, using time-dependent emission and/or relaxation kinetic methods (18–23). The strength and popularity of the method are due, in part, to the fact that the necessary instrumentations are conventional and accessible in a single laboratory. Although usually the measurements are limited to one or several distances, the results are sufficient to deduce structural and mechanistic aspects of the examined systems at low concentrations in solution, i.e., in conditions, where the limiting thermodynamic and kinetic laws apply. There is a plethora of commercially available fluorescent markers, which can serve as suitable donor–acceptor pairs. In the case of the nucleic acid, the markers can be specifically introduced in practically any location of the molecule, allowing the experimenter to access multiple distances (1, 6–9, 11–13, 16, 17). Recombinant DNA techniques permit similar, although not as universal as in the case of the nucleic acids, applications of the specific labeling to the protein molecules, without affecting their structures and interactions (6, 8, 11, 12). Finally, the methodology of measuring the FRET efficiency, $E$, is in most cases relatively easy, though not without possible problems and artifacts (see below). The literature on FRET methodology is very vast. In our discussion, we are not able to do justice to all the excellent works in the field. Instead, we focus on certain aspects of estimating the FRET efficiency in associating macromolecular systems, often ignored in various studies and applications of the FRET measurements, to examine structures of the large protein–nucleic acid complexes, where the presence of known structural constraints allows the experimenter to address the fundamental topology of the complexes. As such, the discussion is limited to the systems where FRET occurs as a result of the formation of the complex between two macromolecules, one containing the fluorescent donor and the other the fluorescence acceptor (6, 8, 11–13, 16). Nevertheless, the discussed methodology can be applied to any associating macromolecular system.

## 2. Materials

All FRET experiments are performed in the same buffers as the thermodynamic binding studies on the corresponding interacting systems. The buffers and, in general, solution conditions are optimized for the system stability over the time period necessary to execute the experiments. The chemical used in the described experiments are AMP-PNP, β,γ-imidoadenosine-5′-triphosphate; Tris, tris(hydroxymethyl)aminomethane; CP, 7-diethylamino-3-(4′-maleimidylphenyl)-4-methylcoumarin; FLM, fluorescein-5-maleimide; Hepes, (*N*-[2-Hydroxyethyl]piperazine-*N*′-[2-ethanesulfonic acid]); DEAE cellulose, diethylaminoethyl cellulose; FITC, fluorescein 5′-isothiocyanate; TRITC, tetramethylrhodamine-6-isothiocyanate.

## 3. Methods

### 3.1. Average Fluorescence Energy Transfer Efficiency from the Donor to the Acceptor

The Förster fluorescence energy transfer efficiency, $E$, is the fundamental quantity, which is accessible from the experiment. The accuracy of the determined distance, $R$, heavily depends on the accuracy of the determination of $E$. The Förster FRET efficiency, $E$, from the donor to the acceptor dipole is related to the distance, $R$, separating the dipoles by the relationship (1–4)

$$E = \frac{R_o^{\,6}}{R_o^{\,6} + R^6} \tag{1}$$

Thus,

$$R = R_o \left[ \frac{(1 - E)}{E} \right]^{\frac{1}{6}} \tag{2}$$

where

$$R_o = 9,790(\kappa^2 n^{-4} \phi_d J)^{1/6} \tag{3}$$

is the so-called Förster critical distance (in angstroms), i.e., the distance, at which the transfer efficiency is 50%, $\kappa^2$, is the orientation factor characterizing the mutual orientation of the fluorescence donor absorption and the fluorescence acceptor emission dipole, respectively, $\phi_d$, is the donor quantum yield in the absence of the acceptor, and, $n$, is the refractive index of the medium ($n = 1.4$). The quantity, $J$, is the overlap integral that characterizes the resonance between the donor and acceptor dipoles and is evaluated by integration of the mutual area of overlap between the donor emission spectrum, $F(\lambda)$, and the acceptor absorption spectrum, $\varepsilon(\lambda)$, as

$$J = \frac{\int F(\lambda)\varepsilon_A(\lambda)\lambda^4 \partial\lambda}{\int F(\lambda)\partial\lambda} \tag{4}$$

Thus, once, $J$ and $\phi_d$, are determined and the possible value of the factor, $\kappa^2$, estimated (see below) the range of values of $R_o$ for a given donor–acceptor pair, in a particular system is obtained. Determination of $E$ provides the distance between the specific elements of the macromolecular system at study (1–4). Nevertheless, at room temperature, any macromolecular system is a dynamic entity with a multitude of different conformational microsubstrates, mutually interchanging at various rates (24). Moreover, the same systems may also exist in equilibrium of several different discrete structural states, with various distances between the examined donor–acceptor pair. The measured FRET efficiency, $E$, and resulting distance, $R$, are then arithmetic averages of all efficiencies and resulting distances, over all such structural states of the examined system.

#### 3.1.1. Apparent Average FRET Efficiencies from the Fluorescence Donor to the Fluorescence Acceptor

The true Förster FRET efficiency, $E$, is a fraction of the photons absorbed by the donor and transferred to the acceptor, in the absence of any additional nondipolar quenching, resulting from the presence of the acceptor (1–4). Two methods of obtaining $E$ are usually applied experimental studies. However, as we discuss it below, both experimental approaches provide only apparent FRET efficiency, not the true Förster FRET efficiency (11–17). The first method relies on the examination of the quenching of the donor fluorescence in the presence of the acceptor. There are important methodological aspects of such measurements, which should be considered (11–17). As an example, we address the FRET efficiency in a complex of two macromolecules with the stoichiometry 1:1, where one macromolecular component of the complex carries the fluorescence donor and the other one the fluorescence acceptor (11–17). A typical case in point is the association of a fluorescent-labeled protein with a fluorescent-labeled DNA oligomer.

The fluorescence of the donor in the presence of the acceptor, $F_{DA}$, is related to the fluorescence of the same donor, in the absence of the acceptor, $F_D$, by (11–17)

$$F_{DA} = (1 - \Theta_D)F_D + F_D\Theta_D(1 - E_D) \tag{5a}$$

or

$$F_{DA} = F_D(1 - \Theta_D E_D), \tag{5b}$$

where $E_D$ is the average FRET efficiency from the donor to the acceptor, determined from the quenching of the donor fluorescence and, $\Theta_D$, is the degree of saturation of the macromolecule, containing the donor, with the macromolecule containing the acceptor. The first term in (5a) is the fluorescence of the free macromolecule containing the donor and the second term is the fluorescence of the macromolecule containing the donor in the complex with the macromolecule containing the acceptor. Notice, in deriving (5a) and (5b), one assumes that the observed quenching of the donor emission is entirely due to the FRET process, but this may not be the case (see below). The average FRET efficiency, $E_D$, determined from the quenching of the donor fluorescence is obtained by rearranging the above expression, providing that

$$E_D = \left(\frac{1}{\Theta_D}\right)\left(\frac{F_D - F_{DA}}{F_D}\right). \tag{6}$$

Thus, besides accurate evaluations of $F_{DA}$ and $F_D$, one must know the value of $\Theta_D$, which can be obtained using binding parameters, determined in the equilibrium thermodynamic studies of the same system (11–17). Notice, the expressions above assume that the degree of the labeling with the fluorescence acceptor is $\gamma_A = 1$. However, if, $\gamma_A < 1$, then the fluorescence of the donor in the presence of the acceptor, $F_{DA}$, is related to the fluorescence of the same donor, in the absence of the acceptor, $F_D$, by

$$F_{DA} = (1 - \Theta_A)F_D + F_D\Theta_D(1 - \gamma_A) + F_D\Theta_D\gamma_A(1 - E_D) \quad (7a)$$

or

$$F_{DA} = F_D(1 - \Theta_D\gamma_A E_D). \quad (7b)$$

The quantity, $F_D\Theta_D(1 - \gamma_A)$, in (7a) is the fluorescence of the macromolecule containing the donor, associated with the fraction of the macromolecule, which should contain the acceptor, but it does not contain the acceptor. The average FRET efficiency, $E_D$, determined from the quenching of the donor fluorescence is then

$$E_D = \left(\frac{1}{\gamma_A\Theta_D}\right)\left(\frac{F_D - F_{DA}}{F_D}\right). \quad (8)$$

Unless it is experimentally determined that $\gamma_A \approx \Theta_D \approx 1$, neglecting the first term in (6) or (8), i.e., assuming a priori ~100% labeling efficiency and complete saturation of the macromolecule containing the donor, with the macromolecule containing the acceptor, sometimes encountered in the literature, may lead to an inaccurate value of $E_D$.

In the second method, the average FRET efficiency, $E_A$, is estimated through the sensitized acceptor fluorescence, i.e., by exclusively measuring the fluorescence intensity of the acceptor, excited at a wavelength where the donor predominantly, but not exclusively, absorbs in the absence and presence of the donor (11–17). This approach is more involved. The fluorescence intensities of the acceptor in the absence, $F_A$, and presence of the donor, $F_{AD}$, are defined as

$$F_A = I_o\varepsilon_A C_{AT}\phi_F^A \quad (9)$$

and

$$F_{AD} = (1 - \Theta_A)F_A + I_o\varepsilon_A C_{AT}\Theta_A\phi_B^A + I_o\varepsilon_D C_{DT}\Theta_D\phi_B^A E_A \quad (10)$$

where, $I_o$ is the intensity of incident light, $C_{AT}$ and $C_{DT}$ are the total concentrations of the acceptor and donor, respectively, $\Theta_A$ is the degree of saturation of the macromolecule, containing the acceptor, with the macromolecule containing the donor, $\varepsilon_A$ and $\varepsilon_D$ are the molar absorption coefficients of the acceptor and the donor at the applied excitation wavelength, respectively, $\phi_F^A$ and $\phi_B^A$ are the quantum yields of the free and bound acceptor, and $E_A$ is the average FRET efficiency determined using the acceptor sensitized emission. Expression (10) assumes that the degree of labeling with the donor and acceptor are $\gamma_D = \gamma_A = 1$.

The first term in (10) is the fluorescence emission from the free macromolecule containing the acceptor. The second term is the fluorescence emission from the macromolecule containing the acceptor in the complex with the macromolecule containing the donor and resulting from direct excitation of the acceptor. These two terms depend on the degree of saturation of the macromolecule containing the acceptor, $\Theta_A$, with the macromolecule containing the donor. The third term is the fluorescence emission from the macromolecule containing the acceptor in the complex with the macromolecule containing the donor, as a result of the fluorescence energy transfer from the donor. Unlike the first two terms, the third term depends on the degree of saturation of the macromolecule containing the donor, $\Theta_D$, with the macromolecule containing the acceptor. Both $\Theta_A$ and $\Theta_D$ can be estimated using the binding parameters determined in the thermodynamic equilibrium studies of the same complex. Moreover, all spectroscopic quantities in (9) and (10) can be experimentally determined. Dividing (9) by (10) and rearranging provides the average FRET efficiency, $E_A$, as

$$E_A = \left(\frac{1}{\Theta_D}\right)\left(\frac{\varepsilon_A C_{AT}}{\varepsilon_D C_{DT}}\right) \times \left\{\left(\frac{\phi_F^A}{\phi_B^A}\right)\left[\left(\frac{F_{AD}}{F_A}\right) + \Theta_A - \Theta_A\left(\frac{\phi_B^A}{\phi_F^A}\right) - 1\right]\right\}. \quad (11)$$

The degree of labeling with the acceptor, $\gamma_A$, is not essential for the determination of $E_A$, as long as the acceptor emission can be observed. However, the degree of labeling with the donor, $\gamma_D$, is essential. Introducing $\gamma_D$ into (10) gives

$$F_{AD} = (1 - \Theta_A)F_A + I_o\varepsilon_A C_{AT}\Theta_A\phi_B^A + I_o\varepsilon_D C_{DT}\gamma_D\Theta_D\phi_B^A E_A. \quad (12)$$

The second term in the right side of (12) represents all macromolecules containing the acceptor associated with the macromolecule, which does or does not contain the donor, directly excited through the acceptor. The third term in the right side of (12) represent all complexes, which contain both the acceptor and the donor excited through FRET from the donor. Correspondingly, (11) becomes

$$E_A = \left(\frac{1}{\gamma_D\Theta_D}\right)\left(\frac{\varepsilon_A C_{AT}}{\varepsilon_D C_{DT}}\right) \times \left\{\left(\frac{\phi_F^A}{\phi_B^A}\right)\left[\left(\frac{F_{AD}}{F_A}\right) + \Theta_A - \Theta_A\left(\frac{\phi_B^A}{\phi_F^A}\right) - 1\right]\right\}. \quad (13)$$

It should be obvious from (8) and (13), that accurate determination of the FRET efficiency in associating macromolecular systems is conditioned by the accurate determination of the degree of labeling of both macromolecules and, to the same extent, by quantitative knowledge of the thermodynamics of the formation of the examined complex (11–17).

#### 3.1.2. Förster Average FRET Efficiencies from the Fluorescence Donor to the Fluorescence Acceptor

As already pointed out above, the FRET efficiencies, $E_D$ and $E_A$, are apparent quantities. The expressions derived for $E_D$ and $E_A$ assume that these quantities represent the true Förster FRET efficiencies. The experimentally determined $E_D$ is a fraction of the photons absent in the donor emission, as a result of the presence of an acceptor, including transfer to the acceptor and the possible nondipolar quenching processes, induced by the presence of the acceptor. An experimental value of $E_A$ is a fraction of *all* photons absorbed by the donor, which are transferred to the acceptor, i.e., it also depends on the possible nondipolar quenching processes affecting the donor emission. Recall, the true Förster FRET efficiency, $E$, is a fraction of the photons absorbed by the donor and transferred to the acceptor, in the absence of any additional nondipolar quenching resulting from the presence of the acceptor (1–4). The value of $E$ is related to the apparent quantities of $E_D$ and $E_A$ by Jezewska and others (11–17):

$$E = \frac{E_A}{(1 - E_D + E_A)}. \tag{14}$$

The expression above was first obtained by Berman et al. (10). It can be derived using the kinetic interpretation of the emission and fluorescence energy transfer processes (13). The true Förster fluorescence energy transfer efficiency is defined as

$$E = \frac{k_T}{k_T + k_F + \sum k_{IQ}} \tag{15}$$

where, $k_T$ is the rate constant describing the transfer of quanta from the donor to the acceptor, $k_F$ is the emission rate constant of the donor, $\sum k_{IQ}$ is the sum of the rate constants characterizing all internal, intrinsic quenching processes in the donor molecule in the absence of the acceptor. This *physical definition of E is, as it should be, independent of the way the FRET efficiency is measured.* It states that the FRET efficiency is a fraction of quanta absorbed by the donor, which is transferred from the donor to the acceptor, while no other quenching processes are affected or induced, as a result of the presence of the acceptor. The fluorescence energy transfer measurements aim at determining this quantity.

If there are no additional quenching processes induced by the presence of the acceptor, then the true FRET efficiency in (15) can be obtained by examining the donor quantum yield in the absence and presence of the acceptor, or, experimentally, the steady-state fluorescence emission of the donor that is proportional to its quantum yield. In the absence of the acceptor, the quantum yield of the donor, $\phi_D$, is

$$\phi_D = \frac{k_F}{k_F + \sum k_{IQ}}. \tag{16}$$

In the presence of the acceptor, the quantum yield of the donor, $\phi_{DA}$, is then

$$\phi_{DA} = \frac{k_F}{k_F + k_T + \sum k_{IQ}}. \tag{17}$$

The apparent fluorescence energy transfer efficiency, $E_D$, determined using the donor quenching is then

$$E_D = \frac{\phi_D - \phi_{DA}}{\phi_D} = \frac{k_T}{k_F + k_T + \sum k_{IQ}}. \tag{18}$$

On the other hand, if besides the intrinsic quenching processes of the donor molecule, the presence of acceptor induces additional quenching of the donor emission, not included in $\sum k_{IQ}$, the quantum yield of the donor in the presence of the acceptor is then

$$\phi_{DA} = \frac{k_T + \sum k_{QAdd}}{k_F + k_T + \sum k_{IQ} + \sum k_{QAdd}} \tag{19}$$

where, $\sum k_{QAdd}$ stands for all additional quenching processes of the donor emission. For the FRET efficiency, $E_D$, one obtains

$$E_D = \frac{\phi_D - \phi_{DA}}{\phi_D} = \frac{k_T + \sum k_{QAdd}}{k_F + k_T + \sum k_{IQ} + \sum k_{QAdd}}. \tag{20}$$

Comparison with (15) indicates that the determined FRET efficiency, $E_D$, is not the true Förster FRET efficiency, $E$.

By the same token, the FRET efficiency, $E_A$, estimated using the sensitized emission of the acceptor, is different from the true Förster FRET efficiency, $E$, if the acceptor introduces additional quenching processes of the donor emission. This is because the absorption process by the donor and the accompanying transfer process from the donor to the acceptor are always measured in the presence of the acceptor. Thus, by measuring $E_A$, one obtains the fraction of all absorbed quanta that is transferred to the acceptor in the presence of all additional quenching processes, including the quenching processes induced by the presence of the acceptor. The FRET efficiency, $E_A$, is then defined as

$$E_A = \frac{k_T}{k_F + k_T + \sum k_{IQ} + \sum k_{QAdd}}. \tag{21}$$

Thus, as discussed above, in the case of $E_D$, the quantity, $E_A$, is not the true Förster FRET efficiency, as defined by (15). However, both,

$E_D$ and $E_A$, can be expressed using the true Förster fluorescence energy transfer efficiency, $E$, as

$$E_D = \frac{1 + \frac{\sum k_{QAdd}}{k_T}}{\frac{1}{E} + \frac{\sum k_{QAdd}}{k_T}} \quad (22)$$

and

$$E_A = \frac{1}{\frac{1}{E} + \frac{\sum k_{QAdd}}{k_T}}. \quad (23)$$

It should be evident that the apparent FRET efficiencies, $E_D$ and $E_A$, do not have to be equal, as is often assumed. Contrary, in many complex systems, they will be different (see below). Only if $\sum k_{QAdd} = 0$ then, as shown by (22) and (23), the value of $E_D = E_A = E$. Combining (22) and (23) provides the true Förster FRET efficiency, $E$, expressed using two experimentally determined, apparent FRET efficiencies by

$$E = \frac{E_A}{1 - E_D + E_A}, \quad (24)$$

which is the same as (14). Therefore, measurements of both apparent fluorescence energy transfer efficiencies, $E_D$ and $E_A$, *are not alternatives* but a necessity to obtain the true Förster FRET efficiency, $E$.

#### 3.1.3. Multiple Fluorescence Donor–Acceptor Approach to Determine Macromolecular Distances in Solution

The FRET efficiency, $E$, determined for a single donor–acceptor pair, depends on the distance between the donor and the acceptor, $R$, and the factor, $\kappa^2$, describing the mutual orientation of the donor and acceptor dipoles (1–4). This results from the fact that $\kappa^2$ determines the value of $R_o$ (3). The general reference values of $R_o$, for different fluorescence donor–acceptor pairs, are estimated using (3), for the complete random orientation of the acceptor and donor dipoles where $\kappa^2 = 0.67$ (1–4). The unusual situation with the factor, $\kappa^2$, is that it cannot be experimentally determined. However, the upper, $\kappa^2_{max}$, and lower, $\kappa^2_{min}$, limits of $\kappa^2$ can be estimated from the measured limiting anisotropies of the donor and acceptor and the calculated axial depolarization factors, using the procedure described by Dale et al. (25). The axial depolarization factors for the donor, $\langle d_D^X \rangle$, and acceptor, $\langle d_A^X \rangle$, respectively, are calculated as square roots of the ratios of the limiting anisotropies and the corresponding fundamental anisotropies (25). When both axial depolarization factors are positive, $\kappa^2_{max}$ and $\kappa^2_{min}$ can be calculated from

$$\kappa^2_{max} = \left(\frac{2}{3}\right)\left(1 + \langle d_D^X \rangle + \langle d_A^X \rangle + 3\langle d_D^X \rangle \langle d_A^X \rangle\right) \quad (25a)$$

and

$$\kappa^2_{\mathrm{min}} = \left(\frac{2}{3}\right)\left[1 - \left(\frac{1}{2}\right)\left(\langle d^{\mathrm{X}}_{\mathrm{D}}\rangle + \langle d^{\mathrm{X}}_{\mathrm{A}}\rangle\right)\right] \quad (25b)$$

which then provide, $R_{\mathrm{omax}}$ and $R_{\mathrm{omin}}$, respectively, using (3). Having, $R_{\mathrm{omax}}$ and $R_{\mathrm{omin}}$, one then obtains the range of the possible distance, corresponding to the measured $E$, $R_{\mathrm{max}}$, and $R_{\mathrm{min}}$, respectively. Although this is a rigorous procedure, unfortunately, it is of little use to the experimentalist, if he/she wants to deduce some functional conclusions from rather involving FRET studies. The estimated ranges of $R_{\mathrm{max}}$ and $R_{\mathrm{min}}$ are often within $\pm 30\%$ of the determined distances using $R_{\mathrm{o}}$ and $R$, corresponding to $\kappa^2 = 0.67$ (see below). Significant conformational changes, even of large macromolecular systems, can be completely obscured by such uncertainty, making any structural assessments unconvincing.

A different and *empirical* procedure to estimate the effect of the orientation factor, $\kappa^2$, is to use the multiple donor–acceptor pairs in examining a given distance in the studied system (11–17). Notice, we want to estimate the effect of $\kappa^2$, not to estimate the value of the factor. Although the factor $\kappa^2$ cannot be experimentally determined, it can only assume a value from 0 to 4 (1–4). However, because the distance between a donor and an acceptor depends upon the 1/6th power of $\kappa^2$, only the two extreme values (0 or 4) significantly affect the determined distance. Notice, the analysis of the possible range of distances between the donor and the acceptor, using the depolarization factors (see above), describes the situation where only a single donor–acceptor pair is used (25). On the other hand, using multiple, different donor–acceptor pairs and/or donors and acceptors placed in interchanged locations in the macromolecular system is an equally rigorous procedure to empirically evaluate the possible effect of $\kappa^2$ on the measured distance (11–17). The different molecular structures of different donors and acceptors and/or interchanged locations introduce an intrinsic randomization of the orientation of the absorption and emission dipoles in the system. The measurement of a similar distance, using different donor–acceptor pairs and/or with interchanged locations, empirically indicates that the orientation of the donor and absorption dipoles is far from the extreme values of 0 or 4, and that the true distance between a donor and an acceptor is represented by the distance obtained using $\kappa^2 = 0.67$ (see below) (11–17, 26). In fact, only two different, donor–acceptor pairs or a single donor–acceptor pair with interchanging locations in the system, between the same sites, is enough to estimate the possible effect of $\kappa^2$. A larger number of donor–acceptor pairs and interchanged locations increase the accuracy of the measurement.

*3.1.4. Hexameric Helicase–ssDNA Complex: Using the Known Structural Constraints of the System to Determine the Location of the ssDNA in the Complex*

The RepA protein is one of the proteins coded by broad host nonconjugative plasmid RSF1010 (16, 27–29). The protein is a hexameric replicative helicase, which unwinds the dsDNA in a 5′–> 3′ direction, in the reaction fueled by the ATP hydrolysis (27). The molecular weight of the RepA monomer is 29,896 Da, i.e., it is one of the smallest known helicases. In the crystal, the RepA hexamer has a ring-like structure with a central cross-channel of ~17 Å in diameter and the structure is depicted in Fig. 1a, b. The hexamer has the largest diameter of ~115 Å and the largest side-length of ~48 Å (28). Similar dimensions of the hexamer can be deduced from the hydrodynamic studies of the protein, although the global shape of the molecule in solution is different from its crystal counterpart (16). The structure of each protomer has an elongated shaped and is built of two domains, small and large. The RepA

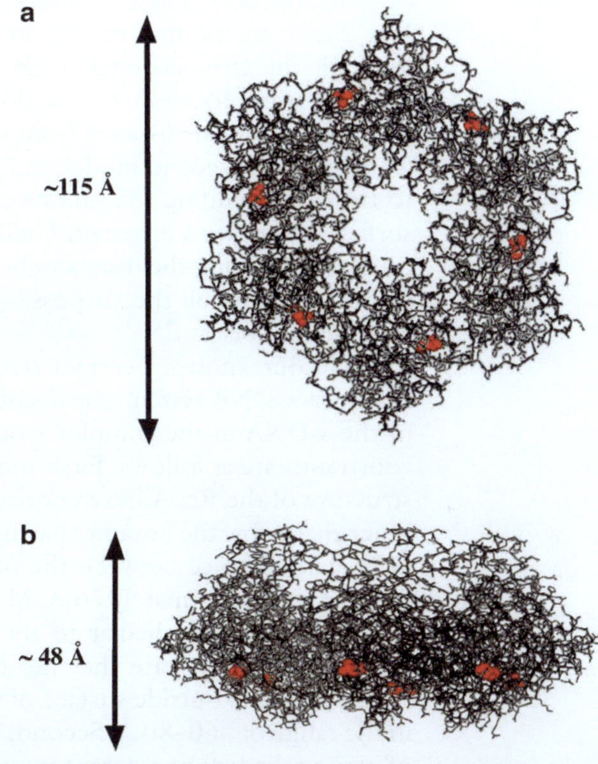


Fig. 1. The crystallographic structure of the RepA helicase hexamer (28). The structure has been generated using data from the RCSB Protein Data Bank, under the code 1G8Y, with ViewerPro (San Diego, CA). (**a**) The six subunits of the hexamer form a ring-like structure with diameter of ~116 Å and with the central cross-channel with a diameter of ~17 Å. (**b**) Side view of the RepA hexamer with the short perpendicular axis, of ~45 Å. Each protomer in the RepA hexamer is built of two domains, small and large. The single cysteine residue in each protomer is located in the large domain and is marked in *red*. All six cysteine residues lie in the plane perpendicular to the short axis of the hexameric structure of the enzyme (Reproduced from *Biochemistry* 2007 (ref. 16) with permission from American Chemical Society).

monomer contains a single cysteine residue (C172), whose location, close to one end of the hexamer within the large domain, is depicted in Fig. 1a, b. For the FRET studies, the cysteine residues were labeled with the fluorescence donor coumarin, CP (16). However, the obtained values of $\gamma_D$ were 0.31 ± 0.05 per hexamer, indicating that only a fraction of the cysteine residues in the RepA hexamer are readily available for modification. The labeled protein is referred to as, RepA-CP.

The site-size of the RepA hexamer–ssDNA has been determined in equilibrium thermodynamic studies and is 19 ± 1 nucleotides (29). Thus, the enzyme binds a single molecule of the ssDNA 20-mer, which occupies the total DNA-binding site of the enzyme. However, there are two fundamentally different structural models, which can generally describe the complex of the RepA hexamer with the bound ssDNA 20-mer, or with the nucleic acid (12, 16, 30, 31). These two models are schematically depicted in Fig. 2a, b. In the first model, in Fig. 2a, the nucleic acid passes through the cross-channel of the RepA hexamer. In this model, every possible location of the donor on the enzyme is approximately at a similar distance from every base of the bound ssDNA 20-mer. In the second model, in Fig. 2b, the DNA oligomer binds to the DNA-binding site on one of the protomers, on the outside surface of the RepA hexamer. Unlike the first model, there are large differences among the distances between a given base in the bound nucleic acid and all the six possible locations of the donor on the enzyme (see below).

Without knowing certain structural constraints of the RepA hexamer–ssDNA system, the decisive determination of the location of the ssDNA in the complex would be extremely difficult. These constraints are as follows. First, the ring-like structure of the crystal structure of the RepA hexamer has a diameter of ~115 Å, with the inner channel of the hexamer having a diameter of ~17 Å (Fig. 1a, b) (28). The distance between the opposing cysteine residues in the hexamer is approximately 70 Å. Moreover, the cysteine residues are in the plane perpendicular to the axis of the hexamer (Fig. 1b). These features indicate that the distance from the center of the hexamer to the outside surface of the protomers in the hexamer is in the range of ~60–80 Å. Second, the Förster critical distances, $R_o$, of the applied donor–acceptor pairs (see below) are known and centered around 50 Å (11, 12, 16). Third, independent FRET measurements of the structure of the ssDNA 20-mer, labeled at its 5′ and 3′ ends with a fluorescence donor and acceptor, in the complex with the RepA hexamer, indicate that the length of the bound 20-mer is ~75 Å. Because the fluorescence markers contribute to the length of the oligomer as two additional bases, this distance is very close to the length of one strand of dsDNA, 22-bp long, in the B form (~78 Å) (32, 33).

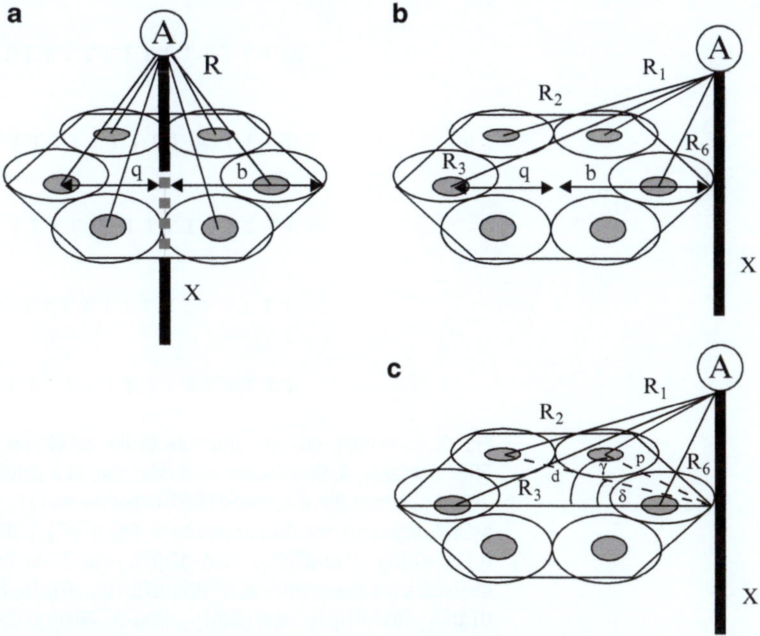


Fig. 2. (**a**) Schematic diagram showing the arrangement of a plane with a ring of six cysteine residues, each being a possible location site of the single FRET donor for the single acceptor located along the ssDNA, bound in the inner channel of the hexamer. The acceptor on the nucleic acid can be located above and below the plane with the donor. The distance from the acceptor to the plane of the ring of donors is designated as, *x*. The distance, *R*, from the acceptor to each donor is the same for all six donors on the hexamer (12, 16). (**b**) Schematic diagram showing the arrangement of a plane with a ring of six cysteine residues, each being a possible location site of a single FRET donor for the single acceptor located along the ssDNA, bound to the outside of one of the protomers of the hexamer. The acceptor on the nucleic acid can be placed above or below the plane with the donor. The distance from the acceptor on the bound DNA to the plane of the ring of donors is designated as, *x*. The possible different distances from the particular location of the donor on the hexamer to the acceptor are $R_1$, $R_2$, $R_3$, and $R_6$ (12, 16). (**c**) Geometrical relationships between the distance from the particular donor location to the center of the hexamer, *q*, radius of the hexamer, *b*, and the particular distances between the donor and the acceptor located along the ssDNA, bound on the outside to one of the protomers of the hexamer *p*, *d*, $R_1$, $R_2$, $R_3$, and $R_6$ (12, 16) (Reproduced from *Biochemistry* 2007 (ref. 16) with permission from American Chemical Society).

Thus, in our FRET analysis, the size of the RepA hexamer, $R_o$ of the applied donor–acceptor pairs, and the length of the nucleic acid impose a set of constraints on the possible distances between the donors and acceptors and the values of the FRET efficiencies (see below). The ssDNA oligomers, which were used in these studies, are depicted in Fig. 3. The FRET acceptor, A, fluorescein, or rhodamine, are placed at the 5′ or 3′ ends of the 20-mer or at different locations along the DNA, separated by five nucleotides. Binding of all ssDNA oligomers to the RepA hexamer is characterized by virtually the same binding constant as the binding of the

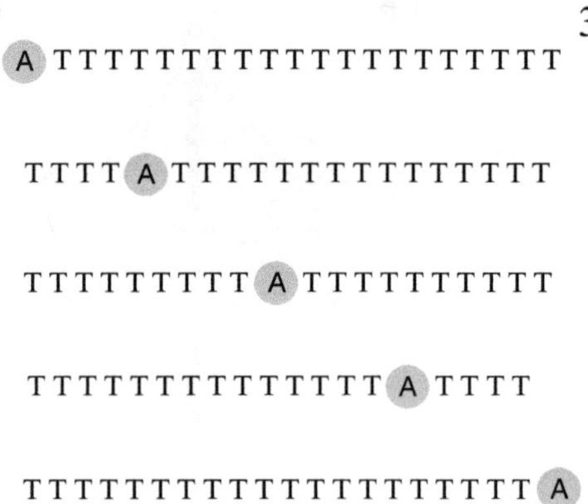


Fig. 3. Schematic primary structure of the ssDNA oligomers, $dT(pT)_{19}$, labeled with the FRET acceptor, A, fluorescein, or rhodamine, at a different specific location on the ssDNA 20-mer used in the discussed FRET experiments (12, 16). In the case of fluorescein, the ssDNA oligomers are designated as 5′-Flu-$dT(pT)_{19}$, $dT(pT)_3$-Flu-$dT(pT)_{14}$, $dT(pT)_8$-Flu-dT$(pT)_9$, $dT(pT)_{13}$-Flu-$dT(pT)_4$, and $dT(pT)_{19}$-Flu-3′. In the case of rhodamine, the ssDNA oligomers are designated as 5′-Rho-$dT(pT)_{19}$, $dT(pT)_3$-Rho-$dT(pT)_{14}$, $dT(pT)_8$-Rho-$dT(pT)_9$, $dT(pT)_{13}$-Rho-$dT(pT)_4$, and $dT(pT)_{19}$-Rho-3′. All oligomers are referred to in the text as 20-mers (Reproduced from *Biochemistry* 2007 (ref. 16) with permission from American Chemical Society).

unmodified oligomers, $dT(pT)_{19}$ or $dT(pT)_{20}$ (16, 29). Knowing the thermodynamic characteristics of the system, we can select the optimal concentrations of the RepA hexamer and the ssDNA oligomer, to perform quantitative FRET studies of the structure of the complex (16, 29).

#### *3.1.5. Theoretical Analysis of the Modes of Binding Using the Structural Constraints of the System*

First, we address the situation where a single acceptor is located in any location along a nucleic acid oligomer, which is bound in the cross-channel of the RepA hexamer (Fig. 2a). In other words, the acceptor can be placed above and below the plane, in which the donor is located on the enzyme. Such a situation corresponds to the fluorescence energy transfer process in the complexes of RepA-CP, with $dT(pT)_{19}$ labeled with fluorescein or rhodamine at different locations along the nucleic acid. The distance between the acceptor placed in an arbitrary location on the ssDNA 20-mer, with respect to the plane of the donor, is designated as $x$. For the DNA bound in the cross-channel of the hexamer (Fig. 2a), the single donor can be located at any of the six specific sites on each protomer. In each of these possible six sites, the donor is at the same distance from the center of the RepA hexamer, $q$, and at the same distance, $R$, from the acceptor placed on the bound nucleic acid. The acceptor can be located at any distance, $x$, from the plane of donor (Fig. 2a). The values of $R$ and $x$ are then described by the Pythagoras formula

$$R = (x^2 + q^2)^{0.5}. \tag{26}$$

The average value of $E$, as a function of $R$, is then described by (2).

The theoretical dependence of the average FRET efficiency, $E$, as a function of the average distance from the acceptor to the plane with the donor, $x$, and for different distances of the donor from the center of the hexamer, $q$, is shown in Fig. 4a. The selected value of $R_o$ is 52 Å, which corresponds to $R_o$ of the CP (donor)–fluorescein (acceptor) system (11, 12). In these calculations, we allowed the ssDNA 20-mer to protrude from both sides of the plane with the donor within the full length of the ssDNA 20-mer, i.e., ~70 Å. This is a practical, although not absolutely necessary generalization, as the binding studies showed that the ssDNA 20-mer assumes a well-defined location within the total DNA-binding site of the enzyme (29). For different values of $q$, as the acceptor is moved along the bound nucleic acid, the plots span different regions of the average distance, $R$ (Fig. 2a) and, correspondingly, different values of the average FRET efficiency, $E$. For $q = 60$ Å, the donor would be as far as the indicated radius of the crystal structure of the RepA hexamer (Fig. 1a). In such a case, the maximum value of the FRET efficiency can only be ~0.29 and corresponds to the acceptor located in the same plane as the donor. For different values of $q$, as the donor is closer to the center of the hexamer, the possible maximum values of $E$ increase. For $q$ in the range of 35–50 Å, which corresponds to the distance from the cysteine residues to the center of the RepA hexamer in the crystal structure of the enzyme (Fig. 1a, b), the maximum values of the average FRET efficiency is in the range of ~0.5–0.9. Another characteristic feature of the plots in Fig. 4a, obtained for $35 < q < 50$ Å, is the sharp decrease of the fluorescence energy transfer efficiency once the distance of the acceptor from the plane with the donor exceeds ~20 Å.

The behavior of the model where the ssDNA 20-mer binds on the outside of the hexamer to one of the protomers, as depicted in Fig. 2b, is very different. The donor can be located at any specific location on each protomer of the RepA hexamer and at the same distance, $q$, from the center of the enzyme molecule. The ssDNA is now tangent to the outside surface of one of the protomers of the hexamer. The geometrical aspects of this model are more complex than the ones previously considered and are shown in Fig. 2c. Inspection of Fig. 2b, c indicates that the distance between the acceptor on the bound nucleic acid and the donor on the protein varies significantly for each particular location of the donor. In general, for the set of $m$ possible locations of the single FRET donor, which is transferring energy to the single acceptor, the average value of $E$ is weighted by the

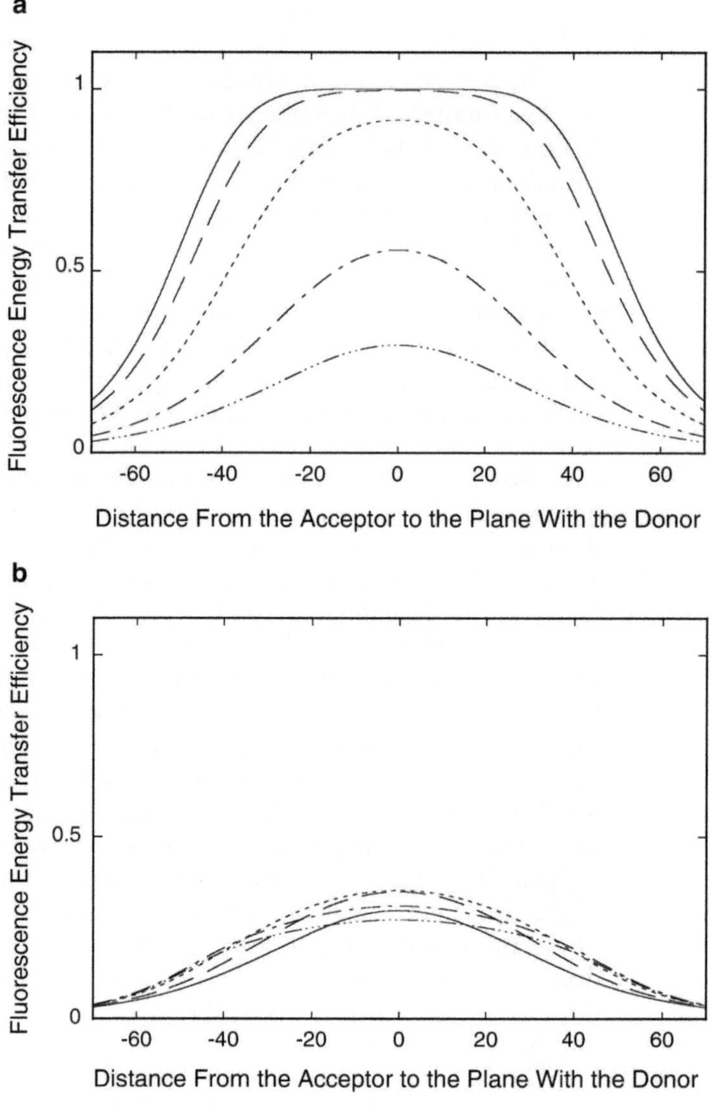


Fig. 4. (**a**) Theoretical dependence of the average Förster FRET efficiency, $E$, from a single donor, placed at any of the six possible cysteine residues of the RepA hexamer, to a single acceptor located along the ssDNA 20-mer bound in the cross-channel of the hexamer, upon the average distance of the acceptor from the plane with the donor, $x$ (Fig. 2a). The plots were generated using (1), (2), and (26), for different distances of the donor from the center of the hexamer, $q$. The maximum distance from the outside surface of the RepA hexamer to its center, i.e., the selected radius of the hexamer, $b = 60$ Å and the maximum selected length of the ssDNA 20-mer, $x = 70$ Å. The critical Förster distance for the selected donor–acceptor pair is 52 Å: $q = 60$ Å (— ··· — ··· —), $q = 50$ Å (— ˉ —), $q = 35$ Å (- - - -), $q = 20$ Å (— — —), $q = 1$ Å (—). (**b**) Theoretical dependence of the average Förster FRET efficiency, $E$, from a single donor, placed at any of the six possible cysteine residues of the RepA hexamer, to a single acceptor located along the ssDNA 20-mer, bound on the outside of the hexamer, upon the average distance of the acceptor from the plane with the donor, $x$ (Fig. 2a). The plots were generated using (1), (2), and (29), for different distances of the donor from the center of the hexamer, $q$. The maximum distance from the outside surface of the RepA hexamer to its center, i.e., the selected radius of the hexamer, $b = 60$ Å and the maximum selected length of the nucleic acid, $x = 70$ Å. The critical Förster distance for the selected donor–acceptor pair is 52 Å: $q = 60$ Å (— ··· — ··· —), $q = 50$ Å (— ˉ —), $q = 35$ Å (- - - -), $q = 20$ Å (— — —), $q = 1$ Å (—) (Reproduced from *Biochemistry* 2007 (ref. 16) with permission from American Chemical Society).

contributions from all possible locations of the donor and is generally defined as (12, 16, 26):

$$E = \left(\frac{1}{m}\right)\sum E_i, \tag{27}$$

where $E_i$ is the transfer efficiency from the particular location, "$i$," of the donor to the acceptor and is defined by (2), for the particular distance, $R_i$. The average distance between the acceptor and the donor in the system is analogously defined as

$$R_{\text{av}} = \left(\frac{1}{m}\right)\sum R_i. \tag{28}$$

In our case, $m = 6$. However, $R_1$, $R_2$, $R_3$, $R_4$, $R_5$, and $R_6$, the particular distances between a given location of the donor on the hexamer and the acceptor placed on the bound ssDNA, are now defined as (Fig. 2c)

$$R_1 = (x^2 + p^2)^{0.5} \tag{29a}$$

$$R_2 = (x^2 + d^2)^{0.5} \tag{29b}$$

$$R_3 = [x^2 + (q + b)^2]^{0.5} \tag{29c}$$

$$R_4 = R_2 \tag{29d}$$

$$R_5 = R_1 \tag{29e}$$

$$R_6 = [x^2 + (b + q)^2]^{0.5} \tag{29f}$$

where $b$ is the radius of the hexamer, $p = (q/2)(3^{0.5}/\sin\gamma)$, $\gamma = 60 + \text{arctg}\{[(q - b)/(q + b)]3^{0.5}\}$; $d = (q/2)(3^{0.5}/\sin\delta)$; $\delta = 30 + \text{arctg}\{[(q - b)/(q + b)]3^{-0.5}\}$ (12, 16).

The theoretical dependence of the average FRET efficiency, $E$, upon the average distance of the acceptor from the plane with the donor, $x$, for different distances of the donor from the center of the hexamer, $q$, is shown in Fig. 4b, with the selected, $b = 60$ Å and $R_o = 52$ Å. Because of the complex relationship between the $R_{\text{av}}$ and the particular donor–acceptor distance, $R_i$, the plots are either very close to or superimpose each other, which is contrary to the model where the ssDNA passes through the inner channel of the hexameric structure (Fig. 4a). In other words, in the considered model, different donor–acceptor configurations may result in the same average FRET efficiencies. However, the most important feature of the plots in Fig. 4b is that, independent of the distance between the donor and the center of the hexamer and independent of where the acceptor is located on the bound ssDNA oligomer, the average FRET efficiency never exceeds the value of ~0.34. This dramatically lower value of $E$, as compared to the FRET efficiency obtained for the model in which ssDNA passes through the inner channel of the hexamer, results from the fact that the average distance between the donor and the acceptor is always significantly larger than $R_o$ (52 Å) of the examined donor–acceptor pair. This is true for any value of $q$, or any location of the acceptor on the bound ssDNA, or any placement of the nucleic acid in the DNA-binding site of the RepA hexamer (Fig. 2b). Moreover, the plots in Fig. 4b, obtained for $35 < q < 50$ Å, show only a gradual decrease of $E$, once the distance of the acceptor from the plane with the donor exceeds ~20 Å, a very different behavior from the model, where the ssDNA passes through the cross-channel of the hexamer (Fig. 4a). Thus, it is evident that the two fundamentally different models of the RepA hexamer–ssDNA complex differ dramatically in the value of the average FRET efficiency for an arbitrary distance of the donor from the center of the protein hexamer, $q$. Moreover, the size of these differences well exceeds the errors due to approximations applied in the analysis (12, 16).

## 4. Experimental Approach

1. Labeling of the cysteine residues of the RepA hexamer, with CP was performed in the buffer T5, which is 50 mM Hepes/HCl, pH 8.1, 100 mM NaCl, 5 mM $MgCl_2$, 10% glycerol, at 4°C (12, 16). The fluorescent label, coumarin (7-diethylamino-3-(4′-maleimidylphenyl)-4-methylcoumarin) further referred as CP, was added from the stock solution to the molar ratio of the dye/RepA hexamer ≈25. The reaction mixture was incubated for ~4 h, gently mixed using a magnetic stirrer. After incubation, the protein was precipitated with ammonium sulfate and dialyzed overnight against buffer T2, 50 mM Tris adjusted to pH 7.6 with HCl at 10°C, 1 mM $MgCl_2$, 10 mM NaCl, and 10% glycerol. The remaining free dye was removed from the modified RepA-CP solution by chromatography on a DEAE cellulose column in buffer T2 containing 500 mM NaCl. The degree of labeling, $\gamma_D$, of the Rep hexamer was determined by the absorbance of a marker using the extinction coefficient, $\varepsilon_{394} = 27 \times 10^3\ M^{-1}\ cm^{-1}$ (16). The obtained values of $\gamma_D$ were $0.31 \pm 0.05$ per RepA hexamer. The concentration of the unlabeled protein was spectrophotometrically determined, using the extinction coefficient $\varepsilon_{280} = 1.656 \times 10^5\ cm^{-1}\ M^{-1}$ (16).
2. The ssDNA 20-mer, $dT(pT)_{19}$, labeled with fluorescein at the 5′ end, or at a different location of the nucleic acid, were synthesized using fluorescein phosphoramidate. Labeling of the ssDNA 20-mers at the 3′ end with fluorescein, or labeling with rhodamine (Rho), was performed by synthesizing dT $(pT)_{19}$ with a nucleotide residue in a given location of the nucleic acid with the amino group on a six-carbon linker and,

subsequently, modifying the amino group with FITC (fluorescein 5′-isothiocyanate) or TRITC (tetramethylrhodamine-6-isothiocyanate). The degree of labeling of the nucleic acids, $\gamma_A$, was determined by absorbance at 494 nm for fluorescein (pH 9) using the extinction coefficient, $\varepsilon_{494} = 7.6 \times 10^4$ $M^{-1}$ $cm^{-1}$, and at 555 nm for rhodamine using the extinction coefficient, $\varepsilon_{555} = 8.0 \times 10^4$ $M^{-1}$ $cm^{-1}$ (12, 16). Concentrations of all ssDNA oligomers have been spectrophotometrically determined (12, 16).

3. Because in the presence of ATP nonhydrolyzable analog, AMP-PNP, the RepA protein acquires a much higher affinity for the ssDNA, fluorescence emission spectra of the RepA-CP, free or in the complex with the labeled nucleic acid, have been recorded in the binding buffer T2, containing 0.5 mM AMP-PNP (16). The measurements were performed in a fluorescence cuvette with an optical path of 0.5 cm. This allowed the use of small volumes (450 μl) of the protein and nucleic acid solutions and decreases the corrections for the inner filter effect. The spectrofluorometer was set in the ratio mode, i.e., the signal coming from the emission photomultiplier (PMT) is always divided by the signal from the reference voltage, which is proportional to the intensity of the light source. This instrumental mode allows the experimenter to eliminate any artifacts resulting from fluctuations, and/or a drift of the light intensity during the spectra recording. Moreover, in order to avoid possible artifacts, due to the fluorescence anisotropy of the sample, polarizers were placed in excitation and emission channels and set at 90° and 55° (magic angle), respectively (12, 16).
4. Concentrations of RepA-CP and the ssDNA oligomer in the sample are $3 \times 10^{-7}$ M (hexamer) and $1 \times 10^{-6}$ M (oligomer), respectively. These concentrations provide optimal concentration of the formed complex and are selected using the equilibrium binding constant determined in independent fluorescence titration experiments (16). Knowing the equilibrium constant, the degree of labeling of both protein and the nucleic acid, allows us to determine the values of $\Theta_A$, and $\Theta_D$, respectively. The excitation wavelength is set at the absorption spectrum of the CP ($\lambda_{ex} = 425$ nm), which is the FRET donor in the discussed measurements.
5. In order to determine the FRET efficiencies, $E_D$, $E_A$, and $E$, as described above, using both fluorescence quenching of the donor emission and the sensitized emission of the acceptor, the following emission spectra were recorded. The fluorescence emission spectrum of the RepA-CP in the presence of unlabeled $dT(pT)_{19}$, which provides the relative quantum yield of the donor, $F_D$, in the presence of the selected concentration of the nucleic acid, but in the absence of the FRET

acceptor. The fluorescence emission spectrum of the RepA-CP, in the presence of the ssDNA oligomer, labeled with the FRET acceptor (fluorescein or rhodamine), which provides the relative quantum yield of the donor, $F_{AD}$, in the presence of the FRET acceptor. The fluorescence emission spectrum of the labeled ssDNA oligomer in the presence of the unlabeled RepA protein, which provides the relative quantum yield of the acceptor, $F_A$, in the absence of the donor at the excitation in the donor absorption band. The fluorescence emission spectrum of the labeled ssDNA oligomer in the presence of the RepA-CP, which provides the relative quantum yield of the acceptor, $F_{AD}$, in the presence of the donor at the excitation in the donor absorption band. Finally, one records the fluorescence emission spectra of the labeled ssDNA oligomer in the absence and presence of the unlabeled RepA protein, with the excitation in the acceptor absorption band. These spectra provide the ratio, $\phi_F^A/\phi_B^A$, for the acceptor. Thus, all parameters in (8) and (13), necessary to obtain the values of the FRET efficiencies, $E_D$, $E_A$, and $E$, are experimentally determined. To increase the accuracy, the relative quantum yields are obtained through integration of the corresponding emission spectra.

## 5. Experimental Analysis of the Modes of Binding of the ssDNA in the Complex with the Hexameric Helicase

Fluorescence emission spectra of the complexes of RepA-CP with the ssDNA oligomer, dT(pT)$_3$-Flu-dT(pT)$_{14}$, the emission spectrum of the RepA-CP complex with unmodified dT(pT)$_{19}$, in the absence of the acceptor, and the emission spectrum of dT(pT)$_3$-Flu-dT(pT)$_{14}$, in the complex with the RepA hexamer in the absence of the donor, are shown in Fig. 5a. The spectra have been recorded with the excitation at $\lambda_{ex} = 425$ nm, i.e., in the major absorption band of the selected donor, CP (12, 16). The spectra in Fig. 5a show that the presence of the acceptor (fluorescein) induces a significant quenching of the donor (CP) fluorescence. Correspondingly, there is an increase of the fluorescence emission of the acceptor in the presence of the donor. Both features, the donor emission quenching and the sensitized emission of the acceptor, indicate that an efficient FRET process occurs in the examined RepA–ssDNA 20-mer system (1–4). Moreover, formation of the complex does not affect the shapes of the donor and the acceptor emission spectra (1–4).

The emission spectrum of the donor, in the complex with the acceptor, has been obtained by normalizing the peak of the spectrum of the donor, recorded in the absence of acceptor, to the peak of the intensity of the same donor recorded in the presence of

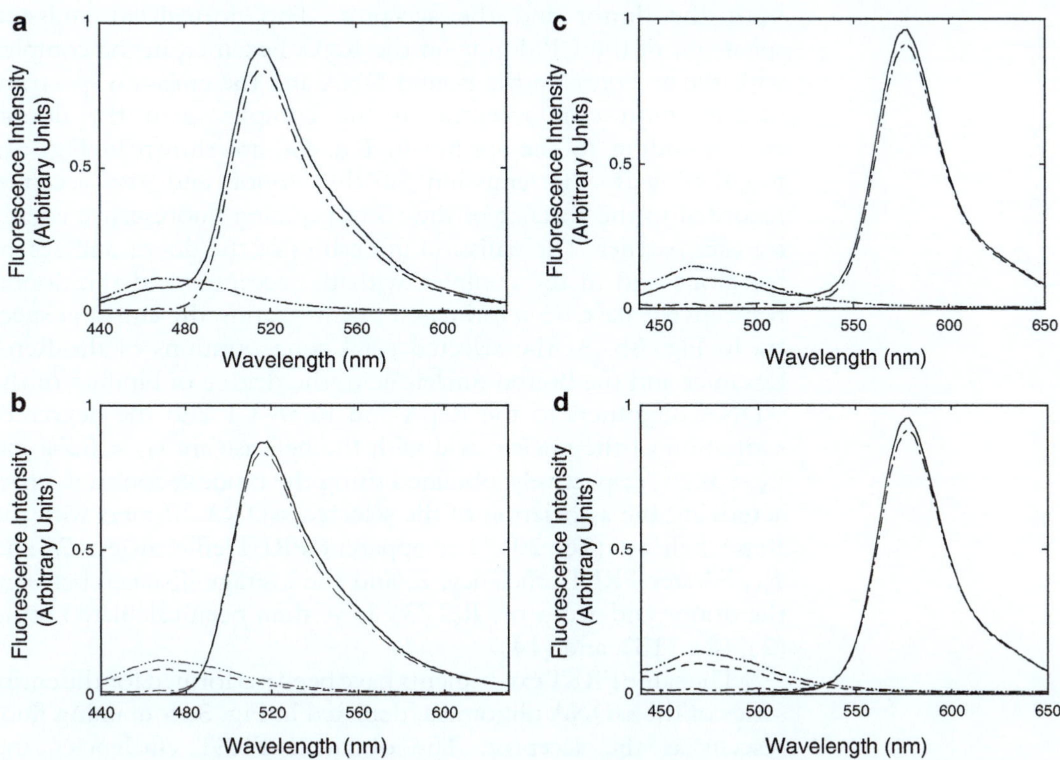


Fig. 5. (**a**) The fluorescence emission spectrum ($\lambda_{ex} = 425$ nm) of the RepA hexamer, RepA-CP, labeled with the coumarin derivative, CP, in the presence of unlabeled dT(pT)$_{19}$ (– · · · –), the fluorescence emission spectrum of the ssDNA 20-mer, dT (pT)$_3$-Flu-dT(pT)$_{14}$ in the presence of unlabeled RepA hexamer (– ˉ –), and the fluorescence emission spectrum of the labeled RepA hexamer, RepA-CP, in the presence of the labeled ssDNA oligomer, dT(pT)$_3$-Flu-dT(pT)$_{14}$ (—) (16). Concentrations of RepA-CP and the ssDNA oligomer are $3 \times 10^{-7}$ M (hexamer) and $1 \times 10^{-6}$ M (oligomer), respectively (16). (**b**) The fluorescence emission spectrum ($\lambda_{ex} = 425$ nm) of the labeled RepA hexamer, RepA-CP, in the presence of unlabeled dT(pT)$_{19}$ (– · · · –), the normalized fluorescence emission spectrum of RepA-CP to the emission of RepA-CP in the complex with dT(pT)$_3$-Flu-dT(pT)$_{14}$ (– – – –), the fluorescence emission of dT(pT)$_3$-Flu-dT(pT)$_{14}$ in the presence of unlabeled RepA hexamer (– ˉ –), and the fluorescence emission spectrum of dT(pT)$_3$-Flu-dT(pT)$_{14}$ in the complex with the RepA-CP hexamer, after subtraction of the normalized spectrum of RepA-CP in the same complex (—). (**c**) The fluorescence emission spectrum ($\lambda_{ex} = 425$ nm) of the RepA hexamer, RepA-CP, labeled with the coumarin, CP, in the presence of the unlabeled dT(pT)$_{19}$ (– · · · –), the fluorescence emission spectrum of dT(pT)$_3$-Rho-dT(pT)$_{14}$ in the presence of unlabeled RepA hexamer (— ˉ —), and the fluorescence emission spectrum of the RepA-CP in the presence of the labeled ssDNA oligomer, dT(pT)$_3$-Rho-dT(pT)$_{14}$ (—) (16). Concentrations of RepA-CP and the ssDNA oligomer are $3 \times 10^{-7}$ M (hexamer) and $1 \times 10^{-6}$ M (oligomer), respectively. (**d**) The fluorescence emission spectrum ($\lambda_{ex} = 425$ nm) of RepA-CP in the presence of unlabeled dT(pT)$_{19}$ (– · · · –), the normalized fluorescence emission spectrum of RepA-CP to the emission of RepA-CP in the complex with dT(pT)$_3$-Rho-dT(pT)$_{14}$ (– – – –), the fluorescence emission of dT(pT)$_3$-Rho-dT(pT)$_{14}$ in the presence of unlabeled RepA hexamer (— ˉ —), and the fluorescence emission spectrum of dT(pT)$_3$-Rho-dT(pT)$_{14}$ in the complex with RepA-CP, after subtraction of the normalized spectrum of RepA-CP in the same complex (—) (Reproduced from *Biochemistry* 2007 (ref. 16) with permission from American Chemical Society).

the acceptor (Fig. 5a). Subsequently, the emission spectrum of the acceptor, in the complex with the donor, has been obtained by subtracting the obtained, normalized emission spectrum of the donor from the emission spectrum of the complex containing both the donor and the acceptor. The normalized emission spectrum of the CP donor on the RepA hexamer, in the complex with the acceptor on the bound DNA and the emission spectrum of the fluorescein acceptor in the complex with the donor, corresponding to the spectra in Fig. 5a, are shown in Fig. 5b, together with the emission of the donor and the acceptor recorded in the absence of the corresponding fluorescence energy transfer partner. The emission intensities of the donor and acceptor alone and in the complex with the acceptor and the donor, respectively, have been estimated by integrating the emission spectra in Fig. 5b. At the selected total concentrations of the RepA hexamer and the bound nucleic acid, the degree of binding of the ssDNA oligomers to the RepA and RepA-CP and the degree of saturation of the nucleic acid with the helicase are $\nu_D \approx 0.74$ and $\nu_A \approx 0.24$, respectively, obtained using the binding constant, characterizing the association of the selected ssDNA 20-mers with the RepA helicase (16, 29). The apparent FRET efficiencies, $E_D$ and $E_A$, Förster FRET efficiency, $E$, and the average distance between the donor and acceptor, $R(2/3)$, have then been calculated using (2), (8), (13), and (14).

The same FRET experiments have been performed for the entire series of the ssDNA oligomers, depicted in Fig. 3, containing fluorescein as the acceptor. The obtained FRET efficiencies and corresponding average distances, $R(2/3)$, are provided in Table 1. The values of $E$ sharply increase from ~0.23 for 5′-Flu-dT$(pT)_{19}$, to ~0.42 and ~0.63 for the dT$(pT)_3$-Flu-dT$(pT)_{14}$, and dT$(pT)_8$-Flu-dT$(pT)_9$, respectively. The corresponding average distances from the center of mass of the donor to the acceptor are $R(2/3)$ ~63.4 Å for 5′-Flu-dT$(pT)_{19}$ and $R(2/3)$ ~47.6 Å for dT$(pT)_8$-Flu-dT$(pT)_9$ (Table 1). However, a shift of the acceptor by an additional five nucleotides along the bound ssDNA in dT$(pT)_{13}$-Flu-dT $(pT)_4$ causes, not an increase, but a drop in the value of $E$ from ~0.63 to ~0.54 and an increase of the average distance to $R(2/3) \approx 51$ Å. A further shift of the acceptor to the 3′ end of the bound DNA in dT$(pT)_{19}$-Flu-3′ results in a dramatic decrease of the FRET efficiency, $E$, to ~0.32 and a strong increase of the average distance to $R(2/3) \approx 59$ Å (12, 16).

Recall, the measurement of the similar distance between the donor and the acceptor for different donor–acceptor pairs provides empirical evidence that the value of the orientation factor, $\kappa^2$, is close to the value of 2/3, which corresponds to the complete random orientation of the donor absorption and acceptor emission dipoles, respectively (12, 14–16). In other words, in such a case, $\kappa^2$ does not affect the obtained specific values of the spatial separation of the donor and the acceptor (see above). To address this issue, analogous studies, as described above for fluorescein as the FRET acceptor, have been performed using rhodamine as the

**Table 1**
**FRET parameters of the complexes of the hexameric helicase, Rep A protein of plasmid RSF1010, labeled with the fluorescence energy transfer donor, coumarin, with different ssDNA oligomers, containing the FRET acceptor, fluorescein (16)**

| ssDNA oligomer | $E_D$ | $E_A$ | $E$ | $R_o$ (Å) | $R(2/3)$ (Å) | $R(2/3)_{av}$ (Å)[a] | $x_{av}$[a] |
|---|---|---|---|---|---|---|---|
| 5′-Flu-dT $(pT)_{19}$ | 0.61 ± 0.04 | 0.12 ± 0.02 | 0.23 ± 0.03 | 52 | 63.4 ± 3.1 | 65.4 ± 3.1 | 44.3 ± 3.1 |
| $dT(pT)_3$-Flu-$dT(pT)_{14}$ | 0.49 ± 0.04 | 0.40 ± 0.03 | 0.42 ± 0.03 | 52 | 54.8 ± 2.5 | 57.7 ± 2.5 | 32.9 ± 2.9 |
| $dT(pT)_8$-Flu-$dT(pT)_9$ | 0.40 ± 0.03 | 1.00 ± 0.03 | 0.63 ± 0.03 | 52 | 47.6 ± 2.5 | 48.1 ± 2.3 | 0 ± 2.3 |
| $dT(pT)_{13}$-Flu-$dT(pT)_4$ | 0.33 ± 0.03 | 0.80 ± 0.03 | 0.54 ± 0.03 | 52 | 51.0 ± 2.5 | 50.5 ± 2.3 | 15.4 ± 2.5 |
| $dT(pT)_{19}$-Flu-3′ | 0.37 ± 0.03 | 0.30 ± 0.03 | 0.32 ± 0.03 | 52 | 59.0 ± 2.8 | 62.2 ± 3.0 | 31.9 ± 2.9 |
| 5′-Flu-$dT(pT)_9$ | 0.24 ± 0.03 | 0.16 ± 0.02 | 0.17 ± 0.03 | 52 | 67.5 ± 3.1 | 67.6 ± 3.1 | 47.8 ± 2.9 |
| $dT(pT)_9$-Flu-3′ | 0.61 ± 0.03 | 0.45 ± 0.03 | 0.53 ± 0.03 | 52 | 50.7 ± 2.8 | 47.6 ± 2.9 | 0 ± 2.3 |

[a] Averaged over two measurements using fluorescein and rhodamine as the FRET acceptor (16) (Reproduced from *Biochemistry* 2007 (ref. 16) with permission from American Chemical Society)

acceptor. Fluorescence emission spectra of the complexes of RepA-CP with ssDNA oligomers, $dT(pT)_3$-Rho-$dT(pT)_{14}$, the emission spectrum ($\lambda_{ex} = 425$ nm) of the RepA-CP complex with unmodified $dT(pT)_{19}$, in the absence of the acceptor, and the emission spectrum of $dT(pT)_3$-Rho-$dT(pT)_{14}$, in the complex with the RepA helicase in the absence of the donor, are shown in Fig. 5c. The selected total concentrations of the RepA helicase, RepA-CP, the labeled and unlabeled ssDNA oligomers are the same as in Fig. 5a. As observed for the ssDNA oligomer labeled with fluorescein, the presence of rhodamine induces a significant quenching of the CP emission and there is a corresponding increase of the fluorescence emission of rhodamine in the presence of the CP.

The analysis of the emission spectra was analogous to the analysis described in the case of fluorescein as FRET acceptor (see above). The obtained FRET efficiencies and corresponding average distances, $R(2/3)$, for the entire series of the ssDNA oligomers, schematically depicted in Fig. 3, containing rhodamine as the acceptor, are provided in Table 2. The behavior of $E$ is very similar to the behavior observed for the analogous ssDNA oligomers containing fluorescein (Table 1). Nevertheless, the values of $E$ are lower than observed for the fluorescein–CP pair, due to the fact that the Förster critical distance, $R_o = 47$ Å, of the CP–rhodamine pair,

**Table 2**
**FRET parameters for the complexes of the hexameric helicase, Rep A protein of plasmid RSF1010, labeled with the fluorescence energy transfer donor, coumarin, with different ssDNA oligomers, containing the FRET acceptor, rhodamine (16)**

| ssDNA oligomer | $E_D$ | $E_A$ | $E$ | $R_o$ (Å) | $R(2/3)$ (Å) | $R(2/3)_{av}$ (Å)[a] | $x_{av}$[a] |
|---|---|---|---|---|---|---|---|
| 5′-Rho-dT $(pT)_{19}$ | 0.32 ± 0.03 | 0.08 ± 0.02 | 0.10 ± 0.02 | 47 | 67.4 ± 3.1 | 65.4 ± 3.1 | 44.3 ± 3.1 |
| $dT(pT)_3$-Rho-$dT(pT)_{14}$ | 0.28 ± 0.04 | 0.14 ± 0.03 | 0.18 ± 0.03 | 47 | 60.5 ± 2.5 | 57.7 ± 2.5 | 32.9 ± 2.9 |
| $dT(pT)_8$-Rho-$dT(pT)_9$ | 0.50 ± 0.03 | 0.41 ± 0.03 | 0.45 ± 0.03 | 47 | 48.6 ± 2.3 | 48.1 ± 2.3 | 0 ± 2.3 |
| $dT(pT)_{13}$-Rho-$dT(pT)_4$ | 0.54 ± 0.03 | 0.32 ± 0.03 | 0.41 ± 0.03 | 47 | 50.0 ± 2.3 | 50.5 ± 2.3 | 15.4 ± 2.5 |
| $dT(pT)_{19}$-Rho-3′ | 0.34 ± 0.03 | 0.09 ± 0.03 | 0.12 ± 0.03 | 47 | 65.3 ± 3.1 | 62.2 ± 3.0 | 31.9 ± 2.9 |
| 5′-Rho-dT $(pT)_9$ | 0.25 ± 0.03 | 0.09 ± 0.03 | 0.10 ± 0.02 | 47 | 67.7 ± 3.1 | 67.6 ± 3.1 | 47.8 ± 2.9 |
| $dT(pT)_9$-Rho-3′ | 0.86 ± 0.03 | 0.20 ± 0.03 | 0.58 ± 0.03 | 47 | 44.4 ± 2.1 | 47.6 ± 2.3 | 0 ± 2.3 |

[a] Averaged over two measurements using fluorescein and rhodamine as the FRET acceptor (16) (Reproduced from *Biochemistry* 2007 (ref. 16) with permission from American Chemical Society)

is lower than $R_o = 52$ Å of the CP–fluorescein pair (12, 16). The value of the FRET efficiency sharply increases from $E$ ~0.1 for 5′-Rho-dT$(pT)_{19}$, to $E$ ~0.45 for $dT(pT)_8$-Rho-$dT(pT)_9$. The corresponding average distance from the center of mass of the donor to the acceptor are $R(2/3)$ ~67.4 Å for 5′-Rho-dT$(pT)_{19}$ and $R(2/3)$ ~48.6 Å for $dT(pT)_8$-Rho-$dT(pT)_9$ (Table 2). Similar to the complexes containing fluorescein as the acceptor, a shift of the rhodamine by an additional five nucleotides along the bound ssDNA oligomer, i.e., in the complex of the RepA helicase with dT $(pT)_{13}$-Rho-$dT(pT)_4$, causes, not an increase, but a drop in the value of $E$ to ~0.41 and an increase of the average distance to $R(2/3)$ ~50 Å. A further shift of rhodamine to the 3′ end of the bound ssDNA 20-mer, in $dT(pT)_{19}$-Rho-3′, is accompanied by a dramatic decrease of the FRET efficiency, $E$, to ~0.12 and the strong increase of the average distance to $R(2/3)$, is ~65.3 Å. It is evident that both FRET acceptors provide very similar values of the average distances between the donor and acceptor in the RepA hexamer complexes with the examined ssDNA oligomers, indicating that $\kappa^2$ does not affect these measurements.

## 6. Topology of the RepA Helicase–ssDNA Complex

For a ring-like hexameric structure of the RepA helicase, with the radius of the hexamer ~60–80 Å and $R_o = 52$ Å, the average FRET efficiency, $E$, cannot reach the value of ~0.54–0.63, if the acceptor is located on the outside surface of the hexamer on one of the protomers (Fig. 2b). However, such high values of $E$ are experimentally observed for the RepA hexamer–dT(pT)$_8$-Flu-dT(pT)$_9$ and RepA hexamer–dT(pT)$_{13}$-Flu-dT(pT)$_4$ complexes, respectively (Table 1). By the same token, for the critical Förster distance, $R_o = 47$ Å, the value of $E$ cannot reach the ~0.41–0.45 observed experimentally for the RepA hexamer–dT(pT)$_8$-Rho-dT(pT)$_9$ and RepA hexamer–dT(pT)$_{13}$-Rho-dT(pT)$_4$ complexes (Table 2). Thus, the high values of the FRET efficiencies in these complexes provide the first strong and very convincing evidence that, in the complex of the RepA helicase, the bound ssDNA passes through the cross-channel of the hexamer.

Among all examined locations of the acceptor, the highest values of the FRET efficiency, $E$, is observed for the ssDNA oligomers with the fluorescein or rhodamine placed in the middle of the nucleic acid molecule, dT(pT)$_8$-Flu-dT(pT)$_9$ and dT(pT)$_8$-Rho-dT(pT)$_9$, respectively (Tables 1 and 2). Slightly lower values of $E$ are observed for the oligomers, dT(pT)$_{13}$-Flu-dT(pT)$_4$ and dT (pT)$_{13}$-Rho-dT(pT)$_4$. Thus, the FRET acceptor, placed at the tenth nucleotide of the nucleic acid molecule (Fig. 3), is at the nearest distance from the plane with the donor on the hexamer, among all examined acceptor locations along the nucleic acid lattice (Tables 1 and 2). Recall, the donor, CP, is at one of the six possible specific locations on the RepA hexamer, in a plane, which is perpendicular to the side axis of the helicase molecule (Fig. 1a, b). The shortest distance from the donor to fluorescein or rhodamine in dT(pT)$_8$-Flu-dT(pT)$_9$ and dT(pT)$_8$-Rho-dT(pT)$_9$ and the second shortest distance, observed for dT(pT)$_{13}$-Flu-dT(pT)$_4$ and dT (pT)$_{13}$-Rho-dT(pT)$_4$, would only result if the plane with the donor passes the nucleic acid axis around the tenth nucleotide residue of the bound ssDNA 20-mer. In such a structure, fluorescein or rhodamine, attached at the 5′ end of the nucleic acid, would be on one side of the plane with the donor, while fluorescein or rhodamine at the 3′ end of the nucleic acid, would be on opposite sides of the plane with the donor, resulting in the lowest and similar FRET efficiencies, as well as the largest distances for these two locations. This is exactly what is experimentally observed (Tables 1 and 2).

Because the acceptor, in the case of dT(pT)$_8$-Flu-dT(pT)$_9$ and dT(pT)$_8$-Rho-dT(pT)$_9$, bound to the RepA hexamer, is in the plane with the donor and the DNA is bound in the cross-channel, the distance from the acceptor to the donor in these complexes is equal to the distance from the donor to the center of the hexamer, $q$. Averaging two independent FRET measurements, using fluorescein or rhodamine as FRET acceptors, provides $R(2/3) = q = 48.1 \pm 2.3$ Å. This distance is significantly larger than the distance from the cysteine residues to the center of the hexamer (~35 Å), as suggested by the crystal structure, indicating the global structure of the RepA helicase is different in solution, as compared to the structure seen in the crystal (Fig. 1a). Subsequent hydrodynamic studies, using the combined application of the analytical ultracentrifugation and the dynamic light scattering techniques, confirmed this finding (34). Analogous average distances between the donor and the acceptor, which is placed at different locations on the ssDNA, averaged over two independent determinations, $R(2/3/)_{av}$, are included in Tables 1 and 2. Knowing these averaged distances and the average distance from the donor to the center of the hexamer (~48.1 Å), allowed us to obtain the average distances between the acceptor, $x$, placed in given locations on the bound DNA, and the plane with the donor on the hexamer, using the Pythagoras formula, defined by (26).

The dependence of the experimentally determined FRET efficiency upon the corresponding distance, $x$, from the plane with the donor, CP, averaged over two donor–acceptor systems, for fluorescein as the FRET acceptor, placed in different locations along the ssDNA 20-mer, is shown in Fig. 6a. The corresponding plot for the ssDNA 20-mer labeled with rhodamine as the FRET acceptor is shown in Fig. 6b. The plots have a characteristic bell-shape with sharp slopes on both sides of the plots, as the FRET acceptor approaches the plane with the donor, with the maximum of the plots centered at the location of the acceptor in the plane with the donor, CP, i.e., at $x = 0$. Only the model in which the ssDNA passes through the cross-channel of the hexamer can describe these data. For plotting purposes, we selected the distances from the 5′ end as negative, although this selection does not affect the character of the plots. To generate plots in Fig. 6a, b, the average distance from the donor to the center of the hexamer, $q \sim 48.1$ Å, was used. However, this does not mean that such an averaged value is the optimal average distance for the specific donor–acceptor pair. The solid lines in Fig. 6a, b are nonlinear least squares fits of the data, using the distance from the CP donor to the center of the hexamer, $q$, as the fitting parameter, which provide $q = 46.3 \pm 2.4$ Å and $q = 49.3 \pm 2.5$ Å, for fluorescein and rhodamine, respectively. For comparison, the dashed lines in Fig. 6a, b correspond to the values of $q$, 3 Å shorter or 3 Å longer than the optimal values of $q$ for a given acceptor–donor pair. The deduced location of the ssDNA in the complex with the RepA hexamer, based on the described above FRET studies is schematically shown in Fig. 7 (16).

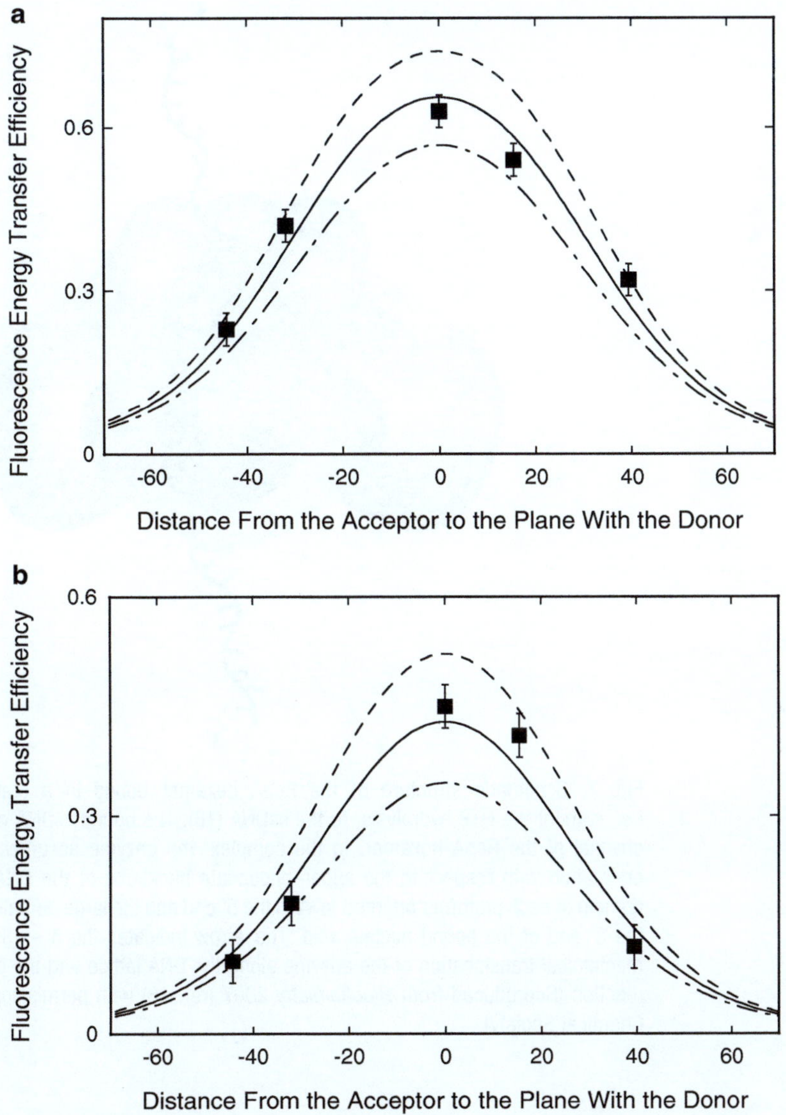


Fig. 6. (**a**) The dependence of the average Förster FRET efficiency from the donor, CP on the labeled RepA hexamer, RepA-CP, to the acceptor, fluorescein, placed at different locations of the ssDNA 20-mer, as a function of the distance of the acceptor from the plane with the donor, *x*, averaged over two independent measurements of the distances between the donor and the acceptor (16). *Solid line* is the nonlinear least squares fit using (1) and (26) with the distance of the donor to the center of the hexamer, *q*, as a single fitting parameter and the Förster distance for the CP–fluorescein donor–acceptor pair, $R_o = 52$ Å. The plane with the donor, CP, passes the axis of the bound ssDNA 20-mer at the middle of the nucleic acid molecule (16). (**b**) The dependence of the average Förster fluorescence energy transfer from the donor, CP on RepA-CP, to the acceptor, rhodamine, placed at different locations of the ssDNA 20-mer, as a function of the distance of the acceptor from the plane with the donor, *x*, averaged over two independent measurements of the distances between the donor and the acceptor (16). *Solid line* is the nonlinear least squares fit using (1) and (26) with the distance of the donor to the center of the hexamer, *q*, as a single fitting parameter and the Förster distance for the CP–rhodamine donor–acceptor pair, $R_o = 47$ Å. The plane with the donor, CP, passes the axis of the nucleic acid at the middle of the nucleic acid molecule (Reproduced from *Biochemistry* 2007 (ref. 16) with permission from American Chemical Society).

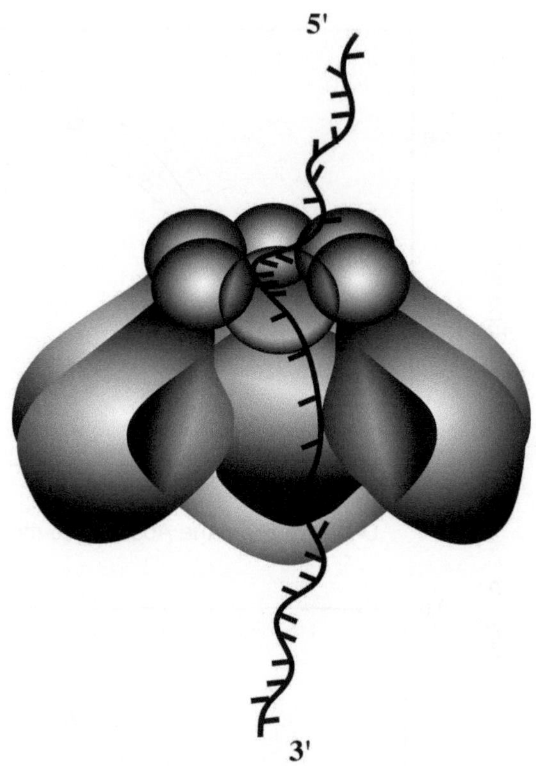


Fig. 7. Schematic structure of the RepA hexamer bound in a stationary complex, i.e., without the NTP hydrolysis, to the ssDNA (16). The bound ssDNA passes the cross-channel of the RepA hexamer. In the complex, the enzyme assumes a strictly single orientation with respect to the sugar–phosphate backbone of the DNA, with the small domain of each protomer oriented toward the 5′ end and the large domain oriented toward the 3′ end of the bound nucleic acid. The *arrow* indicates the 5′–>3′ direction of the mechanical translocation of the enzyme along the DNA lattice and the dsDNA unwinding reaction (Reproduced from *Biochemistry* 2007 (ref. 16) with permission from American Chemical Society).

## Acknowledgments

This work was supported by NIH Grants GM46679 and GM58565 (to W.B.). We wish to thank Gloria Drennan Bellard for a careful reading of the manuscript.

### References

1. Clegg RM (1992) Fluorescence resonance energy transfer and nucleic acids. Methods Enzymol 211:353–388
2. Cheung CH (1991) Resonance energy transfer. In: Lakowicz J (ed) Topics in fluorescence spectroscopy, vol 2. Plenum Press, New York, pp 128–176
3. Lakowicz JR (1999) Principles of fluorescence spectroscopy, 2nd edn. Kluwer Academic, New York, pp 347–367

4. Valeur B (2002) Molecular fluorescence. Principles and applications. Wiley-VCH Weinheim, New York, pp 247–272
5. Yang M, Millar DP (1997) Fluorescence resonance energy transfer as a probe of DNA structure and function. Methods in Enzymol 278:417–444
6. Bailey MF, Thompson EH, Millar DP (2001) Probing DNA polymerase fidelity mechanisms using time-resolved fluorescence anisotropy. Methods 25:62–77
7. Vamosi G, Clegg RM (1992) The helix–coil transition of DNA duplexes and hairpins observed by multiple fluorescence parameters. Biochemistry 37:14300–14316
8. Trakselis MA, Alley SC, Able-Santos E, Benkovic SJ (2001) Creating a dynamic picture of the sliding clamp during T4 DNA polymerase holoenzyme assembly by using fluorescence resonance energy transfer. Proc Natl Acad Sci U S A 98:8368–8375
9. Parkhurst LJ (2004) Distance parameters derived from time-resolved Forster resonance energy transfer measurements and their use in structural interpretations of thermodynamic quantities associated with protein–DNA interactions. Methods Enzymol 379:235–262
10. Berman HA, Yguerabide J, Taylor P (1980) Fluorescence energy transfer on acethylcholinesterase: special relationship between peripheral site and active center. Biochemistry 19:2226–2235
11. Jezewska MJ, Rajendran S, Bujalowski W (1998) Functional and structural heterogeneity of the DNA binding of the *E. coli* primary replicative helicase DnaB protein. J Biol Chem 273:9058–9069
12. Jezewska MJ, Rajendran S, Bujalowska D, Bujalowski W (1998) Does ssDNA pass through the inner channel of the protein hexamer in the complex with the *E. coli* DnaB helicase? Fluorescence energy transfer studies. J Biol Chem 273:10515–10529
13. Galletto R, Jezewska MJ, Bujalowski W (2003) Interactions of the *Escherichia coli* DnaB helicase hexamer with the replication factor the DnaC protein. Effect of nucleotide cofactors and the ssDNA on protein–protein interactions and the topology of the complex. J Mol Biol 329:441–465
14. Jezewska MJ, Galletto R, Bujalowski W (2003) Tertiary conformation of the template-primer and gapped DNA substrates in complexes with rat polymerase β fluorescence energy transfer studies using the multiple donor–acceptor approach. Biochemistry 42:11864–11878
15. Bujalowski W, Klonowska MM (1994) Structural characteristics of the nucleotide binding site of the *E. coli* primary replicative helicase DnaB protein. studies with ribose and base-modified fluorescent nucleotide analogs. Biochemistry 33:4682–4694
16. Marcinowicz A, Jezewska MJ, Bujalowski PJ, Bujalowski W (2007) The structure of the tertiary complex of the RepA hexameric helicase of plasmid RSF1010 with the ssDNA and nucleotide cofactors in solution. Biochemistry 46:13279–13296
17. Jezewska MJ, Bujalowski PJ, Bujalowski W (2007) Interactions of the DNA polymerase X of African Swine fever virus with double-stranded DNA. Functional structure of the complex. J Mol Biol 373:75–95
18. Grinvald A, Haas E, Steinberg IZ (1972) Evaluation of the distribution of distances between energy donors and acceptors by fluorescence decay. Proc Natl Acad Sci USA 69:2273–2277
19. Gryczynski I, Wiczk W, Johnson ML, Cheung HC, Wang C, Lakowicz JR (1988) Resolution of the end-to-end distance distribution of flexible molecules using quenching-induced variations of Förster distance for fluorescence energy transfer. Biophys J 54:577–586
20. Ishmael FT, Trakselis MA, Benkovic SJ (2003) Protein–protein interactions in the bacteriophage T4 replisome. The leading strand holoenzyme is physically linked to the lagging strand holoenzyme and the primosome. J Biol Chem 278:3145–3152
21. Tims HS, Widom J (2007) Stopped-flow fluorescence resonance energy transfer for analysis of nucleosome. Dynam Methods 41:296–303
22. Yang M, Millar DP (1997) Fluorescence resonance energy transfer as a probe of DNA structure and function. Methods Enzymol 278:417–444
23. Wigelsworth DJ, Krantz BA, Christensen KA, Borden KAD, Juris SJ, Collier RJ (2004) Binding stoichiometry and kinetics of the interaction of a human anthrax toxin receptor, CMG2, with protective antigen. J Biol Chem 279:23349–23356
24. Gavish B (1986) Molecular dynamics and the transient strain model of enzyme catalysis. In: Welch GR (ed) The fluctuating enzyme. Wiley, New York, pp 263–339
25. Dale RE, Eisinger J, Blumberg WE (1979) The orientation freedom of molecular probes. The orientation factor in intramolecular energy transfer. Biophys J 26:161–194

26. Bujalowski W, Klonowska MM (1994) Close proximity of tryptophan residues and ATP binding site in *Escherichia coli* primary replicative helicase DnaB protein. Molecular topography of the enzyme. J Biol Chem 269: 31359–31371
27. Scherzinger E, Ziegelin G, Barcena M, Carazo JM, Lurz R, Lanka E (1997) The RepA protein of plasmid RSF1010 is a replicative DNA helicase. J Biol Chem 272:30228–30236
28. Niedenzu T, Roleke D, Bains G, Scherzinger E, Saenger W (2001) Crystal structure of the hexameric helicase RepA of plasmid RSF1010. J Mol Biol 306:479–487
29. Jezewska MJ, Galletto R, Bujalowski W (2004) Interactions of the RepA helicase hexamer of plasmid RSF1010 with the ssDNA. Quantitative analysis of stoichiometries, intrinsic affinities, cooperativities, and heterogeneity of the total ssDNA-binding site. J Mol Biol 343:115–136
30. SenGupta DJ, Borowiec JA (1992) Strand-specific recognition of a synthetic DNA replication fork by the SV40 large tumor antigen. Science 256: 1656–1661
31. Geiselmann J, Wang Y, Seifried SE, Von Hippel PH (1993) A physical model for the translocation and helicase activities of *Escherichia coli* transcription termination protein Rho. Proc Natl Acad Sci U S A 90:7754–7758
32. Saenger W (1984) Principles of nucleic acid structure. Springer-Verlag, New York, p 255–305
33. Bloomfield VA, Crothers DM, Tinoco I (1999) Nucleic acid. Structures, properties, and functions. University Science Books, California, p 79–110
34. Marcinowicz A, Jezewska MJ, Bujalowski W (2008) Multiple global conformational states of the hexameric RepA helicase of plasmid RSF1010 with different ssDNA-binding capabilities are induced by different numbers of bound nucleotides. Analytical ultracentrifugation and dynamic light scattering studies. J Mol Biol 375:386–408

# Chapter 8

# Illuminating Allostery in Metal Sensing Transcriptional Regulators


## Nicholas E. Grossoehme and David P. Giedroc



## Abstract

The intracellular availability of all biologically required transition metal ions in bacteria, e.g., Zn, Cu, Fe, as well as the detoxification of nonbiological heavy metal pollutants, is controlled at the molecular level by a panel of metalloregulatory or "metal sensor" proteins. Metal sensor proteins are specialized allosteric proteins that regulate the transcription of genes linked to transition metal homeostasis as a result of direct binding of a single metal ion or two closely related metal ions, to the exclusion of all others. In many cases, the binding of the cognate metal ion induces a structural change in a metal sensor oligomer that either activates or inhibits operator DNA binding. A quantitative measure of the degree to which a particular metal drives metalloregulation of transcription is the allosteric coupling-free energy, $\Delta G_c$. In this chapter, we outline detailed spectroscopically derived methods for measuring metal binding affinity, $K_{Me}$, as well as $\Delta G_c$ independent of $K_{Me}$, presented in the context of a simple coupled equilibrium scheme. Studies carried out in this way provide quantitative insights into the degree to which a particular metal ion is capable of driving allosteric switching, and via ligand substitution, the extent to which individual coordination bonds establish structural linkage of allosteric metal and operator DNA-binding sites.



**Key words:** Metalloregulation, Metal sensor protein, Metals in biology, Fluorescence anisotropy, Allosteric coupling-free energy


## 1. Introduction

Allostery encompasses the simple idea that the binding of a ligand at one site can influence the binding or chemical reactivity of the same or different ligand at a distinct, often distant, site. In the classical Monod–Wyman–Changeux or two-state model originally developed for oligomeric proteins or enzymes, initial ligand binding in one subunit triggers a structural change that propagates to other subunits, resulting in a much higher affinity for all other ligands (1). This model is exemplified by the textbook case of heterotetrameric ($\alpha_2\beta_2$) hemoglobin. Here, a structural change is thought to occur in the three empty protomers upon $O_2$ binding to


Wlodek M. Bujalowski (ed.), *Spectroscopic Methods of Analysis: Methods and Protocols*, Methods in Molecular Biology, vol. 875, DOI 10.1007/978-1-61779-806-1_8,

a single protomer, resulting in a dramatic increase in $O_2$ affinity for these subunits. The allosteric activation is such that hemoglobin is thought to be limited to two structural states, which resemble the ligand-free and ligand ($O_2$)-saturated quaternary structural states. Alternatively, in the sequential model or allostery, individual subunits within an oligomer do not necessarily adapt the same conformation (2). Experimental data for many allosteric proteins are interpreted in the context of one of these two allosteric models.

Transcriptional regulators are specialized allosteric proteins that sense cellular concentrations of metabolites and other small molecular effectors in order to allow for an appropriate response to changing growth conditions. These proteins function through a specific interaction with the operator/promoter region DNA just upstream of the regulated gene or operon. Ligand binding to the protein–DNA complex, typically to a site distinct from the DNA binding site, drives a structural or dynamic change in conformation that modulates the affinity or structure of the regulatory protein–DNA complex. Metalloregulatory proteins are a further subclassification of transcriptional regulators that have evolved to balance the expression of cellular metal uptake and detoxification systems (3, 4). These specialized proteins have evolved from a number of transcriptional regulator families and within a single family, the metal selectivity of individuals can vary significantly. For example, individual members of the ArsR (*ars*enic *r*epressor) family (5) have been described that regulate metal detoxification systems in response to Co, Ni, Cu, Zn, Cd, Hg, Pb, Bi, Sb, or As, via a classical derepression mechanism. In contrast, MerR (*mer*curic ion *r*epressor) family members are known or predicted to function through transcriptional activation mechanism triggered by the binding of Co, Cu, Zn, Ag, Cd, Au, Hg, or Pb (6, 7).

In this methods review, we outline a general spectroscopic method to quantify allosteric regulation of metalloregulatory protein function by metal ions. This methodology is, however, perfectly general and can be extended to investigate ligand-mediated regulation of operator DNA binding by any transcriptional regulatory protein, provided the ligand affinity can be measured directly. When the ligand is a metal ion, multiple spectroscopic approaches are available to accurately measure this affinity (see Subheading 3.4). As presented below, elucidation of the affinity of the protein and the protein–DNA complex for metal ion allows for the direct determination of the coupling-free energy ($\Delta G_c$), a quantitative reporter of the magnitude of the allosteric driving force (8). Alternatively, $\Delta G_c$ can be obtained by measuring the DNA binding affinity of apo vs. liganded protein. Depending on the nature of the regulator and characteristics of the system, one approach may well be preferential over the other, but each should give rise to the same value of $\Delta G_c$ (9). We first present the generalized coupling scheme and the simple mathematical construct behind it, followed by common experimental considerations and finally, a detailed description of how to successfully perform and quantitatively analyze the results from these experiments.

### *1.1. Allosteric Coupling Scheme*

The thermodynamic cycle presented in Fig. 1a represents a closed system ( $\sum_{i=1}^{4} \Delta X_i = 0$, where $X$ is any thermodynamic state function) that is inclusive of all four possible "end" states that a dimeric metalloregulatory protein ($P_2$) can adopt in equilibrium with a single DNA duplex operator (D) and $n$ total metal ions (M): apo ($P_2$), metal-bound ($P_2{\cdot}M_n$), DNA bound ($P_2{\cdot}D$), and the "ternary" metal–protein–DNA complex ($(P_2{\cdot}M_n){\cdot}D$). Note that $P_2$ is in equilibrium with free monomer P as well, defined by $K_{\text{dimer}}$, and the model assumes that P has negligible affinity for D. Each side of this thermodynamic box represents a measurable transition between two of the four states ($K_1$–$K_4$). This simplistic view of the macroscopic chemical transitions allows for a generic approach to quantify and normalize the allosteric response of a metalloregulatory protein for its DNA binding partner upon metal binding. Note that this scheme can be expanded across the top and bottom equilibria to expressly consider intermediate ligation states with $i$ ligands bound, e.g., where $1 < i < n$; in this case, the macroscopic parameters $K_1$ and $K_2$ would be replaced with the appropriate stepwise binding constants (9). Likewise, this scheme can be expanded to include an additional DNA-binding step, and/or oligomeric assemblies larger than dimers (10).

The magnitude of allosteric regulation, $K_c$, is simply defined as the ligand exchange equilibrium presented (Fig. 1b), where

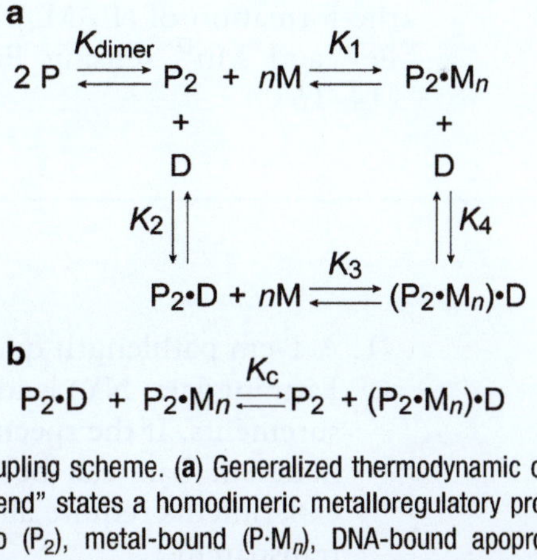

**a**

$$2\,P \underset{}{\overset{K_{\text{dimer}}}{\rightleftharpoons}} P_2 + nM \overset{K_1}{\rightleftharpoons} P_2{\cdot}M_n$$

$$P_2 + D \;\; \overset{K_2}{\updownarrow} \;\; P_2{\cdot}D \qquad\qquad P_2{\cdot}M_n + D \;\; \overset{K_4}{\updownarrow} \;\; (P_2{\cdot}M_n){\cdot}D$$

$$P_2{\cdot}D + nM \overset{K_3}{\rightleftharpoons} (P_2{\cdot}M_n){\cdot}D$$

**b**

$$P_2{\cdot}D + P_2{\cdot}M_n \overset{K_c}{\rightleftharpoons} P_2 + (P_2{\cdot}M_n){\cdot}D$$

Fig. 1. Allosteric coupling scheme. (**a**) Generalized thermodynamic cycle accounting for the four allosteric "end" states a homodimeric metalloregulatory protein ($P_2$) can hypothetically adopt apo ($P_2$), metal-bound ($P{\cdot}M_n$), DNA-bound apoprotein ($P_2{\cdot}D$), and a "ternary" protein–metal–DNA complex ($(P_2{\cdot}M_n){\cdot}D$). Each equilibrium ($K_1$, $K_2$, $K_3$, $K_4$) describes a direct transition from one configurational state to another as shown. Note also that $P_2$ and thus the entire scheme is in equilibrium with free P monomer, defined by $K_{\text{dimer}}$, which has no affinity for the DNA (3, 9). (**b**) Ligand exchange equilibrium, defined by the unitless parameter $K_c$, that dictates the degree of allostery between the metal binding and DNA binding sites (9).

$$K_c = [P_2][(P_2 \cdot M_n) \cdot D]/[P_2 \cdot D][P_2 \cdot M_n] = K_3/K_1 = K_4/K_2. \tag{1}$$

In other words, the allosteric response of a protein is dictated by the stability of the $P_2{\cdot}D$ and $P_2{\cdot}M_n$ states relative to the $(P_2{\cdot}M_n){\cdot}D$ and $P_2$ states. Thermodynamically, this can be thought of as the difference in ***metal affinity*** between the $P_2{\cdot}D$ and $P_2$ states *or* the difference in *DNA binding affinity* between the $P_2{\cdot}M_n$ and $P_2$ states. Therefore, measurement of $K_1$ and $K_2$ or $K_3$ and $K_4$ can provide a quantitative determination of the unitless coupling equilibrium constant $K_c$, which can then be converted to free energy using the standard thermodynamic function

$$\Delta G_c = -RT \ln K_c. \tag{2}$$

For repressors in which metal binding induces dissociation of the repressor of the DNA operator, the $(P_2{\cdot}M_n){\cdot}D$ state is substantially destabilized relative to $P_2{\cdot}M_n$ and free D, and access to the previously occluded promoter by RNA polymerase results in upregulation of the transcription of downstream genes in the operon. In this case, $K_3 < K_1$ (and $K_4 < K_2$) and $\Delta G_c > 0$; that is, the ligand exchange reaction (Fig. 1b) is not favorable and the two biologically relevant "end" states are $P_2{\cdot}M_n$ and $P_2{\cdot}D$. This is most common for regulation of metal detoxification mechanisms such as *Staphylococcus aureus* CzrA and *S. aureus* pI258 CadC in response to $Zn^{II}/Co^{II}$ and $Cd^{II}/Pb^{II}/Bi^{III}$, respectively (11–13). Alternatively, when $K_3 > K_1$ (and $K_4 > K_2$), $\Delta G_c < 0$ and $P_2$ and $(P_2{\cdot}M_n){\cdot}D$ are the two biologically relevant states. In this case, excess cellular metal represses downstream gene transcription by the formation of $(P_2{\cdot}M_n){\cdot}D$ complex as is the case for the largely $Fe^{II}$ and $Mn^{II}$ sensing Fur family and DtxR family repressors (14–16).

## 2. Materials

1. A 1-cm pathlength quartz cuvette (NSG Precision Cells, Inc., Farmingdale, NY) is ideal for carrying out spectroscopic measurements. If the spectrometer has a stirring mechanism, magnetic stir bars can be used for optimal mixing. Fluorescence experiments require all four sides of the cuvette to be optically transparent.
2. Reaction buffer is dependent on experimental requirements as discussed in Subheading 3.1. A commonly used buffer is MOPS (2-(*N*-Morpholino)propanesulfonic acid) (>99%) at pH 7.0 and 100 mM NaCl.

3. An extensive collection of fluorescent metal chelating dyes are available from Molecular Probes. Prepare and store 100 μL aliquots of 1–2 mM concentrations as recommended by the manufacturer.
4. To carry out anaerobic experiments, a variety of gastight syringes are available from Hamilton Company (Reno, NV) and cuvettes are commercially available from a variety of suppliers including NSG Precision Cells, Inc. (Farmingdale, NY).
5. Stock metal solutions should be prepared from ultrapure metal salts, e.g., zinc(II) sulfate, lead(II) chloride, cobalt(II) chloride, etc. (Johnson-Matthey) at neutral pH such that the metal salts are stable. For example, $Zn^{II}$ is stable under atmospheric conditions in a wide range of buffers, while $Fe^{II}$ must be prepared and stored under strictly anaerobic or acidic conditions to prevent air oxidation to $Fe^{III}$. $Pb^{II}$ and $Bi^{III}$ salts must be prepared in a weakly chelating buffer, e.g., bis–Tris, to avoid precipitation of insoluble metal hydroxides (17). We recommend preparing 100 mM metal stocks for use as metal titrants. The metal concentration of these stock solutions are then accurately determined by atomic absorption (18, 19) or atomic emission spectroscopy using National Institute of Science and Technology (NIST) approved standard solutions (Alfa Aesar, Ward Hill, MA) for standard curve calibration.
6. DNA binding assays require a fluorophore covalently bound to one of the strands. Deoxyribooligonucleotides are commercially available in high purity from a number of commercial sources including Operon (Huntsville, AL) and IDT (Coralville, IA). One of the two complementary DNA strands should contain a bright (high fluorescence quantum yield) fluorophore, e.g., fluorescein, covalently attached to the 3′ or 5′ terminus; the complementary strand should be unmodified. Following purification (see Note 1), 100 μM dsDNA stock solutions are prepared from the individual ssDNA by mixing a 1:1 unlabeled:labeled molar ratio of DNA strands. The 10% molar excess of the unlabeled DNA ensures that all labeled DNA (the experimental reporter molecule) is fully complexed, and that there are no unanticipated experimental complications that arise from the slight excess of unlabeled ssDNA. Annealing is accomplished by heating to 95°C followed by slowly cooling to room temperature, with DNA duplex formation confirmed by native polyacrylamide gel electrophoresis. Care must be taken to avoid fold-back intramolecular DNA hairpin structures that might arise from the palindromic or self-complementary nature of the individual ssDNA strands (see Note 2). DNA duplexes prepared in this way are stored in the dark at −20°C and are stable indefinitely.

## 3. Methods

### 3.1. Solution Conditions

The selection of appropriate experimental solution conditions is central to any experimental design, since one or more of these variables may dictate the success or failure of an experiment.

1. *pH.* Metals are positively charged ions and thus tend to associate with anionic or neutral ligands; as a result, the protonation state of these ligands can significantly influence the measured affinity of a regulatory protein for a metal ion. Accordingly, a more acidic environment will generally reduce the apparent affinity ($K_{app}$) according to (3) where $K$ is the pH-independent binding constant and $K_H$ is the macroscopic proton affinity for a ligand. A pH range from 6 to 8 is a common for conducting these experiments.

$$K_{app} = \frac{K}{(1 + K_H[H^+])}. \tag{3}$$

2. *Buffer.* In addition to the major obvious criterion for buffer selection, i.e., to maintain a constant pH, the potential for metal–buffer interactions should also be carefully considered. Very few common biological buffers are actually innocent in this regard and nearly all associate with metal ions to some degree (20). However, several common buffers containing tertiary amines, many of which are derived from the original Good series of biological buffers (21), are very weakly coordinating and span the physiological pH range. These include MES ($pK_a = 6.19$), MOPS ($pK_a = 7.09$), PIPES ($pK_a = 6.77$), and PIPPS ($pK_a = 7.96$) (22). When working with redox active metals, it should also be noted that some buffers promote undesired redox solution chemistry between the metal and buffer. Particularly notable in this regard is HEPES, which should be avoided when studying $Cu^{II}$ interactions in the presence of ligands that stabilize $Cu^{I}$, as redox chemistry may well occur (23). Although phosphate buffer has been commonly used to measure protein–metal interactions, this buffer should be avoided as metal-phosphate chemistry is metal-specific, complicated, often unquantified, and typically ignored. Table 1 lists known metal affinities, $\beta_i$, for several common biological buffers and chelators; in addition, a unitless competition parameter, $\Omega$, is given for these molecules under one set of standard solution conditions, for illustration. $\Omega$ is readily calculated from (4) where $L$ is the deprotonated form of the ligand and $\beta_i$ is the $i$th sequential equilibrium constant:

$$\Omega = 1 + \sum_i \beta_i [L]^i. \tag{4}$$

**Table 1**
**$Zn^{II}$, $Ni^{II}$, and $Cu^{II}$ stability constants (log $\beta$ $M_iL_k$) for common experimental buffers and chelating agents with competition values (Ω) given for common experimental conditions**

| | $\beta M_iL_k$ | H+ | $Zn^{II}$ | $Ni^{II}$ | $Cu^{II}$ |
|---|---|---|---|---|---|
| Tris | ML | 8.1 | 2.24 | 2.63 | 4.05 |
| | $ML_2$ | | | 4.5 | 7.6 |
| | $ML_3$ | | | | 11.1 |
| | $ML_4$ | | | | 14.1 |
| | Ω (50 mM, pH 8) | | 5 | 25 | $3.2 \times 10^7$ |
| Bis–Tris | ML | 6.54 | 2.38 | 3.59 | 5.27 |
| | Ω (50 mM, pH 7) | | 10 | 145 | 6,900 |
| HEPES | ML | 7.52 | n/d[a] | n/d[a] | 3.22 |
| | Ω (50 mM, pH 7) | | | | 20 |
| EDTA | ML | 9.52 | 16.5 | 18.4 | 18.8 |
| | $ML_2$ | 15.65 | | | |
| | Ω (1 mM, pH 7) | | $8.4 \times 10^{10}$ | $6.7 \times 10^{12}$ | $1.7 \times 10^{13}$ |
| NTA | ML | 9.46 | 10.5 | 11.5 | 12.7 |
| | ML | | 14.24 | 16.3 | 17.4 |
| | Ω (1 mM, pH 7) | | $1.0 \times 10^5$ | $1.0 \times 10^6$ | $2.0 \times 10^7$ |

$\beta\, M_iL_k = [M_iL_k]/[M]^i[L]^k$ where L is the deprotonated form of the indicated substance. All values are NIST reviewed (27) for 100 mM NaCl, 25°C
[a] Values not experimentally determined; however, metal association is likely

3. *Ionic strength.* The apparent affinity measured in metal binding equilibria is dependent upon the concentration of ions in solution (ionic strength, *I*) according to the Debye–Hückel relationship where

$$I = \frac{1}{2}\sum\nolimits_i Z_i^2[\mathrm{i}] \quad (5)$$

*I* is a function of the total ionic content and scales with the square of the valency for each species, i.e., 1 μM $Fe^{III}$ ($z = 3$) contributes $9 \times 10^{-6}$ M while 1 μM $Fe^{II}$ ($z = 2$) contributes $4 \times 10^{-6}$ M to *I*. This value directly influences the activity coefficient, and hence the measurable equilibrium constant. However, since typical metal binding experiments with metal sensor proteins are carried out in the 50–500-mM range in monovalent salt concentration (MX, NaCl or KCl), metal salts typically make a negligible contribution to *I*. Elevated [MX] in these systems is often required to enhance protein solubility and/or ensure that the affinity of protein–DNA interactions are within the measurable range (see Subheading 3.4), the latter given the substantial electrostatic contribution to $K_a$ for virtually all protein–nucleic acid binding equilibria (24–26). For a direct comparison of binding affinities between different systems, the solution conditions must obviously be identical; a monovalent salt concentration of 100 mM has been widely used for metal binding assays (27) and provides a point of reference for quantitative comparisons of metal–protein affinities.

4. *Temperature*. The experiments outlined below allow measurement of equilibrium constants that are inherently temperature-dependent thermodynamic quantities. While ambient temperature experiments are common, the temperature in standard laboratories can fluctuate over the course of hours or days. Therefore, a mechanism to maintain constant temperature, such as a circulating temperature bath, is necessary. Common experimental temperatures are 25°C or 37°C to mimic ambient or physiological temperatures, respectively.

5. *Oxygen sensitivity*. Intracellular environments are largely reducing, hence the oxidation states of metal ions and surface cysteines tend to be in a reduced state, i.e., $Fe^{II}$ and Cys-SH or Cys-$S^-$ as compared to $Fe^{III}$ and Cys-S–S-Cys. It is therefore necessary to determine the sensitivity of a system to oxygen. For the DNA binding assays (see Subheading 3.4), it may not be necessary to use stringently anaerobic conditions if a strong reducing agent is present in the buffer. These might include dithiothreitol (DTT), dithionite, or tris(2-carboxyethyl)phosphine (TCEP), which are commonly used for this purpose (see Table 2). However, extreme caution is urged when using DTT in the presence of metal ions since many make high affinity metal–DTT complexes (27) that may ultimately out-compete metal–protein interactions, particularly when considering the large molar excess that will likely be present in metal-binding experiments. For these reasons, we routinely perform metal-binding experiments under anaerobic conditions in the absence of reducing agents (see Subheading 3.3).

### 3.2. Metal-Free Buffers

Since the goal of these studies is to determine the affinity of a protein for a specific metal ion, an undesirable and often overlooked competition from other metals in the buffer needs to be avoided. Adherence to these simple guidelines will ensure the minimization of background metal contamination.

1. *Preparation of glassware*. Standard laboratory glassware is silicate ($SiO_2$) and simple electrostatics predicts that metal cations will nonspecifically adhere to this anionic surface. This is particularly important for metalloregulatory proteins since many possess extremely high affinity for their cognate metal (26, 28), and thus can potentially "leach" metals from contaminated surfaces. Since protons will out-compete any residual metal

**Table 2**
**Common experimental reducing agents used in metal-binding studies**

| Reducing agent | $E^\circ$ (mV, pH 7.0) | log $\beta_{ZnL}$ | log $\beta_{ZnL2}$ | log $\beta_{Zn3L4}$ | log $\beta_{ZnHL}$ |
|---|---|---|---|---|---|
| Dithiothreitol | −330[a] | 11.1[b] | 17.95[b] | 50.9[b] | – |
| TCEP | N/A | 2.91 | – | – | 9.00 |
| Glutathione | −263[c] | n/d[d] | n/d[d] | n/d[d] | n/d[d] |
| Dithionite | −660[e] | – | – | – | – |

$\beta$ $Zn_iH_jL_k = [Zn_iH_jL_k]/[Zn]^i\,[H]^j\,[L]^k$ where L is the deprotonated form of the indicated substance.
N/A, not available
[a]Ref. 54
[b]Ref. 55
[c]Ref. 56
[d]Zn affinity for glutathione is currently not rigorously quantified, although significant affinity is likely via Zn-thiolate coordination
[e]Ref. 57

ions on this surface, all glassware is typically soaked in 1% nitric acid ($HNO_3$) to significantly reduce residual metal contamination derived from the glassware. Following acid treatment, the glassware should be rinsed exhaustively (≥3 times) with metal-free water (see below) to avoid an unwanted change in the pH buffer solutions.

2. *Metal-free water.* Reverse-osmotic treatment and deionization of water (RODI), standard in most research laboratories, is not sufficient to remove metal ions to the degree required for metal binding assays. Additional metal removal can be provided by numerous standard purification systems that are capable of deionization to a resistance ≥18 MΩ cm. Alternatively, strong metal chelators conjugated to solid styrene beads are commercially available, i.e., Chelex, and can be used to treat laboratory grade RODI water in order to produce operationally defined "metal-free" water. This can be accomplished in two ways. The first option is to pass water through a column containing chelating resin and collect in an acid-washed container. Alternatively, the resin can be added directly into the water and shaken for several hours. Separate the phases by centrifugation and careful decanting.
3. *Buffer preparation.* Buffer salts, as provided by the manufacturer, are commonly contaminated with small amounts of divalent metal ions. While very high purity buffer solids can be obtained commercially, removal of residual metal can easily be accomplished by treating the prepared buffer with Chelex, as described just above. Note that $Na^+$ or $H^+$ ions (depending on the regeneration protocol used for the Chelex resin) will replace the metals to maintain electrostatic neutrality and,

depending on the amount of metal removed from the buffer, this may be significant. Determination of the pH and conductivity of the buffer solution following Chelex treatment is therefore strongly suggested.

### 3.3. Anaerobic Preparations

If oxygen reactivity is a concern, additional steps must be taken to ensure a rigorously anaerobic environment since thoroughly degassed or deoxygenated buffers and solutions are required by these experiments.

1. Prepare the buffer solution using metal-free water and remove residual metal as necessary (see Subheading 3.2).
2. Deoxygenate all solvents using one of two standard protocols. The first is a more rapid method (see Note 3) while the second is more thorough and minimizes solvent evaporation (see Note 4).
3. Stock metal solutions should be prepared and stored under an inert atmosphere. The simplest method is to dissolve a known mass of metal salt in an anaerobic chamber (from Vacuum Atmospheres or Coy). If an anaerobic chamber is not available, deoxygenation can be accomplished suboptimally by extensive bubbling of argon or nitrogen from a cylinder of compressed gas through a metal stock solution.
4. The final step is to thoroughly buffer exchange the purified protein into an oxygen-free buffer. Concentrate the protein stock to ~1–2 mL, transfer to an anaerobic chamber, and dialyze at least 4 h in 500 mL of the buffer to be used to metal binding experiments. The dialysis buffer should be exchanged four times to ensure a thorough removal of metal chelators and reducing agents that may have been used during protein purification.

### 3.4. Metal Binding Assays

Biological function of metalloregulatory proteins is dictated, at least in part, by metal selectivity, which is a governed by the relative affinities for cognate vs. noncognate metal ions (3). Therefore, a quantitative measure of metal affinity can provide critical insight into the thermodynamic driving forces behind these sensors. Using metal binding assays described here, the $K_i$ defined by the two *horizontal* equilibria in Fig. 1a ($K_1$ and $K_3$) can be used to determine $K_c$ and $\Delta G_c$ using (1) and (2), respectively. However, if $\Delta G_c > 0$, the ternary $(P_2 \cdot M_n) \cdot D$ complex may not be stable under the experimental conditions since M may well dissociate the complex, i.e., shift the equilibrium to $P_2 \cdot M_n$ and free D. As a result, the integrity of the so-formed $(P_2 \cdot M_n) \cdot D$ complex must be independently verified using size exclusion chromatography, for example (9). A good rule of thumb is to employ a concentration of $P_2 \cdot D$ that is $\geq$50-fold larger than $1/K_4$ (see Fig. 1a). A number of spectroscopic approaches are available to measure $K_1$ and $K_3$; however, a number of considerations will dictate which of these approaches are most directly applicable. The examples given assume an aerobic environment is suitable, but nearly all applications are easily adapted for anaerobic conditions using the guidelines discussed (see Subheading 3.2).

#### 3.4.1. Direct Titrations

Under ideal circumstances, a direct titration of metal into protein can provide an isotherm that can be fit to an appropriate binding model to obtain a unique binding constant(s). Requirements for this approach are twofold: (1) the metal of interest must induce a measurable change in the spectroscopic signal, $S$, e.g., absorbance, fluorescence, or fluorescence resonance energy transfer (FRET); and (2) the apparent affinity must be within a measurable range since $K_i$ is determined by free, not total, metal concentration. For example, $Cd^{II}$ and $Pb^{II}$ binding to *S. aureus* CadC can be directly monitored via intense $S^- \rightarrow Me^{II}$ ligand-to-metal (or metal-to-ligand) charge transfer (LMCT) transitions as a result of metal coordination with cysteine residues with $\lambda_{max}$ observed at 238 and 352 nm, respectively (18, 29). Alternatively, if there is an endogenous protein-derived fluorophore located near the metal binding site, i.e., Trp or Tyr, metal binding may induce a fluorescent enhancement or quenching of the fluorophore. In this case, measurement of the change in total fluorescence may be used as a reporter for metal binding (30). The optimal experimental concentrations will depend on the molar extinction coefficient and apparent metal affinity for the system under study.

The measurable range of metal affinities can easily be extended to much higher $K_i$ values by the addition of a competing ligand of known affinity, e.g., ethylenediaminetetraacetic acid (EDTA) or nitriloacetic acid (NTA) (see Table 1). This approach requires that a known concentration of competitor ligand is added and the affinity of this ligand be known *under the experimental conditions* of the assay. Such competition experiments can be analyzed using the general approach presented below (see Subheading 3.5) provided one ensures that all relevant equilibria are included in the model. This includes the pH-dependent competition parameter, $\Omega$, which incorporates the effects of pH and ligand p$K_a$ values into the fit. Since the chelator ligand EDTA forms a 1:1 complex with most transition metals, $\Omega$ for $Zn^{II}$-EDTA at 1.0 mM EDTA, pH 7.0 will be used here to illustrate this calculation. In most cases, it is the fully deprotonated form of the competing ligand, denoted L, that forms a stable complex with a metal ion; as a result, one needs to calculate the concentration of free L, [L], from $[L]_{total}$ and other protonated forms of L, denoted $H_xL$ according to (6)

$$[L]_{total} = [L] + [HL] + [H_2L]. \quad (6)$$

For EDTA, the two relevant $pK_a$ values that need to be considered are 9.52 and 6.13. From $K_{H_xL} = 10^{-pK_a}$, one obtains (7), from which [L] can be calculated (8):

$$[L]_{total} = [L] + K_1[H][L] + \beta_2[H]^2[L], \quad (7)$$

$$[L] = \frac{[L]_{tot}}{1 + K_1[H] + \beta_2[H]^2}. \quad (8)$$

Note that the denominator in (8) is the binding polynomial of EDTA for protons, denoted $Q$, when taking free L as the reference state. Substituting into (7) and (8) the known values for EDTA (27), $K_1 = 10^{9.52}$ $M^{-1}$, $\beta_2 = K_1K_2 = 10^{9.52+6.13}$ $M^{-2}$, $[H^+] = 10^{-7}$ M and $[L]_{total} = .001$ M, gives $[L] = 3.05 \times 10^{-6}$ M. This value, along with the $Zn^{II}$-L affinity constant, $\beta_{1,Zn} = 10^{16.5}$ $M^{-1}$ (see Table 2) is input directly into (4) (see Subheading 3.1) to obtain the pH-dependent competition parameter, $\Omega$, at pH 7.0. This gives $\Omega = 8.4 \times 10^{10}$ (see Table 2). This value is then used to calculate the affinity of the protein for metal, $K_{Me}$, from an apparent equilibrium constant, $K_{app}$, derived from a fit to the data that does *not* take into account competition with a competitor ligand (see Subheading 3.5), from (9):

$$K_{Me} = \Omega K_{app}. \quad (9)$$

Alternatively, one can simply calculate a conditional stability constant, $K'$, for the $Zn^{II}$–EDTA complex which is given by $K' = \beta_{1,Zn}/Q = 9.6 \times 10^{13}$ $M^{-1}$ at pH 7.0. Figure 2 demonstrates the dramatic influence that a competitor ligand, 50 mM Tris in this case, can have on the determination of $K_{Me}$ for a protein–metal complex.

For the method that follows, it is assumed that 20 μM total protein will lead a spectral signal, $S$, that is readily measured and that $S$ is absorbance.

1. Prepare 1 mL of experimental buffer in a quartz cuvette. Ensure that the cuvette has been acid washed, extensively rinsed with metal-free water and dried with spec-pure methanol, and is free of any optical interferences (such as fingerprints) by cleaning with a Kimwipe. Blank the spectrophotometer over a broad range (200–800 nm).
2. In an optically identical quartz cuvette, prepare 1 mL of 20 μM metal sensor protein.
3. Acquire an initial wavelength scan of the ultraviolet and visible spectral regions (240–700 nm).
4. Prepare a 1.0-mM solution of metal in an identical buffer spiked with 20 μM protein. The addition of protein to the titrant solution avoids the otherwise inevitable dilution of the protein concentration over the course of the experiment.

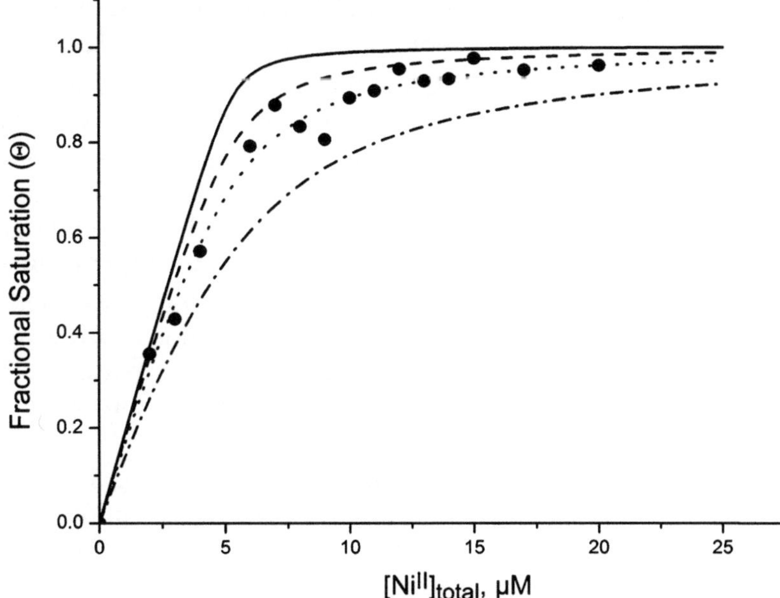


Fig. 2. Representative titrations of a 100-μM Ni$^{II}$ stock solution into Ni$^{II}$/Zn$^{II}$ binding protein, Tm0439, from *Thermotoga maritime* (*filled circles*) (30). The best fit to this data ($K_{Ni}$ = 1.47 × $10^7$ $M^{-1}$) is shown in the figure as a *dotted line*, which includes competition from the buffering substance, 25 mM Tris, pH 8.0 (Ω = 10). The *solid line* represents a simulated curve that corresponds to the same $K_{Ni}$ but without consideration of buffer competition in the fitting model (Ω = 1), while the *dashed* and *hashed* lines represent competitor concentrations of 12.5 and 50 mM Tris, with the same $K_{Ni}$. Table 1 compiles relevant competition values for Tris at pH 8.0. Without consideration of Ω, the value of $K_{Ni}$ would be erroneously determined.

This greatly simplifies data analysis since the most simple fitting procedures require a constant protein concentration (see Subheading 3.5). Alternatively, if this is not possible, i.e., excess metal leads to precipitation of the protein, adjust the titrant stock concentration to ensure that the total dilution of the protein over the course of the experiment does not exceed 5–10%.

5. Make a 1–2-μL injection of 1.0 mM metal into the 20-μM protein solution (corresponding to 1–2 μM metal). Allow appropriate mixing time. If a stirring mechanism is not available, manual mixing may be necessary. Note that some metals, such as Ni$^{II}$ and Cu$^{I}$ have very slow ligand exchange kinetics (31). A kinetic experiment may be required prior to the equilibrium titration in order to ensure that adequate mixing time is allowed for the system to come to equilibrium.
6. Scan the UV–vis region. Note small changes in the electronic spectra. LMCT tend to have sizable changes in absorbance due to a high molar absorptivity (ε > 1,000 $M^{-1}$ $cm^{-1}$); on the other hand, changes in the *d*– > *d* ligand-field region can be

very small given the very low molar intensities of these transitions ($\varepsilon < 200\ M^{-1}\ cm^{-1}$) (32).

7. Repeat the previous two steps until no additional change in $S$ is observed, outside of that expected for dilution.
8. Thoroughly clean the cuvette and repeat the experiment using identical metal aliquots, but in the absence of protein. This titration provides the background signal that can be subtracted point-by-point from the experimental data.
9. These data are commonly presented as absorbance (or molar absorptivity) as a function of wavelength with all aliquots superimposed on the same $x, y$ coordinate. This representation makes it particularly easy to recognize spectral changes as a function of metal concentration.
10. For data fitting (Subheading 3.5), select a wavelength at which maximal change in the absorption spectrum $S$ is observed and plot the $S_i$ as a function of total concentration of added metal titrant.

#### 3.4.2. Surrogate Metal Competition

In some cases, an alternative, or noncognate, metal can be used to determine the binding affinity of the cognate metal ion via a competition experiment. This is particularly applicable when the two metals bind with identical coordination geometries. One common example of this approach is to substitute $Co^{II}$ for $Zn^{II}$ in tetrahedral coordination sites (33). Unlike spectroscopically silent $Zn^{II}$ ($d^{10}$), $Co^{II}$ ($d^7$) has a distinct and measurable spectroscopic signature which enables direct determination of $Co^{II}$ affinity (see Subheading 3.3.1), with a subsequent competitive displacement of $Co^{II}$ by $Zn^{II}$ (34). The experimental details laid out below assume that $Co^{II}$ is used as the surrogate reporter metal for $Zn^{II}$.

1. Carry out a $Co^{II} \rightarrow$ protein experiment as described above (see Subheading 3.3.1). Determine the $Co^{II}$ affinity using the data fitting procedures in Subheading 3.5. This is often straightforward because $K_{Co} \leq 10^7\ M^{-1}$ for many $Co^{II}$ binding sites and a direct titration generates a change in signal $S$ that is nonstoichiometric in the protein concentration range suitable for measurement (20–50 μM).
2. Prepare a 1-mL solution of 20 μM protein and 100 μM $Co^{II}$ in a clean quartz cuvette. The concentration of protein and $Co^{II}$ can be adjusted to optimize $S$. If the affinity for $Co^{II}$ is high enough, stoichiometric equivalence may be adequate. Blank the spectrophotometer over a broad range.
3. Prepare a solution containing 1.0 mM $Zn^{II}$, 100 μM $Co^{II}$, and 20 μM protein. Note that if the concentrations of $Co^{II}$ and protein are adjusted in the previous step, they should be adjusted here as well. This is to ensure that $Zn^{II}$ is the only variable throughout the metal-displacement experiment.

4. Make 1–2 μL injections of $Zn^{2+}$ into the 1-mL protein solution. Allow adequate mixing time. $Co^{2+}$ and $Zn^{2+}$ both have rapid ligand exchange kinetics (31); however if the metal binding site is buried, this can severely impair the exchange. Upon equilibrium, scan the UV–vis region and monitor changes in the spectra. The expected response is directly reverse as the surrogate metal addition.
5. Repeat the previous step as necessary until an adequate baseline is reached. Note that full displacement of the spectroscopically active metal is not necessary to provide a complete data set for rigorous fitting.
6. Clean the cuvette and repeat the experiment in the absence of protein. This provides a point-by-point background for subtraction from the experimental data.

*3.4.3. Spectroscopically Active Metal Chelators*

In many situations, it may not be possible to use a surrogate metal as a spectroscopic reporter or monitor metal binding directly. In these cases, metal chelators with known metal affinities and sensitive spectroscopic signatures are available that can be used to determine the metal affinity, $K_{Me}$, for the sensor protein. This type of competition experiment is based on the relative affinities of the metal (Me) for the chelator (C (10)) and for the protein (P (11)).

$$\mathrm{Me} + \mathrm{C} \rightarrow \mathrm{Me} \cdot \mathrm{C} \qquad K_b \qquad (10)$$

$$\mathrm{Me} + \mathrm{P} \rightarrow \mathrm{Me} \cdot \mathrm{P} \qquad K_{Me} \qquad (11)$$

Since these two equilibria will be in direct competition over the course of this experiment, selection of chelator characterized by an appropriate $K_b$ is critical to the experimental design. Note that this experiment is exactly analogous to the ligand (L) competition experiment discussed above (see Subheading 3.3.1), except the change in $S$ is coming from the ligand C rather from P. As shown in Fig. 3, there is a dynamic range of $K_{Me}$ over which each chelator probe is applicable. When $K_{Me}/K_b < 0.01$, the protein does not effectively compete with the chelator for metal binding and the titration curve is not distinguishable from a direct titration of metal into the free chelator. Alternatively, when $K_{Me}/K_b > 100$, a unique fit for $K_{Me}$ cannot be obtained due to lack of competition from the chelator (35). For example, mag-fura-2 has an estimated affinity of $K_b = 5.0 \times 10^7\ M^{-1}$ for $Zn^{II}$ (pH 7.0 and 25.0°C) (36) and can also be used for a range of divalent cations since their affinities have been measured (37). In the case of $Zn^{II}$, this results in a measurable range of $Zn^{II}$–protein affinity of $5 \times 10^5\ M^{-1} < K_{Me} < 5 \times 10^9\ M^{-1}$. For higher affinity interactions, quin-2 can be used since it has a much higher affinity for $Zn^{II}$ ($K_b \approx 2.7 \times 10^{11}\ M^{-1}$) (38) which extends the dynamic range over which $K_{Me}$ can be measured by this technique to $\geq 10^{13}\ M^{-1}$

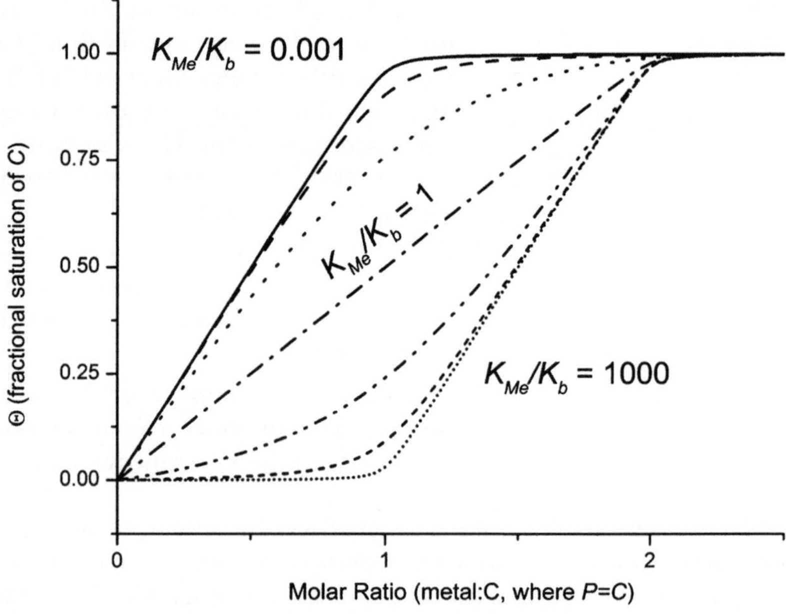


Fig. 3. Simulated data for the titration of a metal into a equimolar solution of protein and spectroscopic metal chelator. Each line corresponds to a tenfold change in the $K_{Me}/K_b$ ratio, as described in (4) and (5). When $K_{Me}/K_b < 0.01$, the protein does not effectively compete with the chelator and the titration cannot be distinguished from a direct metal → chelator titration. When $K_{Me}/K_b > 100$, the chelator does not effectively compete with the protein. In both of these cases, a unique fit to the experimental data cannot be obtained.

(39). Note that parallel experiments done under the same solution conditions should provide internally consistent values of $K_{Me}$ (40). Lower affinity chelators are also commercially available (Molecular Probes); however, metalloregulatory proteins tend to make specific and high affinity interactions with the metals that they sense. Although the experimental protocol described below assumes that the $Zn^{II}$ affinity will be determined by mag-fura-2 competition, the approach is perfectly general and can be used directly with other metals or chelators.

1. Blank the spectrophotometer with the experimental buffer prepared in a clean quartz cuvette.
2. Prepare a 1.0-mM $Zn^{II}$ solution in the experimental buffer.
3. Prepare 1 mL of 20 μM mag-fura-2 in a clean quartz cuvette (lower concentrations of 1–5 μM can be used in the fluorometer). Measure the absorbance minimally from 300 to 400 nm.
4. Make a 1–2-μL injection of $Zn^{II}$ into the mag-fura-2 solution. Allow 1–2 min for mixing or sufficient time for *S* to stabilize. If mechanical mixing is not available, gently aspirate the solution with a micropipette. Measure *S* over the same spectral window.

5. Repeat step (4) until no further change is observed. The absorbance at 325 nm is maximal when $Zn^{II}$ is bound to the chelator and minimized at 366 nm when the chelator is not metallated. These data can then be globally fit to a single equilibrium binding model (see Subheading 3.5) to determine $K_b$.
6. Blank the spectrophotometer with the experimental buffer prepared in a clean quartz cuvette.
7. In a clean quartz cuvette, prepare a 1-mL solution of 20 μM mag-fura-2 and 20 μM protein in the appropriate experimental buffer. Repeat steps (3)–(5). Using $K_b$ determined from the direct titration of $Zn^{II}$ into mag-fura-2, these data can be fit to a simple competition model to fit for $K_{Me}$ (see Subheading 3.5).

### 3.5. DNA Binding Assay

Fluorescence polarization ($P$) or anisotropy ($r$) experiments are most commonly used to determine the affinity of a metalloregulatory protein for duplex DNA operator sequence (41). This hydrodynamic approach, which reports on the size and shape of molecules, utilizes vertically polarized light to selectively excite a subpopulation of fluorophores. The emission intensity is subsequently measured through a polarizer and mathematically translated to anisotropy, $r$, based on the intensities of horizontally and vertically polarized emission (42). The experiments described in this section provide a guide to determining the vertical equilibria in Fig. 1 ($K_2$ and $K_4$). We note however, that for large values of $|\Delta G_c|$, it can be challenging to measure the DNA-binding affinity of the apo (for $\Delta G_c < 0$, or allosteric activation) or the $P_2{\cdot}M_n$ complex (for $\Delta G_c > 0$, or allosteric inhibition), since $K_2$ or $K_4$, respectively, may be very small or difficult to distinguish from nonspecific binding. In the case of allosteric inhibition, our experience is that it is usually possible to measure $K_4$ provided high protein concentrations are achievable in the cuvette ($\geq$10 μM) (43), and excess total metal (0.05–0.1 mM) can be added to the solution without precipitation of the protein. This ensures that all metal sensor is in the $P_2{\cdot}M_n$ form which allows for direct determination of $K_4$ (see Fig. 1). In order to minimize non- or weakly specific binding of the metalloregulatory protein to adventitious or cryptic binding sites on the DNA, the use of a dsDNA of the minimal length that retains all of the known or projected intermolecular contacts that maximize binding energy is desirable (41, 44); even under these conditions, however, the binding of additional dimers to the dsDNA may not be avoidable (9).

The length of the operator-containing duplex DNA also impacts the ability to easily use a change in $r$ as the basis of detection of protein binding. Longer dsDNA molecules are characterized by a larger intrinsic anisotropy, $r_o$, with the change in $r$ smaller over the course of the experiment; this will necessarily result in larger error in data fitting. The minimal length requirement of a dsDNA that maintains high affinity binding varies with each metalloregulator; a DNase I footprint, if available, can be used as a guide for duplex DNA design (26). Alternatively, carrying out binding experiments with a collection of oligonucleotides of incrementally shorter lengths can provide a direct indication of the minimal length as well (43). A 28–32 base pair duplex is, however, a reasonable starting point for most dimeric metalloregulatory proteins for which high resolution structures are known or predicted from functional orthologs (4). The experiment described below assumes a fluorescein-labeled dsDNA.

1. Ensure the spectrofluorometer is configured to measure anisotropy. Polarizers are needed for excitation as well as emitted light. For fluorescein, a 530-nm cut-off filter is often used to remove background light scatter from incident radiation.
2. In a four-sided quartz cuvette, prepare 2 mL of 4–10 nM fluorescein-labeled dsDNA in a appropriate experimental buffer. If a stirring mechanism is equipped, insert a clean magnetic stir bar in the cuvette.
3. Prepare a 2-μM sample of protein in an identical experimental buffer.
4. Measure the anisotropy of the DNA duplex. A typical intrinsic or starting anisotropy ($r_o$) for a fluorescein-labeled $\approx$30 base pair duplex dsDNA is $\approx$ 0.11. Single-stranded DNAs of this length will have $r_o \approx 0.07$, with longer duplexes often $r_o \geq 0.15$. We note that $r_o$ is strongly dictated by the nature of the fluorophore and is also influenced by the degree to which the probe intercalates or stacks against the end of the DNA helix, with $r_o$ values of $\geq$0.2 for many commonly used fluorescent dyes (Cy5, Cy3, rhodamine derivatives, coumarin, etc.) (44–46).
5. Add 1–2 μL of 2-μM protein stock. Stir for at least 1 min to allow adequate mixing. Measure the anisotropy once the signal stabilizes.
6. Repeat step (5) until no change is observed. Avoid addition of >200 μL (10%) as fluorophore concentration is considered to be a constant in simple fitting algorithms. Alternatively, 10 nM labeled dsDNA can be added to the 2-μM protein solution to avoid the effects of dilution.
7. Plot $r_i$ as a function of total protein concentration as shown in Fig. 4 (see Note 5). Fit these data to an appropriate model as outlined below and extract a value for $K_2$ (see Subheading 3.5).
8. Repeat steps (1)–(7) using a preloaded metallated sensor stock as the titrant ($P_2{\cdot}M_n$). If the affinity of the sensor protein for metal when bound to DNA ($K_3$ in Fig. 1) is low, an excess of the same metal salt (50–100 μM) can be added to the binding

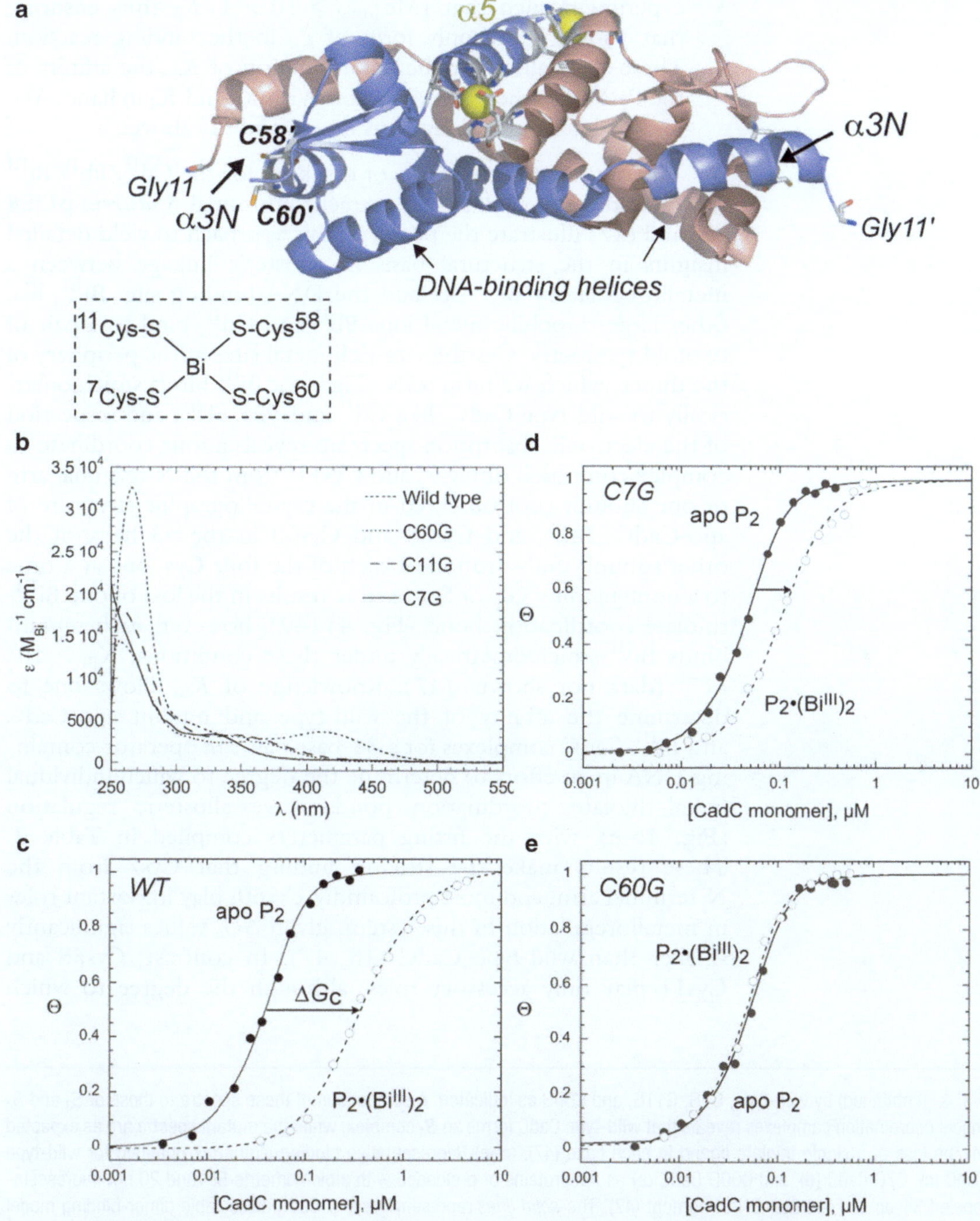


Fig. 4. Representative example of the elucidation of the residue-specific contributions of individual cysteine residues to the allosteric coupling-free energy, $\Delta G_c$. (**a**) Ribbon representation of the crystallographic structure of apo-C11G CadC (residues 11–117 or 11–119), with structural $Zn^{II}$ ions bound in the nonregulatory (18) α5 sites (*yellow spheres*) (48). One protomer is shaded *salmon* and one is shaded *blue*, with the empty regulatory $Cys_4$ α3N sites and the putative DNA-binding helices on each protomer indicated. $Bi^{III}$ is known to form a four-coordinate $S_4$ complex (47), while $Cd^{II}$ forms a distorted $S_4$ complex (18, 29) with Cys11 weakly coordinated (49), while $Pb^{II}$ adopts a trigonal planar complex that excludes coordination by Cys11 (18). (**b**) Molar electronic absorption spectra of the stoichiometric $Bi^{III}$ complexes formed

experiment such that $[\mathrm{Me}]_{\mathrm{total}} \geq 10 \times 1/K_3$ thus ensuring that $P_2{\cdot}M_n$ is the only form of $P_2$ in the binding reaction. These steps allow for the determination of $K_4$, the affinity of the $P_2{\cdot}M_n$ for the DNA operator. With $K_2$ and $K_4$ in hand, $\Delta G_c$ can be determined directly using (1) and (2) above.

The results of experiments of this kind for the $Cd^{II}/Pb^{II}/Bi^{III}$ sensor CadC encoded on the extrachromosomal *S. aureus* pI258 plasmid (47) illustrate the power of this approach to yield detailed insights in the structural basis of allosteric linkage between a metalloregulatory $Bi^{III}$ site and the DNA-binding site. $Bi^{III}$, like other large thiophilic metal ions $Pb^{II}$ and $Cd^{II}$, bind to a pair of twofold symmetric Cys-thiolate-rich metal sites at the periphery of the dimer, which we term α3N (Fig. 4a). $Bi^{III}$ binds stoichiometrically to wild-type CadC, like $Cd^{II}$ and $Pb^{II}$ (18), and inspection of the electronic absorption spectrum reveals a four-coordinate $S_4$ complex composed of Cys7′ and Cys11′ from the N-terminal arm of one subunit (not observed in the crystallographic structure of apo-CadC) (48), and Cys58 and Cys60 in the α3 helix of the other subunit. Substitution of each of the four Cys, one at a time to a nonliganding Gly or Ser residue results in the loss of one $Bi^{III}$-thiolate coordination bond (Fig. 4) (49); however, each mutant binds $Bi^{III}$ stoichiometrically under these conditions, $K_{Bi} \geq 10^9$ $M^{-1}$ (data not shown) (47). Knowledge of $K_{Bi}$ allows one to determine the affinity of the wild-type and mutant apo-CadC and $Bi^{III}$-CadC complexes for a 34-base pair *cad* operator-containing DNA in an effort to determine the degree to which individual metal–thiolate coordination bonds drive allosteric regulation (Fig. 4c–e), with the fitting parameters compiled in Table 3. These results make the striking finding that Cys7 from the N-terminal arm, and most profoundly Cys60, play important roles in metalloregulation in this system, given $\Delta G_c$ values significantly smaller than wild-type CadC (18, 47). In contrast, Cys58 and Cys11 play only accessory roles, although the degree to which

Fig. 4. (continued) by wild-type, C7G, G11G, and C60G as indicated. A comparison of these spectra to those of $S_3$ and $S_4$ model coordination complexes reveals that wild-type CadC forms an $S_4$ complex, while the mutant spectra are as expected for the loss of a single thiolate ligand in each case (47). (**c–e**) Representative binding isotherms obtained for wild-type CadC (**c**), C7G CadC (**d**), and C60G CadC (**e**) as apoproteins or preloaded with stoichiometric $Bi^{III}$ and 20 nM fluorescein-labeled 34-bp *cad* operator DNA fragment (47). The *solid lines* represent fits to a CadC dissociable dimer-binding model (see Subheading 3.5), with $K_{dimer}$ fixed at the values determined from analytical equilibrium ultracentrifugation under the same solution conditions and $K_2$ (for apo-CadC) or $K_4$ (for $Bi^{III}$-loaded CadC) (see Fig. 1) optimized during the fit (29, 47). $Cd^{II}$ has at most a twofold effect on $K_{dimer}$ (29). Solution conditions: 10 mM Bis–Tris, pH 7.0, 0.4 M NaCl, 1 mM DTT, 25.0°C with 50 μM EDTA added to the apo-CadC titrations only. $\Theta = (r_i - r_o)/(r_{complex} - r_o)$, where $r_i$ is the measured anisotropy after each addition of protein titrant (expressed in monomer CadC units), $r_o$ is the starting anisotropy of the free DNA, and $r_{complex}$ is the fitted value for the saturated $P_2{\cdot}D$ complex. $r_o$ for a duplex of this number of base pairs should be 0.13 (±0.01), with the total change in signal ($\Delta r = r_{complex} - r_o$) ranging from ≈0.020 to 0.025 for a 27.6-kDa dimer–34-bp DNA complex (≈50 kDa total) (18, 47).

**Table 3**
**Thermodynamic parameters that define the negative allosteric regulation of wild-type and mutant *S. aureus* pI258 CadCs by various metal ions[a]**

| CadC variant | $K_2$ ($\times 10^9$ $M^{-1}$)[b] | $K_4$ ($Bi^{III}$)[b] ($\times 10^9$ $M^{-1}$) | $\Delta G_c$ ($Bi^{III}$) (kcal/mol) | $\Delta G_c$ ($Cd^{II}$) (kcal/mol) | $\Delta G_c$ ($Pb^{II}$) (kcal/mol) |
|---|---|---|---|---|---|
| Wild type | 1.1 (±0.2) | 0.0067 (±0.0006) | 3.0 (±0.2) | 3.2 (±0.1) | 3.4 (±0.2) |
| C7G | 0.83 (±0.06) | 0.072 (±0.009) | 1.5 (±0.1) | 1.2 (±0.2) | 0.9 (±0.1) |
| C11G | 0.58 (±0.05) | 0.021 (±0.004) | 1.9 (±0.2) | 2.8 (±0.2) | 3.4 (±0.3) |
| C52G[c] | 1.0 (±0.3) | 0.0073 (±0.0008) | 2.9 (±0.2) | 3.3 (±0.3) | 3.4 (±0.4) |
| C58S | 0.11 (±0.02) | ND | ND | 2.7 (±0.6) | 1.8 (±0.2) |
| C60G | 1.0 (±0.6) | 1.0 (±0.1) | 0 (±0.3) | 0.1 (±0.4) | 0 (±0.4) |

[a] Data taken from refs. 18, 47

[b] Fitted parameters from the binding curves shown in Fig. 4 with each $K_i$ corresponding to those in the linkage scheme shown in Fig. 1

[c] Cys52 is not conserved in other CadCs (18) and is not a ligand to the metal ion in any case, and thus represents a control substitution. $\Delta G_c$ determined using eq vv with $T = 298.15$ K

each contributes is clearly metal-ion dependent (Table 3). This reflects the distinct coordination complexes formed in each case, which is a function of both the charge, size, and specific characteristics of the ions. In fact, $Bi^{III}$ is the only metal for which substitution of Cys11 results in an attenuated $\Delta G_c$; for $Pb^{II}$, which forms an $S_3$ complex that excludes Cys11 from the coordination sphere (18) and $Cd^{II}$, which forms a weak coordination bond to Cys11 (49), the C11G substitution is functional silent in vitro and in vivo. Substitution of Cys60 in particular, effectively uncouples metal binding from DNA binding ($\Delta G_c = 0$ kcal/mol), despite the fact that a stoichiometric metal complex is formed in each case (Table 3).

#### 3.6. Data Fitting Using Nonlinear Least Squares Methods

A number of programs are available that are capable of analyzing spectroscopic data to obtain parameters of interest. In the following discussion, we briefly illustrate the use of DynaFit (50) (http://www.biokin.com/dynafit/), a program free to academic users that utilizes a Levenberg-Marquardt "least squares" method to produce estimated parameters from a set of experimental data. The program utilizes a simple symbolic notation in a script or text file to completely define a chemical equilibrium model through a collection of stacked equilibria, and thus does not require that the user-defined closed-form expression that fully encompasses the model. These script files can be as simple (a single chemical equilibrium) or as complex as required by a model. Once the model is set up, it is important that the user consult the cross-correlation matrix if solving for multiple estimated parameters, since all parameters that are optimized will be correlated (or anticorrelated) to some degree (51) (see Note 6). Here, we focus on the appropriate mechanisms to generate estimates of binding constants for simple (direct, no competition) and complex (competition) data sets.

It is also important to point out that the user recognize that, like all programs of this type, it is assumed that the change in signal $S_i$ is directly proportional to a fractional saturation, $\Theta$, of the macromolecule from which $S$ derives, i.e., the relationship between $S_i$ and $\Theta$ is linear:

$$\Theta = (S_i - S_o)/(S_{complex} - S_o)Z \tag{12}$$

where $S_o$ is the signal from the free or uncomplexed macromolecule, and $S_{complex}$ is that associated with a saturated complex. This may not be the case (see an example of $Ni^{II}$ binding to the zinc metalloregulator, *Synechococcus* SmtB in ref. 39), but well-established methods are available that can be used to verify a linear relationship, or if not, explicitly define the relationship using a general method of analysis (52). Although there is no reason to believe, a priori, that the incremental change in the anisotropy of the fluorescence is linearly dependent on fractional saturation by bound sensor protein, we have found in at least one case that this is indeed the case, as determined under solution conditions where the DNA-binding is stoichiometric (26).

1. *Units.* When preparing a script file for DynaFit, a common error occurs in the units. Ensure that the concentration units used throughout the script file matches those used in the data file. For example, the *x*-axis of Fig. 4 is μM monomer protein, therefore the concentration of dsDNA must be scripted in μM and the resulting binding constant will have units of $\mu M^{-1}$. In this case, a least-squares fit that generates $K = 0.1\ \mu M^{-1}$ corresponds to $K = 1 \times 10^5\ M^{-1}$.
2. *Response.* Each independent chemical species that generates a measurable signal, $S$, i.e., unbound dsDNA labeled with a fluorescent probe, as well as the protein–DNA complex, must be included in the response of the script file. The magnitude of the response is dictated by the individual experiment weighted by the concentration. For example, in the fluorescence polarization experiments shown in Fig. 4, the anisotropy of the fluorescein-labeled 34-base pair duplex DNA (D) ($r_o$) is $\approx$0.125 while that for the protein–DNA complex ($r_{complex}$) is $\approx$0.145. The total concentration of dsDNA is 20.0 nM (0.02 μM) in each case; this gives response factors of $r_i/0.02$ of 6.25 and 7.25 for the free DNA and the saturated protein–DNA complex, respectively.

3. *Mechanism.* The mechanism is the actual chemical equilibrium model that describes the binding experiment. DynaFit is capable of fitting data to a theoretically infinite number of coupled equilibria, and obviously many different mechanisms (or models) can adequately describe a given data set. Therefore, existing information regarding the system as well as chemical intuition are needed to guide the data fitting process. For a simple one-site model in which apoprotein binds to a ligand, e.g., $P_2$ binding to D to form a 1:1 complex, the mechanism input into DyanFit is

   P + D <==> PD : K1 assoc.

   where P is protein dimer $P_2$ (note this is one-half the total monomer concentration that is directly measured by UV-absorbance, and assumes a nondissociable dimer), D is DNA, PD is the $P_2{\cdot}$DNA complex, and K1 is the association equilibrium constant (as designated by assoc.).
4. Most metalloregulatory proteins are homodimers and likely only interact with dsDNA in this form (see Fig. 1a). However, the monomer–dimer equilibrium (defined by $K_{dimer}$) may come into play if the total protein monomer concentration is used in the experiment $[P]_{total} \leq 1/K_{dimer}$. In this case, the value of $K_{dimer}$ must be known under the same solution conditions from an independent experiment, e.g., analytical equilibrium ultracentrifugation (29, 53), and used as a fixed parameter in a fitting algorithm that explicitly incorporates this linkage as an additional line of script in the mechanism. We refer to this mechanism as the "dimer-linkage model":

   P + P <==> P2 : K1 assoc.

   P2 + D <==> P2D : K2 assoc.

   If $K_{dimer}$ is not known, one can use the same script file and set K1 assoc. to a very large value (e.g., $10^6$ $\mu M^{-1}$); this has the effect of assuming a nondissociable dimer in the fit. Both $K_1$ and $K_{dimer}$ cannot be simultaneously optimized because the two parameters are nearly infinitely inversely correlated (see Note 6). It is sometimes possible to detect linkage to the monomer–dimer equilibrium (although the extent of that linkage cannot be determined; see Note 5) because the binding curve will appear detectably sharper (more sigmoidal) than a binding curve that is not linked to the equilibrium, due to the fact that $[P]_{total} > 1/K_{dimer}$. This is in fact apparent in the binding curves shown in Fig. 4c–d. Note how the apo-$P_2$ binding curve is "sharper" than the $P_2{\cdot}Bi^{III}$ curve; this is entirely due to the fact that in the former case $[P]_{total} < 1/K_{dimer}$, while in the metallated complex, $[P]_{total} \approx 1/K_{dimer}$.

5. In more complex systems, such as titrating $Zn^{II}$ into a mixture of mag-fura-2 and protein (Fig. 3), the mechanism must include two equilibria that describes $Zn^{II}$ binding to each of the potential ligands.

$$P + Z <==> PZ : K1 \text{ assoc.}$$

$$M + Z <==> MZ : K2 \text{ assoc.}$$

In this example, P is the protein monomer concentration, Z is $Zn^{II}$, M is mag-fura-2, while PZ and MZ are the 1:1 $Zn^{II}$–protein monomer and mag-fura-$Zn^{II}$ complexes, respectively. When scripting this mechanism, the two response elements are M and MZ since PZ is optically transparent; K2 is directly determined from a background experiment (see Subheading 3.3.3). Therefore, the only variable is K1, which directly competes with K2 for the binding of $Zn^{II}$. We note that for most metalloregulatory proteins, metal complexes bridge subunits of dimers or tetramers (4); in this case, P monomer does not actually bind metal to any appreciable degree. As a result, this script has the effect of fitting for two *identical and independent* binding sites on $P_2$ or more generally $n$ such sites on a $n$-oligomeric sensor (10). A more detailed mechanism (or model) that takes these linkages into account can be used to detect negative cooperativity of binding metal ions to a homodimer (32, 44), and is shown below.

$$P + P <==> P2 : K1 \text{ assoc.}$$

$$P2 + Z <==> P2 \cdot Z : K2 \text{ assoc.}$$

$$P2 \cdot Z + Z <==> P2 \cdot Z2 : K3 \text{ assoc.}$$

$$M + Z <==> MZ : K4 \text{ assoc.}$$

In the case of negative homotropic cooperativity, $K3 < K2$. Note this scheme can be readily expanded to include the binding of additional metal ions.

## 4. Notes

1. Synthesis of DNA oligonucleotides on the 200-nmol scale is sufficient to generate adequate material for these experiments. We routinely further purify DNAs by denaturing PAGE followed by electroelution ($l > 20$ nts) or high-resolution anion exchange chromatography ($l \leq 20$ nts), and ethanol precipitation. In the case of denaturing PAGE-purified DNAs, complete removal of acrylamide and urea is ensured by a final reverse phase clean up step using prepacked C18 columns (Alltech) and elution with 50% methanol. Dry to completeness with a speedvac.

2. Some duplex DNA operator sequences are highly palindromic, in which case a ssDNA hairpin may be thermodynamically favored over dsDNA. Annealing under high salt concentration (0.5–1 M NaCl) promotes duplex formation. Additionally, increased strand concentration may be needed to favor the intermolecular complex formation. Note that rapid cooling should be avoided as this process favors hairpin formation.
3. Transfer/prepare the buffer in a 2–3-L round-bottomed vacuum flask (a reaction flask from Kontes works well for this purpose). Attach the flask to a dual line manifold with one dedicated vacuum line and the other attached to a cylinder of argon. Situate the flask on a magnetic stirring mechanism and stir under high vacuum for at least 1 h/L of buffer (2 h/L is recommended). Back-fill with argon for transfer to an anaerobic chamber. Note this method will lead to a small increase in buffer concentration as a result of unavoidable solvent evaporation.
4. Prepare the buffer in a vacuum flask (typically available up to 500 mL) leaving at least 1/3 of the flask volume empty. Submerge the flask into liquid nitrogen or an isopropanol–dry ice slurry until completely frozen. While still frozen, expose to a high vacuum for 10–20 min. Close the flask and warm until completely melted. Submersion in tepid water can help this process; however, caution is urged to avoid fracturing the glassware as a result of a rapid temperature change. Repeat this process three times followed by backfilling the flask with Argon for transfer to anaerobic chamber.
5. Note that Fig. 4 is plotted on a log [CadC monomer] scale which *visually* expands the range of total [protein] used in these experiments. Data fitting should done on a linear scale since a Gaussian distribution of error assumed by these methods does not scale with logarithmic functions.
6. The degree to which two parameters are correlated or anticorrelated is quantified in the cross-correlation matrix associated with any fitting mechanism or model. For example, in the dissociable dimer model discussed above, the cross-correlation coefficient for K1 ($K_{dimer}$) and K2 ($P_2$ DNA-binding affinity) is −0.89 which reflects nearly complete inverse correlation. This means that unique values of K1 and K2 cannot be extracted from the fit, since changing K1 by tenfold and reducing K2 by tenfold will give an imperceptible change in the "goodness of fit" (a $\chi^2$-value). A cross-correlation coefficient of 0 means that the two parameters are independent of one another, with a range of −1 to 1 (51).

## 5. Conclusions

The experimental approach described here outlines a general strategy to experimentally determine the allosteric coupling-free energy, $\Delta G_c$, for any ligand-modulated transcriptional regulator by measuring the DNA-binding affinity of apo-metallated forms of the protein or, alternatively, the metal binding affinity of apo vs. DNA-bound protein (Fig. 1) (9). The approach therefore takes advantage of the ability to measure defined equilibria independently under one set of solution conditions without complications from competing equilibria within a complex linkage scheme. A detailed discussion of how to conduct these experiments is presented, along with consideration of the particular challenges associated with the study of transition metal ions as allosteric ligands. It should be stressed that the nature of the solution conditions required by these titrations can have an enormous impact on the information content of these experiments, and in most cases, can only be determined empirically.

## Acknowledgements

The authors would like to thank Drs. Zhen Ma and Faith Jacobsen and other members of the Giedroc laboratory for comments on the manuscript. This work was supported by grant R01 GM042569 from the NIH.

### References

1. Monod J, Wyman J, Changeux J-P (1965) On the nature of allosteric transitions: a plausible model. J Mol Biol 12:88–118
2. Koshland DE, Nemethy G, Filmer D (1972) Comparison of experimental binding data and theoretical models in proteins containing subunits. Biochemistry 5:365–385
3. Giedroc DP, Arunkumar AI (2007) Metal sensor proteins: nature's metalloregulated allosteric switches. Dalton Trans 29:3107–3120
4. Ma Z, Jacobsen FE, Giedroc DP (2009) Coordination chemistry of bacterial metal transport and sensing. Chem Rev 109:4644–4681
5. Busenlehner LS, Pennella MA, Giedroc DP (2003) The SmtB/ArsR family of metalloregulatory transcriptional repressors: structural insights into prokaryotic metal resistance. FEMS Microbiol Rev 27:131–143
6. Hobman JL, Wilkie J, Brown NL (2005) A design for life: prokaryotic metal-binding MerR family regulators. Biometals 18:429–436
7. Brown NL, Stoyanov JV, Kidd SP, Hobman JL (2003) The MerR family of transcriptional regulators. FEMS Microbiol Rev 27:145–164
8. Reinhart GD (2004) Quantitative analysis and interpretation of allosteric behavior. Methods Enzymol 380:187–203
9. Grossoehme NE, Giedroc DP (2009) Energetics of allosteric negative coupling in the zinc sensor *S. aureus* CzrA. J Am Chem Soc 131: 17860–17870. doi: 10.1021/ja906131b
10. Ma Z., Cowart DM, Ward BP, Arnold RJ, DiMarchi RD, Zhang L et al (2009) Unnatural amino acid substitution as a probe of the allosteric coupling pathway in a mycobacterial Cu

(I) sensor. J Am Chem Soc 131:18044–18045. doi: 10.1021/ja908372b

11. Singh VK, Xiong A, Usgaard TR, Chakrabarti S, Deora R, Misra TK et al (1999) ZntR is an autoregulatory protein and negatively regulates the chromosomal zinc resistance operon znt of Staphylococcus aureus. Mol Microbiol 33:200–207
12. Xiong A, Jayaswal RK (1998) Molecular characterization of a chromosomal determinant conferring resistance to zinc and cobalt ions in Staphylococcus aureus. J Bacteriol 180:4024–4029
13. Sun Y, Wong MD, Rosen BP (2001) Role of cysteinyl residues in sensing Pb(II), Cd(II), and Zn(II) by the plasmid pI258 CadC repressor. J Biol Chem 276:14955–14960
14. Lee JW, Helmann JD (2007) Functional specialization within the Fur family of metalloregulators. Biometals 20:485–499
15. Andrews SC, Robinson AK, Rodriguez-Quinones F (2003) Bacterial iron homeostasis. FEMS Microbiol Rev 27:215–237
16. Stoll KE, Draper WE, Kliegman JI, Golynskiy MV, Brew-Appiah RA, Phillips RK et al (2009) Characterization and structure of the manganese-responsive transcriptional regulator ScaR. Biochemistry 48:10308–10320
17. Payne JC, ter Horst MA, Godwin HA (1999) Lead fingers: $Pb^{2+}$ binding to structural zinc-binding domains determined directly by monitoring lead-thiolate charge-transfer bands. J Am Chem Soc 121:6850–6855
18. Busenlehner LS, Weng TC, Penner-Hahn JE, Giedroc DP (2002) Elucidation of primary (a3N) and vestigial (a5) heavy metal-binding sites in *Staphylococcus aureus* pI258 CadC: evolutionary implications for metal ion selectivity of ArsR/SmtB metal sensor proteins. J Mol Biol 319:685–701
19. Busenlehner LS, Giedroc DP (2006) Kinetics of metal binding by the toxic metal-sensing transcriptional repressor Staphylococcus aureus pI258 CadC. J Inorg Biochem 100 (5–6):1024–1034
20. Magyar JS, Godwin HA (2003) Spectropotentiometric analysis of metal binding to structural zinc-binding sites: accounting quantitatively for pH and metal ion buffering effects. Anal Biochem 320:39–54
21. Good NE, Winget GD, Winter W, Connolly TN, Izawa S, Singh RMM (2002) Hydrogen ion buffers for biological research*. Biochemistry 5:467–477
22. Yu Q, Kandegedara A, Xu Y, Rorabacher DB (1997) Avoiding interferences from good's buffers: a contiguous series of noncomplexing tertiary amine buffers covering the entire range of pH 3–11. Anal Biochem 253:50–56
23. Hegetschweiler K, Saltman P (1986) Interaction of copper(II) with N-(2-hydroxyethyl) piperazine-N′-ethanesulfonic acid (HEPES). Inorg Chem 25:107–109
24. Record MT Jr, Ha JH, Fisher MA (1991) Analysis of equilibrium and kinetic measurements to determine thermodynamic origins of stability and specificity and mechanism of formation of site-specific complexes between proteins and helical DNA. Methods Enzymol 208:291–343
25. Chen X, Agarwal A, Giedroc DP (1998) Structural and functional heterogeneity among the zinc fingers of human MRE-binding transcription factor-1. Biochemistry 37:11152–11161
26. Ma Z, Cowart DM, Scott RA, Giedroc DP (2009) Molecular insights into the metal selectivity of the copper(I)-sensing repressor CsoR from *Bacillus subtilis.* Biochemistry 48:3325–3334
27. National Institutes of Standards and Technology (2003) NIST Standard Reference Database 46, Version 7.0
28. Eicken C, Pennella MA, Chen X, Koshlap KM, VanZile ML, Sacchettini JC et al (2003) A metal-ligand-mediated intersubunit allosteric switch in related SmtB/ArsR zinc sensor proteins. J Mol Biol 333:683–695
29. Busenlehner LS, Cosper NJ, Scott RA, Rosen BP, Wong MD, Giedroc DP (2001) Spectroscopic properties of the metalloregulatory Cd (II) and Pb(II) sites of *S. aureus* pI258 CadC. Biochemistry 40:4426–4436
30. Zheng M, Cooper DR, Grossoehme NE, Yu M, Hung LW, Cieslik M et al (2009) Structure of Thermotoga maritima TM0439: implications for the mechanism of bacterial GntR transcription regulators with Zn2 + -binding FCD domains. Acta Crystallogr D Biol Crystallogr 65:356–365
31. Cotton AW, Geoffrey M, Carlos A, Bochman M (1999) Advanced inorganic chemistry, 6th edn. Wiley, New York
32. Pennella MA, Arunkumar AI, Giedroc DP (2006) Individual metal ligands play distinct functional roles in the zinc sensor Staphylococcus aureus CzrA. J Mol Biol 356:1124–1136
33. Liu T, Golden JW, Giedroc DP (2005) A zinc (II)/lead(II)/cadmium(II)-inducible operon from the Cyanobacterium anabaena is regulated by AztR, an alpha3N ArsR/SmtB metalloregulator. Biochemistry 44:8673–8683
34. Frankel AD, Berg JM, Pabo CO (1987) Metal-dependent folding of a single zinc finger from transcription factor IIIA. Proc Natl Acad Sci U S A 84:4841–4845

35. VanZile ML, Cosper NJ, Scott RA, Giedroc DP (2000) The zinc metalloregulatory protein Synechococcus PCC7942 SmtB binds a single zinc ion per monomer with high affinity in a tetrahedral coordination geometry. Biochemistry 39:11818–11829
36. Walkup GK, Imperiali B (1997) Fluorescent chemosensors for divalent zinc based on zinc finger domains. Enhanced oxidative stability, metal binding affinity, and structural and functional characterization. J Am Chem Soc 119:3443–3450
37. Golynskiy MV, Gunderson WA, Hendrich MP, Cohen SM (2006) Metal binding studies and EPR spectroscopy of the manganese transport regulator MntR. Biochemistry 45:15359–15372
38. Jefferson JR, Hunt JB, Ginsburg A (1990) Characterization of indo-1 and quin-2 as spectroscopic probes for $Zn^{2+}$–protein interactions. Anal Biochem 187:328–336
39. VanZile ML, Chen X, Giedroc DP (2002) Structural characterization of distinct alpha3N and alpha5 metal sites in the cyanobacterial zinc sensor SmtB. Biochemistry 41:9765–9775
40. Liu T, Chen X, Ma Z, Shokes J, Hemmingsen L, Scott RA et al (2008) A Cu(I)-sensing ArsR family metal sensor protein with a relaxed metal selectivity profile. Biochemistry 47:10564–10575
41. VanZile ML, Chen X, Giedroc DP (2002) Allosteric negative regulation of smt O/P binding of the zinc sensor, SmtB, by metal ions: a coupled equilibrium analysis. Biochemistry 41:9776–9786
42. Lakowicz JR (1999) Principle of fluorescence spectroscopy. Kluwer Academic, New York
43. Arunkumar AI, Campanello GC, Giedroc DP (2009) Solution structure of a paradigm ArsR family zinc sensor in the DNA-bound state. Proc Natl Acad Sci U S A 106:18177–18182
44. Lee S, Arunkumar AI, Chen X, Giedroc DP (2006) Structural insights into homo- and heterotropic allosteric coupling in the zinc sensor *S. aureus* CzrA from covalently fused dimers. J Am Chem Soc 128:1937–1947
45. Chen X, Chu M, Giedroc DP (1999) MRE-Binding transcription factor-1: weak zinc-binding finger domains 5 and 6 modulate the structure, affinity, and specificity of the metal-response element complex. Biochemistry 38:12915–12925
46. Grossoehme NE, Li L, Keane SC, Liu P, Dann CE 3rd, Leibowitz JL et al (2009) Coronavirus N protein N-terminal domain (NTD) specifically binds the transcriptional regulatory sequence (TRS) and melts TRS-cTRS RNA duplexes. J Mol Biol 394:544–557
47. Busenlehner LS, Apuy JL, Giedroc DP (2002) Characterization of a metalloregulatory bismuth(III) site in *Staphylococcus aureus* pI258 CadC repressor. J Biol Inorg Chem 7:551–559
48. Ye J, Kandegedara A, Martin P, Rosen BP (2005) Crystal structure of the *Staphylococcus aureus* pI258 CadC Cd(II)/Pb(II)/Zn(II)-responsive repressor. J Bacteriol 187:4214–4221
49. Apuy JL, Busenlehner LS, Russell DH, Giedroc DP (2004) Ratiometric pulsed alkylation mass spectrometry as a probe of thiolate reactivity in different metalloderivatives of *Staphylococcus aureus* pI258 CadC. Biochemistry 43:3824–3834
50. Kuzmic P (1996) Program DYNAFIT for the analysis of enzyme kinetic data: application to HIV proteinase. Anal Biochem 237:260–273
51. Johnson ML, Faunt LM (1992) Parameter estimation by least-squares methods. Methods Enzymol 210:1–37
52. Lohman TM, Bujalowski W (1991) Thermodynamic methods for model-independent determination of equilibrium binding isotherms for protein–DNA interactions: Spectroscopic approaches to monitor binding. Methods Enzymol 208:258–290
53. Pennella MA, Shokes JE, Cosper NJ, Scott RA, Giedroc DP (2003) Structural elements of metal selectivity in metal sensor proteins. Proc Natl Acad Sci U S A 100:3713–3718
54. Cleland WW (1964) Dithiothreitol, a new protective reagent for SH groups. Biochemistry 3:480–482
55. Krezel A, Lesniak W, Jezowska-Bojczuk M, Mlynarz P, Brasuñ J, Kozlowski H et al (2001) Coordination of heavy metals by dithiothreitol, a commonly used thiol group protectant. J Inorg Biochem 84:77–88
56. Millis KK, Weaver KH, Rabenstein DL (1993) Oxidation/reduction potential of glutathione. J Org Chem 58:4144–4146
57. Mayhew SG (1978) The redox potential of dithionite and SO-2 from equilibrium reactions with flavodoxins, methyl viologen and hydrogen plus hydrogenase. Eur J Biochem 85:535–547

# Chapter 9

# Fluorescence-Based Biosensors

Maria Strianese, Maria Staiano, Giuseppe Ruggiero, Tullio Labella, Claudio Pellecchia, and Sabato D'Auria

## Abstract

The field of optical sensors has been a growing research area over the last three decades. A wide range of books and review articles has been published by experts in the field who have highlighted the advantages of optical sensing over other transduction methods. Fluorescence is by far the method most often applied and comes in a variety of schemes. Nowadays, one of the most common approaches in the field of optical biosensors is to combine the high sensitivity of fluorescence detection in combination with the high selectivity provided by ligand-binding proteins.

In this chapter we deal with reviewing our recent results on the implementation of fluorescence-based sensors for monitoring environmentally hazardous gas molecules (e.g. nitric oxide, hydrogen sulfide). Reflectivity-based sensors, fluorescence correlation spectroscopy-based (FCS) systems, and sensors relying on the enhanced fluorescence emission on silver island films (SIFs) coupled to the total internal reflection fluorescence mode (TIRF) for the detection of gliadin and other prolamines considered toxic for celiac patients are also discussed herein.

**Key words:** Ligand-binding proteins, Biosensors, Fluorescence, FRET, Celiac disease, Porous silicon, FCS, Silver island films, Total internal reflection fluorescence

## 1. Introduction

Biosensors and their associated techniques is a rapid-growing field which combines biochemistry, biology, chemistry, physics, electronics, and computer science (1, 2). The term "biosensor" appeared in the scientific literature in the late 1970s (3). The first biosensor was reported by Clark in 1956. IUPAC defined a biosensor as a specific type of chemical sensor comprising a biological or biologically derived recognition element either integrated within or intimately associated with a physicochemical transducer (1, 2, 4).

Aiming at a general classification of optical biosensors, they can be divided in two groups, biosensors of the catalytic type and biosensors of the affinity type (2). The catalytic biosensors make

Wlodek M. Bujalowski (ed.), *Spectroscopic Methods of Analysis: Methods and Protocols*, Methods in Molecular Biology, vol. 875, DOI 10.1007/978-1-61779-806-1_9, © Springer Science+Business Media New York 2012

use of biocomponents capable of recognizing biochemical species and transforming them into a product through a chemical reaction. The affinity biosensors exploit the specific capabilities of an analyte to bind the bio-recognition element.

Compactness, portability, high specificity, and sensitivity represent some reasons why biosensors are considered as having a high potential for replacing current analytical practices (5).

Fluorescence is by far the method most often applied and is the dominant analytical approach in a large variety of schemes in the fields of medical testing, biotechnology, and drug discovery (2, 5, 6). In its numerous variations, it has become the most powerful bioanalytical and diagnostic tool and—in terms of versatility—seems to be second only to NMR spectroscopy, but with an entirely different field of application. This development reflects the transition of fluorescence spectroscopy from a merely academic area of research into a highly practical tool and was paralleled by a gradual move of fluorescence spectroscopy from physics to chemistry and biology (2, 6). One of the reasons of the tremendous increase in the popularity of fluorescence spectroscopy is probably the fact that in the 1980s the first chemically synthesized fluorescence probes for specific analytes became available (5, 6). These early probes were designed so as to include in the same molecule both the specific affinity for the ligand and the capacity to change some intrinsic fluorescent property upon ligand binding. Some of these sensing fluorophores are relatively simple, as illustrated by quinoline probes which are collisionally quenched by chloride (7). However, the complexity of the sensors quickly increases if one requires analyte binding to cause a spectral change. The development, via chemical synthesis, of specific sensors for biochemically relevant analytes is even more challenging. In fact, it is difficult to imagine how one would design a fluorescent probe which specifically binds pyruvate or lactate, or creatinine. Even if suitable structures could be designed and synthesized, there is no guarantee that the final molecules would display adequate water solubility, suitable affinity constants, and a spectral change upon ligand binding. In recent years, however, we have observed significant progress in the synthesis of new chemical probes as fluorescence sensors (8–10). In addition, new spectroscopic methods have been developed that make use of new components like (diode) lasers and LEDs, fiber optics, fast imaging devices, data loggers, and intelligent software which also contributed to the high interdisciplinarity of fluorescence spectroscopy (2).

To circumvent some of the limitations connected with the use of fluorescent probes, modern biotechnology has resorted to the idea of using proteins and enzymes as components of sensors for biochemical analytes. The idea is to exploit the extremely wide range of selective affinities sculpted into the various proteins by biological evolution. The number of potential ligands specifically recognized by different proteins is very large and ranges from small molecules to macromolecules (including protein themselves) (1, 5).

The advantages of using proteins as components of biosensors are many and include relatively low costs in design and synthesis, the fact that proteins are, at least in general, soluble in water, and finally, with the progresses of molecular genetics, the possibility of improving/changing some of the properties of the proteins by genetic manipulation. Many of the ligands that are important in clinical medicine and in the food control industry are relatively small (MW up to 1,000 Da). In these cases the enzymes appear to be the class of proteins endowed with the highest specificity and affinity. Site-directed mutagenesis has allowed alterations in amino acid sequences, resulting in changes in protein-binding constants and insertion of new positions for reporter group labeling. Such amino acid changes have allowed the signal transduction of the binding event to be evaluated by using a variety of physical and chemical techniques (11, 12). In particular, optical methods of detection using fluorescence energy transfer, polarization, and solvent sensitivity have been shown to offer high signal-to-noise ratios and the potential to construct simple and robust devices (13, 14). As a result, the use of such methods has allowed for the development of highly sensitive optical protein biosensors for a variety of analytes, including amino acids, sugars, and metabolic byproducts. Other classes of proteins, such as receptors, transporters, antibodies, etc., often present lower specificity although they offer other advantages such as the fact that they can specifically recognize a wide range of much larger ligands. However, for a broad use of proteins as probes for the development of sensors there are still some problems to be addressed. Protein stability, which is one of the problems, has been improved using a variety of protocols in the preparation of the sensor (immobilization and/or cross linking of the proteins, addition of stabilizing agents, etc.). Finally the use of proteins for the development of sensors requires that conformational changes of the protein, occurring upon ligand binding, can be monitored quantitatively. Fluorescence detection and/or polarization measurements can in principle be used. However, the intrinsic fluorescence of proteins is often low and requires excitation light in the UV range of the spectrum. Thus to develop compact, light weight, portable sensors considerable attention has been given to the derivatization of proteins with high quantum yield fluorophores that can be excited by higher wavelength, lower energy light (5).

Concerning the above considerations, a further classification needs to be done when classifying optical biosensors: the difference between the label-free and label-bound sensors (15). The first exploits any changes that can occur in the intrinsic optical properties of the biomolecule as a result of its interaction with the target analyte. Such changes can occur in absorbance, emission, polarization, or luminescence decay time of a receptor. The label-bound sensors make use of optical dyes and probes of various kinds. This requires the biomolecules to be covalently labeled with a fluorescent probe but enables to shift the analytical wavelength to be monitored to almost any desired value (2, 15).

The 1990s, in turn, saw a tremendous progress in the areas of imaging and single molecule detection. Optical probing of individual binding events represents the ultimate level of sensitivity and it has been a long-standing goal of analytical methods. Recent advances in optical spectroscopy and microscopy have made it possible not only to detect and identify freely diffusing or immobilized molecules but to make spectroscopic measurements and monitor dynamic processes of single molecules as well (16). Furthermore, single-enzyme studies have revealed numerous hidden aspects of enzyme behavior (17, 18) and distinctive functions of individual molecules in real time.

The majority of existing single-molecule enzymatic assays are based on fluorescence, because of its high sensitivity and because a bright signal appears against a dark background. Other exciting novel areas include fluorescence correlation spectroscopy which enables the detection of single molecules, multiphoton excitation with its inherent advantages over conventional excitation, and fluorescence imaging. Nowadays, one of the most commercially successful application of fluorometry is in luminescence immunoassay, followed by the diverse applications of fluorescence activated cell sorting (FACS) and other studies on the function of cells (2).

Numerous kinds of biosensors do exist, but this chapter is confined to sensors where the information is gathered by the measurement of photons (rather than electrons as in the case of electrochemical sensors). More specifically, this chapter focuses on the implementation of affinity sensors based on fluorescence for monitoring environmentally hazardous gas molecules (19, 20) (e.g. hydrogen sulfide and nitric oxide). Reflectivity-based sensors, fluorescence correlation spectroscopy-based systems, and sensors relying on the enhanced fluorescence emission on silver island films (SIFs) coupled to the total internal reflection fluorescence mode (TIRF) for the detection of gliadin and other prolamines considered toxic for celiac patients (21–24) are also discussed herein.

## 2. Discussion

### *2.1. Detection of Environmentally Relevant Gas Molecules*

Our studies deal with the monitoring of hydrogen sulfide ($H_2S$), nitric oxide (NO), and oxygen ($O_2$).

$H_2S$ is a colorless, flammable gas with a characteristic odor of rotten eggs (25). It is considered as one of the most dangerous environmentally toxic and crude metabolic poisons. It is more toxic than hydrogen cyanide and exposure to as little as 300 ppm in air for about 30 min can be fatal in humans (26–28). Many natural sources produce $H_2S$, including volcanic gases, oil, natural gas, and coal reserves, sulfur springs, putrefying vegetable, and animal matter. Several industrial processes also generate $H_2S$ as a by-product. This gas is a particular hazard for workers in the oil drilling and refining industries (29). As a consequence, there is a pressing need for methods allowing NO and $H_2S$ detection in both aqueous and gaseous media.

NO is a ubiquitous by-product of high-temperature combustion (30) and one of the hazardous exhaust gases generated by motor vehicles (31). Emissions of this gas can cause environmental problems such as acid rain, greenhouse effects, destruction of the ozone layer, and air pollution (32).

A different key issue in environmental analysis is the monitoring of $O_2$ levels. Oxygen measurements provide an indispensable guide to the overall condition of the ecology. Nowadays, it is routine practice to detect oxygen levels continuously in the atmosphere and in water (33). To better understand dynamics and drivers of oxygen depletion there is a substantial interest in the development of new, superior techniques for oxygen detection (33).

#### 2.1.1. FRET-Based Sensors

The sensing scheme builds on the translation of the binding event occurring at the enzyme cofactor-binding site into a change in the emission of a fluorescent label covalently attached to the protein through Förster Resonance Energy Transfer (FRET). For engineering such biosensors, a FRET pair for which the emission of the donor fluorophore (the dye label used to functionalize the protein) overlaps with the excitation of the acceptor fluorophore (the enzyme cofactor) is required. In this way, the label acts as a passive "beacon" whose fluorescence is tuned by the presence (or absence) of the analyte. This principle has been successfully advanced in the past few years for a number of applications (34–40). To design an $H_2S$ sensor, we extended the same FRET-based approach to myoglobin from horse skeletal muscle (Mb), exploiting the fact that its optical characteristics are different in the presence and absence of $H_2S$ (19).

Specifically, while the absorption spectrum of Mb exhibits a characteristic Soret band at 409 nm and two less-intense bands centered at 503 nm and 636 nm, $H_2S$ addition shifts the Soret band to 421 nm and quenches the 503- and 636-nm bands leading to the appearance of three new bands centered at 543 nm, 581 nm, and 617 nm (41) (see Fig. 1a). This change of the absorption spectrum of Mb will strongly modulate the fluorescence properties of a FRET donor–acceptor pair, where the Fe site is the energy acceptor and a dye label is the fluorescent donor. The idea is that using a dye label emitting in the 550–570-nm range, when the

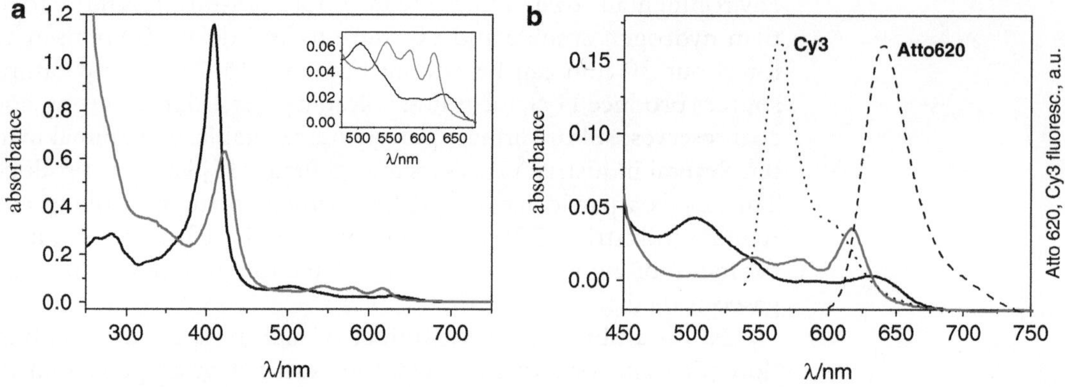


Fig. 1. (**a**) Absorption spectrum of $H_2S$-free (*black*) and $H_2S$-bound (*gray*) Mb. *Inset*: 500–800-nm region of the spectrum with enlarged vertical scale of the $H_2S$-free Mb (*black*) and of the $H_2S$-bound form (*gray*). (**b**) Absorption spectrum of the $H_2S$-free Mb (*dark line*) and of the $H_2S$-bound form (*gray*) and emission spectra of Cy3 ($\lambda_{max} = 570$ nm) and of Atto620 ($\lambda_{max} = 645$ nm). All spectra measured at room temperature. Protein concentration: 9 μM in 100 mM potassium phosphate buffer (pH = 6.8).

protein is in the $H_2S$-free state, the fluorescence of the dye is essentially uninhibited, whereas as soon as $H_2S$ binds to Mb, the fluorescence of the label is quenched as a result of energy transfer to either the 543-nm or the 581-nm band (depending on the emission of the dye label chosen for labeling Mb). In other words, the $H_2S$ binding to labeled Mb turns off the dye fluorescence. Differently, using a dye label emitting around 640 nm when the protein in the $H_2S$-free state, the fluorescence of the dye is initially quenched as a consequence of energy transfer to the 636-nm band, whereas as soon as $H_2S$ binds to Mb, the 636-nm band disappears and all the energy absorbed by the label is emitted as fluorescence. It follows that the fluorescent label acts as a sensitive reporter of the $H_2S$ bound to the iron center. The overlap of the Mb absorption bands ($H_2S$-free and $H_2S$-bound forms) with the emission spectrum of either Cy3 or Atto620 is shown in Fig. 1b.

To test the system, the fluorescence intensity of labeled Mb was monitored as a function of time during a change from an $H_2S$-saturated to an $H_2S$-free environment. Figure 2 shows a typical time trace of a solution containing 100 nM of dye-labeled Mb when excited at the absorption maximum ($\lambda = 550$ nm) of the dye (Cy3). In this particular experiment each cycle was started by adding $H_2S$ to an end concentration of 50 μM (i.e., in excess over the Mb concentration) and completed by passing argon through the solution for the complete removal of $H_2S$. The dye emission was followed at 570 nm.

A fluorescence quenching of label emission was clearly observed upon each $H_2S$ addition. When bubbling argon through the sample solution to displace the $H_2S$, the initial fluorescence intensity of the label was restored; the cycle could be repeated many

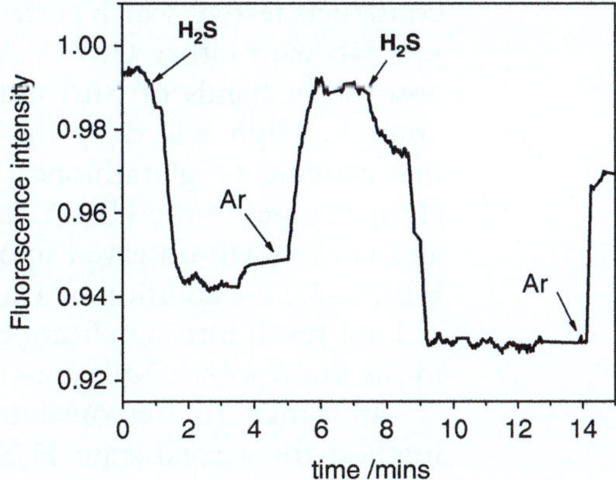


Fig. 2. Room temperature fluorescence intensity time trace observed at 570 nm (exc 550 nm) of Cy3-labeled Mb, upon addition and removal of $H_2S$. Protein concentration: 100 nM in 100 mM potassium phosphate buffer (pH = 6.8). In this particular experiment a standard solution of $(NH_4)_2$ S was used as $H_2S$ source. The $H_2S$ content of the $(NH_4)_2$ S standard solution was quoted by a coulombometric titration experiment.

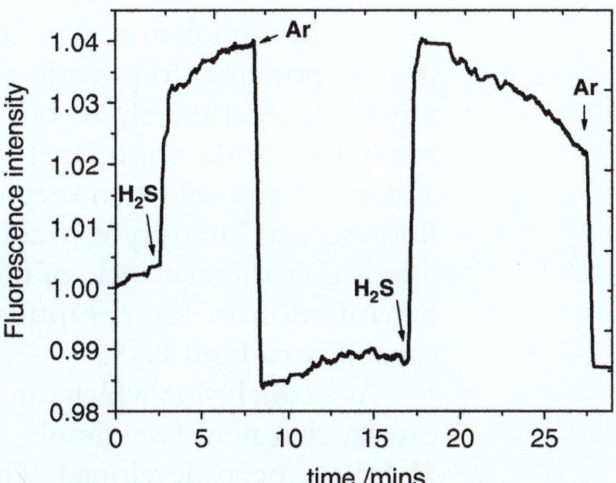


Fig. 3. Room temperature fluorescence intensity time trace observed at 645 nm (exc 620 nm) of Atto620-labeled Mb, upon addition and removal of $H_2S$. Protein concentration: 120 nM in 100 mM potassium phosphate buffer (pH = 6.8). In this particular experiment a standard solution of $(NH_4)_2$ S was used as $H_2S$ source. The $H_2S$ content of the $(NH_4)_2$ S standard solution was quoted by a coulombometric titration experiment.

times. This finding showed that, in the experimental conditions tested, the $H_2S$-binding process is reversible, which is crucial for practical sensing applications. When monitoring the Atto620 labeled Mb sensing construct under the same conditions, an opposite trend was observed (Fig. 3). An increase in label emission was observed upon each $H_2S$ addition. When displacing the $H_2S$ by bubbling through argon, the fluorescence intensity of the label diminished again.

Despite the relatively low amplitude observed for the fluorescence signals, the different trend observed for the two sensing constructs tested (which perfectly matches the fact that when labeling Mb with either Cy3 or Atto620 we are looking at different absorption bands of Mb) indicates that our method is sensitive enough. High selectivity for detecting $H_2S$ against other thiols like cysteine or glutathione (GSH) was found. Contrary to the clean changes in the UV-Vis and fluorescence spectra of unlabeled and labeled Mb observed upon addition of hydrogen sulfide (see Figs. 1–3), the addition of a large excess of either cysteine or GSH did not result into significant changes in the UV-Vis spectrum and in the fluorescence intensities of the system.

In order to demonstrate the potential of this Mb-based method for a solid-state $H_2S$-sensing device, we combined our approach with an immobilization technique. Mb labeled with either Cy3 or Atto620 was entrapped in a silica (TMOS) matrix and immobilized on a quartz support. Binding of $H_2S$ to the immobilized and entrapped Mb was found to be reversible: insertion of the Mb/Atto620 construct into an $H_2S$-containing solution resulted in a fluorescence increase. The bubbling of argon through the solution surrounding the sample drove the removal of $H_2S$ to completion. The immobilization of the sensing device has the potential to provide a more stable and, above all, reusable sensor. Unfortunately in our investigations the condition of the reusability could not be met. After the first complete cycle of $H_2S$ addition/removal when keeping on adding $H_2S$ to the system, the fluorescence intensity was not affected. Most likely the lack of an effect after the first cycle of the experiment could be due to partial denaturation of the encapsulated Mb and consequent loss of its capability to bind $H_2S$.

As a conclusion which can be pointed out from these results, we can say that new, fast, simple, and cost-effective sensing devices for $H_2S$ have been developed. The flexible mode of operation of the sensor not only in solution but also when applied in the form of solid-state devices, make it suited for sensing applications.

#### *2.1.2. Molecular Light-Gating Sensors*

The sensing scheme we have devised builds on the possibility of modulating the excitation intensity of a fluorescent probe used as a transducer and a sensor molecule whose absorption is strongly affected by the binding of an analyte of interest used as a filter (20). There are two simple conditions that have to be fulfilled for our method to work which are both equally important. The first one is that the absorption spectrum of the sensor placed inside the cuvette, and acting as the recognition element for the analyte of interest, should strongly change upon the binding of the analyte. The second one is that the fluorescence dye transducer should exhibit an excitation band which overlaps with one or more absorption bands of the sensor. The absorption band of the sensor affected by the binding of the specific analyte should overlap with the excitation band of the transducer.

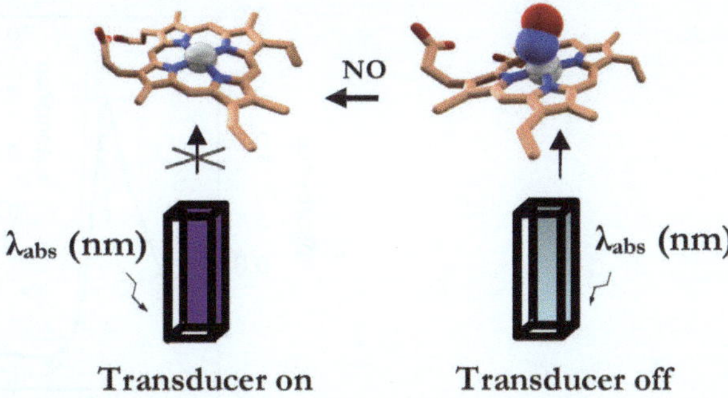


Fig. 4. Schematic representation of the sensing methodology. *Left*: Dye-labeled cuvette filled with CcP-NO free; *Right*: Dye-labeled cuvette filled with CcP-NO bound.

The proposed method uses a commercial cuvette filled with the sensing protein and whose external surface, downstream of the sensor on the light path, is uniformly covered with the fluorescence transducer (Fig. 4). The photodetector reads the fluorescence intensity emission of the transducer. The reading is proportional to the amount of excitation light which in turn depends on the amount of the analyte bound to the protein.

Herein we discuss our recent results obtained with two different pairs of sensor transducers: (1) cytochrome c peroxidase from baker's yeast (CcP) paired with fluorescein for detection of NO, and (2) myoglobin from horse skeletal muscle (Mb) paired with fluorescein for detection of oxygen.

Different than the FRET-based NO sensing device we have recently described (38), in the present case no direct labeling of the CcP was performed. The fluorescence intensity of the dye-label is now modulated by the different amounts of light absorbed by the CcP in the NO-free and NO-bound states. That is to say, the different absorption features of the CcP in the NO-free and NO-bound states change the amount of light reaching the dye transducer, thus tuning its fluorescence emission. In particular, upon NO binding to CcP, the Soret band shifts from 409 nm to 420 nm and its intensity increases (Fig. 5).

By choosing fluorescein (absorption maximum $\lambda = 490$ nm) as dye transducer to functionalize the cuvette, this change in absorption upon NO binding can be translated into a change of fluorescence intensity of the label. To test the system, the fluorescence intensity of the dye transducer (fluorescein, absorption maximum $\lambda = 490$ nm) was monitored during a change from an NO-free environment to an NO-saturated environment. Figure 6 shows a typical fluorescence trace of a solution containing 5 μM of CcP when excited at 450 nm. When adding NO to the system (in considerable excess over the CcP concentration) a 20% decrease of the dye-label fluorescence was detected (Fig. 6). This finding is in accordance with the enhancement of the intensity of the Soret

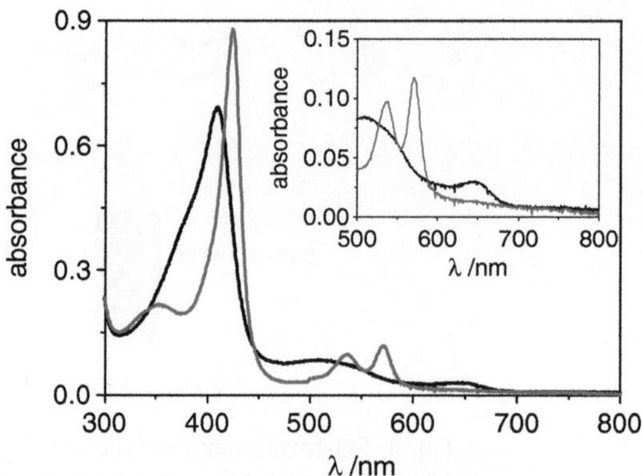


Fig. 5. Absorption spectrum of NO-free (*black*) and NO-bound (*gray*) CcP. *Inset*: 500–800 nm region of the spectrum with enlarged vertical scale showing the 645-nm high spin marker band of the NO-free CcP (*black*) and its disappearance in the NO-bound form (*gray*). All spectra measured at room temperature. Protein concentration: 8 μM in 100 mM potassium phosphate buffer (pH = 6.8).

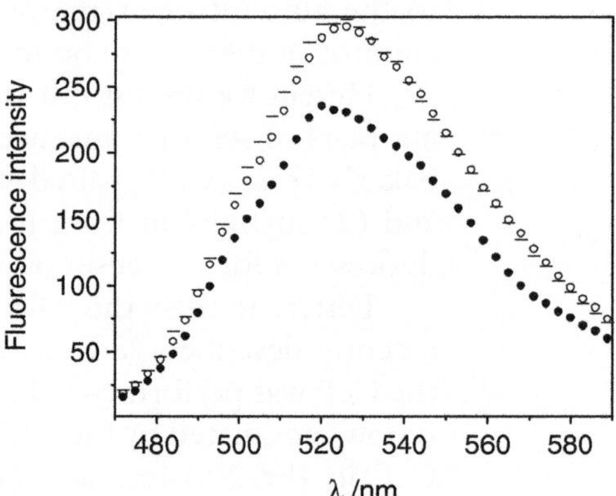


Fig. 6. Room temperature emission spectrum of CcP paired with fluorescein (exc 450 nm) (*dash* trace), upon addition (*filled circle* trace) and removal (*open circle* trace) of NO. Protein concentration: 5 μM in 100 mM potassium phosphate buffer (pH = 6.8). In this particular experiment the molar ratio NO to CcP was about 40 to 1.

band of the CcP in the NO-bound form (see Fig. 5) which leads the protein to absorb more photons. Subsequent bubbling of argon through the solution removed the NO again and made the fluorescence of the dye-label to go back to the initial level (Fig. 6). This showed that the NO-binding process is reversible and our system works successfully when applied for monitoring NO.

Aiming at implementing an oxygen-sensing device by exploiting the potential of our methodology, we considered myoglobin perfectly suited for proof-of-principle experiments. Myoglobin is an 18-kDa protein which is commonly acknowledged as an oxygen-storage protein. It reacts with molecular oxygen by a simple association

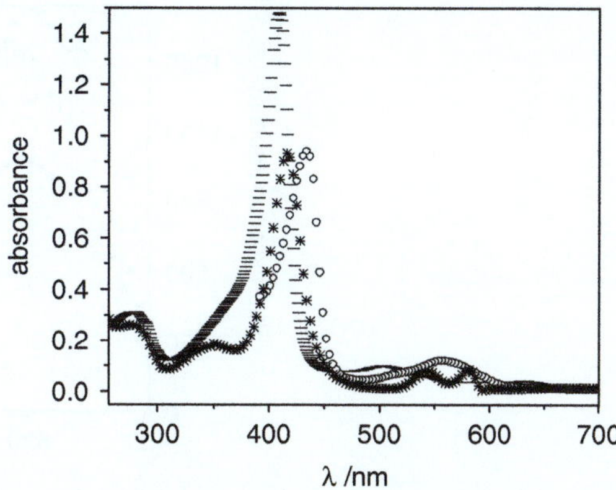


Fig. 7. Absorption spectrum of metMb(Fe3+) (*dash* trace), $O_2$-free($Fe^{2+}$)Mb (*open circle* trace) and $O_2$ bound($Fe^{2+}$)Mb (*asterisk* trace). All spectra measured at room temperature. Protein concentration: 10 μM in 100 mM potassium phosphate buffer (pH = 6.8).

reaction, most likely due to its monomeric form (42–44). Since $O_2$ has a strong preference for binding Mb in the $Fe^{2+}$ state, in all our experiments we first reduced Mb to the $Fe^{2+}$ form. Figure 7 illustrates the absorption spectrum of Mb in the metMb($Fe^{3+}$), oxygen-free($Fe^{2+}$), and oxygen-bound($Fe^{2+}$) states (45–47). When adding a 3-mM sodium dithionite excess over the Mb concentration, a clear shift of the Soret band to 434 nm and the appearance of a band centered around 560 nm occur. These bands are diagnostics of the oxygen-free($Fe^{2+}$)Mb form (45–47). Upon oxygen binding the Soret band shifts again (to 417 nm), whereas the 560-nm band is quenched and two new absorption bands centered at 545 and 580 nm appear (Fig. 7) (45–47).

This change of the absorption spectrum strongly modulates the fluorescence intensity of the dye transducer (fluorescein) used for functionalizing the cuvette. Figure 8 shows a typical fluorescence trace of a solution containing 10 μM of Mb when excited at 450 nm. Upon Mb reduction to the oxygen-free ($Fe^{2+}$) state, a 35% quenching of the dye fluorescence is observed as a result of the fact that more light is absorbed by the Mb (see Fig. 8). As soon as oxygen binds Mb the fluorescence of the label increases (40%), (see Fig. 8). In the oxygen-bound state the fluorescein fluorescence is essentially uninhibited since the Soret band of Mb is quenched if compared to metMb($Fe^{3+}$) and shifts its maximum intensity to 417 nm (Fig. 7). When adding $K_3Fe(CN)_6$ (0.1 mM) the oxygen-bound Mb($Fe^{2+}$) turns to the initial metMb($Fe^{3+}$) and the initial fluorescence intensity of fluorescein is restored (see Fig. 8).

It follows that the dye fluorescence is a sensitive reporter of the oxygen bound to the iron center. The cycle could be repeated several times. This finding showed that the $O_2$-binding process is reversible.

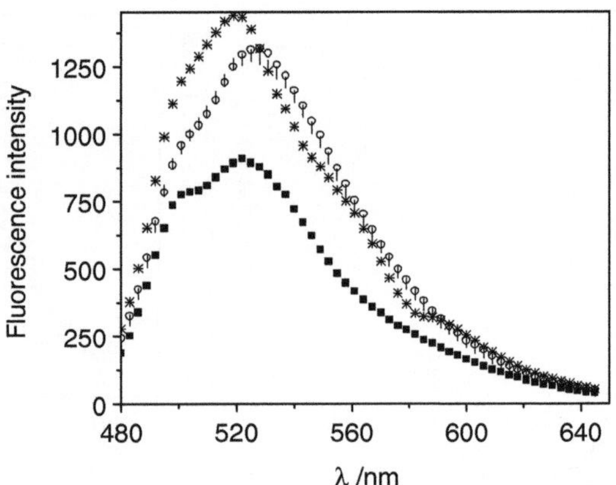


Fig. 8. Room temperature emission spectrum of Mb paired with fluorescein (exc 450 nm) (*open circle* trace), upon reduction to $O_2$-free($Fe^{2+}$)Mb (*filled square* trace), upon addition (*asterisk* trace) and removal (*vertical bar* trace) of $O_2$. Protein concentration: 10 μM in 100 mM potassium phosphate buffer (pH = 6.8).

As a conclusion for this molecular light-gating sensors section we can say that a notable advantage of the new methodology we are presenting herein is its broad applicability. In principle it is a sensitive way to follow any event of interest given that it affects the absorption spectrum of the molecule acting as the recognition element. Advantages like simplicity, rapidity in setting up the system and broad applicability make the developed device a possible alternative to the often time-consuming and expensive conventional screening assays used in the field of environmental analysis.

### 2.2. Detection of Gliadin and Related Toxic Prolamines

Celiac disease is an inflammatory disease of the small intestine affecting genetically susceptible people (48, 49). This pathogenesis is related to inappropriate intestinal T-cell activation in HLA-DQ2 and -DQ8 individuals triggered by peptides from wheat gliadin and related prolamines from barley and rye. At present, a strict gluten-free lifelong diet is mandatory for celiacs for both intestinal mucosal recovery and prevention of complicating conditions such as lymphoma and refractory sprue. However, dietary compliance has been shown to be poor in most patients mainly because of inadvertent gluten consumption and also because even very low amounts of gluten can trigger the disease.

Gluten proteins of wheat have been grouped into two classes according to their solubility: gliadins, soluble in aqueous alcohols, and insoluble glutenins. While gliadins occur in flour predominantly as monomers, glutenins are present in an aggregated state with molecular weights (mw) up to several millions. Gliadin proteins are primarily responsible of CD. As gliadins are a complex mixture of proteins difficult to solubilize and to extract from food, it is difficult to develop an assay capable of accurate quantization of gliadin in food for celiac patients. Reversed-phase HPLC

(RP-HPLC) can separate gliadin in more than 30 components according to their polarity. The amino acid compositions of single components together with the determination of N-terminal amino acid sequences indicated the existence of only three protein types in the gliadin fraction: ω-, α-, and γ-gliadins. The high content of proline residues makes gliadins very resistant to gastric, pancreatic, and intestinal proteases. The gastrointestinal stability of certain gluten peptides can have implications for their immunoactivity in CD (50).

On the basis of these data, to develop immunoassays able to exactly determine the content of gluten remains a hard task. More in particular, such a method should determine gluten and related toxic prolamines in a wide range of food, irrespective of processing, and should be directly related with toxicity. Until now, none of the produced methods are considered to be fully satisfactory.

#### 2.2.1. FRET-Based Sensing

Drawing upon these consideration we recently designed an optical assay for the detection of gliadin, a very well-characterized toxic prolamine (easily extendable to other related toxic prolamines) based on the *Escherichia coli* glutamine binding protein (GlnBP) (21). GlnBP is a monomeric protein that binds with high-affinity glutamine (gln) and poly-gln residues (51). Since gluten proteins are rich of gln residues, we questioned if GlnBP was able to bind amino acid sequences present in gluten proteins such as gliadin, a protein toxic for celiac patients. To check it, we labeled GlnBP and PT-gliadin with fluoresceine isothiocyanate and rhodamine isothiocyanate, respectively. The emission spectra of fluoresceine-labeled GlnBP alone and upon addition of rhodamine-labeled PT-gliadin showed the presence of FRET between fluorescein and rhodamine. The resonance energy transfer process observed upon the addition of rhodamine-labeled PT-gliadin indicates a close interaction between GlnBP and PT-gliadin. Figure 9 shows the effect of rhodamine-labeled PT-gliadin addition on fluoresceine-GlnBP.

Figure 10 shows the variation of the fluorescence intensity ratio ($I_{520\ \mathrm{nm}}/I_{572\ \mathrm{nm}}$) at different concentrations of rhodamine-PTgliadin. The obtained results indicate that the sensitivity of this assay is up to 1.0 μg of PT-gliadin addition; that means 33 nM if we consider the average molecular weight of gliadin before enzyme digestion (30,000 Da) and the volume of the FRET assay (1 mL).

Figure 10 also shows the effect of the unlabeled PT-gliadin on the fluorescein-GlnBP-rhodamine-PT-gliadin complex fluorescence emission. In the presence of 250 nmol of unlabeled PT-gliadin, a marked reduction of the FRET efficiency is observed as a consequence of the competition between unlabeled PTgliadin and rhodamine-labeled PT-gliadin. This result suggests the use of this assay for determining the presence of gliadin in food. In addition, the specificity of the FRET assay for PT-gliadin based on the utilization of GlnBP is confirmed by the fact that GlnBP does not

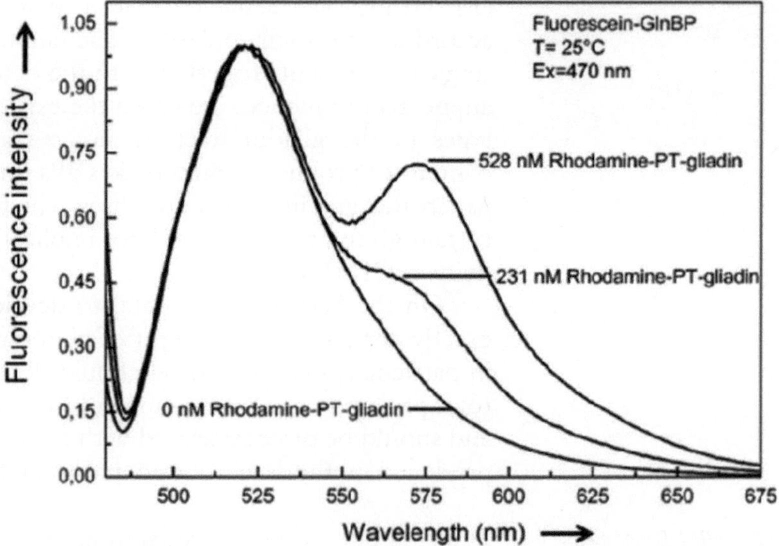


Fig. 9. Normalized fluorescence emission spectra at different concentrations of rhodamine-labeled PT-gliadin.

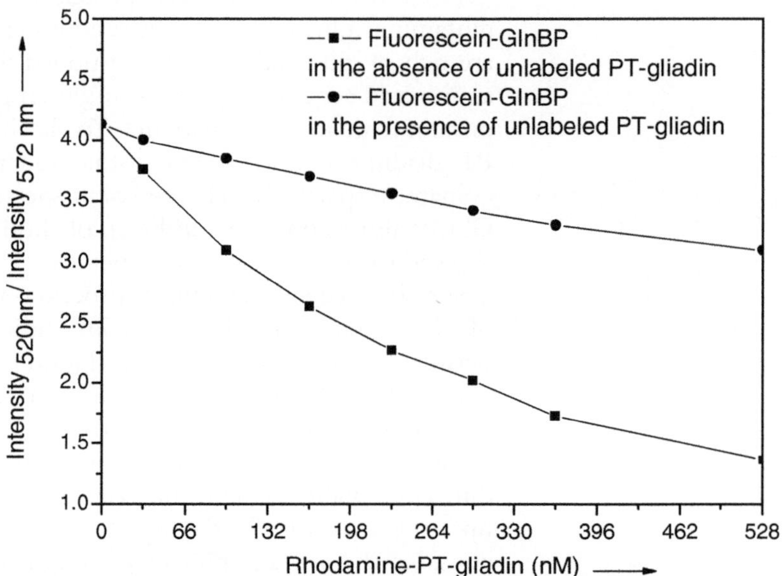


Fig. 10. Effect of rhodamine-labeled PT-gliadin on the emission of fluorescein-labeled GlnBP in the absence (*closed squares*) and in the presence (*closed circles*) of 250 nmol unlabeled PT-gliadin.

bind a peptic-tryptic digest of zein, the corresponding prolamine from corn that, together with rice, is a safe cereal for celiac patients. This result also indicates that the increased intensity of the acceptor molecule is due to energy transfer and it is not a side effect of the excitation of the increased concentration of the acceptor.

The rationale of the proposed strategy is that gluten proteins (gliadins and glutenins) are characterized by a high content of glutamine (gln) residues and, consequently, they represent a substrate for the GlnBP. As a consequence, the relevance of these results for the topic of gluten determination has been addressed.

#### 2.2.2. Reflectivity-Based Sensing

The sensing scheme builds on the possibility of following the ligand binding event by means of a protein linked to the surface of a porous silicon wafer (PSi) which acts as a transducer. Because of its spongelike structure, PSi is an almost ideal material to act as transducer. PSi optical sensors are based on changes of photoluminescence or reflectivity when exposed to the target analytes which substitute the air into the PSi pores (52). The effect depends on the chemical and physical properties of each analyte, so that the sensor can be used to recognize the pure substances. Because of the sensing mechanism, these kinds of devices are not suitable to identify the components of a complex mixture. To enhance the sensor selectivity through specific interactions, the most common approach is to chemically or physically modify the PSi surface. Generally, a covalent bond between the porous silicon surface and the biomolecules which specifically recognize the analytes under investigation is created (51, 53, 54). Among different probes of biological nature, ligand-binding proteins are considered particularly good candidates in designing highly specific biosensors for small analytes.

Recently we took advantage of the principles outlined above to implement a system for monitoring gliadin (Gli) in food (22). The system exploits the binding of the recombinant glutamine-binding protein (GlnBP) from *E. coli* to gliadin peptides (toxic for celiac patients). The GlnBP, which acts as the molecular probe for the gliadin in our assay, had been covalently linked to the surface of the porous PSi by a proper passivation process.

First, we immobilized GlnBP to a porous PSi and tested the strength of the covalent bond between the labeled GlnBP and the porous silicon surface by washing the chip in a demi-water flux. In the presence of gliadin PT, GlnBP undergoes a large conformational change in its global structure to accommodate the ligand inside the binding site. The ligand-binding event was detected as a fringe shift in wavelength, which corresponds to a change in the optical path $nd$. Since the thickness $d$ is fixed by the physical dimension of the PSi matrix, the variation is clearly due to changes in the average refractive index. In Fig. 11, the shifts induced in the fringes of the reflectivity spectrum are reported. A well-defined redshift of $2.6 \pm 0.1$ nm all over a wide range of wavelengths, due to protein–ligand interaction, was registered.

We also measured the signal response to the protein concentration after the ligand interaction. Figure 12 shows the dose-response curve as a function of the PT-gliadin concentration.

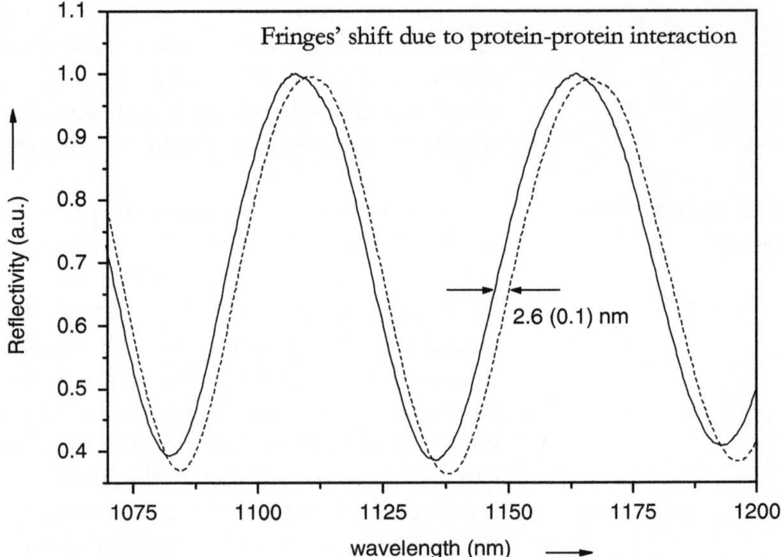


Fig. 11. Fringes shift due to protein–protein interaction. *Continuous curve*: porous silicon + GlnBP. *Dashed curve*: porous silicon + GlnBP + PT-gliadin. Wavelength shift between the continuous curve and the *dashed curve* is 2.6 (0.1) nm.

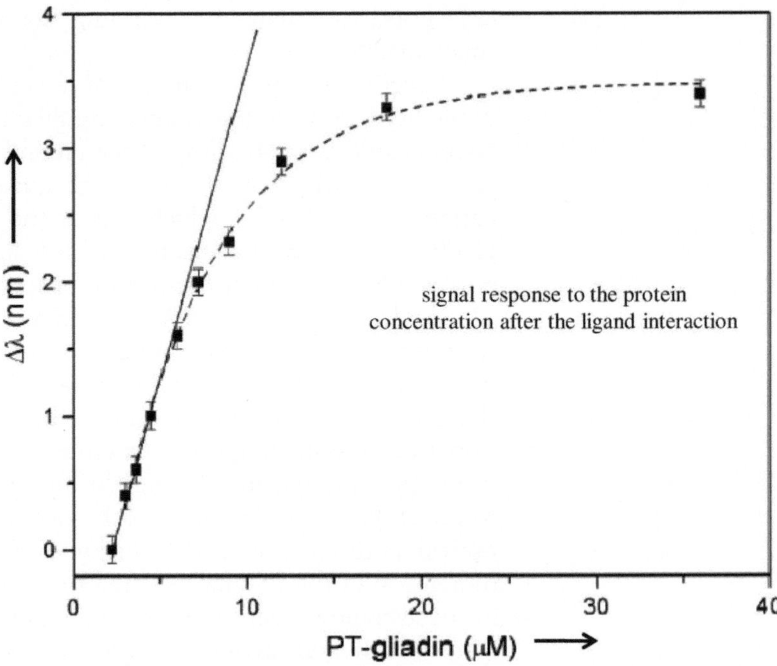


Fig. 12. Dose-response curve. Optical variations as a function of the PT-gliadin concentration.

Since the sensor displays a linear response between 2.0 and 8.0 μM, it is possible to calculate the sensitivity of the optical interferometer to the concentration of PT-gliadin by estimating the slope of the curve: $s_{Gli} = 421\ (13)\ \text{nm}/\mu\text{M}$. The sensor response saturates approximately at 35 μM, which means that about 45% of the spotted proteins have bound the respective peptide.

The rationale of the proposed strategy is that gluten proteins (gliadins and glutenins) are characterized by a high content of glutamine (Gln) residues and, consequently, they represent a substrate for the GlnBP. As a consequence, the relevance of these results for the topic of gluten determination has been addressed.

*2.2.3. Fluorescence Correlation Spectroscopy-Based Sensing*

We recently devised an assay based on the ability of the fluorescence correlation spectroscopy methodology to detect and to investigate the molecular properties of biomolecules at the level of a single-molecule for the detection of traces of gliadin in food.

Detection of a single molecule represents the ultimate level of sensitivity and it has been a long-standing goal of analytical methods. Because of its high sensitivity, and because a bright signal appears against a dark background, fluorescence is one obvious choice for single molecule detection. There are several methods for detection of single or small numbers of fluorophores. Fluorescence correlation spectroscopy (FCS) measures fluctuations in the small number of molecules in a focused laser beam. FCS is based on the analysis of time-dependent intensity fluctuations which are the result of some dynamic process, typically translation diffusion into and out of a small volume defined by a focused laser beam and a confocal aperture. When the fluorophore diffuses into a focused light beam, there is a burst of emitted photons due to multiple excitation–emission cycles from the same fluorophore. If the fluorophore diffuses rapidly out of the volume photon burst is short lived. If the fluorophore diffuses more slowly, the photon burst displays a longer duration. By correlation analysis of the time-dependent emission one can determine the diffusion coefficient of the fluorophore (55).

More specifically the assay builds on the measurement of the fluctuations of fluorescein-labeled gliadin peptides (GP) in a focused laser beam in the absence and in the presence of anti-GP antibodies (high-affinity antibodies raised in mice reared for several generations on a gluten-free diet). A competitive assay based on the utilization of unlabeled GP was developed. The study of interaction between fluorescein-labeled GP and unlabeled anti-GP antibodies was carried out in a three-step procedure. In particular, in the first step we studied the diffusion of the fluorescein-labeled-GP alone (10 nM). We registered a diffusion coefficient value of 120,935 $\mu\text{m}^2/\text{s}$. In the second step, we measured the diffusion coefficient of the fluorescein-labeled-GP upon addition of

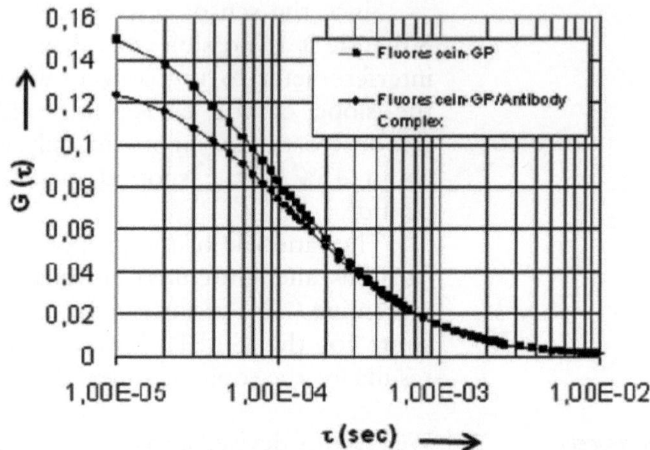


Fig. 13. Fluorescence correlation curves of fluorescein-labeled-GP (*square curve*), and fluorescein-labeled-GP/anti-GP antibodies complex (*circle curve*).

unlabeled-anti-GP mouse polyclonal antibodies. The formation of the complex fluorescein-labeled-GP/unlabeled-anti-GP Abs was detected as dramatic variation of the diffusion coefficient that passed from 120,935 μm$^2$/s to 50,307 μm$^2$/s. The results of the analysis are shown in Fig. 13 where the fluorescence correlation curve of fluorescein-labeled GP and the fluorescence correlation curve of fluorescein-labeled GP/anti-GP Abs complex are reported.

The last step of the assay was realized by the addition of unlabeled GP to the complex fluorescein-labeled-GP/unlabeled-anti-GP Abs. In particular, we were looking for a competition reaction between the fluorescein-labeled-GP and unlabeled-GP.

It appears that the addition of 0.006 ppm of un-labeled GP causes a large variation of the diffusion coefficient complex from 50,307 μm$^2$/s to 654,322 μm$^2$/s. Further additions of unlabeled GP results in a marked increase of the diffusion coefficient complex indicating a significant presence of free-labeled GP.

In conclusion, these results show that a new generation analytical tools to evaluate the presence of trace quantities of gluten in food has been developed. The combination of high-avidity IgG antibodies together with the innovative immunoassay strategy in the proposed system resulted in a gluten-detection limit of 0.006 ppm lower than 3.2 ppm recently reported for R5-ELISA (56) and much lower than the existing threshold of 20–200 ppm. Our method is fully compatible with the quantitative cocktail extraction procedure for heat-processed foods (55).

#### 2.2.4. Nanostructured Silver Surfaces-Based Sensing

The method builds on utilizing enhanced fluorescence emission on silver island films (SIFs) coupled to the TIRF. The process of fluorescence enhancement near silver nanoparticles has been

Fig. 14. Schematic representation of EFLISA methodology for the detection of gluten.

known for several decades (57). There are two general factors responsible for the enhancement. The first is an enhanced local field generated near metallic nanoparticles. The second factor is an interaction of the excited molecule with metallic nanoparticles, the effect known as radiative decay engineering (RDE) (58). The rapid transfer of the excitation to the metallic nanoparticle is followed by the far field radiation. This effect increases molecules' brightness and decreases the lifetime (this is only possible if the radiative rate of deactivation increases). The total enhancement is a product of these two effects, the enhanced local field and RDE. We recently devised a system to detect the presence of gliadin (Gli) by combining the method of enhanced fluorescence on silver island films (SIFs) to the total internal reflection mode (TIRF) (24).

High-affinity antibodies were raised in mice reared for several generations on a gluten-free diet. Under these conditions, mice do not develop immunological tolerance to the food antigen gliadin; consequently, parenteral immunization elicits a much higher specific IgG titer. The methodology is based on the utilization of silver islands coupled to fluorescence emission, namely, enhanced fluorescence linked immunosorbent assay (EFLISA) (Fig. 14) (24).

Our assay was performed as follows: (1) gliadin was first captured on surfaces coated with anti-Gli antibodies; (2) the surfaces were then incubated with fluorophore-labeled anti-Gli antibodies; (3) the signal from the fluorophore labeled anti-Gli antibody bound to the antigen was detected by TIRF. The system was examined on glass surfaces and on SIFs. We observed a relevant enhancement of the signal from SIFs compared to the signal from the glass substrate not modified with a SIF.

Our goal was to compare the gliadin immunoassay utilizing the TIRF detection on SIF-coated surface to the control immunoassay at identical conditions (same reagents, incubation times and temperatures, and detection method) and to demonstrate the signal enhancement due to the SIF coating. We used glass slides because it is very difficult to get stable uniform SIF coating inside the 96-well plate.

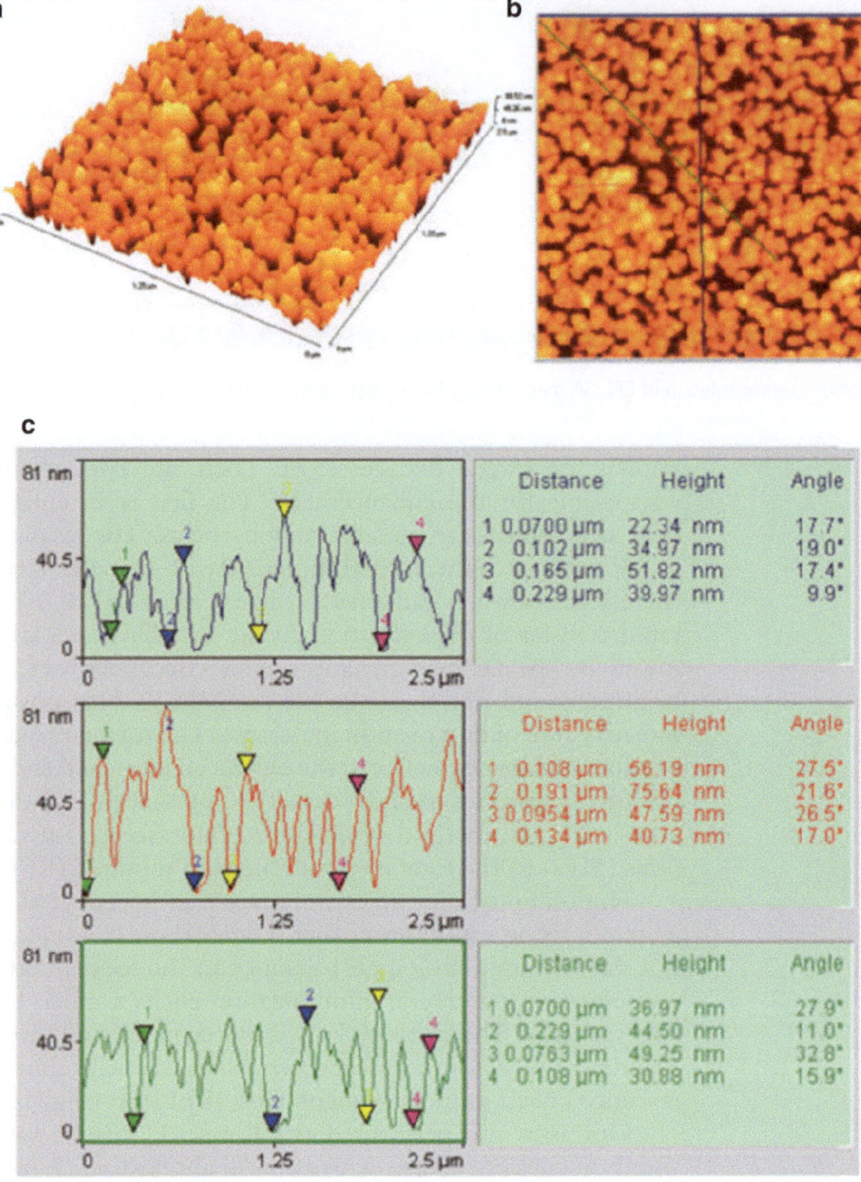


Fig. 15. AFM surface topography of a SIF-coated glass slide: (**a**) 3D image (25 × 25 $\mu m^2$); (**b**) same 2D image showing three profile lines; (**c**) profile analysis data.

Figure 15 presents an example of an AFM image of an SIF-coated slide. As we can see from the image (Fig. 15a), silver islands vary in height and width even within a small AFM scan area (25 × 25 µm). According to the profile analysis (59), the height of the silver islands varies between 25 and 30 and 80 nm, and the width varies between 140 and 300 nm results (first top, second middle, and third bottom profiles from Fig. 15c correspond to vertical blue, horizontal red, and inclined green lines from Fig. 15b).

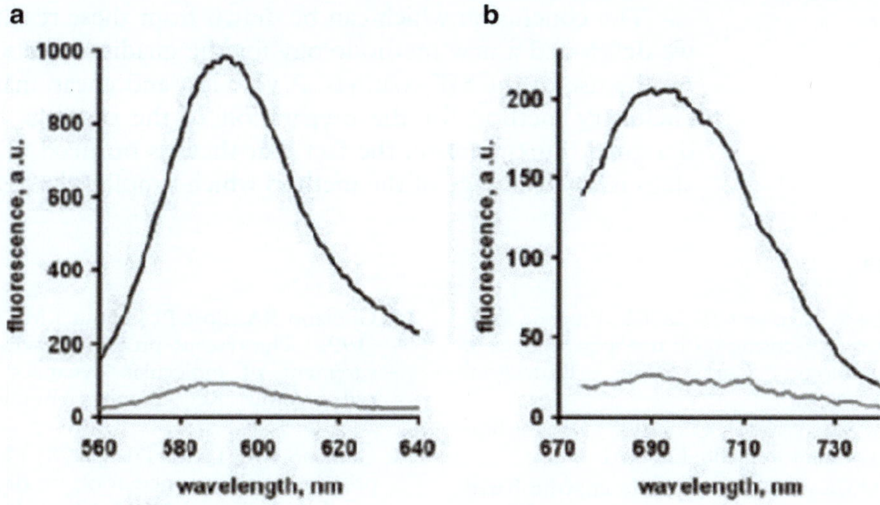


Fig. 16. Gliadin immunoassay response: emission spectra of (**a**) Rhodamine anti-gliadin antibodies and (**b**) Seta anti-gliadin antibodies captured by Gli on the surface in the absence (*gray traces*) and in the presence (*black traces*) of silver islands.

Typical examples of the fluorescence spectra from the assay samples in a sandwich format collected from SIF-modified or non-modified glass substrates are presented in panels *a* and *b* in Fig. 16.

Spectra represent labeled anti-Gli antibodies bound to the surface immobilized antigen (Gli bound to the capture anti-Gli Ab). For both conjugates (Rhodamine Red-X labeled or Seta-670 labeled), there is enhancement of the signal on SIF surface, compared to the bare glass surface. This enhancement may depend on numerous factors, such as the density of the SIF, the type of the fluorophore, the wavelengths of the excitation and emission (60), and in our case of the gliadin sandwich assay, it may also depend on the Gli concentration. Because of large variation in size of the SIF nanoparticles (even at similar average optical density of the SIF-coated slide), we performed averaging in order to calculate the immunoassay signal. The sensitivity of the assay is enhanced by a factor of about five to ten times when the assay is performed in the presence of SIFs. The enhancement factor depends on the SIF surface properties, such as optical density, size, shape, and uniformity of the particles.

In addition, the estimated detection limit (EDL) of this methodology is 60 ng/mL (or lower) that represents the clinical cut-off for Gli presence in food for celiac patients (61).

Clearly, these manually prepared SIF coatings, as used in this research, cannot be used as is for commercial devices without a drastic improvement in uniformity of the particles (which should result in much smaller deviations). Highly uniform coatings similar to SIFs are currently not commercially available; such coatings can be custom-ordered from nanolithography facilities.

The conclusion which can be drawn from these results is that we developed a new methodology for the gliadin detection based on the use of the SIF coatings. A very fast and cheap manual wet chemistry method for the preparation of the coatings has been designed. Furthermore, the fact that there is no need of washing steps is an advantage of the method which simplify the test.

## References


1. D'Auria S, Lakowicz JR (2001) Enzyme fluorescence as a sensing tool: new perspectives in biotechnology. Curr Opin Biotechnol 12:99–104
2. Borisov SM, Wolfbeis OS (2008) Optical biosensors. Chem Rev 108:423–461
3. Choi MMF (2004) Progress in enzyme-based biosensors using optical transducers. Microchim Acta 148:107–132
4. McDonagh C, Burke CS, Maccraith BD (2008) Optical chemical sensors. Chem Rev 108:400–422
5. Staiano M, Bazzicalupo P, Rossi M, D'Auria S (2005) Glucose biosensors as models for the development of advanced protein-based biosensors. Mol Biosyst 1:354–362
6. Wolfbeis OS (2000) Fiber-optic chemical sensors and biosensors. Anal Chem 72:81R–89R
7. Verkman AS, Sellers MC, Chao AC, Leung T, Ketcham R (1989) Synthesis and characterization of improved chloride-sensitive fluorescent indicators for biological applications. Anal Biochem 178:355–361
8. Rudat B, Birtalan E, Thome I, Kolmel DK, Horhoiu VL, Wissert MD, Lemmer U, Eisler HJ, Balaban TS, Brase S (2010) Novel pyridinium dyes that enable investigations of peptoids at the single-molecule level. J Phys Chem B 114:13473–13480
9. Popov AV, Mawn TM, Kim S, Zheng G, Delikatny EJ (2010) Design and synthesis of phospholipase C and A(2)-activatable near-infrared fluorescent smart probes. Bioconjug Chem 21:1724–1727
10. Nandhikonda P, Begaye MP, Cao Z, Heagy MD (2010) Frontier molecular orbital analysis of dual fluorescent dyes: predicting two-color emission in N-aryl-1,8-naphthalimides. Org Biomol Chem 8:3195–3201
11. Ramsden JJ (1997) Optical biosensors. J Mol Recognit 10:109–120
12. Sorochinskii VV, Kurganov BI (1997) Biosensors for detecting organic compounds. 1. Sensors for amino acids, urea, alcohols, and organic acids (review). Appl Biochem Microbiol 33:515–529
13. Giuliano KA, Post PL, Hahn KM, Taylor DL (1995) Fluorescent protein biosensors: measurement of molecular dynamics in living cells. Annu Rev Biophys Biomol Struct 24:405–434
14. Giuliano KA, Taylor DL (1998) Fluorescent-protein biosensors: new tools for drug discovery. Trends Biotechnol 16:135–140
15. Fan XD, White IM, Shopoua SI, Zhu HY, Suter JD, Sun YZ (2008) Sensitive optical biosensors for unlabeled targets: a review. Anal Chim Acta 620:8–26
16. Sauer M (2003) Single-molecule-sensitive fluorescent sensors based on photoinduced intramolecular charge transfer. Angew Chem Int Ed 42:1790–1793
17. Nie S, Zare RN (1997) Optical detection of single molecules. Annu Rev Biophys Biomol Struct 26:567–596
18. Nie S, Chiu DT, Zare RN (1994) Probing individual molecules with confocal fluorescence microscopy. Science 266:1018–1021
19. Strianese M, De MF, Pellecchia C, Ruggiero G, D'Auria S (2011) Myoglobin as a new fluorescence probe to sense H(2)S. Protein Pept Lett 18:282–286
20. Strianese M, Varriale A, Staiano M, Pellecchia C, D'Auria S (2011) Absorption into fluorescence. A method to sense biologically relevant gas molecules. Nanoscale 3:298–302
21. Staiano M, Scognamiglio V, Mamone G, Rossi M, Parracino A, Rossi M, D'Auria S (2006) Glutamine-binding protein from *Escherichia coli* specifically binds a wheat gliadin peptide. 2. Resonance energy transfer studies suggest a new sensing approach for an easy detection of wheat gliadin. J Proteome Res 5:2083–2086
22. De Stefano L, Rossi M, Staiano M, Mamone G, Parracino A, Rotiroti L, Rendina I, Rossi M, D'Auria S (2006) Glutamine-binding protein from *Escherichia coli* specifically binds a wheat gliadin peptide allowing the design of a new porous silicon-based optical biosensor. J Proteome Res 5:1241–1245
23. Varriale A, Rossi M, Staiano M, Terpetschnig E, Barbieri B, Rossi M, D'Auria S (2007)

Fluorescence correlation spectroscopy assay for gliadin in food. Anal Chem 79:4687–4689

24. Staiano M, Matveeva EG, Rossi M, Crescenzo R, Gryczynski Z, Gryczynski I, Iozzino L, Akopova I, D'Auria S (2009) Nanostructured silver-based surfaces: new emergent methodologies for an easy detection of analytes. ACS Appl Mater Interfaces 1:2909–2916
25. Li L, Moore PK (2008) Putative biological roles of hydrogen sulfide in health and disease: a breath of not so fresh air? Trends Pharmacol Sci 29:84–90
26. Reiffenstein RJ, Hulbert WC, Roth SH (1992) Toxicology of hydrogen sulfide. Annu Rev Pharmacol Toxicol 32:109–134
27. Smith RP (1978) Hydrogen sulfide poisoning. Can Med Assoc J 118:775–776
28. Smith RP (2010) A short history of hydrogen sulfide. Am Sci 98:6–9
29. Choi MM, Hawkins P (2003) Development of an optical hydrogen sulphide sensor. Sens Actuators, B 90:211–215
30. Ball JC, Hurley MD, Straccia AM, Gierczak CA (1999) Thermal release of nitric oxide from ambient air and diesel particles. Environ Sci Technol 33:1175–1178
31. Dooly G, Fitzpatrick C, Lewis E (2007) Optical sensing of hazardous exhaust emissions using a UV based extrinsic sensor. Energy 33:657–666
32. Dooly G, Fitzpatrick C, Lewis E (2007) Hazardous exhaust gas monitoring using a deep UV based differential optical absorption spectroscopy (DOAS) system. J Phys Conf Ser 76 (012021)
33. Amao Y (2003) Probes and polymers for optical sensing of oxygen. Microchim Acta 143:1–12
34. Kuznetsova S, Zauner G, Schmauder R, Mayboroda OA, Deelder AM, Aartsma TJ, Canters GW (2006) A Forster-resonance-energy transfer-based method for fluorescence detection of the protein redox state. Anal Biochem 350:52–60
35. Zauner G, Lonardi E, Bubacco L, Aartsma TJ, Canters GW, Tepper AW (2007) Tryptophan-to-dye fluorescence energy transfer applied to oxygen sensing by using type-3 copper proteins. Chemistry 13:7085–7090
36. Zauner G, Strianese M, Bubacco L, Aartsma TJ, Tepper AW, Canters GW (2008) Type-3 copper proteins as biocompatible and reusable oxygen sensors. Inorg Chim Acta 361:1116–1121
37. Strianese M, Zauner G, Tepper AW, Bubacco L, Breukink E, Aartsma TJ, Canters GW, Tabares LC (2009) A protein-based oxygen biosensor for high-throughput monitoring of cell growth and cell viability. Anal Biochem 385:242–248
38. Strianese M, De Martino F, Pavone V, Lombardi A, Canters GW, Pellecchia C (2010) A FRET-based biosensor for NO detection. J Inorg Biochem 104:619–624
39. Kuznetsova S, Zauner G, Aartsma TJ, Engelkamp H, Hatzakis N, Rowan AE, Nolte RJ, Christianen PC, Canters GW (2008) The enzyme mechanism of nitrite reductase studied at single-molecule level. Proc Natl Acad Sci USA 105:3250–3255
40. Schmauder R, Alagaratnam S, Chan C, Schmidt T, Canters GW, Aartsma TJ (2005) Sensitive detection of the redox state of copper proteins using fluorescence. J Biol Inorg Chem 10:683–687
41. Nicholls P (1961) The formation and properties of sulphmyoglobin and sulphcatalase. Biochem J 81:374–383
42. Wang M-YR, Hoffman BM, Shire SJ, Gurd FRN (1979) Oxygen binding to myoglobins and their cobalt analogues. J Am Chem Soc 101:7394–7397
43. Jurgens KD, Papadopoulos S, Peters T, Gros G (2000) Myoglobin: just an oxygen store or also an oxygen transporter? News Physiol Sci 15:269–274
44. Wilson MT, Reeder BJ (2008) Oxygen-binding haem proteins. Exp Physiol 93:128–132
45. Chung KE, Lan EH, Davidsson MS, Dunn BS, Selverstone VJ, Zink JI (1995) Measurement of dissolved oxygen in water using glass-encapsulated myoglobin. Anal Chem 67:1505–1509
46. Millar SJ, Moss BW, Stevenson MH (2010) Some observations on the absorption spectra of various myoglobin derivatives found in meat. Meat Sci 42:272–288
47. Yamazaki I, Yokota KN, Shikama K (1964) Preparation of crystalline oxymyoglobin from horse heart. J Biol Chem 239:4151–4153
48. Maki M, Collin P (1997) Coeliac disease. Lancet 349:1755–1759
49. Tursi A, Giorgetti G, Brandimarte G, Rubino E, Lombardi D, Gasbarrini G (2001) Prevalence and clinical presentation of subclinical/silent celiac disease in adults: an analysis on a 12-year observation. Hepatogastroenterology 48:462–464
50. Shan L, Molberg O, Parrot I, Hausch F, Filiz F, Gray GM, Sollid LM, Khosla C (2002) Structural basis for gluten intolerance in celiac sprue. Science 297:2275–2279
51. D'Auria S, Scire A, Varriale A, Scognamiglio V, Staiano M, Ausili A, Marabotti A, Rossi M,

Tanfani F (2005) Binding of glutamine to glutamine-binding protein from *Escherichia coli* induces changes in protein structure and increases protein stability. Proteins 58:80–87

52. Canham L (1997) Properties of porous silicon. IEE Book, Inspec Publisher, London, UK, 4 (1):23–26
53. Lin VS, Motesharei K, Dancil KP, Sailor MJ, Ghadiri MR (1997) A porous silicon-based optical interferometric biosensor. Science 278:840–843
54. Dattelbaum JD, Lakowicz JR (2001) Optical determination of glutamine using a genetically engineered protein. Anal Biochem 291:89–95
55. Magde D, Elson EL, Webb WW (1974) Fluorescence correlation spectroscopy. II. An experimental realization. Biopolymers 13:29–61
56. Valdes I, Garcia E, Llorente M, Mendez E (2003) Innovative approach to low-level gluten determination in foods using a novel sandwich enzyme-linked immunosorbent assay protocol. Eur J Gastroenterol Hepatol 15:465–474
57. Das P, Metiu H (1985) Enhancement of molecular fluorescence and photochemistry by small metal particles. J Phys Chem 89:4680–4687
58. Lakowicz JR (1999) Principles of fluorescence spectroscopy, 2nd edn. Kluwer Academic/ Plenum Publisher, New York, USA
59. Lakowicz JR, Shen Y, D'Auria S, Malicka J, Fang J, Gryczynski Z, Gryczynski I (2002) Radiative decay engineering. 2. Effects of silver island films on fluorescence intensity, lifetimes, and resonance energy transfer. Anal Biochem 301:261–277
60. Matveeva E, Gryczynski Z, Malicka J, Gryczynski I, Lakowicz JR (2004) Metal-enhanced fluorescence immunoassays using total internal reflection and silver island-coated surfaces. Anal Biochem 334:303–311
61. Sorell L, Lopez JA, Valdes I, Alfonso P, Camafeita E, Acevedo B, Chirdo F, Gavilondo J, Mendez E (1998) An innovative sandwich ELISA system based on an antibody cocktail for gluten analysis. FEBS Lett 439:46–50

# Chapter 10

# Metal-Enhanced Immunoassays

**Ignacy Gryczynski, Rafal Luchowski, Evgenia G. Matveeva, Tanya Shtoyko, Pabak Sarkar, Julian Borejdo, Irina Akopova, and Zygmunt Gryczynski**

## Abstract

The surface-confined assay format is one of the most convenient detection formats used in many immunoassays. Fluorescence emission from monolayers of dyes requires a strong excitation and good detection system. Such samples are susceptible to artifacts due to background fluorescence from substrates. We demonstrate that using silver nanostructures deposited on the slide substrate can significantly enhance measured fluorescence, reduce unwanted background and increase photostability of the used probes. Using thin layers of polymer doped with fluorescein, we tested two nanostructures—silver island films (SIFs) deposited on glass slides and self-assembled colloidal structures (SACS) deposited on thin silver film. The SACS surfaces show extraordinary fluorescence enhancements: over 100-folds in hot spots. We applied these surfaces for enhanced Alexa488 model immunoassay.

**Key words:** Fluorescence enhancement, Alexa488, Silver nanoparticles, Immunoassay, Enhanced detection, Plasmonics

## 1. Introduction

### *1.1. Metal-Enhanced Fluorescence*

Discovery of surface-enhanced Raman scattering (SERS) (1, 2) revitalized Raman spectroscopy and resulted in many new applications of the technology. In contrast to SERS, observed enhancement of fluorescence signal is not significant and often controversial due to strong quenching in a close proximity to the metal surface.

Metal-enhanced fluorescence (MEF) was demonstrated already three decades ago (3–6). Recently, several groups reported fluorescence enhancements on various silvered surfaces, including silver island films (SIFs) (7, 8), deposited colloids (9, 10), photodeposited structures (11), and electron beam-deposited nanostructures (12). Although the enhancements were modest, usually in the order of tenfold as compared to many thousandfolds in SERS, the dependence

Wlodek M. Bujalowski (ed.), *Spectroscopic Methods of Analysis: Methods and Protocols*, Methods in Molecular Biology, vol. 875, DOI 10.1007/978-1-61779-806-1_10, © Springer Science+Business Media New York 2012

of the enhancement on the distance from the metallic surface to fluorophores has been established (9, 13). In general, the strongest enhancements are observed between 40 and 200 Å away from the metallic surface. At shorter distance the quenching dominates the fluorophore–metal interactions and at longer distances the enhancement gradually decreases. There are two enhancement effects: first is the enhanced local field, which is responsible for a higher excitation rate; and a second enhancement effect is due to an interaction of an excited molecule dipole with the nanoparticles, known as radiative decay engineering (RDE), which is responsible for a decrease of fluorescence lifetime. Total enhancement is the product of these two effects. The recent reviews (14, 15) contain more details on MEF.

### 1.2. Model Immunoassays

To demonstrate a metal-enhanced fluorescence immunoassay on the glass surface coated with SIF, we used a model immunoassay format shown in Scheme 1. A model antigen, mouse Immunoglobulin G (IgG), was immobilized on the surfaces (SIF-coated surface and glass surface used as a reference), and corresponding specific antibodies labeled with a fluorophore were allowed to bind to the antigen. We incubated samples with antibodies for about 1 h to complete the binding reaction. Then the non-bound labeled antibodies were removed, a buffer was added, and fluorescence signal was measured in front-face configuration.

The measured fluorescence signal was a result of a specific interaction. To verify this, we performed the immunoassay using the same labeled antibodies, but with a nonspecific antigen, rabbit IgG, immobilized on the surface instead of the specific antigen. When a nonspecific antigen was used for the model immunoassay, no significant increase of the signal (non-specific binding) was observed. For more information about fluorescent immunoassays we refer reader to the recent review (16).

We found that silver colloids deposited on metallic mirrors show stronger enhancements than those deposited on glass substrates (17, 18). It has been established that enhancements on sharp metallic edges are significantly more efficient (19–21). Recently, we reported exceptionally strong local enhancements on self-assembled colloidal structures (SACS) formed on metallic films (22).

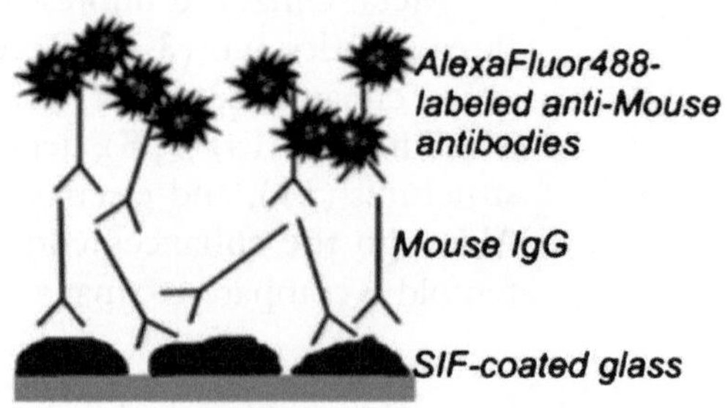


Scheme 1. Alexa488 model immunoassay.

This effect allows a significant reduction of the excitation intensity and effectively eliminates unwanted background from the regions away from the surface. The SACS looks similar to fractal-like structures with sharp edges. The interaction of local plasmons with metallic film results in strong local fields. A similar effect was recently described for a nanowire deposited on a metal film as an efficient platform for SERS (23).

In this chapter we will compare the fluorescence enhancements on SIFs and SACS. We will also emphasize the advantages of SACS in a model Alexa488 immunoassay. The immunoassays are ideal systems for observing strong fluorescence enhancements because the fluorescent probe is located at a close distance from the metallic surface.

## 2. Materials

### 2.1. PVA Films Doped with Fluorescein

Laser Grade Disodiumfluorescein was from Exction, Inc. Low molecular weight (13,000–23,000 MW) poly(vinyl alcohol) (PVA) was from Aldrich. Fluorescein-doped PVA samples were prepared on 1 × 1 in. microscope cover slips, either SIF coated or with SACS prepared on silver film.

### 2.2. Alexa488 Model Immunoassay

Mouse and rabbit IgGs, buffer components and salts (such as bovine serum albimun, sucrose) were from Sigma-Aldrich. Bocking solution was 1% bovine serum albumin, 1% sucrose, 0.05% $NaN_3$, 0.05% Tween-20 in 50 mM Na-phosphate buffer, pH 7.3. Goat-anti-mouse antibodies labeled with AlexaFluor-488 were from Invitrogen, Inc. (dye/protein ratio 7). Microscope glass slides, 3 × 1 in., and 1 mm thick, were from VWR. Water was purified by Milli-Q system.

### 2.3. SIFs and SACS

Trisodium citrate was supplied by Spectrum. Silver nitrate and glucose were from Sigma-Aldrich. Microscope slides were coated by EMF Corp. (Ithaca, NY). A 52-nm thick layer of silver was deposited on the slide with about 2-nm chromium undercoat. The silver films were protected with 5 nm layer of silica. Poly-L-lysine solution was freshly prepared solution: 8 ml water + 1.0 ml Na–phosphate buffer, 50 mM, pH 7.4, +1.0 ml poly-L-lysine solution (0.1%, Sigma).

## 3. Methods

### 3.1. Preparation of Fluorescein-Doped PVA Films

The 0.5% PVA solution containing disodiumfluorescein was spin-coated on slide substrates at 3,000 rpm (about 100 g). The thickness of the sample obtained with this preparation was about 20 nm (24).

### 3.2. Preparation of SIFs and SACS

#### 3.2.1. SIFs Preparation

SIFs-coated surfaces were formed by an incomplete silver mirror reaction as described earlier (25). Glass slides were cleaned by soaking in $H_2SO_4$ (98%) for 10–60 min and then were rinsed with Milli-Q water and dried on air before use. Slides were then coated with poly-L-lysine, with approximately 1.2 ml of the poly-L-lysine solution being added to each slide, incubated for 40–60 min and washed with water. Silver deposition was performed in a glass 100-ml beaker as follows. At intensive stirring, 20 drops of NaOH solution (5 M) were added to $AgNO_3$ solution (500 mg) in Milli-Q water (60 ml), and dark brown precipitate formed immediately. Approximately 2 ml of 30% $NH_4OH$ solution was added to dissolve the precipitate (at continuous intense stirring), and the clear solution was cooled with ice to 10–15°C (10 min). A fresh glucose solution (720 mg D(+)glucose in 15 ml Milli-Q water) was added to the mixture, and glass slides were immediately half-inserted into the mixture. Soaking of slides was performed for pairs of slides, so only one surface of each slide was exposed to the reaction mixture. Stirring of the mixture was continued in an ice bath for 2 min, and then ice was removed and the solution was stirred at warming (i.e., at medium heating) until 30°C for approximately 2 min. Then the heating was turned off, and the solution continued to be mixed intensively for an additional 3–4 min (the temperature increased to 35°C). After the color of the slides became greenish-brown and the solution became opaque, the slides were removed from the beaker and washed with water two times with sonication for approximately 25 s.

#### 3.2.2. SACS Preparation

Prior to SACS preparation, silver colloids were made as previously described (22). Briefly, all necessary glassware were soaked in a base bath overnight and washed scrupulously with deionized water. The solution of 0.18 mg/ml silver nitrate (200 ml) was heated and stirred in a 250 ml Erlenmeyer flask at 95°C. The first 0.5-ml aliquot of 34 mM trisodium citrate solution was added dropwise. The solution was stirred for 20 min and warmed to 96–98°C. Then five aliquots (0.7 ml each) of 34 mM trisodium citrate were added dropwise to the reaction mixture every 15–20 min. Stirring was continued for 25 min until the milky yellow color remained. Then the mixture was cooled in an ice bath for 15 min. After that silver colloids were used to prepare SACS as described earlier (22). Briefly, slide surfaces were cleaned and drop coated with silver colloids. The slides were air-dried to form SACS. The dry slides with self-assembled nanoparticles were stored and used within a month.

### 3.3. Alexa488 Model Immunoassay

Model immunoassay was performed on the slide surface as described earlier (25). Briefly, mouse IgG was non-covalently immobilized on the sample slide, or rabbit IgG on the control slide. Slides were covered with 20 μg/ml IgG solutions in 50 mM

Na-phosphate buffer (pH 7.3) and incubated at room temperature overnight. Next, slides were rinsed with water and covered with blocking solution for 2 h at room temperature. Then, after rinsing with water, AlexaFluor-488 labeled anti-mouse antibody conjugate (33 nM in blocking solution) was added to the sample slides (with mouse IgG) or control slides (with rabbit IgG) and after 1 h incubation and washing, slides were covered with 50 mM Na-phosphate buffer (pH 7.3) and the fluorescence signal was measured.

### 3.4. Atomic Force Microscopy

Atomic Force Microscopy (AFM) was performed on an Explorer Scanning Probe Microscope (ThermoMicroscopes). Scanning was acquired in contact mode with Non-Conductive Silicone Nitride cantilever T: 0.59–0.61 μm (Veeco; MSCT-EXMT-A). Image data was processed and analyzed on Veeco DI SPMLab software.

### 3.5. Photographs

The photographs, showing difference in intensity of Alexa488 signal on glass, SIF, and SACS, were taken with Canon 300D® digital SLR camera. The illumination source was Ti: Sapphire laser (Mai Tai, Newport Corp, Irvine CA) coupled with SCG-800 photonic crystal fiber (Newport Corp, Irvine CA) for supercontinuum generation. 470-nm band pass filter was used to select appropriate wavelength from the supercontinuum and the photographs were taken through 495-nm long-pass filter. For all the photographs, exposure time of one-fourth of a second and aperture of "14" was used. All images were cropped equally to one-fourth of the original image using ZoomBrowser® software (Canon Inc.).

### 3.6. Fluorescence Measurements

Fluorescence spectra and lifetimes were measured on FluoTime200 fluorometer (PicoQuant, GmbH). This instrument is equipped with a monochromator and a microchannel plate photomultiplier (MCP) detector on the observation path. With a 470-nm laser pulsed laser diode (pulse duration 68 ps) excitation, this fluorometer is capable of resolving sub-nanosecond intensity decays.

### 3.7. Microscopy Measurements

The Time-Correlated Single Photon Counting (TCSPC) MicroTime 200 confocal system (PicoQuant, GmbH., Berlin, Germany) coupled to Olympus IX71 microscope was used for measuring fluorescence enhancements. The fluorescence was excited with pulses from picosecond laser diode 470 nm (LDH-P-C-470B) worked at 20 MHz repetition rate and detected by Photon Avalanche Diode (SPAD) Perkin-Elmer (SPCM-AQR-14) detector. The fluorescence was observed confocally from the sample spin-coated on cover slips. Special, non-fluorescing cover slips were used (Menzel-Glasser #1). Sample was placed on the microscope stage, and light was focused (Olympus water immersion objective NA1.2, 60× magnification) in the sample plane. The observation path was equipped with three long-pass filters 500 nm to block excitation light and count fluorescence of the dye only. Photons were counted by TCSPC PicoHarp 300 board. The data was stored in the time-tagged time-resolved mode, which allowed the recording of every detected photon with its individual timing and detection channel information. All measurements and data analysis were performed using SymPho Time Software v. 5.0.

### 3.8. Interpretation of Measurements

#### 3.8.1. Fluorescein-Doped PVA Films

The magnitude of a fluorescence enhancement depends on the distance of fluorophores from the metallic surface as well as on a quantum yield of the probe. The strongest dependence, however, is on the metallic surface structure. The most popular and easy for the preparation are SIFs. Figure 1 shows typical SIFs deposited on a glass substrate evaluated by AFM. The surface is relatively homogeneous. The SACS prepared on a silver film show a different surface morphology (Fig. 2). The silver elongated nanoparticles form fractal-like structures (Fig. 2, left), making the metallic surface very heterogeneous. The expectation is to observe on these surfaces non-uniform enhancements, in contrast to glass and SIFs surfaces.

Emission spectra of fluorescein in PVA spin-coated on glass, SIFs, and SACS substrates are shown in Fig. 3. The spectrum on SACS is not only the strongest but also shifted towards longer wavelengths. Such shift is expected in very strong fields (26) and has been already observed (22). The comparison of the brightness between samples on glass, SIFs, and SACS are shown in the Fig. 4. The photographs were taken at the same excitation and observation conditions.

Next, we collected confocal images from these samples (Fig. 4, top panels).

The fluorescence intensities on SIF and SACS substrates are much stronger than on the glass. Also, the fluorescence on SACS is

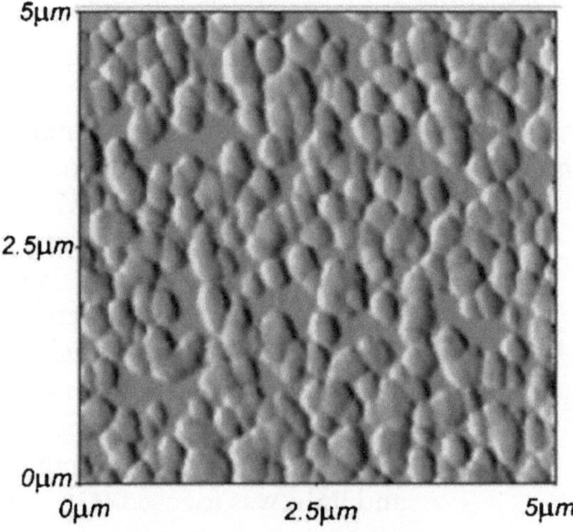


Fig. 1. AFM image of SIFs deposited on a glass substrate.

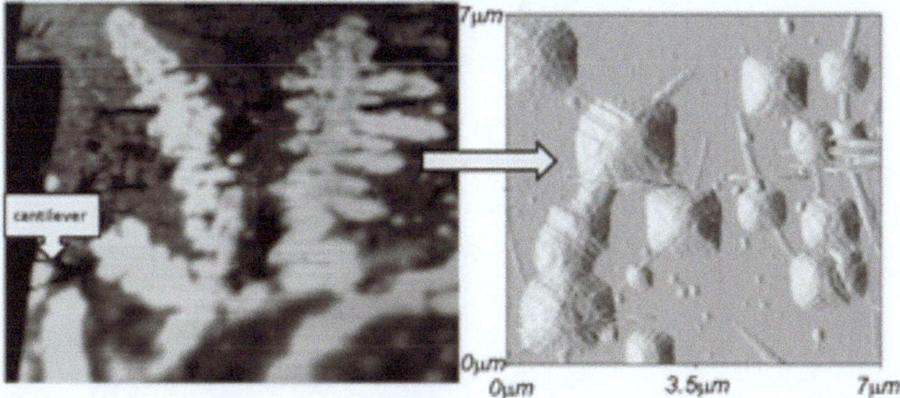


Fig. 2. SACS formed on a silver mirror. *Left*: Photograph of the fractal-like silver nanostructure taken from AFM monitor. *Right* AFM image of SACS.

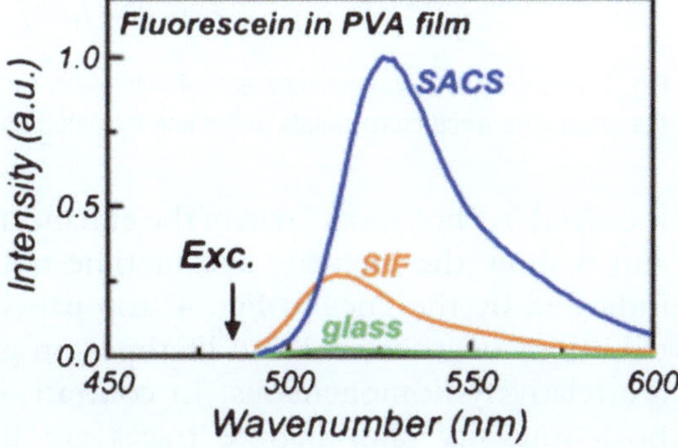


Fig. 3. Fluorescence emission spectra of fluorescein in PVA deposited on glass, SIFs and SACS. The layers (about 20 nm thickness) of PVA doped with fluorescein were spin-coated on the substrates.

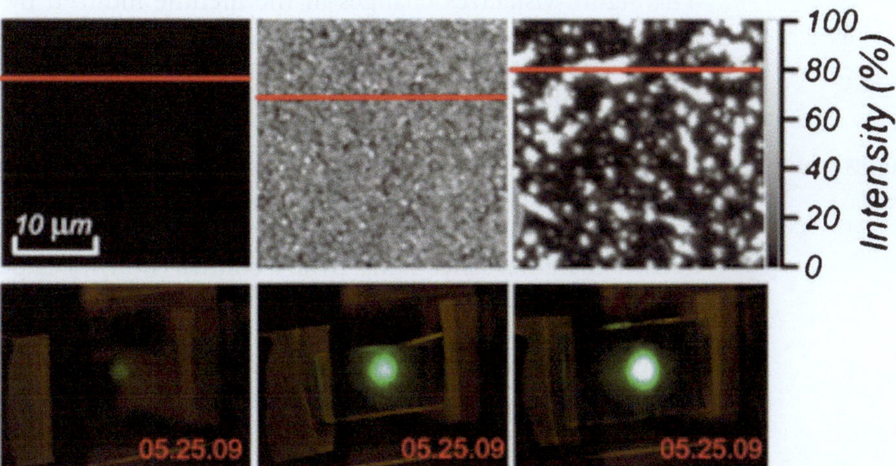


Fig. 4. *Top*: Confocal images collected for fluorescein-doped PVA thin layers on glass, SIFs, and SACS. *Bottom*: Photographs of these samples taken with a 480-nm laser excitation.

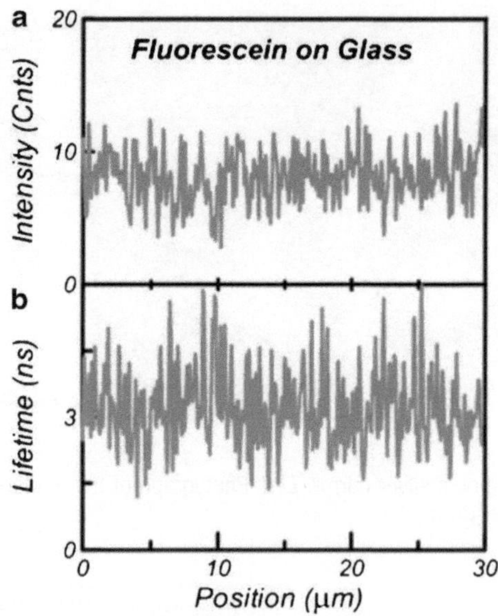


Fig. 5. Distribution of fluorescence-emission intensity for fluorescein in PVA deposited on the glass. The trace corresponds to the line indicated on Fig. 4, *top*.

localized in "hot spots" where the enhancement is higher. Figures 5 and 6 show the intensity and lifetime traces across the images as indicated by the lines in Fig. 4, top panels. As expected, the distributions of intensities and lifetimes on glass and SIFs substrates are relatively homogeneous. In contrast, on the SACS substrate, both intensity and lifetime traces are heterogeneous. Also, as expected, the stronger brightness corresponds to the shorter lifetime, see arrows on Fig. 6. This indicates that RDE effect plays an important role in the total enhancement. The lifetime images and distributions from entire $30 \times 30\ \mu m^2$ areas are shown in Fig. 7. This figure visualizes changes in the lifetime induced by metallic nanoparticles. More precise lifetime measurements are shown on Fig. 8. The amplitude averaged lifetimes are significantly shorter on metallic surfaces than on the glass.

#### *3.8.2. Alexa488 Model Immunoassay*

We performed a model immunoassay on glass and SACS substrates using an Alexa488 fluorescent probe. Emission spectra (Fig. 9) show more than tenfold enhancement on SACS compared to the glass. The fluorescence signal from a control (non-specific binding) was about 10% of the sample signal in both glass and SACS substrates. The distributions of intensities and lifetimes on SACS substrate are shown on Fig. 10. Again, higher intensity corresponds to shorter lifetime, and in the less bright spots lifetimes are longer. Although the average enhancement is not high (see Fig. 9), in the local "hot spots" (Fig. 10, top), the brightness is about 100-fold higher than in the absence of silver a nanostructure.

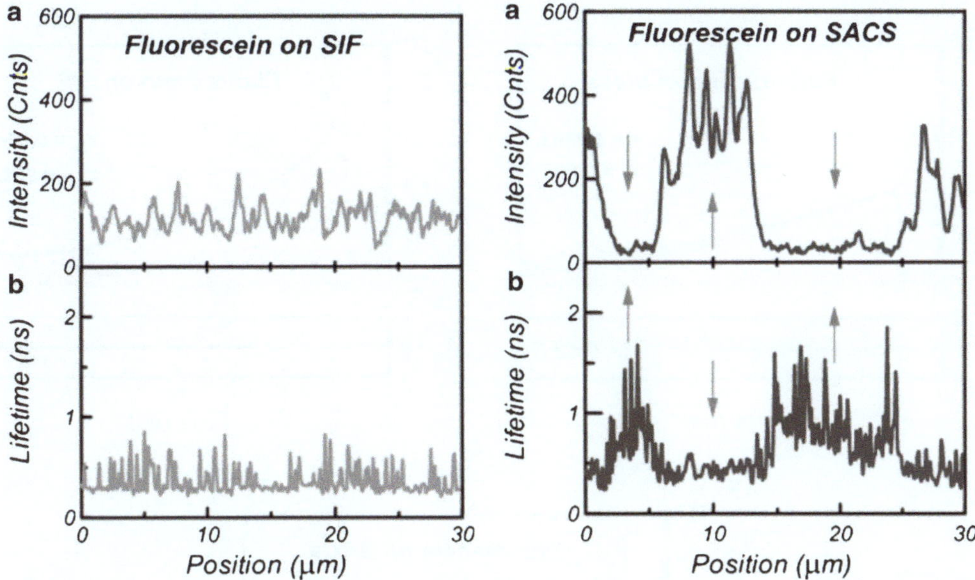


Fig. 6. Distributions of fluorescence-emission intensities for SIFs and SACS. The traces correspond to lines indicated on Fig. 4, *top*.

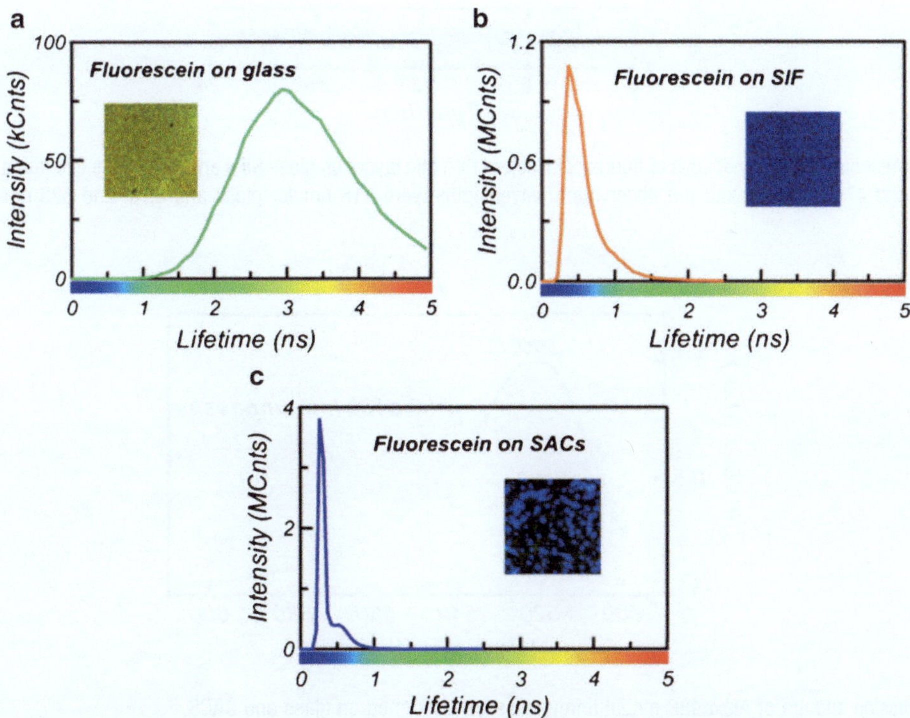


Fig. 7. The distribution of lifetimes measured for the fluorescein-doped PVA thin films on glass, SIFs, and SACS. The distribution corresponds to the *arrows* indicated on lifetime images inserted into figures.

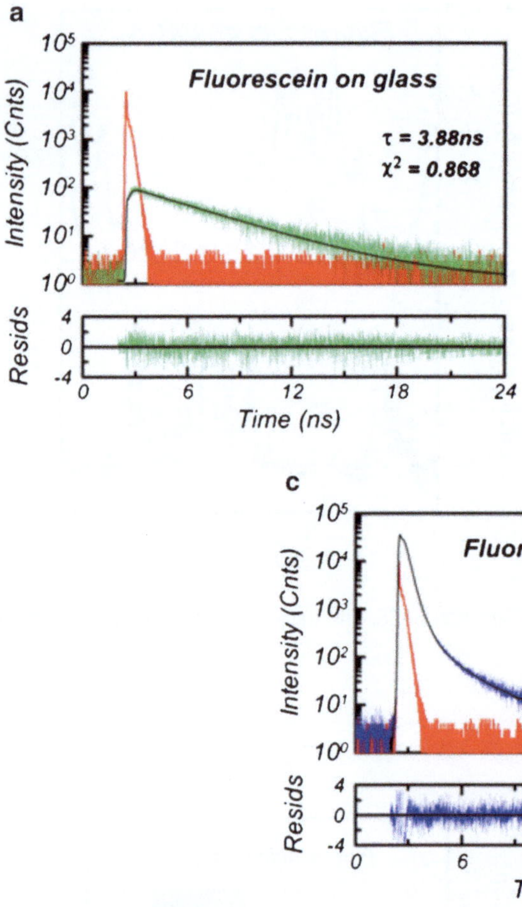

Fig. 8. Fluorescence intensity decays of fluorescein-doped PVA thin layers on glass SIFs and SACS. The excitation was from a picosecond 470-nm laser and the observation wavelengths were 515 nm for glass and SIFs, and 525 nm for SACS substrates.

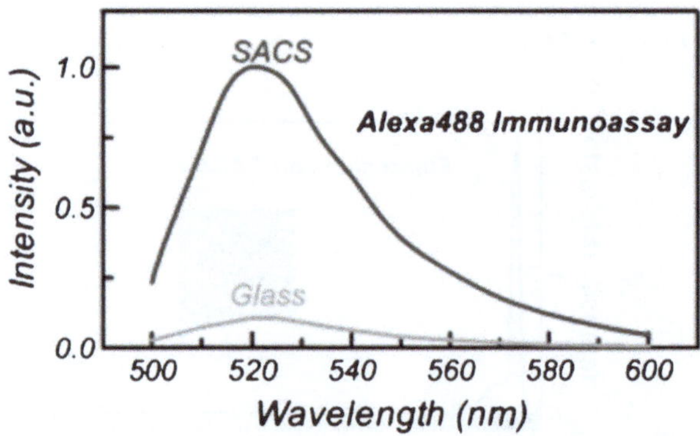


Fig. 9. Emission spectra of Alexa488 model immunoassays performed on glass and SACS.

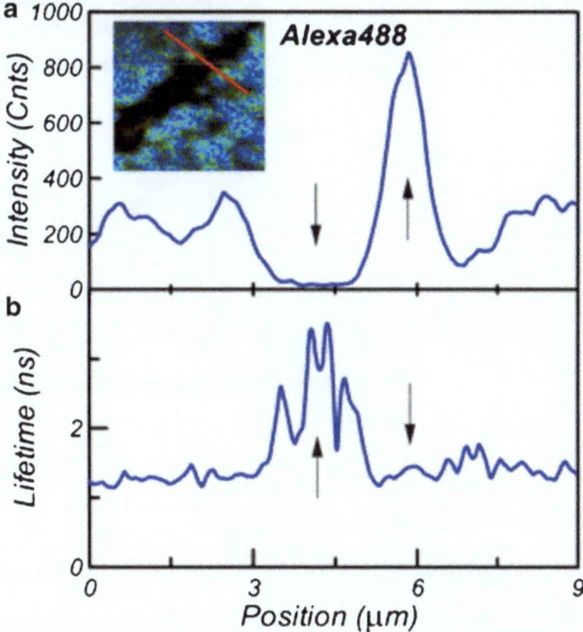


Fig. 10. Intensity (*top*) and lifetime (*bottom*) distributions of Alexa488 immunoassay performed on SACS. The traces correspond to the line indicated on the *inserted image*.

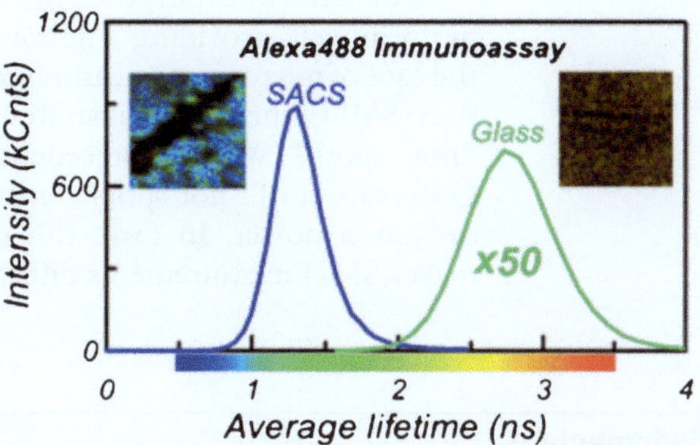


Fig. 11. Lifetime distribution of Alexa488 model immunoassays performed on glass and SACS.

The Alexa488 lifetime distributions on glass and SACS are shown in Fig. 11. The average lifetime on SACS is a few times shorter than on glass. This observation was confirmed in lifetime measurements with a high-resolution fluorometer (not shown). Finally, we compared the photostabilities of Alexa488 immunoassays on glass and SACS (Fig. 12). The shorter lifetimes are responsible for higher photostabilities on metallic surfaces. The photodegradation occurs mostly in the excited state, and near metallic nanostructure molecules can emit many photons before they are bleached.

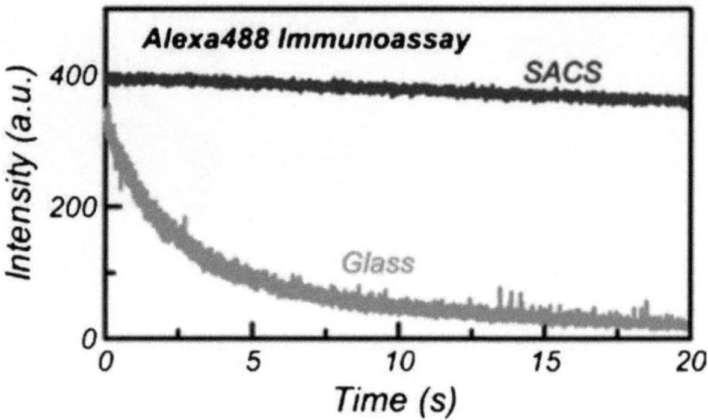


Fig. 12. Photostabilities measured for Alexa488 model immunoassays performed on glass and SACS.

## 4. Notes

### *4.1. Detection of Single Molecules on SACS*

The surface preparation is crucial in MEF and has to be done very carefully.

For detection purposes in macroscopic devices SIFs surfaces perform well, providing a uniform enhancement in large areas. In the case of microscopy measurements, like in single molecule detection (SMD), the SACS substrates are more useful because they offer "hot spots" with unprecedented fluorescence enhancements. Observation of "hot spots" enables a significant reduction of the excitation power. In turn, this reduces background which often makes SMD measurements difficult or impossible.

## Acknowledgments

This work was supported by Texas Emerging Technologies Fund Grant, NIH HG 004364 and R01AR048622, NSF DBI-0649889, and ARPATP Project 000130-0042-2007. RL is the recipient of the Research Mobility program from the Polish Ministry of Science and Higher Education. TS was supported by Research Corporation for Science Advancement (CCSA 7748) and The Welch Foundation (BP-0037).

## References


1. Fleischmann M, Hendra PJ, McQuillan AJ (1974) Raman spectra of pyridine adsorbed at a silver electrode. Chem Phys Lett 26:163–166
2. Jeanmaire DL, Van Duyne RP (1974) Surface Raman spectrochemistry: I. Heterocyclic, aromatic andaliphatic amines adsorbed on the anodized silver electrode. J Electroanal Chem 84:1–20
3. Philpott MR (1975) Effect of surface plasmons on transition in molecules. J Chem Phys 62:1812–1817
4. Weitz DA, Garoff S, Hanson CD, Gramila TJ (1982) Fluorescent lifetimes of molecules on silver-islands films. Opt Lett 7:89–91
5. Das P, Metju H (1985) Enhancement of molecular fluorescence and photochemistry by small metal particles. J Phys Chem 89: 4680–4687
6. Leitner A, Lippitsch ME, Draxler S, Riegler M, Aussenegg FR (1985) Fluorescence properties of dyes adsorbed to silver islands, investigated by picosecond techniques. Appl Phys B 36: 105–109
7. Tarcha PJ, DeSaja-Gonzales J, Rodrigues-Liorente S, Aroca R (1999) Surface-enhanced fluorescence on $SiO_2$-coated silver island films. Appl Spectrosc 53:43–48
8. Lakowicz JR, Shen B, D'Auria S, Malicka J, Fang J, Gryczynski Z, Gryczynski I (2002) Radiative decay engineering: II. Effects of silver island films on fluorescence intensity, lifetimes and resonance energy transfer. Anal Biochem 301:261–277
9. Sokolov K, Chumanov G, Cotton TM (1998) Enhancement of molecular fluorescence near the surface of metal films. Anal Chem 70: 3898–3905
10. Lukomska J, Malicka J, Gryczynski I, Lakowicz JR (2004) Fluorescence enhancements on silver colloid coated surfaces. J Fluoresc 14:417–423
11. Geddes CD, Parfenov A, Roll D, Fang J, Lakowicz JR (2003) Electrochemical and laser depositionof silver for use in metal-enhanced fluorescence. Langmuir 19:6236–6241
12. Corrigan TD, Guo S, Phaneuf RJ, Szmacinski H (2005) Enhanced fluorescence from periodic arrays of silver nanoparticles. J Fluoresc 15:777–784
13. Malicka J, Gryczynski I, Gryczynski Z, Lakowicz JR (2003) Effect of fluorophore-to-silver distance on the emission of cyanine dye-labeled oligonucleotides. Anal Biochem 315:57–66
14. Geddes CD, Aslan K, Gryczynski I, Malicka J, Lakowicz JR (2004) Noble-metal surfaces for metal-enhanced fluorescence. In: Geddes CD, Lakowicz JR (eds) Reviews in fluorescence. Kluwer Academic/Plenum Publishers, New York, Boston, Dordrecht, London, Moscow
15. Lakowicz JR (2006) Principles of fluorescence spectroscopy, 3rd edn. Springer, New York, NY, ch 25, pp 841–859
16. Matveeva EG, Gryczynski I, Gryczynski Z, Goldys EM (2009) Fluorescence immunochemical detection of analytes. In: Goldys EM (ed) Fluorescence applications in biotechnology and life sciences. Wiley-Blackwell, ch 15, pp 309–326
17. Matveeva EG, Gryczynski I, Barnett A, Leonenko Z, Lakowicz JR, Gryczynski Z (2007) Metal particle-enhanced fluorescent immunoassays on metal mirrors. Anal Biochem 363:239–245
18. Barnett A, Matveeva EG, Gryczynski I, Gryczynski Z, Goldys EM (2007) Coupled plasmon effects for the enhancement of fluorescence immunoassays. Physica B 394:297–300
19. Parfenov A, Gryczynski I, Malicka J, Geddes CD, Lakowicz JR (2003) Enhanced fluorescence from fluorophores on fractal silver surfaces. J Phys Chem 107:8829–8833
20. Goldys EM, Drozdowicz-Tomsia K, Xie F, Shtoyko T, Matveeva EG, Gryczynski I, Gryczynski Z (2007) J Am Chem Soc 129: 12117–12122
21. Shtoyko T, Matveeva EG, Chang I-F, Gryczynski Z, Goldys EM, Gryczynski I (2008) Enhanced fluorescence immunoassays on silver fractal-like structures. Anal Chem 80: 1962–1966
22. Sorensen TJ, Laursen BW, Luchowski R, Shtoyko T, Akopova I, Gryczynski Z, Gryczynski I (2009) Enhanced fluorescence emission of Me-ADOTA by self-assambled silver nanoparticles on a gold film. Chem Phys Lett 476:46–50
23. Yoon I, Kang T, Choi W, Kim J, Yoo Y, Joo S-W, Park QH, Ihee H, Kim B (2009) Single nanowire on a film as an efficient SERS-active platform. J Am Chem Soc 131:758–762
24. Gryczynski I, Malicka J, Nowaczyk K, Gryczynski Z, Lakowicz JR (2004) Effects of sample thickness on the optical properties of surface plasmon-coupled emission. J Phys Chem 108:12073–12083
25. Matveeva EG, Gryczynski Z, Malicka J, Gryczynski I, Lakowicz JR (2004) Metal-enhanced fluorescence immunoassays using total internal reflection and silver island-coated surfaces. Anal Biochem 334:303–311
26. Rabani E, Reichman DR, Geissler PL, Brus LE (2003) Drying mediated self-assembly of nanoparticles. Nature 426:271–274

# Chapter 11

# Initial Stages of Angiosperm Greening Monitored by Low-Temperature Fluorescence Spectra and Fluorescence Lifetimes

**Beata Mysliwa-Kurdziel, Anna Stecka, and Kazimierz Strzalka**

## Abstract

In Angiosperms, the reduction of protochlorophyllide (Pchlide) to chlorophyllide (Chlide), a penultimate reaction of chlorophyll biosynthesis, is catalyzed by a photoenzyme Pchlide oxidoreductase (POR) and completely inhibited in darkness. This reaction plays also a regulatory role in plant morphogenesis. In the case of dark-grown Angiosperms, Pchlide is accumulated, mainly in the form of complexes with NADPH and POR but also as an unbound pigment. Etioplasts that develop in the place of chloroplasts in the dark contain a highly organized lipid structure termed prolamellar body (PLB), which is the main site of accumulation of the ternary Pchlide:POR:NADPH complexes. An illumination triggers the photoreduction of Pchlide molecules which are bound to the ternary complexes. This is followed by a set of biochemical reactions and structural changes leading to Chl synthesis that can be monitored with fluorescence techniques. This chapter describes the application of low-temperature fluorescence spectroscopy and fluorescence lifetime measurements for monitoring the Pchlide to Chlide conversion in isolated prolamellar bodies. These techniques enable the analysis of heterogeneity of accumulated pigments: Pchlide and Chlide that reflect the different organization of pigment–protein complexes.

**Key words:** Chlorophyllide, Fluorescence lifetime, Fluorescence spectra, Prolamellar body, Protochlorophyllide, Protochlorophyllide oxidoreductase, Protochlorophyllide photoreduction

## Abbreviations

| | |
|---|---|
| Chl | Chlorophyll |
| Chlide | Chlorophyllide |
| EDTA | Ethylenediamine tetraacetic acid tetra-sodium salt |
| EPIM | Etioplast inner membranes |
| NADPH | Nicotinamide adenine dinucleotide phosphate |
| PLB | Prolamellar body |
| Pchlide | Protochlorophyllide |
| POR | Light-dependent Pchlide oxidoreductase |
| PT | Prothylakoid |

Wlodek M. Bujalowski (ed.), *Spectroscopic Methods of Analysis: Methods and Protocols*, Methods in Molecular Biology, vol. 875, DOI 10.1007/978-1-61779-806-1_11, © Springer Science+Business Media New York 2012

## 1. Introduction

Development of Angiosperm seedlings is determined by light condition and directly connected to a light-dependent step of chlorophyll (Chl) biosynthesis (1–5). In the dark, seedlings have long hypocotyls; an apical hook; and small, yellowish, and closed cotyledons, which is characteristic for a developmental program known as skotomorphogenesis. On the structural level, proplastids differentiate to etioplasts containing a prolamellar body (PLB), which is a regular paracrystalline lipid structure characteristic for etioplasts (6). The PLB is surrounded by prothylakoids (PTs), a membrane system that resembles primitive thylakoids. Proteomic analysis of purified PLBs showed the oxidoreductase:protochlorophyllide: NADPH (POR, EC 1.3.1.33) as the main protein as well as some proteins of the photosynthetic apparatus, pigment biosynthesis and others (7).

The Chl biosynthesis process in the dark is stopped at the stage of the formation of protochlorophyllide (Pchlide), which accumulates mainly in PLBs in the form of ternary complexes with POR and NADPH. These complexes are usually described as the photoactive Pchlide because this Pchlide pool is immediately conversed to chlorophyllide (Chlide), a direct precursor of Chl, with a flash of ms duration. Some Pchlide also accumulates in PTs but is not attached to POR and cannot be reduced to Chlide with a flash. Thus, it is called non-photoactive. Low temperature (i.e. 77 K) absorption or fluorescence spectroscopy is a useful tool to distinguish between the photoactive and non-photoactive Pchlide due to their different positions of absorption and fluorescence maxima (3, 4).

The light-triggered reduction of the photoactive Pchlide to Chlide can be observed using fluorescence spectroscopy because the fluorescence maximum of the substrate (Pchlide) and the product (Chlide) differs about 30 nm. The Pchlide photoreduction directly triggers a switch between skotomorphogenesis and photomorphogenesis, which is termed deetiolation. This is a complex process including changes on morphological, structural, and molecular levels that finally lead to the formation of functional photosynthetic apparatus. Initial deetiolation steps, i.e., Pchlide to Chlide photoreduction and release of Chlide from complexes with POR, can be investigated in vitro using isolated PLBs.

In this work, we describe a simple method for PLBs isolation, an experiment on Pchlide to Chlide phototransformation and early deetiolation, an analysis of fluorescence properties of Pchlide in isolated PLBs using fluorescence spectroscopy and fluorescence lifetimes.

**Table 1**
**Modified Hoagland medium (8)**

| Ingredients | | Concentration of ingredient in media |
|---|---|---|
| $Ca(NO_3)_2 \times 4H_2O$ | Calcium nitrate | 0.95 g/l |
| $KNO_3$ | Potassium nitrate | 0.61 g/l |
| $MgSO_4 \times 7H_2O$ | Magnesium sulfate | 0.49 g/l |
| $NH_4H_2PO_4 \times 4H_2O$ | Ammonium acid phosphate | 0.12 g/l |
| Fe-tartrate | Iron tartrate | 0.005 g/l |
| $CuSO_4 \times 5H_2O$ | Copper sulfate | 0.05 g/l |
| $ZnSO_4 \times 4H_2O$ | Zinc sulfate | 0.05 g/l |
| $H_3BO_3$ | Boric acid | 0.6 g/l |
| $MnCl_2 \times 4H_2O$ | Manganese chloride | 0.4 g/l |
| $H_2MoO_4$ | Molybdic acid | 0.02 g/l |

## 2. Materials

### 2.1. Plant Material

Seeds of wheat (*Triticum aestivum* cv. Minaret) were soaked in water for 24 h, and then grown hydroponically on modified Hoagland medium (8) (Table 1) in darkness at 295 K for 6–8 days. Seedlings of about 7–10 cm height were cut, 1 cm top fragments were discarded and the following 5 cm pieces were used for PLBs isolation (see Note 1). All manipulations connected with harvesting of seedlings were done in a weak green light that did not produce any detectable amount of Chlide.

### 2.2. Buffers for PLBs Isolation (see Note 2)

*Buffer 1* (isolation medium): 25 mM Hepes–NaOH, pH 7.6, 0.4 M sorbitol, 1 mM $MgCl_2$, 1 mM ethylenediamine tetraacetic acid tetra-sodium salt. *Buffer 2*: 25 mM Hepes–NaOH, pH 7.6, 1 mM $MgCl_2$, 1 mM EDTA. *Buffer 3*: a mixture of buffer 1 with glycerol (80:20 v/v). Stock solution of NADPH (20 mM) prepared and stored in aliquots at 253 K, and then added to the buffer to reach the required NADPH concentration.

## 3. Methods

### 3.1. Isolation of Prolamellar Bodies

Approximately 25 g of 5-cm pieces of etiolated wheat seedlings were cut and homogenized in about 200 ml of buffer 1 (isolation medium) using Moulinex easy power TYPE Y45 blender three times for 5 s. The homogenate was filtered via four layers of commercially available nylon. The suspension was poured into two centrifugation tubes and centrifuged two times as described in the Fig. 1 (I and II centrifugation). The pellet containing mainly etioplasts was suspended in 1 ml of the buffer 1, separately for each centrifugation tube. Just before the suspending the pellet, 70 μl of NADPH from the stock solution was added to each tube. Subsequently, 6 ml of the buffer 2 was added to each tube and left for 2–3 min to ensure an osmotic shock and centrifuged (see Fig. 1, III centrifugation). The resulting pellet of etioplast inner membranes (EPIMs) was then washed once in 5 ml of the buffer 2 containing 0.2 mM NADPH and centrifuged (Fig. 1, IV centrifugation). Next, 50 μl of NADPH from the stock solution was added to each tube and the washed EPIMs were suspended in 5 ml of buffer 3. PLBs were separated from PTs by sonication: 2 × 5 s, 50% duty cycle and amplitude 5 using Ultrasome Homogenizer 4710 equipped with a microtip (Cole-Parmer Instrument Co, USA). PLBs were centrifuged as shown in Fig. 1 (V centrifugation). PLBs were suspended in 1–2 ml of buffer 1 and used for experiment or stored (see Note 3). The above procedure of isolation of PLBs was originally published by Ouazzani Chahdi et al. (9) and with some modification described in ref. (10).

### 3.2. Fluorescence Spectroscopy

Fluorescence emission spectra were measured using a steady-state Perkin Elmer LS-50B Luminescence Spectrometer (UK) equipped with a liquid nitrogen device for measurements at low temperature (77 K). Spectra were collected in the range between 600 and 750 nm for the excitation at 440 nm. Excitation and emission slits were 5 nm. Spectra were corrected for the wavelength-dependent sensitivity of the photomultiplier.

Preparation of sample for spectra measurement: an aliquot (~150 μl) of PLBs diluted (5–10 times) with the buffer 1 was used to fill a capillary (2.5 mm diameter and 7 cm length). The filled capillary was slowly frozen in a small container filled with liquid nitrogen (see Note 4).

### 3.3. Fluorescence Lifetimes

For fluorescence lifetime measurements, PLBs were diluted with the buffer 1 containing 0.2 mM NADPH to reach the optical density lower than 0.15 at the excitation wavelength. Fluorescence lifetimes were measured using phase and modulation K2 ISS spectrofluorimeter (ISS Instruments, Urbana, IL). This instrument was equipped with 300 W Xenon lamp as the excitation source. The intensity of the excitation beam was sinusoidally modulated using a Pockell cell in the range between 2 and 250 MHz. Fluorescence was recorded through a cutoff filter transmitting light with $\lambda > 550$ nm. A single measurement was performed for 12 frequencies of modulation of the excitation beam intensity that were set randomly to avoid any systematic errors caused by bleaching of the sample.

1. Homogenisation and filtration.
2. I Centrifugation 300g/5min.

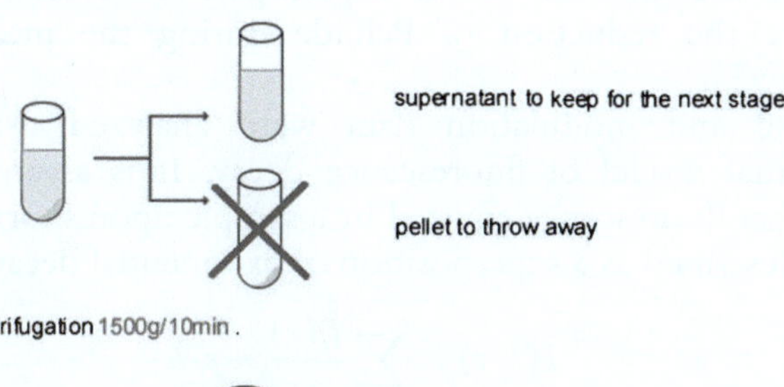


II Centrifugation 1500g/10min.

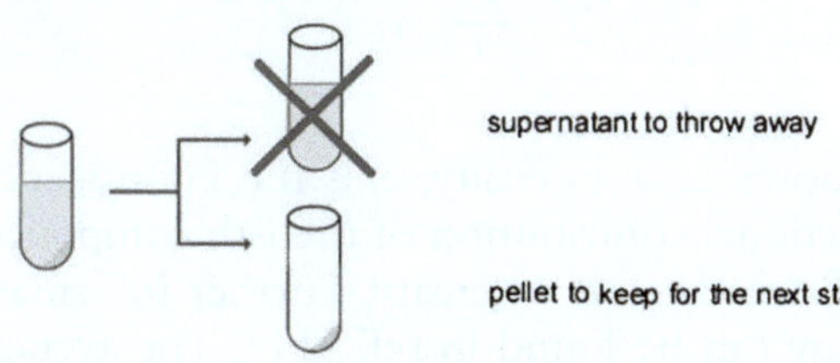


3. Osmotic shock - pellet with 70 μl of NADPH from the stock solution. 1 ml of the buffer 1 and 6 ml of the bufer 2. Incubation 2-3min.

III Centrifugation 7000g/5min.

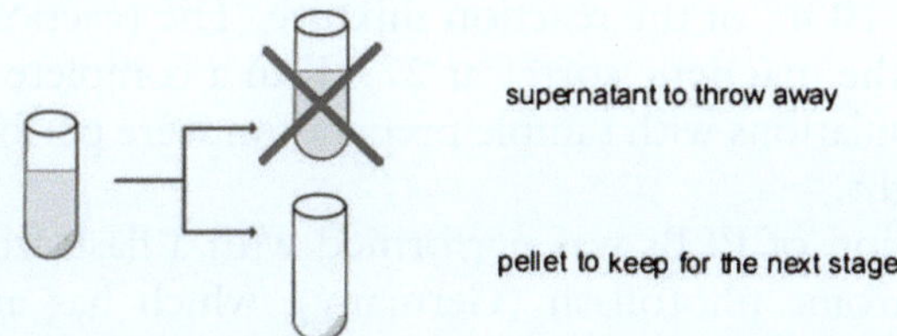


4. Washing of pellet of EPIM in 5 ml of the buffer 2 containing 0,2mM NADPH.
IV Centrifugation 7000g/5min.

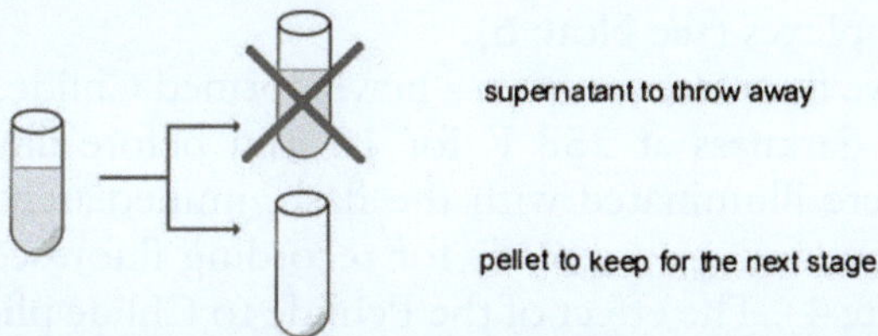


5. Separation of PLB from PT - sonication 2x5s in the isolating buffer mixed with glycerol (80/20 v/v) and containing 0,2mM NADPH.
6. V Centrifugation 7000g/5min.

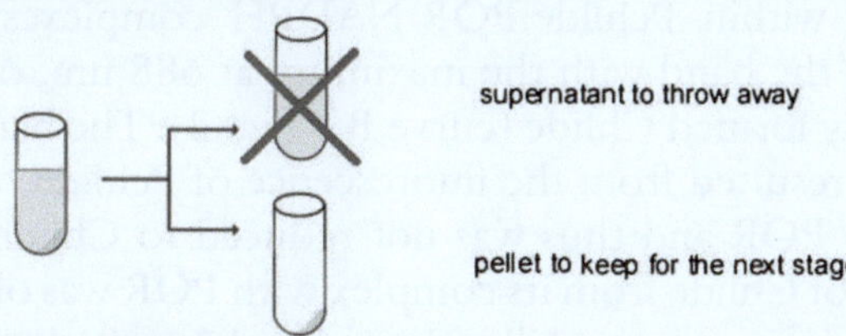


7. Pellet was dissolved in buffer 3.

Fig. 1. Procedure of PLBs isolation.

A diluted glycogen suspension was used as a reference solution of fluorescence lifetime equal to 0 ns. The intensity of the excitation light and the duration of the experiments were set in order to minimize the reduction of Pchlide during the measurements (see Note 5).

Phase and modulation data were analyzed using multi-exponential model of fluorescence decay. It is assumed in this model that fluorescence emitted by a sample upon short excitation can be described as a superposition of exponential decays by

$$I(\lambda, t) = \sum_i \frac{f_i(\lambda)}{\tau_i} \times e^{-\frac{t}{\tau}} \tag{1}$$

where:
$I(\lambda,t)$ is the fluorescence intensity, $\tau_i$ is the fluorescence lifetime, and $f_i$ is the fractional contribution of the i-th component's intensity to the total steady-state intensity. Further information about the methodology can be found in ref. (11). The average random errors in the experimental data, used in fluorescence lifetime analysis, were equal $\pm 0.3°$ in phase angle and $\pm 0.01$ in modulation.

### 3.4. Photoreduction of Pchlide and Early Deetiolation of Isolated PLBs

An aliquot of PLBs was diluted five to ten times with the buffer 1 to obtain about 10 ml of the reaction mixture. The reaction mixture was kept on the magnetic stirrer at 277 K in a complete darkness. All the manipulations with sample preparation were performed in a dim green light.

Illumination of PLBs was performed with a flash from Hama SF-30E electronic photoflash (Germany), which has an average energy output of 70 J ($\pm$10%) in the range of 450–700 nm, in the experimental condition. The lamp was placed about 30 cm from the sample. The flash reduced all the Pchlide within Pchlide:POR: NADPH complexes (see Note 6).

To observe fluorescence from a newly formed Chlide, the PLBs were kept in darkness at 253 K for 10 min before illumination. Then they were illuminated with the flash, immediately placed in liquid nitrogen, and used directly for recording fluorescence spectrum (see Note 4). The effect of the Pchlide to Chlide photoreduction was manifested by disappearance of the fluorescence band having the maximum at 655 nm, (curve A – Fig. 2) originating from Pchlide within Pchlide:POR:NADPH complexes, and the appearance of the band with the maximum at 688 nm, originating from the newly formed Chlide (curve B – Fig. 2). The band around 630–633 nm resulted from the fluorescence of Pchlide, which was not bound to POR and thus was not reduced to Chlide after the flash. Release of Chlide from its complex with POR was observed in the reaction mixture that was illuminated with one flash and kept in the dark at room temperature for 40–50 min before taking sample for recording the fluorescence spectrum. As it can be seen from the

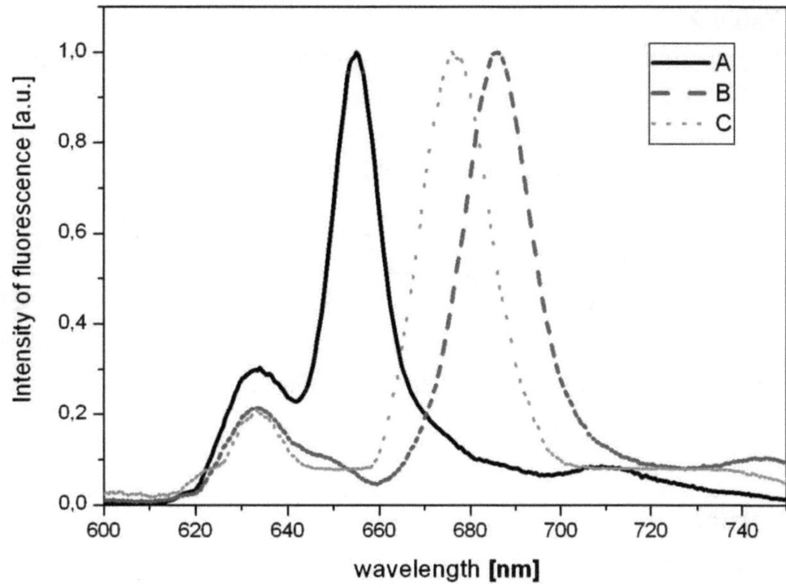


Fig. 2. Fluorescence spectra measured at 77 K for PLBs (**a**), PLBs illuminated with a single flash of ms duration (**b**), PLBs illuminated with a single flash of ms duration and kept in darkness for 30 min (**c**). Excitation: 440 nm. The spectra were normalized at their maxima.

Fig. 2, this time was sufficient for the shift of the fluorescence maximum from 688 nm to 676 nm, which reflects the Chlide release from the complex with POR (curve C – Fig. 2).

Fluorescence lifetimes were measured for the following samples: (1) PLBs in darkness: 2 ml of the reaction mixture was poured into fluorescence cuvette (1 × 1 cm) and transferred in the dark to the fluorimeter. (2) PLBs after the Pchlide to Chlide conversion: 2 ml of the reaction mixture in the fluorescence cuvette was illuminated with the flash and used for fluorescence lifetime measurements. (3) PLBs kept in the dark for 40 min after the illumination. In this case, the fluorescence lifetimes of Chlide released from the PLBs and the residual Pchlide were measured.

Analysis of the results: the best fitting of the phase and modulation data was obtained for three-exponential model of fluorescence decay. In the case of PLBs in darkness, the main fluorescence lifetime component was about 3.5 ns (Table 2); however, the component shorter than 1 ns had contribution to the total fluorescence higher than 25%. After the flash, one main component of 4.8 ns with the fractional intensity higher than 90% was observed. Chlide:POR:NADP$^+$ is the main fluorophore in this sample. The release of Chlide was accompanied by the slight decrease of the main fluorescence lifetime component to 4.5 ns. The value of the longest component, which is of μs–ms range, cannot be precisely determined with the K2 spectrofluorimeter. The fractional intensity of this component is definitely higher in not-illuminated PLBs than PLBs after the Pchlide to Chlide photoreduction.

**Table 2**
**Fluorescence lifetimes (T) and fractional intensities (F) measured for isolated PLBs at 288 K for PLBs' samples described in 3.4. The average results and standard deviation from five repetitions is shown. The error of F did not exceed 0.05. Excitation wavelength: 440 nm**

| Sample | $T_1$ [ns] | $T_2$ [ns] | $T_3$ [ns] | $T_1$ | $F_2$ | $F_3$ |
|---|---|---|---|---|---|---|
| PLBs in the dark | 0.76 ± 0.04 | 3.49 ± 0.15 | 6,345 | 0.25 | 0.62 | 0.13 |
| PLBs after the flash | 0.10 ± 0.24 | 4.86 ± 0.11 | 1,050 | 0.01 | 0.94 | 0.04 |
| PLBs kept in the dark after the flash | 0.05 ± 0.11 | 4.50 ± 0.06 | 488 | 0.03 | 0.93 | 0.05 |

## 4. Notes

1. Etiolated seedlings can be stored wrapped in a wet paper and closed in a metal box for 1–2 days in a fridge (at 277–279 K).
2. All the buffers can be stored at 253 K for several months. The stock solution of NADPH can be stored at 253 K for several weeks. If it becomes oxidized, the solution changes color to yellow.
3. PLBs suspended in buffer 3 with addition 0.2 mM NADPH can be stored at 193 K or in a container filled with liquid nitrogen without any loss of their ability of Pchlide to Chlide photoreduction until used.
4. Frozen PLBs samples in capillaries can be stored at 77 K for several weeks before recording the fluorescence spectra at 77 K.
5. All the manipulations with the setting of the instrument have to be performed on different PLB sample than that used for the measurement. The intensity of the excitation beam has to be kept as low as possible during measurement due to a high probability of photoreduction of Pchlide in PLBs. The degree of the Pchlide photoreduction during the single measurement was calculated according to ref. (12).
6. Photoreduction of Pchlide could also be performed in a continuous light of 100–200 μE for 30 s.

### References

1. Masuda T (2008) Recent overview of the Mg branch of the tetrapyrrole biosynthesis leading to chlorophylls. Photosynth Res 96:121–143
2. Heyes DJ, Hunter CN (2005) Making light work of enzyme catalysis: protochlorophyllide oxidoreductase. Trends Biochem Sci 30: 642–649
3. Schoefs B (2005) Protochlorophyllide reduction – what is new in 2005? Photosynthetica 43:329–343

4. Belyaeva OB, Litvin FF (2007) Photoactive pigment–enzyme complexes of chlorophyll precursor in plant leaves. Biochemistry (Mosc) 72:1458–1477
5. Bollivar DW (2006) Recent advances in chlorophyll biosynthesis. Photosynth Res 90: 173–194
6. Solymosi K, Schoefs B (2008) Prolamellar body: a unique plastic compartment, which does not only occur in dark-grown leaves. In: Schoefs B (ed) Plant cell compartments – selected topics. Research Signpost, Kerala, India, pp 152–202
7. Blomqvist LA, Ryberg M, Sundqvist C (2008) Proteomic analysis of highly purified prolamellar bodies reveals their significance in chloroplast development. Photosynth Res 96:37–50
8. Hoagland DR, Arnon HI (1950) The water-culture method for growing plants without soil. Calif Agric Exp Sta Cir 347: 32p
9. Ouazzani Chahdi MA, Schoefs B, Franck F (1998) Isolation and characterization of photoactive complexes of NADPH: protochlorophyllide oxidoreductase from wheat. Planta 206:673–680
10. Mysliwa-Kurdziel B, Franck F, Ouazzani Chahdi MA, Strzalka K (1999) Changes in endothermic transitions associated with light-induced chlorophyllide formation, as investigated by differential scanning calorimetry. Physiol Plantarum 107:230–239
11. Lakowicz JR (2006) Principles of fluorescence spectroscopy, 3rd edn. Chapter 5, Springer, ISBN-10:0-387-31278-1
12. Sperling U, Franck F, Van Cleve B, Frick G, Apel K, Armstrong GA (1998) Etioplast differentiation in Arabidopsis: both PORA and PORB restore the prolamellar body and photoactive protochlorophyllide-F655 to the cop1 photomorphogenic mutant. Plant Cell 10:283–296

# Chapter 12

# Activation of the Mammalian Cells by Using Light-Sensitive Ion Channels

Mandy Siu Yu Lung, Paul Pilowsky, and Ewa M. Goldys

## Abstract

With advances in molecular biology and gene cloning techniques, it is now possible to selectively stimulate living cells of interest by using an external light source. This is done by transfecting the cells of interest with a plasmid carrying the channelrhodopsin (ChR2) gene. By stimulating these transfected cells with laser, the light-sensitive ion channels ChR2 are opened, followed by an influx of cation resulting in cell activation. This combination of optical and genetic technique is known in the literature as optogenetics. It is particularly useful in the functional studies of excitable cells, such as neurons, muscle and endocrine cells, to mimic the stimulation from action potentials to trigger the release neurotransmitters and hormones. Here, we describe the methods needed to make selected mammalian cells (PC12) respond to light excitation.

**Key words:** Optogenetics, HEK-293, Channelrhodopsin 2, Tranfection, Fluorescent microscope, Laser

## 1. Introduction

This chapter focuses on photostimulation of HEK-293 cells, which is an epithelial cell line derived from the human embryonic kidney. HEK-293 has been used extensively for transient gene expression studies as it is easy to transfect and to grow in culture (1, 2). pcDNA3.1/hChR2(H134R)-mCherry is a plasmid that contains a ubiquitous promoter derived from the cytomegalovirus (CMV), to drive the expression of channelrhodopsin 2 (ChR2). ChR2 is a non-specific light-gated cation channel which, when activated by light, will allow the conduction of positively charged ions $H^+$, $Na^+$, $K^+$, and $Ca^{2+}$. In the pcDNA3.1/hChR2(H134R)-mCherry plasmid, the ChR2 gene is fused to a gene for a fluorescent protein marker mCherry. The plasmid construct is delivered into HEK-293 cells in culture by the method of lipofection. Successfully transfected cells are identified by the expression of mCherry which is

Wlodek M. Bujalowski (ed.), *Spectroscopic Methods of Analysis: Methods and Protocols*, Methods in Molecular Biology, vol. 875, DOI 10.1007/978-1-61779-806-1_12, 

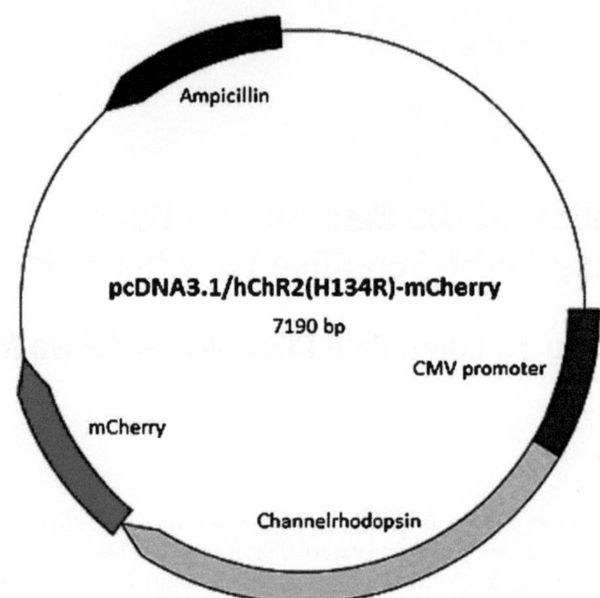


Fig. 1. Genetic structure of the plasmid pcDNA3.1/hChR2(H134R)-mCherry.

visible under a fluorescent microscope with an appropriate filter. The expressed ChR2 is then activated by light, and the level of cell activation can be quantified by various activity assays. The example used in this protocol is an intracellular calcium assay.

The presented protocols can be easily modified to address other cell types by taking advantage of the specificity of different enhancer and promoter elements used to drive the expression of ChR2 within the target cells. An alternative light-sensitive ion channels can also be used to achieve a different stimulation outcome. For example, halorhodopsin (NpHR) which is a light-activated chloride channel can be used to mediate the inhibition of cells.

## 2. Materials

### *2.1. Growing Up of Plasmids from Commercial Stock*

1. pcDNA3.1/hChR2(H134R)-mCherry in DH5α (Addgene) (Fig. 1).
2. Luria Bertani (LB) broth powder (Sigma).
3. Ampicillin sodium salt (Sigma).
4. Bacteriological agar (Sigma).
5. HiSpeed Plasmid Maxi Kit (Qiagen).

### *2.2. Plasmid Transformation into E. coli*

1. Subcloning Efficiency DH5α Competent Cells (Invitrogen).

#### 2.3. Cell Culture

1. HEK-293 human embryonic kidney cells (ATCC).
2. Dulbecco's Modified Eagle's Medium (DMEM) with 4,500 mg/L glucose, L-glutamine, sodium pyruvate, sodium bicarbonate (Sigma).
3. MEM non-essential amino acids solution 10 mM (Gibco).
4. Foetal bovine serum (FBS) (Gibco).
5. PBS solution pH 7.2 (Gibco).
6. Trypsin–EDTA solution (Sigma).
7. Dimethyl sulfoxide (Sigma).

#### 2.4. Transfection

1. Lipofectamine LTX and Plus Reagent (Invitrogen).
2. Opti-MEM Reduced Serum Medium (Invitrogen).

#### 2.5. Fluorescent Microscopy

1. Fluorescence or laser scanning microscope with fluorescence option. The microscope we used was the Leica SP2 confocal microscope.
2. Halogen light source for excitation.

#### 2.6. Photostimulation

1. OptoGeni 473 photostimulation laser (IKECOOL).
2. Function/Arbitrary Waveform Generator, 20 MHz (Agilent 33220A).
3. Laser power meter with readings calibrated to the laser wavelength (Coherent Fieldmate).
4. TPS 2000 oscilloscope for displaying the laser pulses (Textronics).

#### 2.7. Intracellular Calcium Assay

1. 1 mM Fura 2-AM in DMSO.
2. Hanks balanced salt solution (HBSS).
3. HEPES buffer saline (20 mM HEPES, 115 mM NaCl, 5.4 mM KCl, 1.8 mM $CaCl_2$, 0.8 mM $MgCl_2$, 13.8 mM glucose, pH 7.4).

#### 2.8. Action Potential Assay

1. 1 mM RH155 dye in saline.

## 3. Methods

#### 3.1. Growing Up of Plasmids from Commercial Stock

1. To prepare LB broth, suspend 20 g of LB broth powder in 1 L of Milli-Q water and autoclave.
2. To prepare selective LB broth, add 100 μg/mL ampicillin to autoclaved LB broth prior to use. Note that antibiotics should never be added to liquid medium when it is too hot, i.e. immediately after autoclaving. Allow medium to cool down to about 70°C before adding antibiotics.

3. To prepare selective LB agar plates, add an extra 15 g of agar to 1 L of LB broth before autoclaving. Wait for agar solution to cool down to about 70°C after autoclaving, then add 100 μg/mL ampicillin before pouring agar solution into plastic petri dishes. Once the agar has set, the plates can be sealed and stored at 4°C and should be used within 2–3 months.
4. The pcDNA3.1/hChR2(H134R)-mCherry plasmid is supplied in DH5α, in the form of a bacterial stab. To recover plasmid, inoculate a selective LB agar plate with the stab culture and incubate at 37°C overnight.
5. The next day select a single colony and grow it up in selective LB broth.
6. To obtain a large amount of pure plasmids, follow manufacturer's instructions for plasmid purification procedures using the HiSpeed Plasmid Maxi Kit.

### 3.2. Plasmid Transformation into E. coli Cell

1. For plasmid transformation, mix 10–20 ng of plasmid DNA with 20 μL of competent DH5α cells in a 1.5 mL microcentrifuge tube. Incubate on ice for 30 min.
2. Heat-shock the cells for 30 s at 42°C without shaking.
3. Immediately transfer the tubes onto ice.
4. Add 475 μL of LB broth (pre-warmed at 37°C) to the tube.
5. Incubate the tube at 37°C for 1 h, shaking at 200 rpm.
6. Spread 20–100 μL of transformed culture onto pre-warmed selective LB agar plates.
7. Incubate plates overnight at 37°C.
8. The next day, select a single colony from the plate and grow up the selected clone in LB broth with ampicillin.
9. Follow manufacturer's instructions for plasmid purification procedures using the HiSpeed Plasmid Maxi Kit.

### 3.3. Cell Culture

1. The following protocol is for HEK-293 cells maintained in 75 $cm^2$ flasks. Adjust volumes proportionally when using culture flasks or plates with different surface areas, or when working with cells of different sizes.
2. Cultures are maintained in DMEM with 10% FBS in a humidified incubator with 5% $CO_2$.
3. For subculturing, remove and discard culture medium.
4. Briefly rinse the cell layer with PBS to remove all traces of serum that contain trypsin inhibitor.
5. Add 1 mL (or enough to cover the surface of the flask) of trypsin–EDTA solution to flask and observe cells under an inverted microscope until cell layer is dispersed (usually within 5 min). Flask may be incubated at 37°C to facilitate dispersal.

6. Add 9 mL of fresh medium to rinse cells off the surface of the flask. The FBS in the medium acts as an inhibitor to trypsin to stop the digestion of cells. Collect the total volume (10 mL) to a falcon tube.
7. Centrifuge cells at 1,500 rpm (420 × *g*) for 5 min at room temperature.
8. Remove and discard supernatant.
9. Resuspend cell pellet in 9 mL of fresh medium.
10. Add 27 mL of fresh medium to a new 75 $cm^2$ flask and seed flask with 3 mL of cell suspension (1:3 split).
11. Incubate cells and subculture when cells appeared to be confluent under the microscope.
12. Replace exhausted medium (indicated by pH indicator in the medium as a change of colour from red to orange/yellow) with fresh medium, approximately every 2–3 days.
13. The density of cells can be estimated by counting under a microscope using a haemocytometer.

### 3.4. Transfection

1. This protocol (Invitrogen Lipofection LTX protocol) is for plasmid transfection of HEK-293 cells in a 24-well tissue culture plate. Adjust volumes proportionally when using culture flasks or plates with different surface areas, or when working with cells of different sizes.
2. The day before transfection, count and plate cells ($2 \times 10^5$ cells per well) in 500 μL of complete growth medium. Cell density should be 90–95% confluent on the day of transfection.
3. For each well dilute 500 ng of plasmid DNA into 100 μL of Opti-MEM Reduced Serum Medium without serum.
4. Mix PLUS Reagent gently before use, and then add 0.5 μL to diluted DNA. Mix gently and incubate for 5–15 min at room temperature.
5. For each well add 0.75 μL of Lipofectamine LTX into the above diluted DNA solution. Mix gently and incubate for 25 min at room temperature to form DNA–liposome complexes.
6. Remove growth medium from cells and replace with 500 μL of complete growth medium. Add 100 μL of DNA–liposome complexes directly to each well and mix gently by rocking the place back and forth.
7. Incubate cells at 37°C in a 5% $CO_2$ incubator for 24 h post-transfection before assaying for transgene expression.
8. Because the cells are meant to express mCherry which should not be bleached, carefully minimize the exposure of cells to light.

### 3.5. Fluorescent Microscopy

1. The confirmation that the cells have been successfully genetically engineered is performed by the assessment of the expression level of the fluorescent protein mCherry.
2. The standard imaging conditions for exciting mCherry in cells is to excite the sample at 594 nm and to image cells at around 630 nm. This can be accomplished by using an excitation filter at 560 nm and emission filter at 630 nm (similar to filters for Texas Red).

### 3.6. Photostimulation

ChR2 is a light-activated cation channel which is opened upon exposure to blue light, which depolarizes the membrane potential. Thus in a cell expressing ChR2, light excitation can be used to produce precisely timed spike trains (10 Hz routine, ≥30 Hz readily achievable in fast spiking cells) with temporal jitter of 1–3 ms. Light excitation is best used for precise activation of neurons on the ms timescale, but it can be used to evoke single spikes or defined trains of action potential over a range of frequencies (3, 4). The optimum excitation wavelength for ChR2 is about 480 nm, but other wavelengths are also effective and may produce somewhat different results, this area is in need of further research.

#### 3.6.1. Producing Desired Laser Pulses

1. High power light emitting diodes are eye-safe (although uncomfortably bright), but lasers normally require eye protection (in form of special laser goggles with high neutral density) when working with exposed light beams.
2. The first task when working with a diode laser capable of external modulation is to actually implement that modulation and verify that it is occurring, and establish the modulation characteristics before the cell experiment starts. The process would then typically be repeated regularly during cell experiments, for checking and when modifying the stimulation parameters.
3. Put on laser goggles for eye protection while using lasers outside the enclosure.
4. Turn on the laser and turn down its power to 3 mW. Measure it with the power meter located in front of the laser.
5. If the power meter is not fast enough, replace the power meter with a photodiode or another semiconductor light detector. Their active element may be small so make sure that the laser beam is falling onto this active element.
6. Connect the detector output to the Y channel of an oscilloscope, working in a Y-time mode. One should see the light detector signal on the oscilloscope screen. Confirm that the laser light should be blocked with an opaque object which causes the detector signal becoming zero.

7. Unblock the beam and connect the function generator to the relevant input on the power supply of the laser. Select some pulse pattern, for example rectangular. It should be observable on the oscilloscope as a rectangular pulse train. Keep adjusting the oscilloscope controls (start with putting it on "auto") until the train of pulses stops. This means the oscilloscope is properly triggered. Keep modifying the pulse train on the signal generator until you reach the desired pulse shape. Typical pulse duration for stimulation is ~0.5–50 ms (to ensure millisecond time scale precision). A typical repetition rate is below 10 Hz. Make sure the pulse meets these requirements.

*3.6.2. Launching Laser Light into the Optical Fibre*

1. Optical fibres are frequently fragile and need to be handled with care, especially near terminations.
2. Check the ends of your optical fibre and make sure they look perfectly flat and perpendicular to the fibre axis.
3. Place the fibre in a "fibre chuck"—a round piece of metal with a fine groove holding the fibre. Lock the chuck with the fibre in place.
4. Put the laser goggles on. Turn on the laser and turn down laser power to a safer level of $<3$ mW.
5. By using a strip of paper locate the place where the microscope objective focuses the light. Remove the paper, place the chuck in the x-y-z translation stage mounted with the microscope objective. The end of the fibre should be located near the location where the microscope objective focuses light.
6. Place the power meter at the other end of the fibre. Use the x-y-z controls of the translation stage to focus the laser light exactly at the end of the fibre. The first sign of progress is the appearance of a bright blue star at the end of the fibre, indicating that the laser focus has found the fibre end. The meter should now show some power reading.
7. Keep adjusting the x-y-z controls until this power reading is maximized. One should aim for the power through a fibre to be at least 50% of the total power measured previously without a fibre. With a well-terminated fibre ~70–80% can be obtained.

*3.6.3. Producing a Desired Power Density at the Cells*

1. After coupling the laser to the fibre as described previously, observe the light spreading out from the other end of the fibre. Put up a paper screen rectangular to the fibre and find a distance $d$ from the fibre to the screen where the illuminated area is comparable to the cell area $A$ you wish to illuminate. Draw a line around this area and measure it.
2. Calculate the laser power providing a desired power density. Power density is defined as the ratio of power leaving the output fibre divided over the illuminated area $A$.

3. The stimulating laser beam needs to be normal to the cell surface for best power uniformity. Uniformity can be improved by using a narrow illumination cone.
4. The minimum threshold power for photostimulation is ~5 mW/mm$^2$. Make sure that your power remains within the safe limit for the cells. Maximum reported power density for harmless sustained light stimulation >100 mW/mm$^2$.
5. Set up the cell illumination system comprising the laser coupled to an optical fibre as described above and a signal generator. In order to achieve specified power density at the location of the cells the end of the fibre needs to be located at a specified distance $d$ from the cell surface found previously so that the illuminated area is approximately $A$.

### 3.7. Intracellular Calcium Assay

1. The most popular calcium dye for intracellular calcium assay is Fura-2, which shows a spectral response upon binding $Ca^{2+}$ and enables the measurement of intracellular-free $Ca^{2+}$ concentrations using ratiometric fluorescent microscopy.
2. Ratio-imaging is performed by alternating excitation at the wavelengths of 340 and 380 nm, while monitoring emission at the wavelength of 510 nm.
3. The imaging conditions should be such that the excitation light for the calcium assay should neither interfere with the ChR2 present on cell surfaces nor with mCherry expressed within the transfected cells.

#### 3.7.1. Protocol for Staining with Fura-2

1. Culture cells on a glass-bottom dish.
2. Dilute 1 mM Fura 2AM DMSO solution with HEPES buffer saline to prepare 1 mM Fura 2-AM working solution.
3. Remove the culture medium, and add 500 μL of Fura 2-AM working solution to the cells.
4. Incubate for 20 min then remove the Fura 2-AM working solution.
5. Wash cells once with HEPES buffer saline. Incubate cells for 1 h in the HEPES buffer saline.
6. Use the cells for fluorescent calcium ion detection.
7. Monitor the excitation spectra at 380 nm (free calcium) and 340 nm (calcium complex) with fixed emission at 510 nm.

### 3.8. Action Potential Assay

The action potentials induced by photostimulation can also be detected using optical methods. These voltage transients are much faster than calcium transients and they need to be detected using fast-response voltage-sensitive dyes.

1. The example demonstrated in this protocol is to detect voltage transients by measuring absorbance changes of cells stained with RH155 (5).
2. RH155 is an anionic oxonol dye and its adsorption to the cell membrane increases upon depolarization.
3. Responses are detected at the wavelength of approximately 720 nm, as the absorption of 720 nm light is increased as transmembrane voltages become more positive.

The signals from voltage-sensitive dyes should be presented as the fractional intensity change ($\Delta I/I$) because this normalizes the differences in the amount of dye in the cells. These signals give information about the time course of the potential change but no direct information about its magnitude.

#### 3.8.1. Protocol for Staining with RH155

1. Prepare RH155 as a solution of 1 mg/mL in saline. It is critical to wrap foil around staining container and keep in dark. Make fresh solution for each experiment.
2. Expose the cells to the solution for about 15 min in the dark.
3. For culture dishes with higher density of cells, it may be necessary to expose for longer, up to 3 h. If long exposure times are needed, provide conditions similar to the growth conditions of the cells.
4. Ensure cells are completely immersed in the staining solution. Consistent timing, dye concentration, and staining area are crucial.
   Wash cells with high-Ca saline.

#### 3.8.2. Optics Setup for Registering Action Potentials

1. On an upright microscope, set up the stimulation light source to interfere with the epifluorescence port. The light source or a shutter needs to be in place to allow precisely timed flashes of stimulation light.
2. Set up the filters by placing the imaging filter cube carrying both the dichroic (reflect < 665 nm light to the cell sample) and emission filter (710 nm bandpass).
3. Connect the halogen lamp to the diascopic lamphouse port and use this light source for illumination of RH155-stained cells.
4. Place the illumination filter (710 nm) into a slot in the condenser. This filter isolates 710 nm light from the diascopic lamphouse for illumination of RH155-stained cells.

### 3.9. Data Processing for Signal Transients and Images

A key challenge of imaging is that the speed required to monitor Ca-related signals and the action potential. Their duration may be too short for high quality imaging of large areas in a conventional laser scanning microscope, so a compromise may be needed between image quality and high frame rate. There are ways to overcome this in a conventional laser microscope but at the expense of the degree of detail imaged, by setting the microscope to image a small spot only, or, as frequently done, by forgoing the imaging completely, in favour of line scans. In each case, the imaging needs to be synchronized with the pulse from the stimulating laser. Using a low magnification objective may be beneficial, with a small spot covering a single cell for example, high frame rates may be possible in fluorescence microscope equipped with a high speed camera, if available. It is necessary for the camera to be gated from the laser pulse. Another problem is the fact that the stimulating laser is producing a powerful light beam some of which may enter into the light detector used for imaging. This unwanted light needs to be filtered out by using a suitable (high neutral density, band stop) laser filter, to prevent light detector damage.

## 4. Notes

1. Voltage stabilized halogen sources are superior for stable measurements of transmission, with stability better than $10^{-5}$.
2. Transmission signals used with voltage sensitive dyes need to be sampled at a couple of kHz, and the low pass filter set at 0.1 ms. Optical signals benefit from such filtering which eliminates high-frequency noise.
3. Transmission or fluorescence signals for calcium indicators need to be sampled at 500 Hz and low pass filter set at 10 ms. Further smoothing with digital filters is often carried out.
4. The signal-to-noise is usually proportional to inverse square root of the signal sue to shot noise contribution. Thus, increasing the light intensity should increase the signal-to-noise ratio (S/N). However, a strong light intensity risks photo-induced damage to the specimen and bleaching of voltage sensitive dye, which prevent stable recording of physiological parameters. For transmission measurements, most authors use a 150 W halogen lamp at maximum power with a bandpass filter.
5. An inferior S/N can be improved if the signal is statistically identical during each trial, allowing the use of an averaging technique.
6. When an imaging device is used, to increase the upper limit of its dynamic range and the frame rate, it is advisable to bin $n \times n$ (for example $4 \times 4$) pixels into one pixel. This reduces the spatial resolution of the collected images though.

7. For measurements of fluorescence, it is advisable to use objectives with high numerical aperture as the amount of collected light is proportional to a fourth power of this quantity.
8. High magnification may not be necessarily beneficial, as low magnification facilitates sample averaging.
9. Statistical analysis is complicated by variation in values of the signal ($\Delta I/I$) because of differences in cell sample thickness, dye staining, and detector sensitivity.

## References


1. Baldi L, Muller N, Picasso S, Jacquet R, Girard P, Thanh HP, Derow E, Wurm FM (2005) Transient gene expression in suspension HEK-293 cells: application to large-scale protein production. Biotechnol Prog 21:148–153
2. Durocher Y, Perret S, Kamen A (2002) High-level and high-throughput recombinant protein production by transient transfection of suspension-growing human 293-EBNA1 cells. Nucleic Acids Res 30:E9
3. Zhang F, Wang LP, Boyden ES, Deisseroth K (2006) Channelrhodopsin-2 and optical control of excitable cells. Nat Methods 3:785–792
4. Zhang F, Gradinaru V, Adamantidis AR, Durand R, Airan RD, de Lecea L, Deisseroth K (2010) Optogenetic interrogation of neural circuits: technology for probing mammalian brain structures. Nat Protoc 5:439–456
5. Zochowski M, Wachowiak M, Falk CX, Cohen LB, Lam YW, Antic S, Zecevic D (2000) Imaging membrane potential with voltage-sensitive dyes. Biol Bull 198:1–21

# Chapter 13

# Detection of Specific Strains of Viable Bacterial Pathogens by Using RNA Bead Assays and Flow Cytometry with 2100 Bioanalyzer

**Philip Butterworth, Henrique T.M.C.M. Baltar, Martin Kratzmeier, and Ewa M. Goldys**

## Abstract

Bead assays are an emerging microbial detection technology with the capability for rapid detection of extremely low levels of viable pathogens. Such technologies are of high value in clinical settings and in the food industry. Here, we perform a bead assay for extracted 16S rRNA from *Escherichia coli* (strain K12) with the flow cytometry readout on a 2100 Bioanalyzer, a highly accurate, small-scale flow cytometer system.

**Key words:** RNA assay, Bead assay, Fluorescent bead, Fluorochrome, Oligonucleotide probe, Flow cytometry

## 1. Introduction

Research into diagnostic methods to detect specific microorganisms is highly significant for areas such as health care and the food industry. The ultimate aim is to develop a simple, rapid, reliable, and reproducible methodology that can accurately detect low levels of viable pathogenic microorganisms in complex matrices such as food matrices or human samples and which can be employed at the point of source, ideally by untrained personnel. Traditional methods of microbial pathogen detection and identification do not meet these requirements. They use a range of microbiological staining and culture techniques and biochemical tests such as immunoassays (1) and flow cytometry (2). These traditional methods can be time-consuming and/or expensive, and/or do not have the sensitivity required to detect the extremely low viable pathogen levels required by the food industry. Recently introduced technologies with the potential to be both rapid and inexpensive include

Wlodek M. Bujalowski (ed.), *Spectroscopic Methods of Analysis: Methods and Protocols*, Methods in Molecular Biology, vol. 875, DOI 10.1007/978-1-61779-806-1_13, 

ultra-violet resonant Raman spectroscopy (3, 4), real-time polymerase chain reaction (PCR) (5, 6), immuno-chromatography (7), capillary electrophoresis (8), autofluorescence (9, 10), and certain applications of nazchnology (11). However, their wider utilisation is yet to be demonstrated.

Lab-on-a-chip (LOC) technology is a recent development that can help automate laboratory processes, as well as to reduce the quantity of sample/reagents used, and thereby lower costs (12). Such LOC systems are already on the market, e.g., the Agilent 2100 Bioanalyzer (Agilent Technologies Inc., USA), a microfluidics-based system that can carry out sizing, quantification, and quality control of DNA, RNA, proteins, and cells. The system is small-scale (desktop computer size) and it can carry out analyses within 30–40 min. Other systems on the market include a LOC construction kit (thinXXS Microtechnology AG, Germany) and LabChips (Caliper LifeSciences Inc., USA) for electrophoretic separation and drug discovery. Owing to small size and feasibility of low cost solutions, it is expected that LOC devices will become widely applicable in clinical applications and for industrial diagnostics.

Molecular detection of specific pathogens important in health and food industries can be carried out by targeting their DNA or RNA. While DNA is relatively stable outside of cells, RNA rapidly degrades once removed from living organisms. Therefore, the detection of RNA provides an indication of viable pathogens. In addition, RNA is highly abundant in cells, e.g., its content in one *Escherichia coli* cell is $58 \times 10^{-15}$ g of which approximately 27% is 16S ribosomal RNA (rRNA) (13).

In this chapter, we describe how to conduct an RNA assay using beads coated with oligonucleotide probes for the capture of the RNA of a specific microorganism; the assay is performed using 2100 Bioanalyzer. The assay targets rRNA because it is highly conserved among closely related species and the differences between these highly conserved rRNA sequences enable a definite identification of target microorganisms. 16S rRNA used here is the most commonly used target for the molecular identification of bacteria. A nonpathogenic *E. coli* (strain K12) has been selected because it is safer/easier to work with than the pathogenic *E. coli* strains of relevance to the food industry. *Bacillus subtilis* is used as a control microorganism to enable the assessment of nonspecific binding.

The procedure is briefly summarised as follows. The first step is to extract and purify RNA from bacterial cultures and then label this RNA with a suitable fluorescent dye—in this case, fluorochrome Alexa Fluor 647 (excitation 650 nm, emission 670 nm). Short oligonucleotides with sequences specific to the target organism (probes) are coupled to synthetic fluorescent microspheres (beads) to form a probe–bead complex. A hybridisation reaction is then carried out with the fluorescently labelled RNA and the

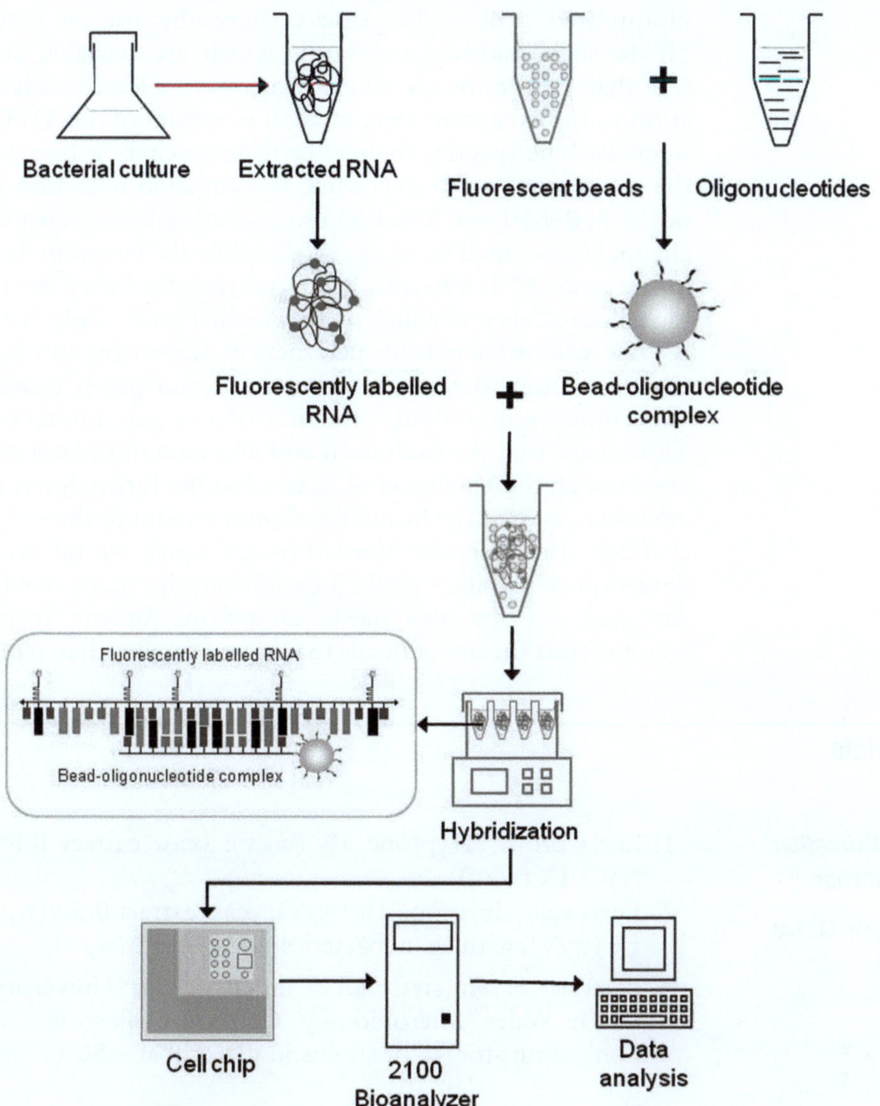


Fig. 1. Pictorial description of the key steps for the RNA assay using beads coated with oligonucleotide probes for the capture of the RNA of a specific microorganism.

probe–bead complexes. The sample is loaded to a cell chip which is placed into the 2100 Bioanalyzer for flow cytometer analysis to determine whether the labelled RNA has hybridised with the probe–bead complexes, i.e., that the target organism has been detected. The key steps of the assay procedure are illustrated in Fig. 1. The isolation of RNA and other assay procedures are carried out using commercial kits with minimal modifications (see below for more details of the assay protocols).

The present protocols describe an assay using carboxylated beads and amino-modified probes. With modified protocols, other types of binding may be used, e.g., streptavidin beads with biotinylated probes. We discuss here the use of commercial, off-the-shelf, carboxylated beads which are available at a lower cost than custom beads. These beads were selected with consideration to the excitation and emission spectrum of the 2100 Bioanalyzer. Its blue spectral channel provides excitation by a LED with the maximum at 470 nm, while the emission detection is carried out at 510–540 nm. The 633 nm excitation wavelength in the red channel is provided by a laser diode while the emission detection is in the range 674–696 nm. Beads are therefore read out using the blue fluorescence channel of the Bioanalyzer, while the labelled RNA is read by the red channel where excitation intensity is stronger. Co-registration (detection of both red and green signals at the same time) indicates the detection of a targeted microorganism. This means that the excitation and emission of the selected beads must match the blue spectral channel of the Bioanalyzer, while the emission curve of the beads should not extend to the red emission channel otherwise the bleed-through signal would prevent the detection of specific signals. In addition, the beads need to be of adequate size for blue diode excitation; Agilent Technologies recommends the use of beads that are no smaller than 6 μm in size.

## 2. Materials

### 2.1. RNA Extraction and Purification

#### 2.1.1. Bacterial Culture Preparation

1. Luria broth: Tryptone 1% (w/v): yeast extract 0.5% (w/v), NaCl 1% (w/v).
2. Luria agar: Tryptone 1% (w/v): yeast extract 0.5% (w/v), NaCl 1% (w/v), with agar (bacteriological) 1.5% (w/v).
3. Bacteria: *E. coli* (strain K12) and *B. subtilis* (University of New South Wales Microbiology Culture Collection, Australia). Long-term storage of strains in glycerol at −80°C.

#### 2.1.2. Lysis of Cells

1. Tubes: 1.5-ml LoBind tubes (Eppendorf, Germany) (see Note 1).
2. TRIzol® (Invitrogen, USA) (active components: phenol and guanidine isothiocyanate). TRIzol® disrupts cellular membranes and dissolves cellular components while maintaining the integrity of the RNA by eliminating enzyme activity. TRIzol® is toxic and caustic; avoid contact with skin, wear suitable gloves and eye protection, and always work under a chemical fume hood.
3. RNaseZAP® (Sigma, USA). RNaseZAP® is used to remove the RNases (which degrade RNA) which may be present on utensils, apparatus, and on work surfaces (see Note 2).

#### 2.1.3. RNA Purification

1. Pure Link™ Total RNA Purification System (Invitrogen, USA).
2. Chloroform (Univar, USA). It is an irritant and potentially carcinogenic/mutagenic; avoid contact with skin, wear suitable

gloves and eye protection, and always work under a chemical fume hood.

3. Ethanol (Univar, USA). Working solution of 70% ethanol freshly prepared before each extraction procedure (see Note 3).

#### 2.1.4. Assessment of RNA Quantity and Quality

1. RNA analysis kit: RNA 6000 Nano Kit (Agilent Technologies, USA).
2. NanoDrop spectrophotometer.

### 2.2. Coupling of Oligonucleotide Probes to Beads

1. Beads: Dragon Green Fluorescent Carboxyl Polymers 5.78 μm (Product No. FC06F/8549, Bangs Laboratories, USA). Store at 2–8°C, do not freeze. Limit exposure to light.
2. Oligonucleotide 5′ amino-modified probes (Sigma) appropriate for targeting *E. coli* (strain K12) as shown in the table below.

| Oligonucleotide | Position[a] | Sequence 5′-3′ | Specificity |
|---|---|---|---|
| Eco16S07C (14, 15) | 447 | ACTTTACTCCC-TTCCTC | *E. coli* spp. *Shigella* spp. |

[a]Relative to the *E. coli* 16S rRNA gene (16)

3. Coupling Buffer (2-(*N*-morpholino)ethansulfonic acid-MES) (Sigma, USA): 0.1 M MES, pH 4.5.
4. EDC (*N*-(dimethylaminopropyl)-*N*′-ethylcarbodiimide) (Sigma, USA): 200 mg/ml working solution; freshly prepare before each coupling reaction. Store powder desiccated at −20°C. Use a fresh aliquot for each coupling reaction and discard after use. This is used to activate the carboxyl groups on the beads' surfaces. EDC is an irritant and potentially carcinogenic/mutagenic; avoid contact with skin, wear suitable gloves and eye protection, and always work under a chemical fume hood.
5. Wash/storage buffer: Phosphate buffered saline tablets (Sigma, USA); 0.02% Tween® 20 (Sigma, USA), pH 7.2. Prepare buffer using RNase-free water and autoclave before use.
6. Tubes: 1.5-ml LoBind tubes (Eppendorf, Germany).

### 2.3. Fluorescent Labelling of RNA

1. RNA fluorescent labelling kit: Ulysis™ Alexa Fluor® 647 Nucleic Acid Labelling Kit (Invitrogen, USA).
2. Sodium acetate, Trihydrate (Sigma, USA): 3 M working solution, adjusted to pH 5.2 using glacial acetic acid (Univar, USA).
3. Ethanol (Univar, USA): Absolute and working solution of 70%.
4. Tubes: 1.5-ml LoBind tubes (Eppendorf, Germany).

#### 2.4. RNA Purification Postlabelling

1. RNA purification kit: RNeasy® MinElute® Cleanup Kit (Qiagen, Netherlands).
2. Ethanol (Univar, USA): absolute and working solution of 80%.
3. Tubes: 1.5-ml LoBind tubes (Eppendorf, Germany).

#### 2.5. Hybridisation and Analysis on 2100 Bioanalyzer

1. Cell assay kit: Bioanalyzer Cell Assay Kit (Agilent Technologies, USA).
2. Hybridisation buffer, 20 ml solution: 18 ml of 5 M TMAC (tetramethylammonium chloride); 0.15 ml of 20% sodium dodecyl sulphate (SDS); 1.5 ml of 1 M Tris–HCl, pH 8.0; 0.24 ml of 500 mM EDTA, pH 8.0; 0.11 ml of RNase-free water. All reagents for hybridisation buffer are from Sigma. TMAC is an irritant to eyes, skin, and mucous membranes, always work under a chemical fume hood.
3. Wash/Storage buffer: phosphate buffered saline tablets (Sigma, USA); 0.02% Tween® 20 (Sigma), pH 7.2. Prepare buffer using RNase-free water and autoclave before use.
4. Tubes: 0.5-ml LoBind tubes (Eppendorf, Germany).

## 3. Methods

#### 3.1. RNA Extraction and Purification

##### 3.1.1. Bacterial Culture Preparation

1. Resuscitate *E. coli* and *B. subtilis* strains from lyophilised cultures.
2. Bacterial cultures are to be grown overnight at 30°C or for a few hours (typically 3 h) at 37°C. Growth conditions are dependent on the particular strain used. Avoid isolating RNA from saturated cultures. Optical density at 600 nm (OD at 600 nm) readings can approximate culture density (a reading of 1 at 600 nm is approximately $10^9$ bacteria). Serial dilutions of culture should be plated to confirm culture density at a particular absorbance reading.

##### 3.1.2. Lysis of Cells

1. Transfer 10 ml of the cell solution to a 15-ml polypropylene centrifuge tube.
2. Centrifuge at 2,000 × *g* for 5 min at room temperature to pellet the cells.
3. Discard the supernatant taking care not to dislodge the bacterial pellet.
4. Immediately add 1 ml of TRIzol® to the tube and vortex until the cell pellet is completely lysed and no visible particulate matter remains.
5. Transfer the lysate to a 1.5-ml LoBind tube.

#### *3.1.3. RNA Purification*

1. Follow manufacturer's protocol provided in Pure Link™ Total RNA Purification System with the following exception: use LoBind tubes in place of tubes provided in the kit, where possible.

#### *3.1.4. Assessment of RNA Quantity and Quality*

1. Follow manufacturer's protocol provided in the RNA 6000 Nano Kit and the operator instructions provided in the manual for the 2100 Bioanalyzer (for RNA quality) and NanoDrop spectrophotometer (RNA quantity).
2. Store purified total RNA at −80°C.

### 3.2. Coupling of Oligonucleotide Probes to Beads

1. Warm beads and reagents to room temperature.
2. Vortex beads to thoroughly resuspend.
3. Pipette 100 μl of beads into a 1.5-ml LoBind tube (see Note 4).
4. Centrifuge the beads for 3 min at 12,000 × *g*, and carefully remove and discard the supernatant so that the pellet is not disturbed. If there are beads spread on the surface of the tube, take less supernatant and repeat the process, then resuspend pellet in 400 μl of coupling buffer and vortex.
5. Repeat the previous step, but now resuspend pellet in 170 μl of coupling buffer.
6. Just before use, prepare a 200 mg/ml EDAC solution by dissolving 10 mg of EDAC in 50 μl of coupling buffer. Immediately, add 20 μl of the EDAC solution to the bead suspension.
7. Mix gently end-over-end (preferred) or briefly vortex.
8. Add 9 nmol (90 μl) of oligonucleotide probes (see Note 5). Mix gently end-over-end (preferred) or briefly vortex.
9. Incubate for 2 h in the dark at room temperature (25°C) with gentle end-over-end mixing.
10. Centrifuge mixture for 3 min at 12,000 × *g*, and resuspend microparticle pellet in 400 μl of wash/storage buffer.
11. Repeat previous step twice but on final centrifugation resuspend in 100 μl wash/storage buffer and store at 4°C.

### 3.3. Fluorescent Labelling of RNA

1. To fluorescently label the RNA, follow the manufacturer's protocol provided in the Ulysis™ Alexa Fluor® 647 Nucleic Acid Labelling Kit with the following exceptions: 1 μl of the labelling reagent.

### 3.4. RNA Purification Postlabelling

1. To purify the RNA postfluorescent labelling follow the manufacturer's protocol provided in the RNeasy® MinElute® Cleanup Kit.

### *3.5. Hybridisation and Analysis on 2100 Bioanalyzer*

1. Pre-heat the hybridisation buffer at 70°C to dissolve any precipitate.
2. Add 0.6 μl of probe immobilised fluorescent bead suspension to a 1.5-ml LoBind tube containing 25 μl of hybridisation buffer and 24.4 μl of RNase-free water (see Note 6).
3. Incubate at 45°C for 10 min.
4. Incubate labelled RNA at 95°C for 5 min and rapidly cool on ice for at least 2 min. This step is designed to avoid the influence of the secondary structure of the RNA.
5. Pulse centrifuge RNA to move solution to the bottom of the tube.
6. Pipette the desired amount of the labelled RNA to the hybridisation mixture and incubate in the dark for 2 h at 45°C on a platform shaker with gentle agitation.
7. Add 100 μl of wash buffer to the hybridisation mixture. The hybridisation buffer is viscose; this dilution step prior to centrifugation facilitates pelleting of the beads.
8. Centrifuge for 3 min at 13,000 × *g*, then aspirate fluid by pipette, resuspend in 100 μl of wash buffer, and vortex.
9. Repeat the previous step twice, but centrifuge at 12,000 × *g*. In the final step, instead of resuspending in wash buffer, resuspend with 20 μl of cell buffer (from Bioanalyzer Cell Assay Kit) to give the desired concentration of beads.
10. Prime cell chip with 10 μl of priming solution (from Bioanalyzer Cell Assay Kit).
11. Add 10 μl of focusing dye and 30 μl of cell buffer to specific wells on the chip (from Bioanalyzer Cell Assay Kit).
12. Add 10 μl of the sample suspension to a selected sample well and run the chip on the Bioanalyzer.
13. The Bioanalyzer produces a report with plots of the intensity of the red fluorescence (from labelled RNA) against the intensity of blue fluorescence (from the beads), histograms, and basic statistics. The key information is the average red fluorescence, referred to as "the signal" (see Note 7). A gating option is available in the Bioanalyzer to eliminate the data outliers and undesired regions of the scattergrams.

## 4. Notes

1. In all procedures, LoBind tubes should be used in order to reduce binding between nucleic acids and tubes.
2. Application of RNaseZAP® (or a similar reagent) to all utensils, apparatus, and work surfaces is recommended to eliminate

the presence of RNases. In addition, RNase-free disposable plastic ware should be used wherever possible and common use glassware should be avoided. Where used, common use glassware and other heat resistant common use items (e.g., metal spatulas) should be placed in a 170°C oven overnight before use.

3. Water used in RNA work should always be RNase-free.
4. The literature suggests that $10^6$ or 1 mg of beads are required (17–21).
5. The amount of oligonucleotide probes used here is excessive. Bangs Laboratories recommend using an excess (3–10×) of the surface titration value reported as a starting place before further optimisation (Dragon Green beads have 0.9 μmol COOH groups/gram of beads). The amount of amino-modified oligonucleotide bound to the microparticles is dependent on the concentration of amino-modified oligonucleotide in solution and on the size of the microparticles.
6. The amount of beads should be calculated aiming to reach a maximum concentration of two million beads/ml, giving a number of events around 1,000 per 4-min assay. If the concentration is higher, a fraction of counted events will be generated by more than one particle in the detection window, affecting the data quality (22).
7. In order to establish the presence of the target pathogen, one needs to carry out control experiments. These quantify the amount of nonspecific binding and the extent of binding to nontarget organisms. To obtain the nonspecific binding, the beads without attached probes need to be hybridised to labelled RNA using the protocols as above (with the probe coupling step omitted). The binding to nontarget organisms is quantified by using probes to *E. coli* as specified earlier, but hybridising with RNA extracted from the nontarget organism (*B. subtilis*). The level of bleed-through signal from the beads also needs to be quantified by examining pure beads on the 2100 Bioanalyzer. The bleed-through signal for the Dragon Green beads used here was very small. This signal is subtracted from assay, nonspecific, and other signals. The percentage binding is determined by subtracting the nonspecific signal from the assay signal and dividing the difference by the assay signal. By using the presented protocols, we have been able to obtain specific binding as high as 96%, and no binding (0%) to the nontarget microorganism. We have been able to detect RNA quantities as small as 125 ng, with limited optimisation.

## References


1. Delehanty JB, Ligler FS (2002) A microarray immunoassay for simultaneous detection of proteins and bacteria. Anal Chem 74: 5681–5687
2. McClelland RG, Pinder AC (1994) Detection of Salmonella typhimurium in dairy products with flow cytometry and monoclonal antibodies. Appl Environ Microbiol 60:4255–4262
3. Jarvis RM, Goodacre R (2004) Ultra-violet resonance Raman spectroscopy for the rapid discrimination of urinary tract infection bacteria. FEMS Microbiol Lett 232:127–132
4. Manoharan R, Ghiamati E, Dalterio RA, Britton KA, Nelson WH, Sperry JF (1990) UV resonance Raman spectra of bacteria, bacterial spores, protoplasts and calcium dipicolinate. J Microbiol Methods 11:1–15
5. Belgrader P, Benett W, Hadley D et al (1990) Infectious disease—PCR detection of bacteria in seven minutes. Science 284:449–450
6. Hinata N, Shirakawa T, Okada H, Shigemura K, Kamidono S, Gotoh A (2004) Quantitative detection of *Escherichia coli* from urine of patients with bacteriuria by real-time PCR. Mol Diagn 8:179–184
7. Pugia MJ, Sommer RG, Kuo HH, Corey PF, Gopual DL, Lott JA (2004) Near-patient testing for infection using urinalysis and immunochromatography strips. Clin Chem Lab Med 42:340–346
8. Desai MJ, Armstrong DW (2003) Separation, identification, and characterization of microorganisms by capillary electrophoresis. Microbiol Mol Biol Rev 67:38–51
9. Bhatta H, Goldys EM, Learmonth RP (2006) Use of fluorescence spectroscopy to differentiate yeast and bacterial cells. Appl Microbiol Biotechnol 71:121–126
10. Giana HE, Silveira L, Zângaro RA, Pacheco MTT (2003) Rapid Identification of bacterial species by fluorescence spectroscopy and classification through principal components analysis. J Fluoresc 13:489–493
11. Basu M, Seggerson S, Henshaw J et al (2004) Nano-biosensor development for bacterial detection during human kidney infection: use of glycoconjugate-specific antibody-bound gold NanoWire arrays (GNWA). Glycoconj J 21:487–496
12. Ateya DA, Erickson JS, Howell PB Jr, Hilliard LR, Golden JP, Ligler FS (2008) The good, the bad, and the tiny: a review of microflow cytometry. Anal Bioanal Chem 391: 1485–1498
13. Neidhardt FC, Umbarger E (1996) In: Neidhardt FC, Curtiss III R, Ingraham JL, Lin ECC, Low KB, Reznikiff WS, Riley M, Schaechter M, Umbarger HE (eds) *Escherichia coli* and Salmonella: cellular and molecular biology, 2nd ed. AMS Press, Washington 1:2822
14. Joachimsthal EL, Ivanov V, Tay STL, Tay JH (2003) Quantification of whole-cell in situ hybridization with oligonucleotide probes by flow cytometry of *Escherichia coli* cells. World J Microbiol Biotechnol 19:527–533
15. Stender H, Broomer AJ, Oliveira K et al (2001) Rapid detection, identification, and enumeration of *Escherichia coli* cells in municipal water by chemiluminescent in situ hybridization. Appl Environ Microbiol 67:142–147
16. Brosius J, Palmer ML, Kennedy PJ, Noller HF (1978) Complete nucleotide sequence of a 16 S ribosomal RNA gene from *Escherichia coli*. Proc Natl Acad Sci USA 75:4801–4805
17. Diaz MR, Fell JW (2004) High-throughput detection of pathogenic yeasts of the genus Trichosporon. J Clin Microbiol 42:3696–3706
18. Fitzgerald C, Collins M, van Duyne S, Mikoleit M, Brown T, Fields P (2007) Multiplex, bead-based suspension array for molecular determination of common Salmonella serogroups. J Clin Microbiol 45:3323–3334
19. Iannone MA, Taylor JD, Chen J et al (2000) Multiplexed single nucleotide polymorphism genotyping by oligonucleotide ligation and flow cytometry. Cytometry 39:131–140
20. Spiro A, Lowe M, Brown D (2000) A bead-based method for multiplexed identification and quantitation of DNA sequences using flow cytometry. Appl Environ Microbiol 66: 4258–4265
21. Wilson WJ, Erler AM, Nasarabadi SL, Skowronski EW, Imbro PM (2005) A multiplexed PCR-coupled liquid bead array for the simultaneous detection of four biothreat agents. Mol Cell Probes 19:137–144
22. Nitsche R (2002) Cell fluorescence assays on the Agilent 2100 Bioanalyzer—general use. Agilent Technologies.

# Chapter 14

# Photosynthetic Antenna Complex LHCII Studied with Novel Fluorescence Techniques

**Wieslaw I. Gruszecki, Rafal Luchowski, Wojciech Grudzinski, Zygmunt Gryczynski, and Ignacy Gryczynski**

## Abstract

LHCII is the largest light-harvesting pigment-protein complex of plants, comprising more than half of photosynthetically active chlorophyll pigments in biosphere. Understanding relationship between the molecular structure of the complex and photophysical processes that undergo in this pigment-protein complex is an aim of numerous current studies. This chapter addresses possibility of the application of single-molecule fluorescence measurements and fluorescence lifetime imaging microscopy (FLIM) in a study of LHCII.

**Key words:** LHCII, Single-molecule spectroscopy, FLIM, Photosynthetic antenna, Photoprotection

## 1. Introduction

Light-harvesting pigment-protein complex LHCII is a main photosynthetic antenna complex of plants, responsible for capturing quanta of Sun radiation and for transferring electronic excitation energy towards the reaction centers, where primary electric charge separation takes place (1, 2). According to recent crystallographic analysis (2), the complex is constituted with three principal α-helical components, spanning the thylakoid membrane, hosting several chlorophyll, and xanthophyll pigment components: eight molecules of chlorophyll *a*, six molecules of chlorophyll *b*, two molecules of lutein, one molecule of neoxanthin, and one molecule of violaxanthin (see Fig. 1). LHCII easily associates into trimers and such an organization form is observed in vivo, in the chloroplast membranes (2, 3). Relatively tight packing of all the accessory pigments in LHCII and unique spectral properties of their chromophores assures very efficient electronic excitation energy exchange. Figure 2a presents absorption spectra of principal accessory pigments of the complex.

Wlodek M. Bujalowski (ed.), *Spectroscopic Methods of Analysis: Methods and Protocols*, Methods in Molecular Biology, vol. 875, DOI 10.1007/978-1-61779-806-1_14, © Springer Science+Business Media New York 2012

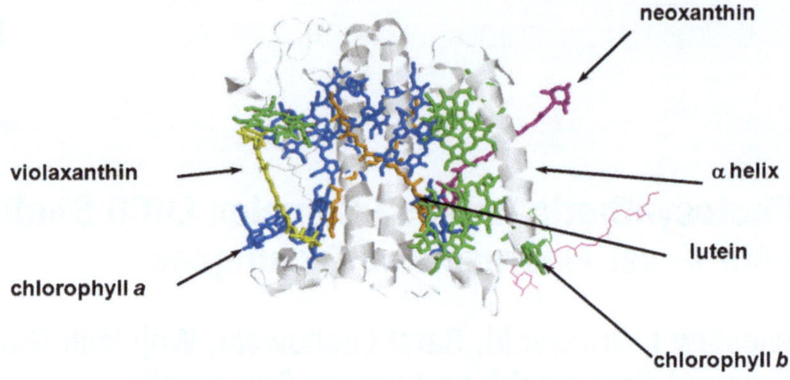


Fig. 1. Structure of the LHCII complex based on crystallographic analysis of Liu et al. (2). The main constituents are labeled.

Excitation of the complex at any wavelength, corresponding to the absorption spectrum of any of the LHCII pigment constituents, gives rise to the fluorescence emission typical of chlorophyll *a* (Fig. 2b). This is a demonstration of very efficient excitonic coupling within the complex, on the one hand, and functional intactness of this protein, on the other hand. Also, the chlorophyll *a* fluorescence excitation spectrum recorded from LHCII (emission at 680 nm) displays a complex structure which reflects relative contribution from all the protein-embedded accessory pigments (Fig. 2c). Those classical experiments, employing fluorescence technique, show very nicely basic photo-physical mechanisms which operate in a functional antenna protein. On the other hand, more complex approach is required to address several open problems regarding detailed molecular mechanisms of the regulation of the excitation energy transfer and conversion in the photosynthetic apparatus. An example of such an open question is: why Nature has chosen the trimeric organization of LHCII as an optimal one? In this chapter, we present the possibility of studying photo-physical processes in LHCII with the application of the single-molecule time-resolved fluorescence spectroscopy and FLIM technique.

## 2. Materials

### 2.1. Isolation of LHCII

1. LHCII was isolated from fresh spinach leaves according to the method elaborated by Krupa et al. (4). Purity and integrity of the complex has been checked by means of HPLC, mass spectrometry, and gel electrophoresis (5).
2. Isolated complex was suspended in Tricine buffer (10 mM, pH 7.6, containing 10 mM KCl). In order to change aggregation status of the protein, a detergent n-Dodecyl-α-D-maltoside (DM) was added to the buffer in concentrations between 0 and 0.4% (6).

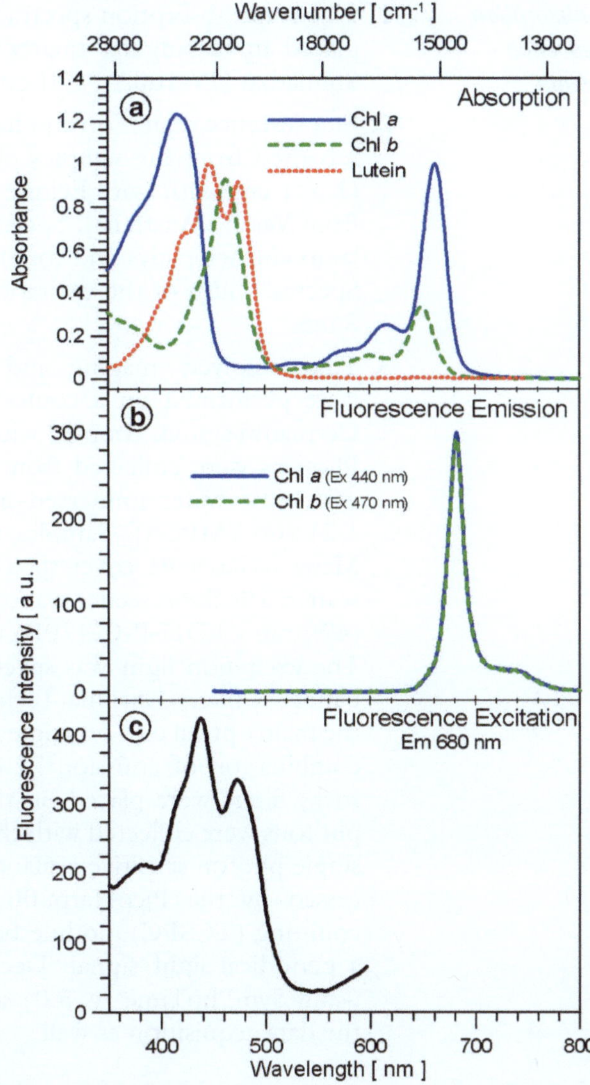


Fig. 2. (**a**) Absorption spectra of the main pigment constituents of LHCII: chlorophyll *a*, chlorophyll *b*, and the xanthophyll lutein, indicated. The spectra recorded in acetonitryl:methanol:water (75:20:5, v:v:v) solvent mixture. (**b**) Fluorescence emission spectra recorded from LHCII suspension in the Tricine buffer containing 0.025% DM. Excitation wavelengths at 440 nm and at 470 nm (indicated) correspond to the main absorption bands in the Soret region of chlorophyll *a* and chlorophyll *b*, respectively. (**c**) Chlorophyll *a* fluorescence excitation spectrum recorded from LHCII solution as in panel (**b**). Emission recorded at 680 nm.

## 3. Methods

### 3.1. Preparation of LHCII Samples

1. In order to carry single-molecule FLIM measurements of LHCII, the complexes were deposited from picomolar suspensions to nonfluorescent Menzel-Glaser #1 cover slips covered with poly-L-lysine (see Note 1).

### 3.2. Light Absorption and Fluorescence Measurements

1. Electronic absorption spectra were recorded from the samples placed in rectangular quartz cells (optical path 1 cm) on a Shimadzu UV-160A-PC spectrophotometer (Japan).
2. Fluorescence excitation and fluorescence emission spectra were recorded from the samples placed in rectangular quartz cells (1 × 1 cm) with Cary Eclipse fluorescence spectrophotometer from Varian (Australia). Spectra were corrected for the Xenon lamp characteristics and for the sensitivity of photomultiplier. Spectral width of the excitation and emission slits were set to 3 nm.
3. Time resolved imaging and single-molecule measurements were performed on a confocal MicroTime 200 (Picoquant, Germany) system coupled with OLYMPUS IX71 microscope. Photons were collected from different places of chosen area with 60× water immersed infinity corrected objective (NA 1.2) (OLYMPUS). Samples were placed on nonfluorescent Menzel-Glaser #1 cover slips and mounted on a stage piezo-scaner. The fluorescence was excited by a solid state pulsed laser (470 nm—LDH-P-C-470B) with repetition rate of 20 MHz. The excitation light was spectrally cleaned by z463/25x-HT bandpass filter (Chroma Technology Corp.) and delivered to the main optical unit by single mode fiber. A 30 μm pinhole and combination of emission 720/240 and 650 longpass (Semrock) filters were placed on the detection path. Fluorescence photons were collected with the Perkin-Elmer SPCM-AQR-14 single photon sensitive avalanche photodiode (APD) and processed by the PicoHarp300 time-correlated single photon counting (TCSPC) module based on detection of photons of a periodical light signal. Decay data analysis was performed using SymPhoTime (v. 5.0) software package that controlled the data acquisition as well.

### 3.3. Interpretation of Measurements

1. LHCII is a largely hydrophobic protein and appears in the aggregated form in a water medium. Aggregated structures of LHCII can be gently decomposed, without pigment extraction and without affecting functional excitation energy exchange between the protein-bound pigments, by the application of mild detergents, such as DM, (6). Application of relatively high detergent concentration (~0.4%) yields samples composed of trimeric and also monomeric LHCII. Figure 3 presents the FLIM image of the polylysine surface with one single LHCII monomer and one single LHCII trimer trapped. The trimeric and monomeric organization of the single particles detected have been assigned as based on the range of the characteristic fluorescence lifetime components of the trimeric complex (7) (aggregated LHCII would have much shorter fluorescence lifetimes) and apparent smaller size of monomeric particle. Moreover, the laser-induced bleaching experiment (see Fig. 4) shows

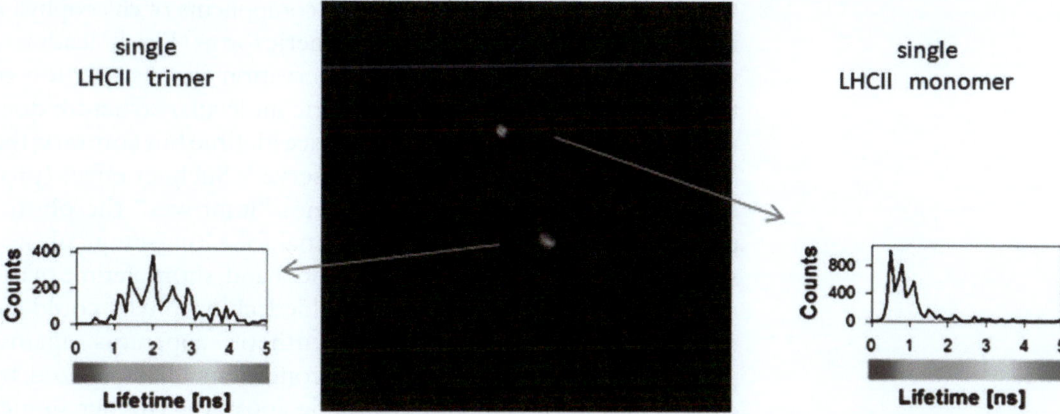


Fig. 3. Fluorescence lifetime imaging microscopy (FLIM) image of the 20 × 20 μm area of the glass slide covered with polylysine, at which the single LHCII monomer and the single LHCII trimer (*indicated*) were trapped from the picomolar suspension of LHCII in the buffer containing 0.4% DM. The *left* and *right panels* show fluorescence lifetime distributions recorded from the monomeric and trimeric LHCII. Excitation wavelength 470 nm, emission was detected in the spectral region above 650 nm. The laser power density was 97 μW/μm$^2$.

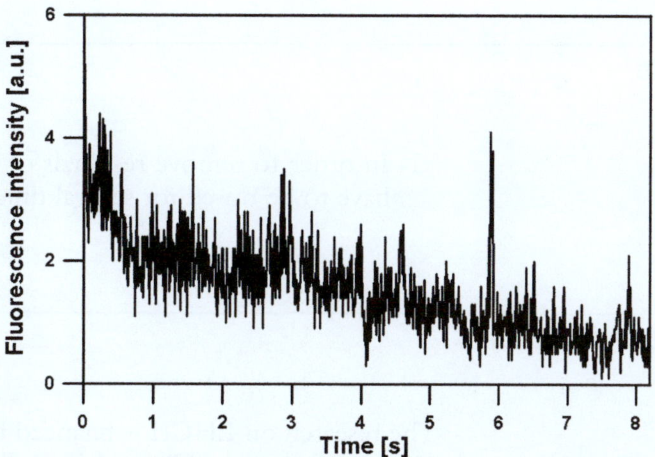


Fig. 4. Time course of fluorescence intensity recorded from the single LHCII monomer (such as visualized in Fig. 3). The decrease steps indicate bleaching of individual chlorophyll pigments in the complex. Excitation wavelength 470 nm, emission was detected in the spectral region above 650 nm. The laser power density was 97 μW/μm$^2$.

that the single particle is composed of several chromophores. This is demonstrated by a stepwise fluorescence decrease, unexpected in the case of a single chlorophyll molecule extracted from the protein. The fact that the excitation was at 470 nm (excitation of chlorophyll *b*) and the fluorescence emission was detected in the spectral window above 650 nm (characteristic of the emission of chlorophyll *a*) assures that the LHCII monomer examined was in a functional physiological state.

2. Comparison of fluorescence lifetimes components of chlorophyll *a* in LHCII in the monomeric and trimeric forms (Fig. 3) leads to a very interesting and surprising observation that association of the protein monomers into a trimeric molecular structure does not result in shortening of fluorescence lifetime but contrary, the longer lifetime components are observed. Such an effect (prolongation of excitation state lifetime) "improves" the photophysical parameters of LHCII essential for a long-range photosynthetic energy transfer. On the other hand, short lifetime of the excited states of the LHCII-embedded chlorophylls could be essential to protect the photosynthetic apparatus against photo-damage, under light stress conditions characterized by overexcitation of the photosynthetic apparatus (higher singlet exciton density than can be utilized by the photochemical processes in the reaction centers). Interestingly, it has been reported that expose of LHCII to strong illumination can induce the trimer to monomer transition (8). Such a light-induced mechanism can be attributed to a photo-protective activity, as based on the fluorescence lifetime analysis presented here.

## 4. Note

1. In order to remove residuals of a protein suspensions samples have to be rinsed for several times with pure buffer.

## Acknowledgments

The research on LHCII is financed by the Ministry of Science and Higher Education of Poland from the funds for science in the years 2008–2011 within the research project N N303 285034. RL acknowledges the postdoctoral fellowship from the Ministry of Science and Higher Education of Poland (grant No.17/MOB/2007/0).

### References

1. Kühlbrandt W (1994) Structure and function of the plant light-harvesting complexes, LHC-II. Curr Opin Struct Biol 4:519–528
2. Liu Z, Yan H, Wang K, Kuang T, Zhang J, Gui L, An X, Chang W (2004) Crystal structure of spinach major light-harvesting complex at 2.72 A resolution. Nature 428:287–292
3. Standfuss R, van Scheltinga ACT, Lamborghini M, Kuhlbrandt W (2005) Mechanisms of photoprotection and nonphotochemical quenching in pea light-harvesting complex at 2.5 A resolution. EMBO J 24:919–928
4. Krupa Z, Williams JP, Khan MU, Huner NPA (1992) The role of acyl lipids in reconstitution

of lipid-depleted light-harvesting complex II from cold-hardened and nonhardened rye. Plant Physiol 100:931–938

5. Gruszecki WI, Gospodarek M, Grudzinski W, Mazur R, Gieczewska K, Garstka M (2009) Light-induced change of configuration of the LHCII-bound xanthophyll (tentatively assigned to violaxanthin): a resonance Raman study. J Phys Chem B 113:2506–2512
6. Voigt B, Krikunova M, Lokstein H (2008) Influence of detergent concentration on aggregation and spectroscopic properties of light-harvesting complex II. Photosynth Res 95:317–325
7. van Oort B, van Hoek A, Ruban AV, van Amerongen H (2007) Aggregation of light-harvesting complex II leads to formation of efficient excitation energy traps in monomeric and trimeric complexes. FEBS Lett 581: 3528–3532
8. Garab G, Cseh Z, Kovacs L, Rajagopal S, Varkonyi Z, Wentworth M, Mustardy L, Der A, Ruban AV, Papp E, Holzenburg A, Horton P (2002) Light-induced trimer to monomer transition in the main light-harvesting antenna complex of plants: thermo-optic mechanism. Biochemistry 41:15121–15129

# Chapter 15

# Analysis of RNA Folding and Ribonucleoprotein Assembly by Single-Molecule Fluorescence Spectroscopy

**Goran Pljevaljčić, Rae Robertson-Anderson, Edwin van der Schans, and David Millar**

## Abstract

To execute their diverse range of biological functions, RNA molecules must fold into specific tertiary structures and/or associate with one or more proteins to form ribonucleoprotein (RNP) complexes. Single-molecule fluorescence spectroscopy is a powerful tool for the study of RNA folding and RNP assembly processes, directly revealing different conformational subpopulations that are hidden in conventional ensemble measurements. Moreover, kinetic processes can be observed without the need to synchronize a population of molecules. In this chapter, we describe the fluorescence spectroscopic methods used for single-molecule measurements of freely diffusing or immobilized RNA molecules or RNA–protein complexes. We also provide practical protocols to prepare the fluorescently labeled RNA and protein molecules required for such studies. Finally, we provide two examples of how these various preparative and spectroscopic methods are employed in the study of RNA folding and RNP assembly processes.

**Key words:** RNA folding, Ribonucleoprotein assembly, Förster resonance energy transfer, Single-molecule fluorescence spectroscopy, RNA labeling, RNA ligation, In vitro transcription of RNA, Protein labeling

## 1. Introduction

Beyond their role as information carriers during gene expression, RNA molecules are involved in a broad range of other activities required for normal cellular function, including pre-mRNA splicing (1–3), translational regulation (4), metabolite sensing (5), protein synthesis (6), and protein targeting (7). RNA molecules also play a key role in viral infection processes (8–10). To execute their biological functions, RNA molecules must fold into specific tertiary structures and/or associate with one or more proteins to form ribonucleoprotein (RNP) complexes. There is often a close interplay between RNA folding and RNP assembly, since binding of

Wlodek M. Bujalowski (ed.), *Spectroscopic Methods of Analysis: Methods and Protocols*, Methods in Molecular Biology, vol. 875, DOI 10.1007/978-1-61779-806-1_15, © Springer Science+Business Media New York 2012

one protein can induce an RNA folding transition that creates the binding site for a second protein. This coupling between protein binding and RNA folding events naturally results in the hierarchical assembly of large RNPs such as the ribosome, spliceosome, and signal recognition particle (3, 6, 11). Once assembled, these large RNPs generally undergo a series of dynamic and reversible conformational changes during the course of their biological functions. Consequently, it is important to elucidate the mechanisms of RNA folding and RNP assembly, as well as the conformational dynamics of RNP complexes, in order to fully understand the diverse biological functions of RNA.

Fluorescence spectroscopy offers a number of advantages as a method for the study of RNA folding and RNP assembly processes. Fluorescence measurements can be performed in solution under physiologically relevant conditions, without restrictions arising from the size of the molecules under study, and dynamic information is obtained over a wide range of time scales, from picoseconds to minutes. Various types of fluorescence parameters can be recorded, including emission intensity, polarization, and lifetime, each of which reports different molecular properties. Owing to technical advances over the past decade, it is now possible to record fluorescence signals from individual biomolecules under physiologically relevant conditions (12–14). Single-molecule measurements are especially powerful, as they directly reveal different RNA conformational subpopulations that are hidden in conventional (ensemble-averaged) measurements. Moreover, kinetic processes can be observed without the need to synchronize a population of molecules. In principle, with these unique capabilities, it is possible to monitor the temporal order of protein-binding events during RNP assembly and the associated RNA conformational changes at each step. Despite this potential, very few such studies have been performed to date (15, 16). Single-molecule fluorescence methods can also be used to monitor autonomous RNA folding processes, such as occur during the catalytic cycle of various small ribozymes (17–19).

The purpose of this chapter is to describe some of the single-molecule fluorescence techniques that are useful in the study of RNA folding and RNP assembly processes. We also provide practical protocols to generate the necessary protein and RNA constructs. In addition, we present two examples from our own laboratory to illustrate how the methods are applied and the type of information that is forthcoming from single-molecule fluorescence studies of RNA folding and RNP assembly. It is hoped that this chapter will encourage the use of single-molecule fluorescence techniques in a broad range of RNA systems in the future.

### 1.1. RNA Folding: Overview of Methods

The spectroscopic phenomenon of Förster resonance energy transfer (FRET) has proven to be very informative as a tool to monitor the folding of individual RNA molecules (20–22). In the FRET process, the excitation energy of a donor fluorophore is transferred non-radiatively through space to a fluorescent acceptor, thereby reducing the fluorescence intensity and lifetime of the donor while giving rise to enhanced emission from the acceptor. The efficiency of FRET ($E$) is strongly dependent on the intervening donor–acceptor distance ($R$), as described by the Förster equation, $E = \left(1 + \left(\frac{R}{R_0}\right)^6\right)^{-1}$, where $R_0$ is the Förster radius for the donor–acceptor pair. Hence, the FRET efficiency can be used to measure donor–acceptor distances (typically in the range from 30 to 70 Å) or to monitor changes in distance during macromolecular folding processes. To monitor the folding of RNA, the donor and acceptor dyes must be appropriately positioned to report on distance changes during the folding process. Suitable positions can often be deduced if the crystal structure of the folded RNA is known. Otherwise, the donor and acceptor can be incorporated in a variety of different positions and the resulting constructs tested under conditions in which the RNA is either completely folded or unfolded. Constructs that exhibit the largest differences in FRET efficiency between the folded and unfolded states are most suitable for detailed studies of the folding process. In addition, the donor and acceptor dye pair has to be selected so that its Förster radius is appropriate to report the anticipated distance change during the folding process.

A limiting factor for FRET experiments is the production of suitable doubly labeled RNA molecules. There are several different ways to obtain an RNA construct that contains donor and acceptor dyes at specific positions. In the case of relatively short RNA molecules (typically less than 50 nt), both dyes can be directly incorporated at defined sites during solid-phase oligonucleotide synthesis, either at strand termini or at internal positions. Alternatively, reactive amino or sulfhydryl groups can be incorporated during solid-phase synthesis and subsequently modified with amine-reactive or sulfhydryl-reactive dye derivatives. In other cases, separate RNA oligonucleotides labeled with either donor or acceptor are annealed to obtain the desired doubly labeled construct for FRET measurements, as in the case of the hairpin ribozyme described later. However, these approaches are not feasible for labeling large RNA molecules that are beyond the practical size limit for solid-phase synthesis (generally longer than 50 nt). In this case, dye-labeled DNA oligonucleotides can be annealed with the unlabeled RNA of interest (generated by in vitro transcription), usually within unstructured loop regions. Such an approach has been used to label ribosomal RNA for FRET studies (23).

In cases where an active, double-labeled RNA molecule cannot be constructed by simple annealing, short RNA oligonucleotides labeled with either donor or acceptor can be covalently joined to create longer molecules (24, 25). The donor and acceptor dyes are directly incorporated into each oligonucleotide during solid-phase synthesis or by post-synthetic labeling of amino- or sulfhydryl-modified oligonucleotides, as described above. The two oligonucleotides are then annealed with a complementary DNA strand (splint), which positions the 5′ end of one oligonucleotide in close proximity to the 3′ end of the other. The two ends are then covalently joined in an enzymatic reaction catalyzed by RNA ligase 2. Since this reaction requires the presence of a 5′ phosphate group and a 3′ hydroxyl group, one of the oligonucleotides is first phosphorylated at the 5′ end using T4 polynucleotide kinase. A detailed protocol is presented later.

Once suitable doubly labeled RNA constructs are in hand, FRET measurements can be performed on individual RNA molecules using either of two different detection formats. In one approach, individual RNA molecules are observed in a confocal microscope setup as they freely diffuse through a tightly focused laser beam (26). Single-molecule events are registered as discrete bursts of fluorescence emission. The number of donor and acceptor photons recorded during each burst is used to calculate the FRET efficiency. The results from many single-molecule events are then compiled in the form of FRET efficiency histograms. These histograms can directly reveal different conformational subpopulations present during the RNA folding process (19). These populations appear as separate peaks in the histogram. Moreover, the shape of the histogram peaks can provide information on the rate of exchange between the different conformers (19). The advantage of this approach is that the RNA molecules are examined under natural solution conditions. However, the observation time is limited by the period of time that the RNA molecule remains in the confocal volume, which is usually on the order of a few to tens of milliseconds. Kinetic information on longer time scales is lost, although equilibrium populations can still be obtained. The necessary instrumentation, data acquisition and data analysis protocols are described below.

In the other approach, individual RNA molecules are tethered to a solid surface and visualized by means of total internal reflection fluorescence (TIRF) microscopy (27, 28). An evanescent field is created by total internal reflection of a laser beam at a quartz–water interface, using either an objective or prism-based optical system (28). Since the evanescent field penetrates just ~100 nm into the solution, only the fluorophores localized in the interfacial region are excited by the field and become capable of emitting fluorescence. Since the RNA molecules are immobilized, the fluorescence from both donor and acceptor can be monitored for relatively long periods of time (typically tens of seconds), until one of the dyes is destroyed by photobleaching. Moreover, many individual RNA molecules can be monitored in parallel using a charge coupled device (CCD) camera, which records a two-dimensional image of the surface. The FRET efficiency for each RNA molecule in the field of view can be tracked over time, directly revealing dynamic folding transitions on a molecule-by-molecule basis. For this approach, biotinylated RNA molecules are immobilized by binding to a streptavidin-coated surface. Generally, the surface is also coated with polyethylene glycol (PEG) to suppress nonspecific adsorption of RNA or protein molecules. The necessary protocols are provided below. We also describe the instrumentation required for TIRF measurements, as well as the data acquisition and analysis protocols.

### 1.2. Ribonucleoprotein Assembly: Overview of Methods

During the process of RNP assembly, one or more proteins bind to a single RNA molecule, usually in a defined temporal order. During assembly, the conformation of the RNA molecule may remain fixed or else it may change in response to each protein-binding event. There are two types of information that can be obtained from single-molecule fluorescence studies of RNP assembly: (1) the temporal order of protein-binding events and (2) the nature of the associated RNA conformational changes (if any). To obtain temporal information, individual steps in the assembly pathway are directly visualized by TIRF microscopy as suitably labeled proteins bind one at a time to an immobilized RNA molecule. As there is no necessity to label the RNA in this approach, arbitrarily long RNA molecules can be employed, which are readily generated in an in vitro transcription reaction. The protocols used for protein labeling and in vitro transcription of RNA are described later. Information on RNA conformational changes can be obtained by single-pair FRET measurements with donor–acceptor RNA, using either freely diffusing or immobilized RNA molecules. In this case, the proteins are usually unlabeled, although more sophisticated three-color FRET experiments can be devised in which both the RNA and proteins are labeled.

## 2. Materials

### 2.1. In Vitro RNA Transcription

1. 100 μg of plasmid DNA containing an insert coding for the desired RNA. To transcribe only the insert of interest, rather than the entire vector, the vector must be completely linearized with a suitable restriction endonuclease prior to the reaction. Following the restriction digestion, the linearized plasmid is extracted with phenol:chloroform:isoamyl alcohol, precipitated with ethanol, and resuspended in nanopure water that has been filtered through a 0.22-μm sterile filter.
2. Transcription buffer (10×): 800 mM K-Hepes (pH 8.1), 10 mM spermidine, Triton X-100 (1% v/v).

3. FPLC buffer A: 50 mM Tris–HCl (pH 7.8).
4. FPLC buffer B: 50 mM Tris–HCl (pH 7.8), 2 M NaCl.
5. 100 mM ATP, divided into 120-μL aliquots and stored at −20°C.
6. 100 mM GTP, divided into 120-μL aliquots and stored at −20°C.
7. 100 mM CTP, divided into 120-μL aliquots and stored at −20°C.
8. 100 mM UTP, divided into 120-μL aliquots and stored at −20°C.
9. 1 M $MgCl_2$, stored at room temperature.
10. 1 M DTT, dissolved in 0.01 M sodium acetate and stored at −20°C.
11. 0.5 M EDTA (pH 8.0), stored at room temperature.
12. T7 RNA polymerase (50 U/mL): divided into 40-μL aliquots and stored at −20°C.
13. 5 mL HiTrap Q High Performance (HP) column (Pharmacia/GE Healthcare).
14. 30,000 NMWL Centriprep centrifugal filter unit (Millipore).
15. TE Buffer: 10 mM Tris–HCl (pH 7.5), 1 mM EDTA.

### 2.2. RNA Ligation

1. Annealing buffer: 50 mM Tris–HCl (pH 7.5) and 50 mM NaCl. Filtered through a 0.22-μm sterile filter and stored at room temperature.
2. T4 polynucleotide kinase buffer (1×): 70 mM Tris–HCl (pH 7.6), 10 mM $MgCl_2$, 1 mM ATP and 5 mM dithiothreitol (DTT). Stored at −20°C.
3. T4 RNA ligase 2 buffer (1×): 50 mM Tris–HCl (pH 7.5), 2 mM $MgCl_2$, 1 mM DTT and 400 μM ATP. Stored at −20°C.
4. DNase I buffer (1×): 10 mM Tris–HCl (pH 7.6), 2.5 mM $MgCl_2$ and 0.5 mM $CaCl_2$. Stored at −20°C.
5. Soaking buffer: 50 mM Tris–HCl (pH 7.5), 300 mM NaCl and 0.5 mM EDTA. Filtered through a 0.22-μm sterile filter and stored at room temperature.
6. Gel running buffer (5×): Tris–borate (54 g tris base and 27.5 g boric acid per 1 L) and 10 mM EDTA (pH 8.0).
7. PAGE loading buffer: 7.3 M urea, 50% formamide and 9 mM EDTA (pH 8.0).
8. Nanopure water, filtered through a 0.22-μm sterile filter.
9. Quick Spin Columns (TE), (G-25 Sephadex Column, Roche Diagnostics). Stored at +2 to +8°C.
10. Illustra NAP-5 columns (Sephadex G25, GE Healthcare).

11. Polyacrylamide gel, such as SequaGel (National Diagnostics). Mini-gel electrophoresis system, such as PROTEAN II (Biorad).
12. Two RNA oligonucleotides. One oligonucleotide is labeled with the FRET donor at a desired location, the other with the FRET acceptor. Synthetic RNA oligonucleotides labeled with fluorescent dyes can be obtained from commercial sources. Alternatively, synthetic oligonucleotides containing reactive amino or sulfhydryl groups can be labeled with succinimidyl ester or maleimide derivatives of the desired fluorescent dyes, respectively. Pure and homogeneous starting material should be used for the ligation. The RNA 5′-end at the ligation site has to be phosphorylated and the RNA 3′-end at the ligation site has to carry a hydroxyl group.
13. DNA oligonucleotide. The DNA splint is selected for efficient annealing with both RNA oligonucleotides. Longer splints might help prevent the formation of undesired RNA secondary/tertiary structures.
14. T4 RNA ligase 2 (10 U/mL, New England Biolabs).
15. RNasin (40 U/mL).
16. DNase I (10 U/mL, Roche).
17. Proteinase K (20 mg/mL).

### 2.3. Biotinylation of RNA

1. 1 nmol RNA.
2. 100 mM $NaIO_4$.
3. 3.3 M KCl.
4. 3 M NaOAc, pH 5.0.
5. 2.5 M NaOAc, pH 5.0.
6. EtOH stored at −20°C.
7. 70% v/v EtOH in $dH_2O$ stored at −20°C.
8. 25 mg biotin hydrazide (Sigma B3770).
9. 1.53 mL DMSO.
10. Refrigerated centrifuge (4°C) with rotor for 1.5-mL vials.

### 2.4. Protein Labeling

1. Labeling buffer: 50 mM $Na_3PO_4$ (pH 7.2) and 500 mM NaCl. Filtered through a 0.22-μm sterile filter, stored at room temperature but kept on ice during the labeling protocol.
2. Elution buffer: 50 mM $Na_3PO_4$ (pH 7.2), 500 mM NaCl, and 250 mM imidazole. Filtered through a 0.22-μm sterile filter, stored at room temperature but kept on ice during the labeling protocol.
3. Storage buffer: 50 mM Tris–HCl (pH 7.2), 1.6 M NaCl, and 1 mM EDTA. Filtered through a 0.22-μm sterile filter and stored at 4°C.

4. Resuspension buffer: 50 mM Tris–HCl (pH 7.2), 1.6 M NaCl, 1 mM EDTA, and 6 M guanidine HCl. Filtered through a 0.22-μm sterile filter and stored at room temperature.
5. Nanopure water, filtered through a 0.22-μm sterile filter.
6. 1 M DTT solution in 0.01 M sodium acetate. Stored at −20°C.
7. Acetic acid.
8. Illustra NAP-10 column (GE Healthcare).
9. HPLC buffer A: 0.1% TFA in $dH_2O$. Filtered through a 0.22-μm sterile filter and stored at room temperature.
10. HPLC buffer B: 0.1% TFA in acetonitrile. Filtered through a 0.22-μm chemically resistant nylon filter and stored at room temperature.
11. 0.8 cm × 4 cm Poly-Prep chromatography column (BioRad).
12. Ni-NTA Superflow nickel-charged resin (Qiagen). The resin is stored in a slurry of 50% resin and 50% storage buffer.
13. 250 mm × 4.6 mm HAISIL 300, C18 5 mm column (Higgins Analytical, Inc.) for HPLC purification.
14. Dialysis membranes: 3.5 K molecular weight cutoff (MWCO) Slide-A-Lyzer, 0.1–0.5 mL capacity (Pierce Biotechnology), 3.5 K MWCO Slide-A-Lyzer, 0.5–3 mL capacity (Pierce Biotechnology), 3.5 K MWCO Slide-A-Lyzer, 3–12 mL capacity (Pierce Biotechnology).
15. Maleimide derivative of the desired fluorophore, such as Alexa Fluor® $555C_2$-maleimide (Invitrogen).

### 2.5. Diffusion spFRET Measurements

1. Annealing buffer: as described in Subheading 2.2.
2. spFRET buffer: A buffer containing an oxygen scavenger should be used (Example: 50 mM Tris–HCl (pH 7.5), 10 mM $MgCl_2$, 1 mM propyl gallate). Each stock solution should be filtered through a 0.22-μm sterile filter. The buffer solution should be prepared freshly before each measurement.
3. 8-chamber micro cuvettes (Nalgene).
4. 10% Tween.

### 2.6. RNA Immobilization and Surface Treatment

1. 30 mm × 40 mm × 1 mm quartz microscope slides (TIRF Solutions), with holes drilled at each end to allow input and output of sample solutions.
2. 3 in. × 1 in. No. 1 glass microscope cover slips (Fisher Scientific).
3. Adhesive spacers (typically double-sided tape).
4. 50 mM $Na_2CO_3$ (pH 8.5). Bring the solution to pH 8.5 with acetic acid and filter through 0.22-μm sterile filter. This solution should be made no more than 1 week in advance due to an unstable pH. Store buffer at 4°C.

5. mPEG-5000 (Laysan Bio, Inc.). Aliquot dry mPEG into ~100 mg quantities in centrifuge tubes. Seal tubes with parafilm and store in a desiccated jar at −20°C.
6. Biotin-mPEG-5000 (Laysan Bio, Inc.). Aliquot into ~2–5 mg quantities in centrifuge tubes. Seal tubes with parafilm and store in a desiccated jar at −20°C.
7. Dry acetone. To prepare, add molecular sieves to a 500-mL beaker and add acetone to fill. Prepare at least 1 day before use.
8. Acetone.
9. 3-Aminopropyltriethoxysilane (Pierce).
10. Immunopure streptavidin. Dissolve dry streptavidin stock in water to 1 mg/mL. Prepare 500 μL of 0.2 mg/mL streptavidin from stock prior to slide preparation.
11. 100 mM propyl gallate. Add propyl gallate to 100 mM in acetonitrile. Separate into 20-μL aliquots and store at −80°C.

## 3. Methods

### 3.1. In Vitro RNA Transcription Protocol

The following protocol is used to generate unlabeled RNA molecules for studies of RNP assembly.

1. This protocol for a 1 mL transcription reaction requires 50 μg of linearized plasmid DNA (prepared as described in Subheading 2.1).
2. Prepare the following reaction mixture in a centrifuge tube. All the RNA nucleotides should be thawed completely before adding to the reaction to ensure proper concentration. T7 polymerase should be kept on ice at all times and added to the reaction last.

| | |
|---|---|
| 100 μL | 10× transcription buffer |
| 60 μL | 100 mM ATP |
| 60 μL | 100 mM GTP |
| 60 μL | 100 mM CTP |
| 60 μL | 100 mM UTP |
| 25.3 μL | 1 M $MgCl_2$ |
| 10 μL | 1 M DTT |
| 20 μL | T7 RNA polymerase |
| 50 μg | Linearized plasmid (volume depends on original concentration) |
| X μL | DEPC-treated distilled $H_2O$ (X is adjusted to bring the total reaction volume to 1 mL) |

3. Incubate the reaction mixture at 37°C for 5 h. After 5 h the mixture should be cloudy, indicating that the reaction was successful.
4. Add 0.5 M EDTA to 50 mM final concentration (111 μL per 1 mL reaction) to quench the reaction. At this point the reaction mixture can be stored at −20°C overnight, if necessary, prior to FPLC purification.
5. Separate the RNA product from the plasmid DNA using FPLC purification as follows. The FPLC column details are listed in Materials #13, Subheading 2.1.
   (a) Wash the column with nanopure water for 10 min.
   (b) Equilibrate with FPLC buffer A for 10 min.
   (c) Set the following program using a flow rate of 1 mL/min:

| | |
|---|---|
| 0–10 min | 0% FPLC buffer B |
| 10–160 min | 0–70% B |
| 160–180 min | 70–100% B |
| 180–190 min | 100% B |
| 190–200 min | 100–0% B |

   (d) Load reaction mixture onto column.
   (e) Run the program set in point (c) in step 5. Collect 3 mL fractions over the entire run. The RNA elutes before the plasmid.
   (f) Following the run, wash the column with 20% ethanol for 10 min. The column is stored in 20% ethanol at 4°C.
   (g) Store all fractions at 4°C before and during step 6.
6. Run an agarose gel (adjust percentage according to RNA length) on all of the fractions surrounding the RNA peak, with a relevant molecular weight marker, to determine which fractions contain pure RNA (no contaminating plasmid).
7. Combine all fractions containing pure RNA.
8. Concentrate and remove the salt from the RNA solution with a Centriprep centrifugal filter unit (Material #14, Subheading 2.1) as follows: Rinse the Centriprep with 15 mL of TE buffer. Load the RNA fractions into the Centriprep and concentrate to 1 mL by centrifugation. Add 15 mL of TE buffer to the Centriprep and concentrate to 1 mL. Repeat twice. Ethanol precipitation can also be used in place of this step.
9. Aliquot the RNA into desired quantities.
10. Store in TE buffer at −20°C, or dry the RNA using a speed vacuum concentrator and store at −20°C.

### 3.2. RNA Ligation Protocol

This protocol is used to create large doubly labeled RNA molecules for FRET studies of RNA folding or RNP assembly. Since the enzymatic ligation reaction requires the presence of a 5′ phosphate group and a 3′ hydroxyl group, one of the oligonucleotides is first phosphorylated at the 5′ end using T4 polynucleotide kinase. One potential drawback of this approach is that the 5′-phophorylated RNA can potentially react with itself to form a cyclic ligation product. A simple way to prevent this undesirable side product is by using a dideoxy modification at the 3′-end.

1. To phosphorylate the 5′-end of RNA, incubate RNA containing 5′-hydoxyl group (up to 300 pmol), T4 polynucleotide kinase buffer, and T4 polynucleotide kinase (20 U) in 50 μL for 1 h at 37°C.
2. To denature T4 polynucleotide kinase, incubate the reaction mixture for 20 min at 65°C.
3. To purify the RNA, use Quick Spin Columns (Roche). Homogenize the resin and let column run dry by gravity. Centrifuge the column for 2 min at 1,100 × *g*. To equilibrate the column with water, add 200 μL nanopure water and centrifuge column for 2 min at 1,100 × *g*. Load 50 μL (maximum) reaction mixture onto the column and centrifuge 4 min at 1,100 × *g*. Collect the sample from flow through. The purified RNA is used in the following ligation reaction.
4. To anneal the two RNAs with the DNA splint, combine the two RNA strands (1 nmol each) and the DNA splint (1 nmol) in annealing buffer and heat the mixture for 3 min at 90°C and let cool to room temperature.
5. To ligate the two RNA strands, add RNasin (660 U/nmol RNA), T4 RNA ligase 2 buffer, and T4 RNA ligase 2 (65 U/nmol RNA) to the solution and incubate for 20 h at 37°C.
6. To remove the DNA splint from the reaction mixture, add DNase I buffer and DNase I (660 U/nmol DNA) and incubate the reaction for 2 h at 37°C.
7. To digest the proteins remaining after the reaction, add Proteinase K (1.3 mg per 1 nmol RNA ligation reaction) and incubate 5 h at room temperature.
8. To prepare the reaction mixture for purification, reduce the volume of the ligation mixture to 500 μL (if necessary) in a speed vacuum concentrator (Savant SC110).
9. To desalt the solution, use a NAP-5 (Sephadex G25, GE Healthcare) column, concentrate the sample with a speed vacuum concentrator and spin until sample is dry.
10. To purify the ligation product, dissolve the pellet in PAGE loading buffer and use a polyacrylamide gel (SequaGel, adjust

polyacrylamide percentage for optimal separation) for separation of products. Pre-run the gel in TBE buffer (1×) for 30 min (300 V for PROTEAN 2 mini-gel apparatus). Load the sample and bromophenol blue dye (to observe progress of gel run) onto the gel and run the gel in running buffer (at 300 V for Mini-gel system) until the dye runs out of the gel. Identify the ligation product either by UV shadowing or fluorescence and excise the corresponding RNA band. Crush and soak in soaking buffer at 4°C overnight.

11. To desalt the ligated RNA, use a NAP-5 column.
12. To concentrate the sample, spin the sample in a speed vacuum concentrator until the sample is dry.
13. Record a UV absorption spectrum to determine the amount of the ligated RNA.

### 3.3. Biotinylation of RNA

The following protocol is used to biotinylate RNA molecules (labeled or unlabeled) at the 3′ end for subsequent attachment to a solid surface (required for TIRF measurements).

1. To oxidize the 3′ cis diol into aldehydes, prepare the following reaction: 5 μM RNA, 100 mM NaOAc, pH 5.0, and 100 mM $NaIO_4$ in a volume of 200 μL. The $NaIO_4$ solution should be made freshly for each reaction.
2. Incubate this reaction for 1 h on ice and protect from light.
3. Add 12.9 μL of 3.3 M KCl to precipitate $IO_4^-$ ions as potassium periodate.
4. Incubate this reaction for 10 min on ice.
5. Spin the reaction mixture in a centrifuge at 4°C for 5 min at a speed of 16,000 × *g*.
6. Following the spin, discard the resultant precipitate.
7. Precipitate the RNA in the supernatant with EtOH as follows:
   (a) Add 20 μL of 3 M NaOAc and 500 μL EtOH to the supernatant.
   (b) Incubate the mixture for 30 min at −20°C.
   (c) Spin the mixture at 4°C for 30 min in a centrifuge at a speed of 16,000 × *g*.
   (d) Discard the resultant supernatant and add 500 μL of cold 70% EtOH.
   (e) Spin the reaction mixture at 4°C in a centrifuge for 5 min at a speed of 16,000 × *g*.
   (f) Discard the resultant supernatant and dry the precipitated RNA. This can be done by incubating the open centrifuge tube at 37°C for ~10 min.

8. Dissolve 25 mg biotin hydrazide in 1.53 μL DMSO and 170 μL $dH_2O$.
9. Add together 79.5 μL of $dH_2O$ and 8 μL of 2.5 M NaOAc, pH 5. Keep this mixture on ice.
10. Keeping this reaction on ice, add 12.5 μL of the biotin hydrazide solution (step 8) and mix.
11. Add this mixture to the dried RNA and swirl to resuspend the RNA.
12. Incubate the reaction mixture for 2 h at room temperature.
13. Add an additional 5 μL of the hydrazide solution and incubate at 4°C overnight.
14. Precipitate the RNA by repeating step 7.
15. Store the dried RNA at −20°C. Alternatively, the RNA can be resuspended in $dH_2O$ or TE buffer (10 mM Tris–HCl pH 7.5, 1 mM EDTA) and stored at −20°C.

### 3.4. Protein Labeling Protocol

The following protocol is used to prepare dye-labeled proteins for studies of RNP assembly. The protocol was developed for the HIV-1 Rev protein but should be applicable to any His-tagged protein containing a single cysteine residue. The HPLC purification and refolding steps are most suitable for relatively small proteins that can be reversibly unfolded and refolded.

1. This protocol is for labeling 1 mL of His-tagged protein containing a single cysteine residue. The protein concentration should be >75 μM.
2. Add 1 M DTT to protein to 5 mM final concentration (5 μL for 1 mL aliquot). Incubate overnight at 4°C.
3. To remove DTT, pass protein solution through an NAP-10 column as follows: To equilibrate the column, add 15 mL (3 column volumes) of labeling buffer to the column 5 mL at a time. Once the labeling buffer has completely passed through column, add the 1 mL protein aliquot and allow to fully enter the gel bed. To elute the protein, add 1.5 mL labeling buffer to the column. Collect elution product in a 1.5-mL centrifuge tube.
4. Dissolve 3× to 5× molar excess, compared to protein, of maleimide dye, such as Alexa Fluor® 555 $C_2$-maleimide (Invitrogen) in DMSO. Dye should be dissolved in DMSO to a final volume of less than 45 μL (DMSO content of the solution must be less than 3%). Add dye solution to protein (elution product from step 3). Wrap in foil and place on a rocker for ~3 h at 4°C.
5. To quench the reaction, add 3 mL β-mercaptoethanol (2 mg/mL).

6. To remove free dye from the protein solution, run a Ni:NTA column as follows:
   (a) To prepare the column, add 1 mL Ni:NTA resin slurry to column. Allow storage buffer to completely pass through column.
   (b) Wrap the column in foil to protect the labeled protein from light.
   (c) To wash the column, add 10 mL sterile-filtered nanopure $H_2O$ to the column. Allow water to fully pass through column.
   (d) Add 4 mL labeling buffer to the column and allow to fully pass through the column.
   (e) Add the labeled protein solution (1.5 mL) to the column and allow to fully enter the resin.
   (f) To remove free dye, add 5 mL labeling buffer. If the flow-through from the column is not clear (no visible fluorescence) after 5 mL, continue adding labeling buffer until flow-through is clear.
   (g) To elute the protein, add 1 mL elution buffer to the column and collect elution product in 1.5-mL centrifuge tube. Repeat until elution product and column resin is clear.
   (h) Combine the elution product from all the tubes with visible fluorescence (typically 2–5 mL).
7. If the labeling efficiency is sufficiently high (typically > 90%), the labeled protein can be used without further purification. In this case, dialyze the protein against storage buffer (point (b) in step 14) and concentrate using Centriprep centrifugal filter unit. Small aliquots of protein are then flash frozen in liquid nitrogen and stored at −80°C.
8. If the labeling efficiency is low, it is necessary to separate the labeled and unlabeled fractions. In the following steps, the protein is denatured, purified by reverse-phase HPLC and then refolded to the native state.
9. Dialyze labeled protein against 2 L of 2% acetic acid in nanopure $H_2O$, contained in a dialysis reservoir (Nalgene) with magnetic stir bar. Load labeled protein solution into a 3.5 K MWCO Slide-A-Lyzer with the appropriate volume capacity, and place in the reservoir. Cover the reservoir in foil to protect the sample from light. Place reservoir on magnetic stirrer and dialyze at 4°C for ~3 h. Switch to fresh dialysis buffer and continue dialysis overnight.
10. Remove protein from Slide-A-Lyzer.

11. Divide protein into 0.5-mL fractions in 1.5-mL centrifuge tubes. Place fractions in a speed vacuum concentrator (Savant SC110) and spin until protein fractions are dry.
12. Purify labeled protein from unlabeled protein using HPLC purification, as follows. The column details are listed in Subheading 2.4.
    (a) Dissolve dry protein fractions in HPLC buffer A to a total volume of 110 μL. Combine all fractions and mix. This volume is sufficient for a 10 μL analytical run, to check for elution time and to ensure everything is working correctly, and 2 × 50 μL preparative runs to purify the labeled protein.
    (b) Equilibrate with 5% HPLC buffer B for 30 min.
    (c) Set the following program:

| | |
|---|---|
| 0–5 min | 5% B |
| 5–20 min | 5–35% B |
| 20–50 min | 35–55% B |
| 50–55 min | 55–80% B |
| 55–65 min | Wash 80% B |
| 65–75 min | 80–5% B |
| 75–90 min | Equilibrate 5% B |

    (d) Load 10 μL of protein solution onto the column and begin the program.
    (e) The labeled protein will elute after the unlabeled protein. Collect the elution product corresponding to the labeled protein. Fluorescence should be visible by eye during collection.
    (f) Repeat points (b)–(e) in step 12 with 50 μL of protein solution. Repeat with remaining protein.
    (g) During the final HPLC run stop the program when 80% B has been reached. The HPLC column should be stored in 80% B.
13. Combine collected labeled protein and dry using a speed vacuum concentrator, as described in step 11.
14. Renature the protein as follows:
    (a) Dissolve the protein in chilled resuspension buffer. The concentration should be adjusted to prevent the protein from precipitating from solution during refolding.
    (b) Prepare 2 L storage buffer for dialysis. Pour storage buffer into 2 L dialysis reservoir with magnetic stir bar. Store at 4°C until buffer is chilled.

(c) Load the protein into a 3.5-K MWCO Slide-A-Lyzer of appropriate volume (typically 0.1–0.5 mL) and place in the dialysis reservoir.

(d) Cover the reservoir in foil, place it on a magnetic stirrer, and dialyze at 4°C for ~3 h. Switch to fresh prechilled storage buffer and continue dialysis overnight.

(e) Remove protein from the Slide-A-Lyzer.

15. Divide protein into 20-μL aliquots and store at −80°C.

### 3.5. Diffusion spFRET Instrumentation

Diffusion single-pair FRET measurements are performed using an epi-illuminated confocal microscope setup (Fig. 1) (26). Lasers selected for excitation of donor and acceptor dyes are used as light sources. For alternating laser excitation (ALEX) (29), electro-optical modulators are used to rapidly switch between donor and acceptor excitation. The ALEX method automatically rejects

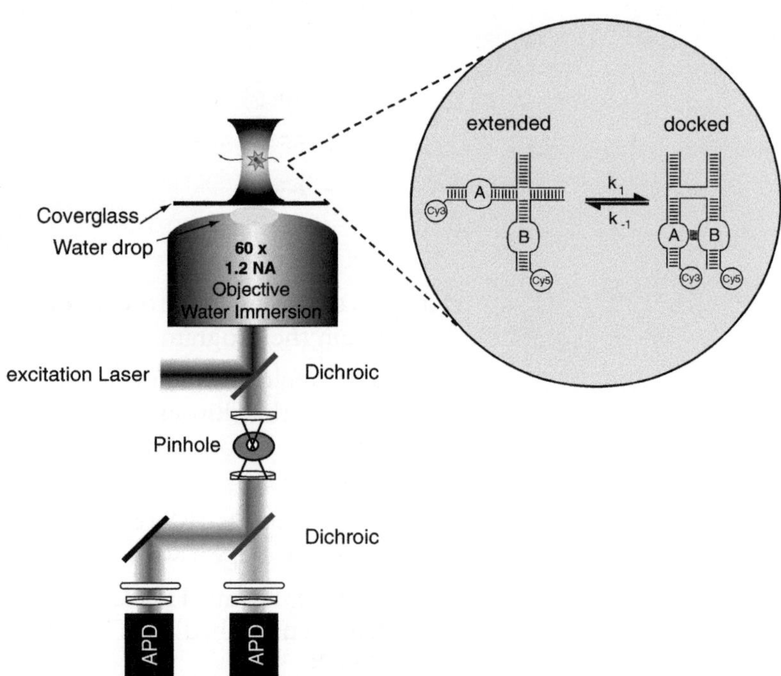


Fig. 1. Experimental setup used for diffusion spFRET measurements. As an example, a single hairpin ribozyme molecule labeled with donor (Cy3) and acceptor (Cy5) dyes is shown traversing the excitation volume of a tightly focused laser beam. The ribozyme can exist in two conformations (extended or docked), which are distinguished by their different FRET efficiencies (see Fig. 2). The fluorescence emitted by both donor and acceptor is collected by the same objective used for laser excitation, transmitted through a small pinhole to reject emission of molecules outside the confocal volume, spectrally separated with dichroic mirrors and bandpass filters, and finally detected on separate avalanche photodiode (APD) detectors. For ALEX measurements, two different lasers are used to alternately excite the donor and acceptor.

RNA molecules labeled with only the FRET donor, which often arise from photobleaching of the acceptor or incomplete acceptor labeling. For ease of alignment the two laser beams are directed into a water immersion objective (60×, 1.2 NA) via an optical fiber. The emitted donor and acceptor fluorescence is collected through the same objective and a pinhole is used to select the light coming from the confocal volume only. A dichroic mirror and cutoff filters are used to spectrally separate the donor and acceptor emission, which are detected simultaneously on separate avalanche photo diodes (APD). The collected signals are acquired with a photon-counting card (National Instruments or Becker & Hickl) under the control of appropriate data acquisition software (e.g., Labview).

### 3.6. Diffusion spFRET Data Acquisition

1. To prepare the cuvettes for spFRET measurements, treat each sample chamber with Tween (10%) for 1 h at room temperature. Clean the sample chambers four times with distilled water and dry the chambers overnight.
2. To anneal the RNA, heat a stock solution of labeled RNA (1 mM) in annealing buffer to 90°C for 3 min and let cool to room temperature.
3. For alignment and measurement, dilute the RNA solution twice in spFRET buffer to a final concentration of 1 nM (setup alignment) and 100 pM (measurement).
4. Load 200 μL of the 1 nM RNA solution into a sample chamber and place the cuvette onto the stage of the confocal microscope.
5. Focus the laser beam on the inner glass surface of the cuvette. Then move the objective so that the focus is located 35 μm into the solution.
6. Reduce the power of the excitation laser to avoid damage to the APDs. Choose a large pinhole (200 μm diameter) to align the APDs for maximum signal intensity. Reduce the pinhole size (75 μm diameter) and position the pinhole for maximum signal intensity.
7. To perform a measurement, load the 100 pM RNA sample and adjust the power of the excitation laser to approximately 200 μW (for Cy3/Cy5 dye pair).
8. To perform an ALEX measurement, the sample is excited with two excitation lasers that alternate rapidly (e.g. every 50 μs). The data acquisition is the same, regardless of the number of excitation lasers.
9. Count photons separately for donor and acceptor. For ALEX experiments an additional synchronization signal is recorded to be able to assign collected photons to the corresponding excitation laser during subsequent data analysis.

### 3.7. Diffusion spFRET Data Analysis

Isolated bursts of donor and acceptor fluorescence corresponding to single-molecule events are identified using a burst search algorithm (30). Only events that exceed a defined background threshold are used for further analysis. The FRET efficiency ($E$) for each single-molecule event is calculated using the equation $E = (1 + \gamma I_d/I_a)^{-1}$, where $I_d$ and $I_a$ represent the donor and acceptor intensities, respectively, and $\gamma$ is a correction factor that accounts for differences in quantum yield and detection efficiency of the donor and acceptor (usually set to 1). A histogram containing the FRET efficiencies from many single-molecule observations (typically thousands of events are recorded during an acquisition time of a few minutes) is compiled (e.g., using Matlab software). Conformational subpopulations can be directly identified as separate peaks in the FRET efficiency histogram (an example is shown in Fig. 2). Gaussian functions are used to fit these peaks and obtain the mean FRET efficiency, the width of the distribution and the area under each peak (using Origin or Kaleidagraph software). The mean FRET efficiencies are related to the average donor–acceptor distance ($R$) for each conformer, through the Förster equation, given above. The peak areas reflect the fractional populations of each conformer, and the width of the peak reflects conformational flexibility of the corresponding subpopulation if the theoretical shot noise width is exceeded. Hence, changes in the conformer populations and conformer properties can be followed in response to changes in solution conditions, such as metal ion concentration (an example is shown in Fig. 2, bottom), ionic strength, temperature, or the presence of specific RNA-binding proteins.

### 3.8. RNA Immobilization and Surface Treatment

The following protocol is used to immobilize biotinylated RNA molecules for TIRF measurements and to passivate the quartz surface with polyethylene glycol (PEG) chains.

1. Place the desired number of quartz slides and glass cover slips in holders designed to separate all surfaces and keep all surfaces open to surrounding air or liquid.
2. Clean the slides and cover slips using a plasma cleaner as follows: Place the holders containing the slides and cover slips in the plasma cleaner (Harrick Scientific). Follow the manufacturer's directions to start cleaning process. Set the RF power to HIGH and run plasma cleaner for ~5 min.
3. Prepare a 2% solution of 3-aminopropyltriethoxysilane (Aminosilane) in dry acetone by adding 2 mL of Aminosilane to 98 mL of dry acetone in a 200-mL glass beaker. Following slide and cover slip cleaning, immerse each holder in the 2% Aminosilane solution for 60 s.
4. To rinse the slides and cover slips, immerse each holder in acetone for 60 s.

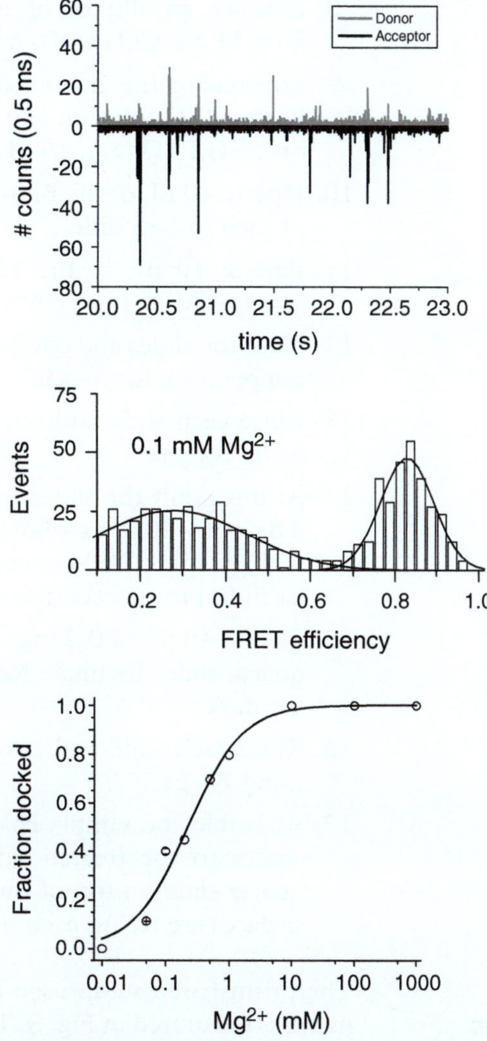


Fig. 2. Example of diffusion spFRET data for a doubly labeled hairpin ribozyme molecule. *Upper panel*: Typical bursts of fluorescence emission from donor and acceptor observed as single ribozyme molecules diffuse through the confocal volume. Photons are counted in a 0.5 ms time bin. The acceptor counts are presented as negative values for ease of presentation. *Middle panel*: FRET efficiency histogram compiled from thousands of single-molecule events. The $Mg^{2+}$ ion concentration is 0.1 mM. The two well resolved peaks correspond to extended ($E \sim 0.3$) and docked ($E \sim 0.85$) ribozyme conformations. The smooth lines are fits to Gaussian functions. *Lower panel*: Fraction of docked ribozymes as a function of $Mg^{2+}$ ion concentration. Increasing $Mg^{2+}$ concentrations favor formation of the docked conformation. The *solid line* is a fit to the Hill equation.

5. Dry all the slides and cover slips using a stream of $N_2$ gas.
6. Place the dry slides and cover slips into individual holders.
7. Dissolve an aliquot of mPEG-5000 (PEG) in 50 mM $Na_2CO_3$ to 150 mg/mL.

8. Dissolve an aliquot of Biotin-mPEG-5000 (Biotin-PEG) in 50 mM $Na_2CO_3$ to 10 mg/mL.
9. To prepare the quartz slide surface treatment, combine the Biotin-PEG solution and PEG solution to a ratio of 1:19 Biotin-PEG:PEG (e.g. 15 μL of Biotin-PEG and 285 μL of PEG).
10. Pipette 40 μL of the Biotin-PEG:PEG solution onto the center of each quartz slide.
11. Pipette 40 μL of the 150 mg/mL PEG solution onto the center of each glass cover slip.
12. Place the slides and cover slips in the dark and incubate at room temperature for 3–4 h.
13. Rinse each slide and cover slip with nanopure water and dry using $N_2$ gas.
14. At this point the slides and cover slips can be stored for up to 3 days at −20°C as follows: Seal each slide and cover slip holder with parafilm. Place sealed holders in a desiccated jar. Cover the jar in foil to protect it from light. Store at −20°C.
15. Pipette 40 μL of 0.2 mg/mL streptavidin on the center of each quartz slide. Incubate for 15–20 min at room temperature in the dark.
16. Rinse each slide and cover slip with nanopure water and dry using $N_2$ gas.
17. Assemble the sample chamber as follows: Attach an adhesive spacer to the treated side of each glass coverslip. Place the quartz slide on top of the cover slip such that the two treated surfaces are facing each other.

### 3.9. TIRF Instrumentation

The prism-based setup used in our laboratory for TIRF measurements is illustrated in Fig. 3. The excitation laser is focused through the prism on a quartz slide. The incident angle is adjusted for total internal reflection at the quartz–water interface, yielding an evanescent field from the upper sample chamber surface on which the RNA molecules are immobilized. Fluorescence emitted by excited fluorophores is collected through a water immersion objective (63×, 1.2 NA) using dichroic and cutoff filters to select the fluorescence of the desired dye and reject scattered laser light. Fluorescence is detected with an intensified CCD camera (Andor model 897). For FRET experiments, the fluorescence from donor and acceptor dyes are spectrally separated and detected on different areas of the CCD camera using a Dual View optical splitter (Photometrics). Individual immobilized molecules appear as fluorescent spots whose intensity is recorded over time in the form of a movie. The time resolution is determined by the frame rate of the CCD camera, which is typically adjustable from 5 ms to several hundred milliseconds per frame.

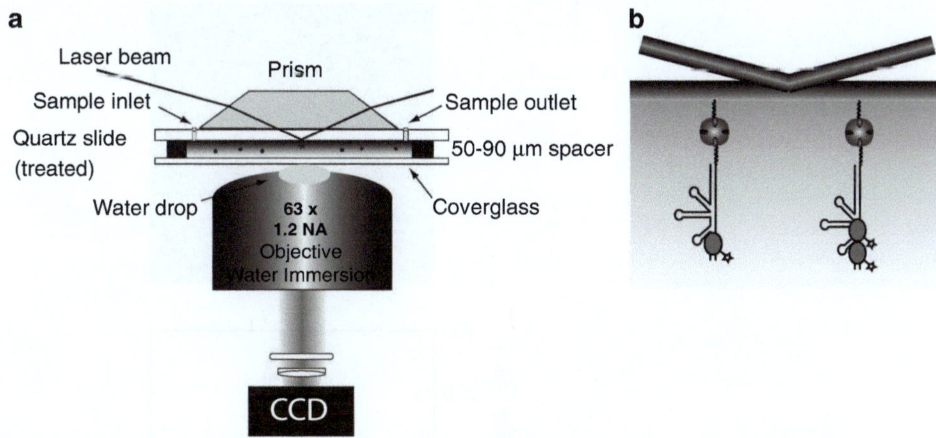


Fig. 3. Experimental setup used for single-molecule TIRF measurements. Panel (**a**): A laser beam enters a prism and impinges on a quartz surface at an incident angle above the critical angle for total internal reflection. The resulting evanescent field transmitted into the aqueous medium excites fluorescently labeled molecules immobilized on the quartz surface. The resulting fluorescence emission is collected though a water immersion objective and imaged on a CCD camera. Panel (**b**): As an example, a close up view of individual RRE molecules immobilized on a quartz slide by means of biotin–streptavidin bonds is shown. One or two fluorescently labeled Rev molecules are shown bound to the immobilized RRE molecules.

### *3.10. TIRF Data Acquisition*

Two different detection formats are typically used. For single-color measurements of protein binding during RNP assembly, unlabeled RNA is biotinylated and immobilized on the quartz surface (as above) and the labeled protein is present in solution. Binding or dissociation of the labeled protein to the immobilized RNA is registered as discrete jumps in fluorescence intensity from individual spots within the TIRF field of view (Fig. 4). For FRET measurements of RNA folding and/or RNP assembly, donor–acceptor labeled RNA molecules are biotinylated and immobilized on the quartz surface and the fluorescence signals from donor and acceptor are monitored over time on separate segments of the CCD camera. Magnesium ions or unlabeled proteins can also be added. The following protocol describes the general procedure to obtain TIRF movies using a prism-based microscope setup.

1. Place the sample chamber onto the microscope stage, fill with buffer and align the excitation laser beam to obtain total internal reflection at the quartz–water interface. Focus the objective on the upper glass surface. Make sure to use glycerol at the contact site of the quartz slide and prism and water at the contact site between objective and cover slip.
2. Load biotinylated RNA molecules into the sample chamber. Generally, 1 mM propyl gallate or another oxygen scavenger is added along with the sample to hinder photobleaching.
3. Turn on the laser light source and adjust the focus to observe the immobilized fluorescent molecules. Adjust the excitation

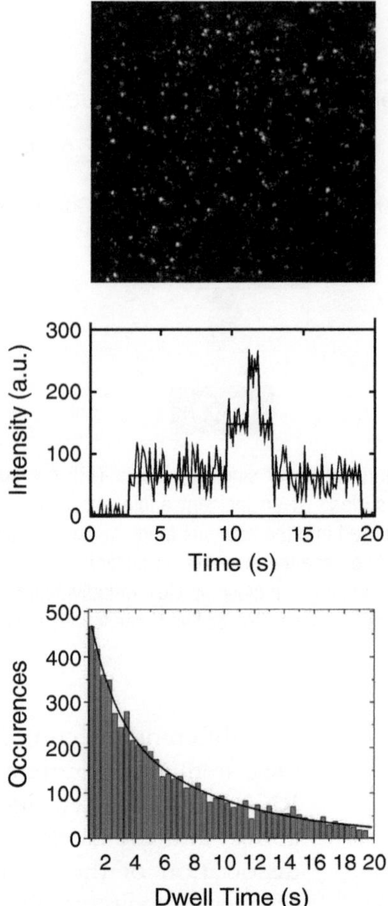


Fig. 4. Visualizing Rev–RRE complex assembly by single-molecule TIRF microscopy. *Upper panel*: Typical TIRF image of surface-immobilized Rev–RRE complexes. The discrete spots correspond to individual Rev–RRE complexes containing various numbers of bound Rev monomers. *Middle panel*: Intensity trajectory for a single fluorescent spot, revealing discrete Rev binding and dissociation events, manifested as abrupt jumps in the fluorescence intensity. *Lower panel*: Histogram of dwell times spent in the intensity state corresponding to zero Rev monomers before transition to the intensity state corresponding to a single Rev monomer. The histogram is fitted to an exponential function to obtain the association rate constant for the first Rev monomer. A similar analysis can be performed for each Rev binding and dissociation event (16).

laser power to yield distinguishable fluorescent spots over background.

4. Record the intensity fluctuations over time as a movie file using the CCD camera software.
5. Introduce other sample components, such as monovalent or divalent ions or specific protein partners (unlabeled), as required, and continue data acquisition.

6. Similar TIRF measurements are usually also performed in the absence of immobilized RNA to assess the extent of nonspecific protein adsorption on the quartz surface (in the case of labeled proteins).

### *3.11. TIRF Data Analysis*

Custom software is used to analyze the recorded TIRF movies. Individual fluorescent spots are first identified in the 2D images and their location is determined. In FRET experiments, individual spots in the donor channel are matched with corresponding spots in the acceptor channel. For each single spot, the time-dependent intensity values are established after suitable background correction, and intensity or FRET efficiency trajectories (time traces) are constructed. These trajectories are analyzed using hidden Markov modeling to deduce the number of distinguishable states, the fluorescence intensity or FRET efficiency for each state and the transition rates connecting them (31).

## 4. Example: Folding of the Hairpin Ribozyme

Here we provide an example of an RNA folding study performed under equilibrium conditions using the diffusion spFRET method. The hairpin ribozyme is a small endonucleolytic RNA molecule responsible for reversible phosphodiester cleavage reactions (32). In order to attain catalytic activity, the ribozyme must fold into a compact structure that positions the two internal loops (A and B) in close physical proximity. We monitored the folding of a hairpin ribozyme construct that contains donor and acceptor dyes on the two loop-bearing arms (Fig. 1) (19). The ribozyme was assembled from four synthetic RNA oligonucleotides, two of which were end-labeled with donor (Cy3) or acceptor (Cy5) dyes. Individual ribozyme molecules traversing a confocal volume give rise to isolated bursts of donor and acceptor fluorescence (Fig. 2, upper panel). A FRET efficiency histogram was constructed from many single-molecule events, revealing separate peaks due to extended and docked conformational subpopulations (Fig. 2, middle panel). The equilibrium distribution of extended and docked conformers is readily determined as a function of Mg-ion concentration (or other solution conditions) from these spFRET measurements, revealing that the docked conformation is stabilized by one or more Mg-ions (Fig. 2, lower panel). Similar measurements performed with ribozymes containing different helical junction geometries (two-way or three-way junctions instead of the natural four-way junction) or specific mutations have identified the structural elements required for efficient docking (19). The same approach could be used to monitor folding of a doubly labeled RNA molecule in the presence of a cognate binding protein (unlabeled).

## 5. Example: Assembly of the HIV-1 Rev–RRE Complex

Here we describe a TIRF-based study of RNP assembly, using the HIV-1 Rev–RRE system as an example. The Rev protein mediates export of unspliced and partially spliced viral mRNAs from the nucleus to the cytoplasm of an infected cell (8). To do so, Rev binds to a conserved region of the viral mRNA, known as the Rev response element (RRE), where it forms an oliogomeric RNP consisting of multiple Rev proteins bound to a single RNA molecule. The detailed assembly mechanism of the Rev–RRE complex was investigated using the methods described in the preceding sections (16). A large fragment of the RRE generated by in vitro transcription was immobilized on a quartz slide. The Rev protein was mutated to change one of the two native cysteine residues to serine. The mutant protein was labeled at the remaining cysteine residue with Alexa Fluor555, purified by HPLC and refolded as described above. Single-color TIRF microscopy was used to monitor the binding of one or more labeled Rev proteins to the immobilized RRE molecules. A two-dimensional image of the surface-immobilized Rev–RRE complexes is shown in Fig. 4 (upper panel) and a typical fluorescence intensity trajectory for a single spot within this image is shown in the middle panel. Abrupt jumps between different fluorescence intensity levels are observed as individual Rev proteins bind to or dissociate from the immobilized RNA. Binding of up to four Rev monomers to a single truncated RRE molecule was observed. A jump size distribution compiled from thousands of such trajectories revealed that more than 90% of transitions correspond to binding or dissociation of single Rev monomers (16). Hence, the Rev–RRE complex assembles by a sequential monomer-binding pathway, rather than by direct binding of preformed Rev oligomers. The binding and dissociation rates for each Rev monomer were obtained from a statistical analysis of the dwell times in each intensity state. An example of a typical dwell time histogram is shown in Fig. 4 (bottom panel). While all the proteins are of the same type in this example, the experimental system could be readily extended to visualize the binding of different proteins, each labeled with a distinct dye, to the same RNA molecule. In this case, multicolor TIRF detection could be used to monitor binding of the different proteins and to establish temporal correlations between individual binding events. In principle, this approach could be applied to a variety of RNPs, limited only by the ability to label the constituent proteins.

## Acknowledgments

Work in the author's laboratory was supported by NIH grants RO1 GM058873 and P50 GM082545.

### References

1. Collins CA, Guthrie C (2000) Nat Struct Biol 7:850–854
2. Valadkhan S (2007) Biol Chem 388:693–697
3. Wahl MC, Will CL, Lührmann R (2009) Cell 136:701–718
4. Winkler WC, Breaker RR (2005) Annu Rev Microbiol 59:487–517
5. Roth A, Breaker RR (2009) Annu Rev Biochem 78:305–334
6. Staley JP, Woolford JL (2009) Curr Opin Cell Biol 21:109–118
7. Keenan RJ, Freymann DM, Stroud RM, Walter P (2001) Annu Rev Biochem 70:755–775
8. Pollard VW, Malim MH (1998) Annu Rev Microbiol 52:491–532
9. Iglesias N, Stutz F (2008) FEBS Lett 582: 1987–1996
10. Boulo S, Akarsu H, Ruigrok RWH, Baudin F (2007) Virus Res 124:12–21
11. Sykes MT, Williamson JR (2009) Annu Rev Biophys 38:197–215
12. Nie S, Zare RN (1997) Annu Rev Biophys Biomol Struct 26:567–596
13. Moerner WE, Orrit M (1999) Science 283:1670–1676
14. Joo C, Balci H, Ishitsuka Y, Buranachai C, Ha T (2008) Annu Rev Biochem 77:51–76
15. Stone MD, Mihalusova M, O'Connor CM, Prathapam R, Collins K, Zhuang X (2007) Nature 446:458–461
16. Pond SJ, Ridgeway WK, Robertson R, Wang J, Millar DP (2009) Proc Natl Acad Sci USA 106:1404–1408
17. Tan E, Wilson TJ, Nahas MK, Clegg RM, Lilley DM, Ha T (2003) Proc Natl Acad Sci USA 100:9308–9313
18. Zhuang X, Kim H, Pereira MJ, Babcock HP, Walter NG, Chu S (2002) Science 296: 1473–1476
19. Pljevaljcić G, Millar DP, Deniz AA (2004) Biophys J 87:457–467
20. Bokinsky G, Zhuang X (2005) Acc Chem Res 38:566–573
21. Pljevaljčić G, Millar DP (2008) Methods Enzymol 450:233–252
22. Alemán EA, Lamichhane R, Rueda D (2008) Curr Opin Chem Biol 12:647–654
23. Dorywalska M, Blanchard SC, Gonzalez RL, Kim HD, Chu S, Puglisi JD (2005) Nucleic Acids Res 33:182–189
24. Moore MJ, Query CC (2000) Methods Enzymol 317:109–123
25. Bullard DR, Bowater RP (2006) Biochem J 398:135–144
26. Mukhopadhyay S, Deniz AA (2007) J Fluoresc 17:775–783
27. Roy R, Hohng S, Ha T (2008) Nat Methods 5:507–516
28. Axelrod D (1989) Methods Cell Biol 30: 245–270
29. Kapanidis A, Lee NK, Laurence TA, Doose S, Margeat E, Weiss S (2004) Proc Natl Acad Sci USA 101:8936–8941
30. Eggeling C, Berger S, Brand L, Fries JR, Schaffer J, Volkmer A, Seidel CA (2001) J Biotechnol 86:163–180
31. McKinney S, Joo C, Ha T (2006) Biophys J 91:1941–1951
32. Walter NG, Burke JM (1998) Curr Opin Chem Biol 2:24–30

# Chapter 16

# Single-Molecule Force Spectroscopy of Polycystic Kidney Disease Proteins

**Liang Ma, Meixiang Xu, and Andres F. Oberhauser**

## Abstract

Atomic force microscopy in its single-molecule force spectroscopy mode is a nanomanipulation technique that is extensively used for the study of the mechanical properties of proteins. It is particularly suited to examine their response to stretching (i.e., molecular elasticity and mechanical stability). Here, we describe protein engineering strategies and single-molecule AFM techniques for probing protein mechanics, with special emphasis on polycystic kidney disease (PKD) proteins. We also provide step-by-step protocols for preparing proteins and performing single-molecule force measurements.

**Key words:** Atomic force microscopy, Single-molecule force spectroscopy, Protein mechanics, Mechanical properties, Protein elasticity, Polycystic kidney disease, Polycystin

## 1. Introduction

Sensing of mechanical forces is mediated by mechanosensitive proteins. Mechanical forces induce conformational changes which result in the modulation of signaling pathways. Mutations in these proteins typically lead to defects in mechanotransduction which have been implicated in the development of various diseases ranging from muscular dystrophies to kidney disease (1). For example, polycystin-1—a large membrane protein found in the kidney—functions as a mechanosensor, receiving signals from the primary cilia, neighboring cells, and extracellular matrix and transduces them into biochemical pathways that regulate proliferation, adhesion, and differentiation that are essential for the control of renal tubules and kidney morphogenesis (2–4). Mutations in polycystin-1 account for 85% polycystic kidney disease (PKD), which is one of the most common life-threatening inherited diseases worldwide. Most domains of the large ectodomain of polycystin-1 are mechanical stable Ig-like domains and have been suggested to function as force transmitters to regulate the multifunction properties of polycystin-1 (5, 6).

Wlodek M. Bujalowski (ed.), *Spectroscopic Methods of Analysis: Methods and Protocols*, Methods in Molecular Biology, vol. 875, DOI 10.1007/978-1-61779-806-1_16, © Springer Science+Business Media New York 2012

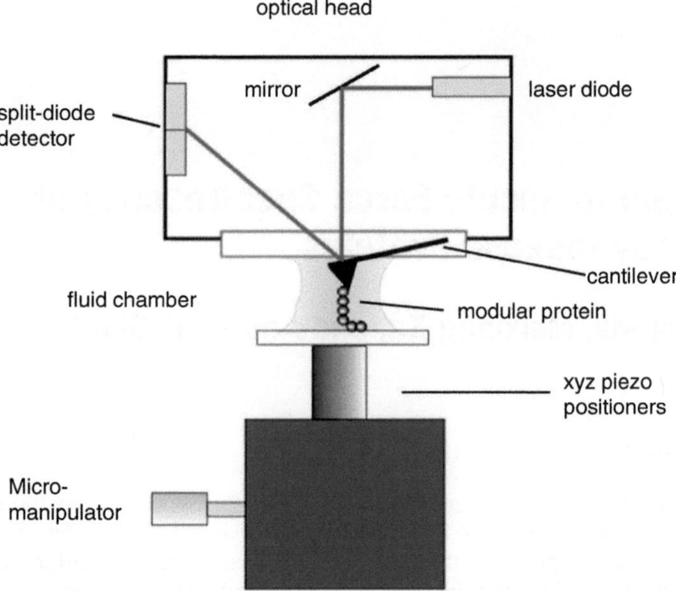


Fig. 1. Diagram of our home-built AFM. The mechanical properties of single proteins are studied using a single-molecule atomic force microscope that consists of a detector head (Digital Instruments) mounted on top of an xyz piezoelectric positioner with capacitive sensors (PicoCube, Physik Instrumente). This system has a *z*-axis resolution of a few nanometers and can measure forces in the range of 10–10,000 pN. The monitoring of the force reported by the cantilever, and the control of the movement of the piezoelectric positioners, are achieved by means of two data acquisition boards (PCI 6052E, PCI 6703, National Instruments) and controlled by custom written software (LabView; National Instruments and Igor, Wavemetrics). With this system, it is possible to measure the force as a function of the extension of the protein (force–extension mode) or measure the elongation of the protein at a constant force (force-feedback mode (31)).

One technique that has been widely employed to investigate the mechanical properties of single proteins is atomic force microscopy (AFM) (7–18). AFM in its single-molecule force spectroscopy mode (for simplicity, referred to as SMFS) is one of the nanomanipulation techniques used for the study of the mechanical properties of proteins (Fig. 1). These proteins appear in a wide range of biological systems and are critical for their survival. Examples include muscle proteins (such as titin and projectin), cytoskeletal proteins (such as spectrin and filamin), proteins that form the extracellular matrix (such as tenascin and fibronectin), and cell adhesion proteins (such as polycystin, cadherin and NCAMs). SMFS can provide unique information not only about the elastic properties of these proteins, but also about other biophysical parameters, such as mechanical stability, unfolding, and refolding rates and free energies for the unfolding and refolding pathways. The aim of this chapter is to provide a quick guide for experimental protocols in this new exciting field. More comprehensive descriptions can be found elsewhere (19–21).

## 2. Materials

### 2.1. Surfaces

Usually, different kinds of surfaces, such as glass, Ni-NTA, silanized glass and gold-coated glass, have been used to adsorb different polymers and proteins in AFM experiments. Here, we describe several methods for functionalizing glass coverslips.

#### 2.1.1. Glass Coverslip

Reagents:

1. MilliQ $H_2O$ (18.2 MΩ).
2. 70% (v/v) ethanol.
3. 10% (v/v) Hellmanex (liquid cleaning concentrate for glass).
4. Parafilm.
5. Transferring pipette.

Equipment:

1. 50 ml glass beakers.
2. Sonicator.
3. Compressed $N_2$ gas.

#### 2.1.2. Silanized Coverslips

Reagents:

1. 0.1 M $H_2SO_4$.
2. Trimethylchlorosilane (TMS-Cl).
3. Dry methanol.
4. Dry acetone.
5. Dry chloroform.
6. MilliQ $H_2O$.

Equipment:

1. Glass substrates (Round Glass Coverslips, 15 mm diameter, 1 oz, Ted Pella, Inc.).
2. 50 ml glass beakers.
3. Nitrogen gas (filtered, compressed).

#### 2.1.3. Ni-NTA Coverslips

To permit tight binding of the proteins to Ni-NTA, a polyhistidine tag is usually fused to the N-terminus of the protein. To attach histidine-tagged proteins to glass coverslips (Glass Coverslips, 15 mm diameter, 1 oz, Ted Pella, Inc.), these must be silanized with a derivative of NTA that coordinates nickel. Below we describe the protocol in detail.

Reagents:

1. Distilled water.
2. MilliQ water (18 MΩ).

3. ~20 ml 20 N KOH.
4. Acetic acid.
5. 3-mercaptopropyltrimethoxysilane (TSL8380, Toshiba GE Silicone, Tokyo).
6. 50 ml 10 mM MOPS-KOH, pH 7.
7. *N*-[5-(3′-maleimidopropylamido)-1-carboxypentyl] iminodiacetic acid (maleimide-$C_3$-NTA (Dojindo); 20 mg/ml in 10 mM MOPS-KOH.
8. 10 ml 100 mM DTT in MilliQ $H_2O$.
9. 50 ml 10 mM $NiCl_2$ in MilliQ $H_2O$.
10. Parafilm.

Equipment:

1. Rocker.
2. 50 ml centrifuge tube.
3. 50 ml beakers.
4. 2 l beaker.
5. Sand bath.
6. Thermometers (110°C range).
7. Pipette and pipette tips (1 ml).

#### 2.1.4. Au Coverslips

Gold-coated coverslips are useful in adsorbing proteins with thiol-containing amino acids. The interaction of gold with cysteine is quite strong, due to the high affinity for gold (22, 23).

Reagents:

1. Chromium/nickel powder (Goodfellow).
2. Gold pellet (99.99%, GoodFellow, GB).
3. Hydrochloric acid (HCl; 1 N).
4. Hydrogen peroxide ($H_2O_2$; 30%).

Equipment:

1. Vacuum metal evaporator.

### 2.2. AFM Cantilevers

There are two key parameters that are used to characterize AFM cantilevers in force spectroscopy measurements: the spring constant, $k_C$ (pN/nm), and the resonant frequency, $f_0$ (Hz). The spring constant of a typical cantilever, $k_C$, is in the range of 10–100 pN/nm (for MLCT or MSNL silicon nitride, Veeco). For a cantilever with a $k_C$ of 60 pN/nm, the thermal noise of the force measurements is calculated to be approximately 15 pN rms (the root–mean–square force fluctuation) (24). Most of the AFM cantilevers are available with metal coatings. Some coatings are intended only to improve the reflectivity of the back side of the cantilevers, and are called reflex coatings, such as aluminum, gold, and platinum. The cantilevers that we typically use are silicon nitride gold-coated (MLCT-AUHW, Veeco Metrology Group, Santa Barbara, CA).

### 2.3. PKD Polyproteins

1. PKD constructs: we focus on PKD domains of polycystin-1.
2. Vectors: T-A cloning vector pGEM-T vector from Promega; Expression vectors include pEQ80L, p202 (vector with MBP) and pAFM (a modified from pRSET vector (25)).
3. Host cell: *E. coli* Top10 (Invitrogen), JM109 (Promega) and Sure-2 (Stratagene) are used for plasmid cloning. *E. coli* BL21 (Stratagene), BLR(DE3) (Novagene), and C41 (Lucigen Corp.) strains are used for protein expression.
4. Media: LB medium (10 g bacto-tryptone, 5 g bacto-yeast extract, 10 g NaCl, in total volume 1 l $ddH_2O$, adjust pH to 7.0), and YT medium (16 g bacto-tryptone, 10 g bacto-yeast extract, 5 g NaCl, in total volume 1 l $ddH_2O$, adjusted to pH 7.2) are used for growing *E. coli*. Before use, antibiotics such as Ampicillin or Kanamicin are supplied into medium at a concentration of about 100 μg/ml.
5. Enzymes: Platinum® Taq DNA Polymerase High Fidelity (Invitrogen) is used in polymerase chain reaction (PCR) to amplify the DNA fragments. Restriction endonucleases, such as BamHI/NheI/EcoRI/NotI/KpnI/SacI/XbaI/SpeI/NdeI (NEB Inc.) are used for enzyme digestion to obtain DNA fragments and prepare the expression vectors. T4 DNA ligase from Promega is used in ligation reactions.
6. Kits and others: QIAprep Spin Miniprep Kit is used for plasmid extraction. QIAquick Gel Extraction Kit (Qiagen) is used to purify DNA fragments and vectors from gels after DNA gel electrophoresis. Homemade competent cells are prepared with Z-Competent™ E. Coli transformation Kit (Zymo Research Corp.). Protease Inhibitor Cocktail Tablets (Roche), Ni-NTA (nickel-nitrilotriacetic acid) resins (Qiagen), Ni-NTA spin columns, and Poly-Prep Chromatography Columns (Bio-Rad) are used during protein purification. PD-10 Columns (GE), Vivaspin series (GE), or Amicon Ultra Centrifugal Filter Devices (Millipore) are used for further protein purification.
7. Protein purification buffers. Cell lysis buffer/column equilibration buffer 1× PBS with 10 mM imidazole, pH 7.4; the wash buffer is 1× PBS with 25 mM imidazole; the elution buffer is 1× PBS with 250 mM imidazole.
8. SDS-PAGE gel analysis: we use phastgel system (GE) for protein analysis.

## 3. Methods

### 3.1. PKD Polyprotein Construction

The ectodomain of polycystin-1 is about 3,000 aa long and it contains multiple PKD domains. We use several recombinant DNA techniques and cloning methods to construct PKD polyproteins. PKD polyproteins are expressed in *E. coli* strains and then purified by $Ni^2$ affinity chromatography.

#### 3.1.1. Cloning and Expression Strategies

1. PKD DNA fragments preparation: Design proper primers to amplify the target PKD DNA fragments and introduce different restriction enzyme sequences on both ends of each target DNA fragment by using PCR. Then, the target fragments are purified using QIAquick Gel Extraction Kit (QIAGEN). The purified DNA fragments with introduced restriction enzyme sequences should be cloned into pGEM-T vector by ligation reaction. The plasmids are transformed into competent cells. The positive colonies are screened using the blue/white screening method. Extract the plasmids from the cell culture of white clones (positive clones) with QIAprep Spin Miniprep Kit (QIAGEN). The purified plasmids are digested with restriction endonucleases and identified by agarose gel electrophoresis (Fig. 2b). The positive clones should be confirmed by sequencing.
2. Cloning and expression of a $(I27\text{-HuPKDd1})_3$ hetero-polyprotein. This polyprotein based on the first PKD domain (HuPKDd1, residues V268 to E354) and the titin immunoglobulin domain #27 (I27). The I27 domain has been extensively studied by force spectroscopy; hence, it can serve as an internal fingerprint (26, 27). We assembled an I27-HuPKDd1 hetero-polyprotein containing three multiples of the I27-HuPKDd1 dimer, by applying a multiple-step cloning technique that makes use of four restriction sequences (*BamHI*, *BglII*, *BstY*, and *KpnI*) (5, 8). This construct was cloned in an *E. coli* recombination-defective strain, Sure-2 (Stratagene), and expressed in the BLR (DE3) strain (Novagene).
3. All the vectors are digested with proper pair of restriction endonucleases in order to match the insert PKD DNA fragments (Fig. 2c). The digested vectors are recovered by QIAquick Gel Extraction Kit (QIAGEN) and stored at −20°C.
4. Ligation and transformation: mix the insert DNA fragment and vector in the range of 3:1 (molar ratios), set up regular ligation reaction with a final volume of 10 μl and incubate overnight at 4°C. The *E. coli* Top10, Sure-2, and JM109 cells are used for cloning and transformation and the BL21, BLR (DE3), and C41 cells are used for protein expression.

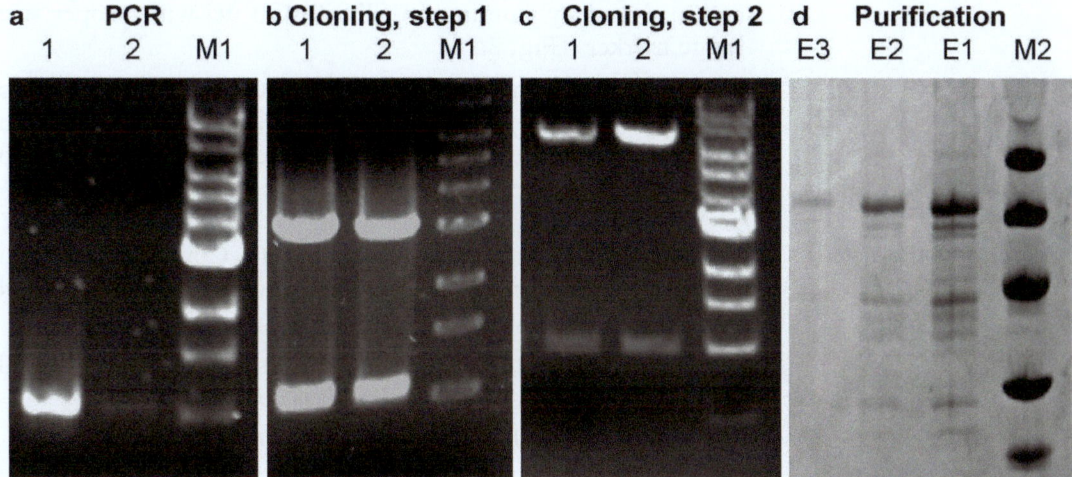


Fig. 2. Cloning and expression of a PKD polyprotein. Example of the different steps in the construction and expression of a PKD polyprotein containing PKD domains 11-14 (PKD11-14) of human polycystin-1. (**a**) Ethidium bromide-stained agarose gel showing PKD11-14 DNA fragments obtained by PCR (*lanes 1* and *2*). *Lane M1* shows the DNA marker (1 kb DNA ladder; Biolabs). (**b**) Ethidium bromide-stained agarose gel showing the PKD11-14 DNA fragment cloned into the T-A cloning vector pGEMT vector and digested with BamHI and NdeI. *Lanes 1* and *2* are positive clones for the PKD11-14 fragment. (**c**) Cloning of the PKD11-14 DNA fragment into the p202 expression vector. The gel shows a digest of the positive clones using BamHI and NdeI. (**d**) SDS-PAGE (12%) stained with Coomassie brilliant blue showing the purification of the PKD11-14 protein fragment. *Lanes E3*, *E2*, and *E1* show different elution fractions. *Lane M2*: low molecular weight-SDS Marker (GE lifescience).

5. PKD polyprotein constructs expression: *E. coli* expression protocol is modified from the manual of pRSET A for high-level expression of recombinant proteins (Invitrogen, Catalog no. V351-20). Incubate by shaking *E. coli* host cells in LB or YT media supplied with antibiotics at 37°C. When OD600 reaches ~0.6, the protein expression is induced by adding IPTG to a final concentration of 1–5 mM (see Note 1).

*3.1.2. PKD Polyprotein Purification*

1. Cell lysis: Disolve the cell pellets in the lysis buffer (with protease inhibitors) in an ice bath. The cells are lyzed by sonication or by using a fluidizer. During the procedure, always keep the samples on ice. Centrifuge the lysate to collect the supernatant for purification.
2. Purification: Ni-NTA resins are used to purify PKD polyproteins. Before use, the resins should be equilibrated with buffer. Then, mix the supernatant of cell lysate with the resin and allow binding for 30–60 min at 4°C with gentle agitation to keep the resin completely mixed with lysate. Settle the resin by gravity and collect the supernatant running through. Wash the resin with wash buffer (containing 25 mM imidazole) several times before adding the elution buffer (containing ~250 mM imidazole; see Note 2). The proteins can be eluted by adding elution buffer and stored at 4°C. The proteins can be identified

and analyzed by running an SDS-PAGE gel with proper size range markers (Fig. 2d).

3. Desalting and concentrating: PD-10 Columns are used to remove the high concentration salt (e.g., imidazole). Vivaspin series or Amicon Ultra Centrifugal Filter Devices are used to concentrate the proteins.

### 3.2. Preparation of Surfaces

#### 3.2.1. Glass Coverslips

1. Take around 20 coverslips (Round Glass Coverslips, 15 mm diameter, 1 oz, Ted Pella, Inc.) and put them into a 50 ml beaker.
2. Spray 70% (v/v) EtOH on the coverslips and pipette to get rid of the dust on the surface of the glass coverslips.
3. Rinse with MilliQ $H_2O$ until the solution is clear.
4. Add 30 ml 10% (v/v) Hellmanex (3 ml Hellmanex into 30 ml MilliQ $H_2O$).
5. Cover the beaker with Parafilm and sonicate for 20 min.
6. Discard the Hellmanex solution and rinse the coverslips with MilliQ $H_2O$.
7. Rinse the coverslips with 70% (v/v) EtOH.
8. Rinse the coverslips with MilliQ $H_2O$ to remove the alcohol.
9. Add 30 ml MilliQ $H_2O$ and sonicate for 20 min.
10. Discard the MilliQ $H_2O$ and add 20 ml fresh MilliQ $H_2O$, covered by parafilm and store at room temperature until further use.
11. Before use, the coverslips should be dried in a stream of $N_2$ gas.

#### 3.2.2. Silanized Glass Coverlips (28)

1. Dip the glass coverslips in 0.1 M $H_2SO_4$ for 20 s and rinse with MilliQ $H_2O$.
2. Dip the glass coverslips for 20 s in each of the following solutions: methanol, dry acetone, and dry chloroform (see Note 3).
3. Dry with nitrogen gas.
4. Dip the glass coverslips in freshly prepared solution of 5% (v/v) trimethylchlorosilane (TMS-Cl) in chloroform for 20 s.
5. Rinse twice in chloroform.
6. Dry with nitrogen gas and store them at room temperature prior to use.

#### 3.2.3. Ni-NTA-Coated Glass Coverslips (Modified from Ref. (29))

1. Put ~20 glass coverslips (15 mm diameter, 1 oz, Ted Pella, Inc.) into a 50 ml centrifuge tube and immerse them in ~20 ml 20 N KOH, rocking for ~13 h.
2. Rinse with distilled water to completely wash off the KOH by using a 2 l glass beaker.

3. Transfer all the coverslips in to a 50 ml glass beaker and immerse the coverslips in a solution containing 0.02% (v/v) acetic acid and 2% (v/v) 3-mercaptopropyltrimethoxysilane, incubate the beaker on sand bath at 90°C for 1 h.
4. Rinse thoroughly with distilled water using the 2 l glass beaker for ~1 h.
5. Collect all the coverslips and bake them in an oven at 120°C for 10 min.
6. After cooling to room temperature, the SH groups of the silane on the glass surface are reduced using 100 mM DTT for 10 min, then rinse the coverslips with distilled water in a 2 l beaker.
7. React with 20 mg/ml *N*-[5-(3′-maleimidopropylamido)-1-carboxypentyl]iminodiacetic acid (Dojindo) in 10 mM MOPS-KOH for 30 min by pipetting a drop of the solution on the top of each coverslip (see Note 4).
8. Gently rinse the coverslips with MilliQ water.
9. Air-dry the coverslips and add a drop of 10 mM $NiCl_2$ on the top side of coverslip and allow the reaction for 10 min.
10. Repeat step 8 to rinse off the free $NiCl_2$.
11. Air-dry the Ni-NTA-coated coverslips and store them in air at room temperature until use.

#### 3.2.4. Au-Coated Glass Coverlips

1. Evaporate a layer of chromium/nickel (1 nm) on the glass coverslips and then a 50 nm layer of high-purity gold, under vacuum, at a pressure of $1–2 \times 10^{-6}$ mbar (see Note 5).

### 3.3. Calibration of Cantilevers

The spring constant values of AFM cantilevers that originate even from the same batch (wafer) can differ quite significantly (up to ±30%). Therefore, each individual cantilever must be calibrated before measurement. In our experiments, we use the so-called thermal method which is based on the energy equipartition theorem (30). When a cantilever system is modeled as a simple harmonic oscillator, the average potential energy of the cantilever, $1/2k_C Z^2$, is equal to the "thermal energy," $1/2k_B T$, where $k_C$ is the cantilever spring constant, $Z$ is the amplitude of its random oscillations in thermal equilibrium, $k_B$ is the Boltzmann constant, and $T$ is the absolute temperature. Thus, calibrating the cantilever involves the determination of $Z^2$.

1. $Z^2$ is not measured directly but is determined from the measurement of $V^2$, where $V$ is the voltage of the split photodiode generated by the laser beam that traces the movement of the cantilever. To convert $V^2$ to $Z^2$, it is necessary to determine the optical lever sensitivity, $S$, of the photodiode voltage, $V$, to the amount of the cantilever bending, $Z$, using the equation $S = V/Z$.

2. To determine $S$ the cantilever is pushed vertically—and therefore bent—by the piezoelectric actuator of the AFM. $Z$ (in nm) and $V$ (in volts), corresponding to the bending, are directly measured.
3. $V^2$ is typically not evaluated in the time domain but is converted to the frequency domain by performing the Fourier transform on the time signal, $V(t)$. This approach permits the evaluation and rejection of low-frequency mechanical (nonthermal) noise that contributes to the motion of the cantilever.

### 3.4. Analysis of Force–Extension Curves

1. Select the most suitable substrates to attach protein molecules (see Subheading 3.2) depending on the reactive group. For proteins containing a thiol group (from Cys residues) in C- or N-termini use Au-coated coverslips. For proteins with a His-tag use Ni-NTA functionalized coverslips. However, sometimes we find that proteins bind well to just clean glass or silanized glass.
2. In the experiment shown in Fig. 3, we used Ni-NTA glass coverslips to immobilize a recombinant protein chimera $(I27\text{-}HuPKDd1)_3$ (see Subheading 3.1.1).
3. Glue the Ni-NTA coverslip on a magnetic sample holder by using a double-stick tape, and mount it on top of a vertical piezoelectric positioner.
4. Mount an AFM cantilever onto the cantilever holder/AFM chamber and measure its spring constant (see Subheading 3.3)
5. Prepare a protein solution to be studied. We use the PBS or Tris-buffer and make the concentration in the range of 10–100 μg/ml.
6. Add 25 μl of the protein solution on the coverslip and incubate ~10 min at room temperature to allow binding of the proteins to surface.
7. Remove the unbound protein molecules by rinsing with 50 μl buffer.
8. Raise the piezoelectric positioner up to allow the cantilever into contact with sample on the surface, by using the *z*-axis macromanipulator as gently as possible. Then adjust the contact between the cantilever and protein layer by using a fine, computer-controlled motion.
9. Start an SMFS experiment under "length-clamp" mode (7, 31). Adjust the time interval where the cantilever tip remains in contact with the sample in order to increase or decrease the probability of picking up single molecule instead of multiple molecules or none.
10. Examine the initial recordings. Look for mechanical fingerprints of individual protein molecules. Typically, unfolding of tandem-repeated Ig-like domains (such as those in the

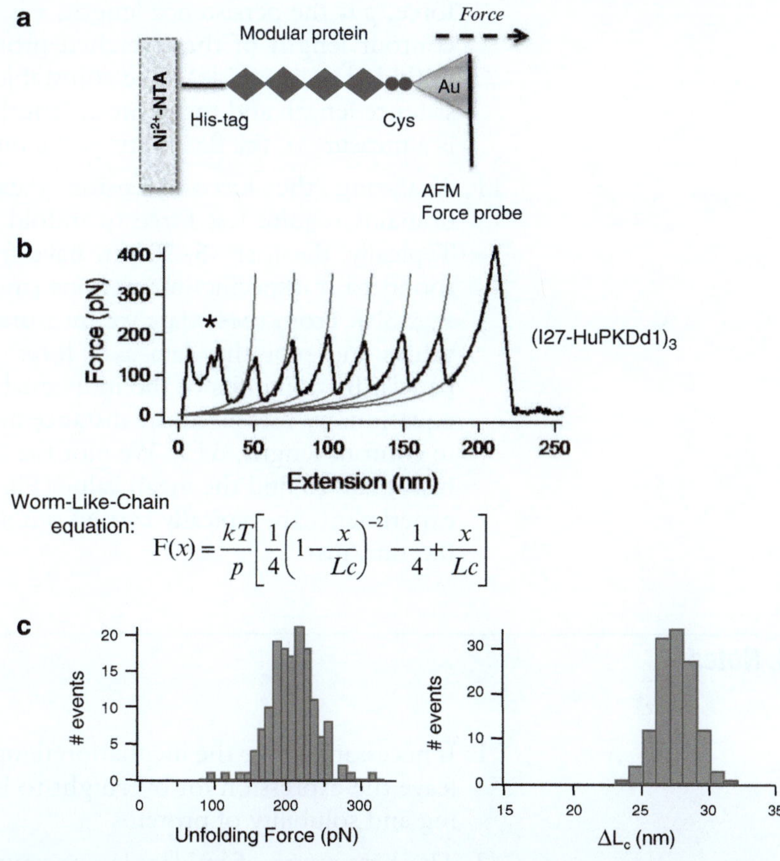


$$F(x) = \frac{kT}{p}\left[\frac{1}{4}\left(1-\frac{x}{Lc}\right)^{-2} - \frac{1}{4} + \frac{x}{Lc}\right]$$

Fig. 3. Analysis of the mechanical properties of a PKD protein using SMFS techniques. (**a**) Experimental design: the proteins used for SMFS experiments are typically made of multiple domains and the ends have tags for specific attachment to surfaces (i.e., His-tag to Ni-NTA and SH groups from Cys residues to Au surfaces). The AFM force probe is typically a silicon cantilever that has a sharp tip coated with Au. (**b**) Representative example of a force–extension curve obtained after stretching a polyprotein that consist of repeats of the human PKD domain 1 and the titin I27 domain $(I27\text{-}HuPKDd1)_3$. Both are ~100 amino acid long Ig-like beta-sandwich domains. The unfolding of repeats of Ig-like is seen as a sawtooth pattern. The initial region of the force–extension typically has nonspecific interaction between the AFM tip and the surface (marked with an *asterisk*). The *thin lines* correspond to fits to the WLC equation. (**c**) We measure to key parameters: the unfolding forces and the increases in contour lengths, $\Delta L_c$, and plot these data as histograms in order to obtain their mean values.

$(I27\text{-}HuPKDd1)_3$ protein) results in a sawtooth pattern containing regular unfolding force peaks with similar amplitude and similar spacing (Fig. 3b). This force–extension relationship can be formally described by the wormlike chain (WLC) model of polymer elasticity (Fig. 3b, thin lines) (32, 33), where $F$ is force, $p$ is the persistence length, $x$ is end-to-end length, $L_c$ is contour length of the stretched protein (total length of the unfolded polypeptide). The adjustable parameters are the persistence length and the contour length. The persistence length is a measure of the flexibility of the molecule.

11. Analyzing the force–extension data: Usually, the weaker domains require less force to unfold and these are seen first. Typically, the first ~5–30 nm have force patterns that correspond to nonspecific interactions (marked with an asterisk in Fig. 3b). From these data we measure the different force peak values and plot the data as a force histogram (Fig. 3c, left panel). By doing fits of the individual force peaks to the WLC equation, we measured the distance between peaks (or increase in contour length, $\Delta L_c$). We plot the $\Delta L_c$ values as a frequency histogram to find the mean value (Fig. 3c, right panel). All the experiments are typically carried out at a pulling speed of 0.4–0.6 nm/ms.

## 4. Notes

1. If necessary, lower the incubation temperature to 25–15°C and leave the expression for overnight to improve the correct folding and solubility of proteins.
2. The best range of imidazole concentration and pH levels for protein elution have to be tested experimentally (150–800 mM and pH 6–8).
3. This process has to be performed in the fume hood.
4. Remember which is the top side of the coverslip since there is only one side being functionalized.
5. The gold will not glue to the glass without the layer of chromium/nickel.

## Acknowledgments

This work was funded by NIH grant R01DK073394, the John Sealy Memorial Endowment Fund for Biomedical Research, and by the Polycystic Kidney Foundation (grant 116a2r).

## References


1. Jaalouk DE, Lammerding J (2009) Mechanotransduction gone awry. Nat Rev Mol Cell Biol 10:63–73
2. Praetorius HA, Spring KR (2003) Removal of the MDCK cell primary cilium abolishes flow sensing. J Membr Biol 191:69–76
3. Praetorius HA, Spring KR (2001) Bending the MDCK cell primary cilium increases intracellular calcium. J Membr Biol 184:71–79
4. Nauli SM, Alenghat FJ, Luo Y, Williams E, Vassilev P, Li X, Elia AE, Lu W, Brown EM, Quinn SJ, Ingber DE, Zhou J (2003) Polycystins 1 and 2 mediate mechanosensation in the primary cilium of kidney cells. Nat Genet 33:129–137
5. Qian F, Wei W, Germino G, Oberhauser A (2005) The nanomechanics of polycystin-1 extracellular region. J Biol Chem 280:40723–40730
6. Forman JR, Qamar S, Paci E, Sandford RN, Clarke J (2005) The remarkable mechanical strength of polycystin-1 supports a direct role in mechanotransduction. J Mol Biol 349:861–871
7. Oberhauser AF, Carrion-Vazquez M (2008) Mechanical biochemistry of proteins one molecule at a time. J Biol Chem 283:6617–6621
8. Carrion-Vazquez M, Oberhauser AF, Fisher TE, Marszalek PE, Li H, Fernandez JM (2000) Mechanical design of proteins studied by single-molecule force spectroscopy and protein engineering. Prog Biophys Mol Biol 74: 63–91
9. Fisher TE, Carrion-Vazquez M, Oberhauser AF, Li H, Marszalek PE, Fernandez JM (2000) Single molecular force spectroscopy of modular proteins in the nervous system. Neuron 27:435–446
10. Fisher TE, Marszalek PE, Oberhauser AF, Carrion-Vazquez M, Fernandez JM (1999) The micro-mechanics of single molecules studied with atomic force microscopy. J Physiol 520 (Pt 1):5–14
11. Fisher TE, Oberhauser AF, Carrion-Vazquez M, Marszalek PE, Fernandez JM (1999) The study of protein mechanics with the atomic force microscope. Trends Biochem Sci 24:379–384
12. Best RB, Clarke J (2002) What can atomic force microscopy tell us about protein folding? Chem Commun (Camb) (3):183–192
13. Rounsevell RW, Steward A, Clarke J (2005) Biophysical investigations of engineered polyproteins: implications for force data. Biophys J 88:2022–2029
14. Linke WA, Grutzner A (2008) Pulling single molecules of titin by AFM – recent advances and physiological implications. Pflugers Arch 456:101–115
15. Muller DJ, Krieg M, Alsteens D, Dufrene YF (2009) New frontiers in atomic force microscopy: analyzing interactions from single-molecules to cells. Curr Opin Biotechnol 20(1):4–13
16. Samori B (2000) Stretching single molecules along unbinding and unfolding pathways with the scanning force microscope. Chemistry 6:4249–4255
17. Mehta AD, Rief M, Spudich JA (1999) Biomechanics, one molecule at a time. J Biol Chem 274:14517–14520
18. Rief M, Grubmuller H (2002) Force spectroscopy of single biomolecules. Chemphyschem 3:255–261
19. Rabbi M, Marszalek PM (2008) Probing polysaccharide and protein mechanics by atomic force microscopy. In: Selvin PR, Ha T (eds) Single-molecule techniques. Cold Spring Harbor Laboratory Press, New York, pp 371–394
20. Rounsevell R, Forman JR, Clarke J (2004) Atomic force microscopy: mechanical unfolding of proteins. Methods 34:100–111
21. Carrion-Vazquez M, Oberhauser AF, Diez H, Hervas R, Oroz J, Fernandez J, Martinez-Martín D (2006) Protein Nanomechanics – as studied by AFM single-molecule force spectroscopy. In: Arrondo JL, Alonso A (eds) Advance techniques in biophysics. Springer, Berlin, pp 163–245
22. Lim II, Ip W, Crew E, Njoki PN, Mott D, Zhong CJ, Pan Y, Zhou S (2007) Homocysteine-mediated reactivity and assembly of gold nanoparticles. Langmuir 23:826–833
23. Petean I, Tomoaia G, Horovitz O, Mocanu A, Tomoaia-Cotisel M (2008) Cysteine mediated assembly of gold nanoparticles. J Optoelectron Adv Mater 10:2289–2292
24. Bustamante C, Macosko JC, Wuite GJ (2000) Grabbing the cat by the tail: manipulating molecules one by one. Nat Rev Mol Cell Biol 1:130–136
25. Steward A, Toca-Herrera JL, Clarke J (2002) Versatile cloning system for construction of multimeric proteins for use in atomic force microscopy. Protein Sci 11:2179–2183
26. Li H, Carrion-Vazquez M, Oberhauser AF, Marszalek PE, Fernandez JM (2000) Point

mutations alter the mechanical stability of immunoglobulin modules. Nat Struct Biol 7:1117–1120

27. Oberhauser AF, Badilla-Fernandez C, Carrion-Vazquez M, Fernandez JM (2002) The mechanical hierarchies of fibronectin observed with single-molecule AFM. J Mol Biol 319: 433–447
28. Sundberg M, Rosengren JP, Bunk R, Lindahl J, Nicholls IA, Tagerud S, Omling P, Montelius L, Mansson A (2003) Silanized surfaces for in vitro studies of actomyosin function and nanotechnology applications. Anal Biochem 323:127–138
29. Sakaki N, Shimo-Kon R, Adachi K, Itoh H, Furuike S, Muneyuki E, Yoshida M, Kinosita K Jr (2005) One rotary mechanism for F1-ATPase over ATP concentrations from millimolar down to nanomolar. Biophys J 88: 2047–2056
30. Florin EL, Rief M, Lehmann H, Ludwig M, Dornmair C, Moy VT, Gaub HE (1995) Sensing specific molecular-interactions with the atomic-force microscope. Biosens Bioelectron 10:895–901
31. Oberhauser AF, Hansma PK, Carrion-Vazquez M, Fernandez JM (2001) Stepwise unfolding of titin under force-clamp atomic force microscopy. Proc Natl Acad Sci USA 98:468–472
32. Bustamante C, Marko JF, Siggia ED, Smith S (1994) Entropic elasticity of lambda-phage DNA. Science 265:1599–1600
33. Marko JF, Siggia ED (1995) Statistical-mechanics of supercoiled DNA. Phys Rev E 52:2912–2938

# Chapter 17

# Single Molecule Detection Approach to Muscle Study: Kinetics of a Single Cross-Bridge During Contraction of Muscle

**Julian Borejdo, Danuta Szczesna-Cordary, Priya Muthu, Prasad Metticolla, Rafal Luchowski, Zygmunt Gryczynski, and Ignacy Gryczynski**

## Abstract

D166V point mutation in the ventricular myosin regulatory light chain (RLC) is one of the causes of familial hypertrophic cardiomyopathy (FHC). We show here that the rates of cross-bridge attachment and dissociation are significantly different in isometrically contracting cardiac myofibrils from right ventricle of WT and Tg-D166V mice. To avoid averaging over ensembles of molecules composing muscle fibers, the data was collected from a single molecule. Kinetics were derived by tracking the orientation of a single actin molecule by fluorescence anisotropy. Orientation oscillated between two states, corresponding to the actin-bound and actin-free states of the myosin cross-bridge. The cross-bridge in a wild-type (healthy) heart stayed attached and detached from thin filament on average for 0.7 and 2.7 s, respectively. In FHC heart, these numbers increased to 2.5 and 5.8 s, respectively. These findings suggest that alterations in myosin cross-bridge kinetics associated with D166V mutation of RLC ultimately affect the ability of a heart to efficiently pump the blood.

**Key words:** Hearth hypertrophy, Cardiac muscle, Single molecule detection, Correlation function

## Abbreviations

| | |
|---|---|
| $t_{ON}$ | The time cross-bridge is strongly attached to actin |
| $t_{OFF}$ | The time cross-bridge is detached from actin |
| $\Psi$ | Duty cycle of the cross-bridge |
| AP | Alexa488-phalloidin |
| RP | Rhodamine-phalloidin |
| UP | Unlabeled-phalloidin |
| DA | Detection area |
| EDC | 1-Ethyl-3-(3′-dimethylaminopropyl) carbodiimide |
| ROI | Region-of-interest |
| DTT | Cleland's reagent |
| APD | Avalanche photodiode |
| FCS | Fluorescence correlation spectroscopy |
| SMD | Single molecule detection |

Wlodek M. Bujalowski (ed.), *Spectroscopic Methods of Analysis: Methods and Protocols*, Methods in Molecular Biology, vol. 875, DOI 10.1007/978-1-61779-806-1_17, © Springer Science+Business Media New York 2012

## 1. Introduction

Muscle contraction results from the cyclical interactions of actin and myosin. During this interaction, myosin cross-bridge delivers force impulses to actin (1). Gross isometric force is the time average of trillions of those impulses and therefore it contains no kinetic information—it has been lost due to temporal averaging. Likewise, gross measurements of orientation using either polarization of fluorescence (1, 2) or electron paramagnetic resonance (3) provide only information about average orientation of cross-bridges. But there are ways of extracting kinetic rates from a signal originating from a large assembly of molecules. The first is to apply a transient of a parameter under study and follow its decay to a steady state. The rationale is that all molecules are suddenly brought to a new and identical state and return to a steady state with a well-defined rate. In the case of muscle contracting ex vivo, the most commonly applied transient is muscle length, which is suddenly decreased just enough to drop tension to near zero (4). This brings all myosin cross-bridges to the same relaxed conformation and the rate of their recovery to tension-generating state is a kinetic measure of force-generating step. This approach yielded invaluable data ex vivo (4). In vitro a rapid mixing is used (5–7). The disadvantage of the transient method is that the displacement from equilibrium may upset steady state. The second method is to measure stochastic fluctuations in a steady-state signal. These fluctuations contain information about kinetics of individual events. The size of fluctuations varies inversely with the number of observed molecules (8). Kinetic information can be extracted from fluctuations when a few molecules contribute to the signal and stochastic fluctuations become important (mesoscopic regime, (9)). Extreme case is when a single molecule is observed. The disadvantage of this approach is that it is technically challenging to observe one molecule among large assembly. It is particularly difficult in muscle where the concentration of proteins is large (10). Nevertheless, with the recent advances in fluorescence technology and single molecule detection (11–14), it has been done in vitro. In the application to muscle, Warshaw and collaborators measured orientation of a single molecule of smooth myosin II (15, 16) and Goldman, Selvin, and collaborators measured the orientation of a single molecule of myosin V (17–19). A more physiologically relevant problem is performing single molecule experiments in tissue, where excluded volume effects become important. At high concentrations, the access to proteins may be limited to only small solvent molecules. As a result, proteins in certain regions of a muscle cell may become over hydrated and behave differently than isolated proteins that are normally hydrated (20, 21).

In this chapter, we describe our latest attempts to measure kinetic behavior of muscle using single molecule detection (SMD). We first describe our efforts to develop necessary methods using skeletal muscle as a model system and then apply them to the study of cardiac muscle and to a particularly malignant disease of the heart—familial hypertrophic cardiomyopathy (FHC).

There are two ways to observe single molecules. The first is to restrict the observational volume so that only one molecule is in the field-of-view, and the second is to limit the concentration of a fluorophore so that only one labeled molecule is in the field-of-view. The conventional wide-field microscope cannot be used for the former approach because its detection volume is much too large (~$10^{-9}$ L). In the case of skeletal muscle, this volume contains ~$10^{11}$ myosins. High aperture objectives forming diffraction-limited illuminated spots and confocal detection made it possible to limit the detection volume to ~0.5 fL ($10^{-15}$ L) and eliminate much background (22), but half of a femtoliter is still far too large to optically isolate just a few myosin molecules. The situation is not much improved by the use of two photon excitation (23). It is necessary to reduce the volume to about an attoliter ($10^{-18}$ L). Attoliter detection volumes have recently been obtained utilizing evanescent field excitation. Zero-mode waveguides consisting of small apertures in a metal film deposited on a coverslip (24) limit the Z-dimension to the depth of an evanescent wave (~100 nm) and the X- and Y-dimensions by the size of the aperture. However, the manufacture of the film with small apertures is complex and expensive. Near-field scanning optical microscopy (NSOM) produces an evanescent field at the end of a tapered fiber optic tip that can be scanned over the sample surface. Excitation field dimensions are likewise on the order of 100 nm (25, 26). However, NSOM probes are expensive and must be near to (within 5 nm) of a muscle fiber, or be inserted into it. This is difficult to accomplish without breaking a fragile NSOM fiber tip. Other techniques utilize the evanescent field produced at a glass/water interface by light undergoing total internal reflection (TIR). This method was previously utilized by us, but the smallest number of molecules we succeeded in observing was ~5–7 (27). Here, we summarize the results obtained by the second method: to limit the concentration of a fluorophore so that only one labeled molecule is in the field-of-view.

In applying this approach, a critical question is what to label and at what concentration. To observe single molecule in muscle, it is preferable to observe actin rather than myosin. Observing actin has five advantages: (1) Labeling actin with phalloidin preserves the regular structure of a myofibril, unlike the less gentle labeling of myosin. To observe myosin cross-bridges directly, it is necessary to label myosin light chain (28) with a fluorophore and then exchange labeled LC with the native LC of myosin—a procedure that involves incubating muscle at 36°C which destroys some of the

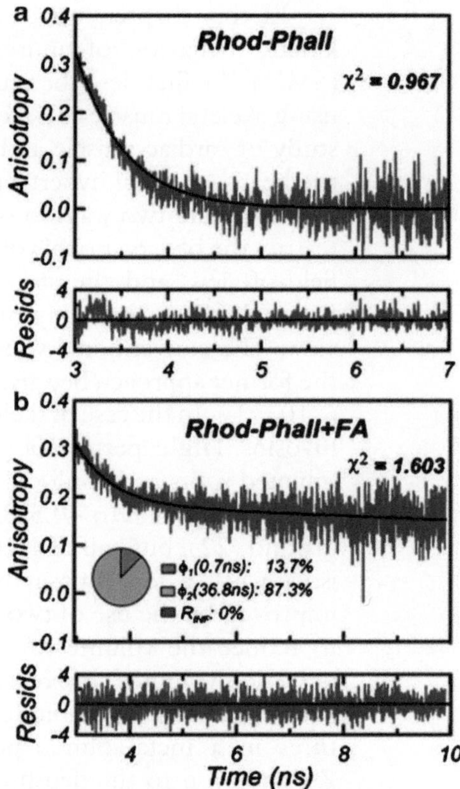


Fig. 1. Decay of anisotropy of the dye alone (**a**) and the dye attached to F-actin (**b**) to show that the rhodamine-phalloidin is rigidly immobilized by F-actin. (**a**) −0.1 μM rhodamine-phalloidin, (**b**) −2 μM F-actin + 0.5 μM rhodamine-phalloidin. The *inset* in (**b**) shows the estimated contributions of the decay components to the anisotropy decay. $R_{INF}$ was fixed at 0. *Bottom panels* show residuals to the least square fit. The goodness-of-fit is reflected by the small value of $\chi^2$. Excitation 470 nm, emission >590 nm.

regular structure of a sarcomere. (2) Phalloidin does not alter enzymatic properties of muscle (29, 30). (3) Phalloidin labels actin stoichiometrically, which allows strict control of the degree of labeling. Such control is impossible with myosin and extensive controls have to be done to assure small degree of labeling (31). (4) Phalloidin attaches to actin noncovalently but strongly. Noncovalent binding is preferable in the case where orientation of dipole moment of a probe is measured—it is rigid (see Fig. 1) because it involves attachment over large surfaces through electrostatic and hydrogen bonding (5) In skeletal muscle, labeling actin with phalloidin allows observing events that occur only in the area where actin and cross-bridges interact. This is because nebulin prevents phalloidin to label central part of a thin filament. It initially labels only the pointed ends of thin filaments (Overlap zone, O-band), precisely the region where filaments overlap and interact (32). This is not true in the case of cardiac muscle, which does not have nebulin.

Our goal is therefore to select a convenient indicator of (a) actin binding and (b) to assure that just one molecule is detected with adequate signal-to-noise ratio (SNR). Regarding point (a), observing orientation of actin is a valid way of observing the effect of cross-bridges, because it has been known for a long time that actin changes orientation in response to cross-bridge binding (30, 33–35) and that those changes parallel the changes of orientation of a cross-bridge (36). Orientation of fluorophores is conveniently monitored by polarized fluorescence (1, 37–40).

## 2. Materials

### 2.1. Chemicals and Solutions

Alexa488 (AP)- and Rhodamine-phalloidin (RP) were from Molecular Probes (Eugene, OR). All other chemicals including 1-ethyl-3-(3′-dimethylaminopropyl) carbodiimide (EDC), dithiotreitol (DTT), creatine phosphate, and creatine kinase were from Sigma. EDTA-rigor solution contained 50 mM KCl, 2 mM EDTA, 1 mM DTT, 10 mM Tris–HCl buffer, pH 7.5. Ca-rigor solution contained 50 mM KCl, 4 mM $MgCl_2$, 0.1 mM $CaCl_2$, 1 mM DTT, 10 mM Tris–HCl buffer, pH 7.5. Mg-rigor solution contained 50 mM KCl, 4 mM $MgCl_2$, 1 mM DTT, 10 mM Tris–HCl buffer, pH 7.5. Contracting solution was the same as Ca-rigor, except that it contained in addition 5 mM ATP. When low concentrations of ATP were used, contracting solution contained also 20 mM creatine phosphate and 10 U/mL creatine kinase (~1 mg/mL).

### 2.2. Preparation of Myofibrils

Thin strips of glycerinated rabbit psoas muscle were incubated in EDTA-rigor solution until they turned white (~1 h). The fiber bundle was then homogenized using a Heidolph Silent Crusher S homogenizer for 20 s (with a break to cool after 10 s) in $Mg^{2+}$-rigor solution (it was important that the fibers were not homogenized in the EDTA rigor buffer to avoid foaming). Myofibrils were always freshly prepared for each single molecule experiment. Labeled myofibrils (25 μL) were applied to a 25 × 60 mm coverslip (Menzel-Glaser 20 × 20 mm #1 or Corning #1). The sample was left on a coverslip for 3 min to allow the myofibrils to adhere to the glass. The bottom cover slip was covered with a 25 × 25 mm coverslip (to prevent drying) and separated from the bottom coverslip by Avery Hole Reinforcement Stickers. Labeled myofibrils were washed with 5 volumes of the $Ca^{2+}$-rigor solution by applying the solution to the one end of the channel and absorbing with #1 filter paper at the other end.

### 2.3. Cross-Linking

To prevent shortening of muscle in contracting solution, myofibrils (1 mg/mL) were incubated with 20 mM EDC for 10 min at room temperature.

### 2.4. Labeling

1 mg/mL myofibrils (~4 μM actin) were mixed with 0.01 nM Alexa488-phalloidin + 10 μM unlabeled-phalloidin or with 0.1 nM rhodamine-phalloidin + 10 μM unlabeled-phalloidin. Unlabeled-phalloidin was necessary to prevent uneven labeling. If it was not there, sarcomeres closest to the tip of the pipette used to add the label would have contained more chromophores than sarcomeres further away from the tip. The degree of labeling was 10 μM/0.01 nM = 1,000,000 or 10 μM/0.1 nM = 100,000, i.e., on the average 1 actin protomer in $10^6$ or $10^5$ was fluorescently labeled.

## 3. Methods

### 3.1. Data Collection

The experiments were done using Micro Time 200 (PicoQuantGmbH, Berlin, Germany) or ISS-Alba-FCS (ISS Co, Urbana, IL) confocal systems coupled to Olympus IX 71 microscopes. For MT200, the excitation was by a 470-nm laser pulsed diode, and the observation was through a 500-nm long-pass filter. Confocal pinhole was 30 μm. The instrument measured fluorescence lifetimes as well as anisotropies. For Alba, the excitation was by a 532 nm CW laser. Confocal pinhole was 50 μm. Fluorescence was collected every 10 μs. Orthogonally linearly polarized analyzers were placed before Avalanche PhotoDiodes (APDs). The laser was polarized vertically (on the microscope stage). Myofibril was also vertical.

The first step in quantitative measurements of orientation is to determine whether the probe is immobilized by the protein so that the transition dipole of the fluorophore reflects the orientation of the protein. Figure 1 shows that this is the case for Rhodamine-phalloidin labeling thin filaments. The decay of anisotropy of free Rhodamine-phalloidin is shown in Fig. 1a. All of the signal was contributed by the fluorophores with decay time of 0.519 ns, consistent with the rotation of a molecule with $M_w = 1{,}250$. No independent rotation of rhodamine moiety was observed. Figure 1b shows the decay of anisotropy of rhodamine-phalloidin coupled to thin filaments. Now the decay curve is best fitted by the two exponents with correlation times of 0.665 and 36.8 ns with the relative contributions of 13.7 and 86.3%, respectively. The short correlation time is due to the rotation of rhodamine-phalloidin and the long one is due to the rotation of F-actin oligomers. Thus, over 86% of fluorescent phalloidin is immobilized by F-actin. This is consistent with the fact that probes attached to proteins through interactions that stretch over large surface areas, such as hydrophobic or Van der Waals interactions, are attached more rigidly than probes that are attached by covalent links. We conclude that

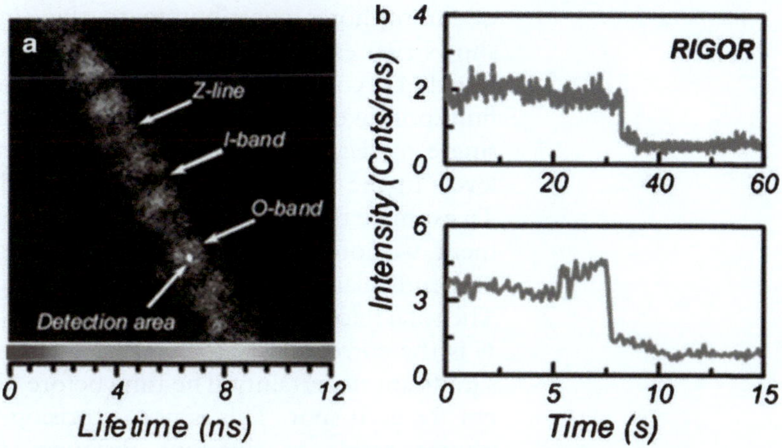


Fig. 2. (**a**) The lifetime image of a myofibril in rigor sparsely labeled with fluorescent phalloidin. The location of various bands is indicated by *white arrows*. The *bottom arrow* points to the area (diameter 0.4 μm) from which the microscope collects the data. Myofibril irrigated with 0.1 nM Alexa488-phalloidin + 10 μM unlabeled-phalloidin. The scale of lifetimes is shown at the *bottom*. Cross-linked myofibril in rigor, excitation at 470 nm, emission viewed through 500 nm long-pass filter. (**b**) Sudden drop of $I_V$ intensity to the level of the background in rigor muscle—behavior characteristic of single molecule bleaching. In the *top and bottom panels*, the intensity suddenly dropped at ~35 s and 10 s. The rate of arrival of fluorescent photons in *top and bottom experiments* was estimated as 4 and 6 photons/ms, respectively (see text).

fluorescent phalloidin is a good dye for monitoring anisotropy of muscle.

A simple calculation, based on the known ratio of phalloidin to actin, gives an order of magnitude estimation of concentration of the dye to achieve SMD: The length, width, and height of a typical half-sarcomere (HS) are 1, 1, and 0.5 μm so its volume is $0.5\ \mu m^3 = 5 \times 10^{-16}$ L. Since the concentration of actin in muscle is 0.6 mM (10), this volume contains ~200,000 actin monomers. If myofibril is labeled with 0.1 nM AP + 10 μM UP, only one actin in 100,000 carries fluorescent phalloidin and there should be, on average, 2.0 fluorophores per half-sarcomere. Since we collect data from ~0.5 $\mu m^2$ ($\pi$ times square of lateral resolution), we detect approximately one molecule. Figure 2a shows a typical image of a rigor skeletal myofibril irrigated with 0.1 nM Alexa488 phalloidin + 10 μM UP.

Figure 2b shows sudden drop of one of the polarized intensities to the level of the background—a test that SMD condition has been achieved. To measure the photon rate associated with this signal, we measured the rate, $\lambda$, at which photons are emitted from a single fluorophore. Providing one works in a saturated regime, where increasing laser power does not lead to an increase in the number of photons emitted by a fluorophore, $\lambda$ is constant. If the total rate at which photons are emitted from muscle is $I_{tot}$, then the number of fluorophores contributing to this signal is $I_{tot}/\lambda$. Figure 2b shows two examples of the time course of the parallel polarized signal (41) collected from rigor muscle. In rigor muscle, it is easy to find spots like these, exhibiting classical symptoms of bleaching of a single molecule—a sudden drop of intensity to the background level. In the top panel, the intensity suddenly dropped at ~35 s. To estimate the rate of arrival of fluorescent photons in this experiment, we note that the $I_V$ was steady at first at ~2 photons/ms. The perpendicular component ($I_H$) of the signal was ~1 photon/ms. The total fluorescence rate was $I_V + 2GI_H \approx 4$ photons/ms, where $G$ is the correction factor (=1.06). However, this number carries significant uncertainty. The time before photobleaching was different for each spot. This is not surprising because different fluorophores reside at different distances from the focus of the illuminating laser beam. They are thus subjected to different illuminating light intensities and take varying amount of time to absorb the number of photons required for photobleaching. For example, the spot shown in Fig. 2a was located closer to the focus because it bleached faster and fluoresced stronger. The total fluorescence rate for this spot was 6 photons/ms. Overall, the time before photobleaching event was within 5–40 s range (average 20 s). We conclude that depending on the position relative to the focus, a single fluorescent phalloidin molecule bound to F-actin contributes photons at the rate of 2–6 (average 4) photons/ms.

Figure 3a shows a typical signal of contracting muscle irrigated with 0.1 nM AP + 10 μM UP. The polarized fluorescence is best characterized by its orthogonal intensities: $I_H$—the fluorescence intensity perpendicular to the axis of a myofibril and $I_V$—the fluorescence intensity parallel to the axis of a myofibril. The average photon rate in $I_V$ and $I_H$ signal was 10 and 4 photons/ms. The inset in Fig. 3b shows the signal from an empty area immediately adjacent to a myofibril. It is due to the fluorescence of glass coverslip/solvent. The average photon rate in $I_H$ and $I_V$ of background was $I_{Hb} = 2$ and $I_{Vb} = 4$ photons/ms, respectively. The total photon rate was $I_{Tot} = (I_V - I_{Vb}) + 2\,G(I_H - I_{Hb}) = 10$ photons/ms.

When there is only a few observed molecules, it should be possible to distinguish individual impulses by inspection of the intensity trace. In practice, this was not possible; the signal was too noisy. A practical way to reduce the noise is to compute autocorrelation function of the signal. Figure 4 (at left) shows the relation between signal and its autocorrelation function.

Correlation function at a given delay time $\tau$ is a sum of the products of a signal multiplied by a signal shifted by a delay time $\tau$. Thus, if $\tau$ is small and there is correlation between signal and its value at time $\tau$ later, the correlation function will be large. As the value of $\tau$ increases, however, a signal has drifted from the original point and it is now of the opposite sign. The product will be negative resulting in smaller correlation. If there is no correlation between signal and its value at time $\tau$ later, the correlation function will be zero. If the

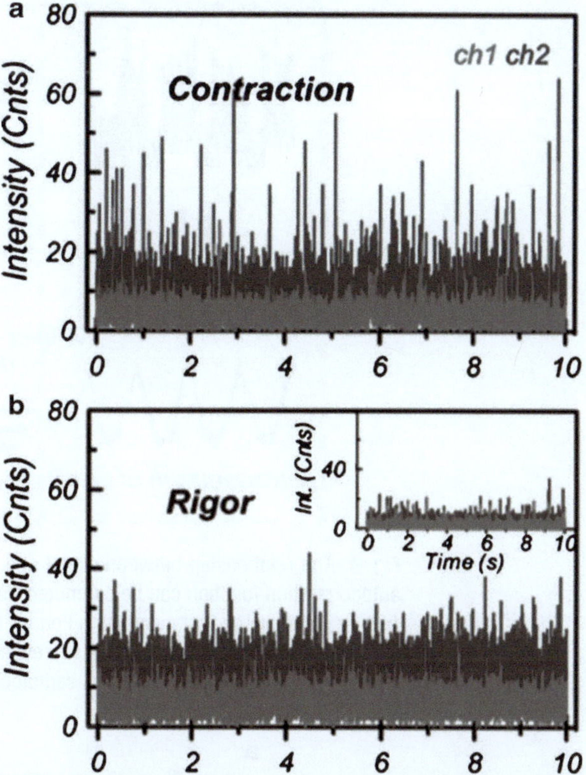


Fig. 3. The time course of polarized intensity of contracting (**a**) and rigor (**b**) myofibril. The data was collected every 10 μs but 1,000 points were pooled together so that the vertical scale is the number of counts during 10 ms. Note that the scale in both panels is the same. Myofibrillar axis is vertical on the microscope stage. Laser polarization is vertical on the microscope stage. Ch1 (*gray*) and ch2 (*dark gray*) are the fluorescence intensities polarized perpendicular and parallel to the myofibrillar axis, respectively. (*Inset*) Signal from an empty area immediately adjacent to the myofibril. Excitation 470 nm, emission $> 590$ nm. Laser power $= 40$ μW, light flux ~143 μW/μm$^2$. Data collected with PicoQuant MT200.

signal is periodic, its correlation function will also be periodic with the same period. The noise is decreased because noise is random and there is no correlation between any two points. This method has the additional advantage that it makes no difference whether we observe one or two molecules. As long as the signal falls within mesoscopic regime [when stochastic fluctuations become important (42)].

The shape of the autocorrelation function corresponding the precise number is not important, i.e. 2 molecules must yield the same result as 10 molecules to the signal of Fig. 3 is shown in Fig. 5. The relationship between the shape of correlation function and the shape of the underlying molecular event is explored below.

$t_{ON}$ was extracted from the correlation function by measuring the time from 0 until the correlation function first changed the slope (for the reasons that will become apparent later). $t_{OFF}$ was measured from the time correlation function changed the slope to the next minimum.

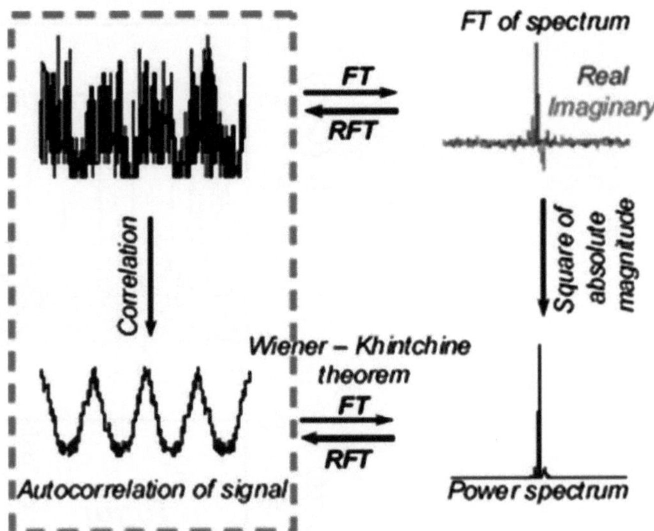


Fig. 4. The relationship between signal and its autocorrelation function (*green box*). The autocorrelation function can be computed from signal directly, as indicated in the *shaded box*, or indirectly by first calculating Fourier transform (FT), squaring its absolute value to obtain power spectrum, and computing reverse Fourier transform (RFT). Power spectrum is a histogram of signal power at each sampled frequency. The indirect route is much faster.

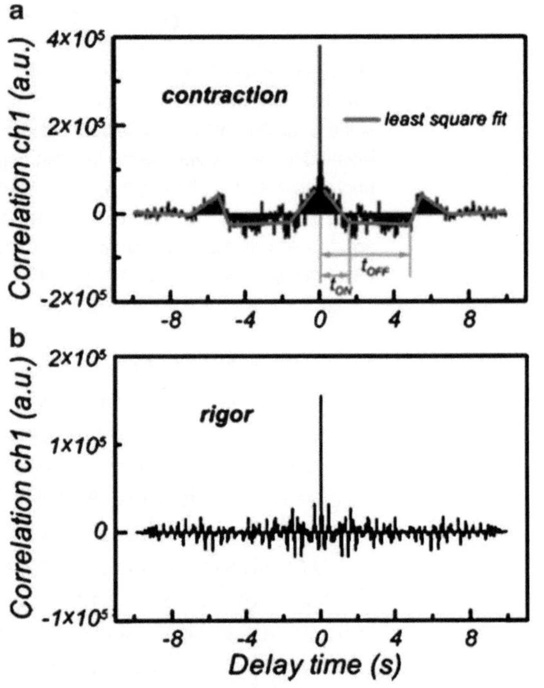


Fig. 5. The correlation function of contracting (**a**) and rigor (**b**) myofibril. Data from ch1 of Fig. 3. Data was fitted to the exponential function, the exponential was subtracted from the data to get zero average value and correlation function was computed. The *red line* in (**a**) is the least square fit to the correlation. The times, delimited by *light gray lines*, are defined in the text. In order to get correlations only between corresponding points, padded zeros have been added before the correlation calculation. 1,000 points averaging was used.

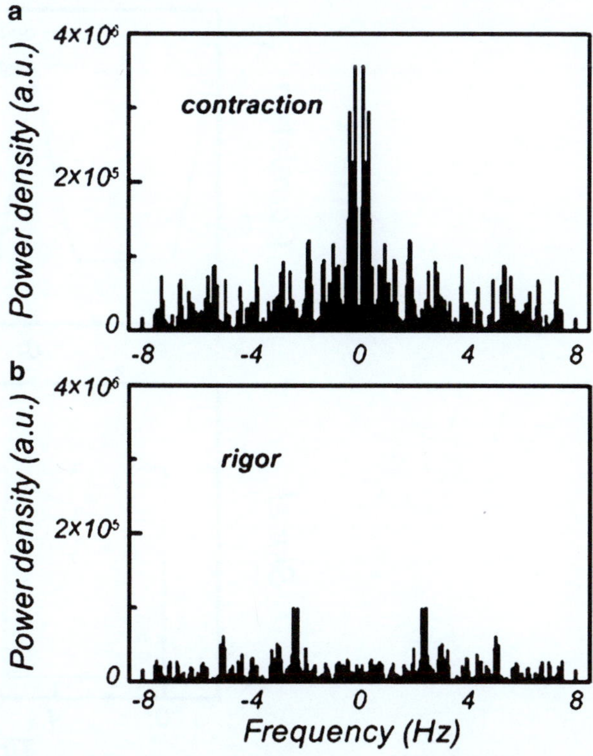


Fig. 6. Power density spectra of contracting (**a**) and rigor (**b**) myofibril. Data from ch1 of Fig. 3.

An alternative way to display kinetic properties is in the form of a power spectrum. The relation of power spectrum to signal is shown in Fig. 4. Figure 6a and b shows power spectra of contracting and rigor myofibril of Fig. 3. Periodic character of the Fourier transform can be seen in the power spectrum.

### 3.2. From Correlation Function to Signal

After calculating correlation function, the next step is to retrieve it from the original signal. However, the counterclockwise operations indicated in Fig. 4 lead to degeneracy, i.e., one correlation function can lead to a number of signals (it is a "price" one has to pay in order to reduce the noise). It is therefore necessary to select a model of a signal (acto-myosin interaction) and to test it by seeing whether it is consistent with the observed correlation function. One such model is the scheme originated by Huxley (43), where a cross-bridge is either attached or detached from actin. The impulse begins when dissociated cross-bridge, containing products of hydrolysis of ATP, attaches to actin. It is believed that a cross-bridge is strongly bound to actin for a period of time $t_{ON}$ (ON time) during which it generates contractile force. The binding and force generation are followed by a release of phosphate, the dissociation of ADP and the onset of rigor state. In rigor state, a cross-bridge assumes a well-defined orientation (44). The end of an impulse is marked by the binding of a fresh

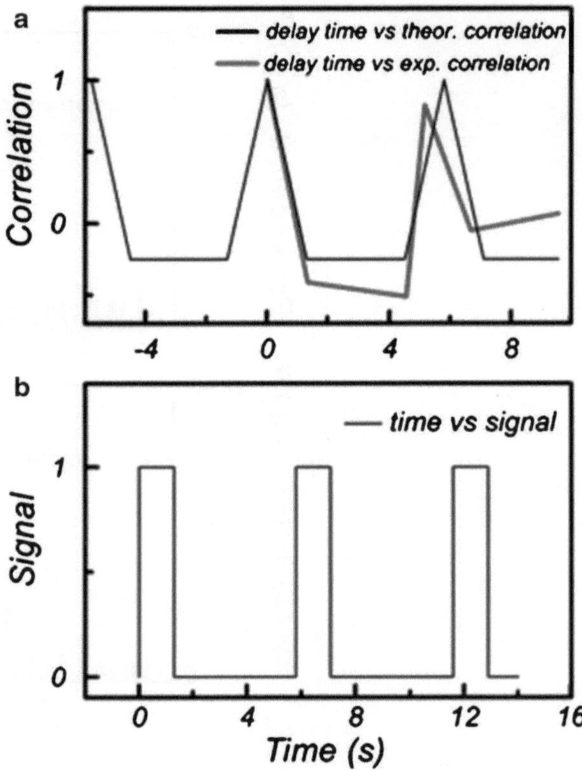


Fig. 7. (**a**) The best fit of the experimental autocorrelation function from Fig. 5 (*red*) to the ideal saw tooth function (*black*). (**b**) *Blue signal* is a wave that gives rise to autocorrelation function in (**a**). In this example, $t_{ON}$ and $t_{OFF}$ were 1.3 s and 4.5 s, respectively.

molecule of ATP and by the dissociation of myosin from actin. Cross-bridge remains detached from actin for a period of time $t_{OFF}$ (OFF time) during which it is idle and disordered (45). Finally, ATP is hydrolyzed to ADP and phosphate, and the cycle repeats with the period $t_{ON} + t_{OFF}$. This model predicts that the autocorrelation function of a signal be a triangular wave. The model is a gross over-simplification but Fig. 7a (gray line) shows that the best fit of the data of Fig. 5a is indeed a triangular wave. This is in turn approximated with a piecewise linear train of triangles (black lines). Such a train of triangles is the correlation function of a train of rectangular waves. The rectangular signal in Fig. 7b was calculated from the fitted triangle train. The nomenclature assigned in Fig. 5a now becomes clear: $t_{ON}$ is the time when the polarization signal is high, and $t_{OFF}$ when the polarization is low.

The results from 17 independent experiments on 6 different myofibrillar preparations show that the average ± SD of ON and OFF times were 1.2 ± 0.40 and 2.3 ± 1.0 s, respectively, suggesting the duty ratio, defined as the fraction of a total cycle time that a cross-bridge remains attached to actin $\Psi = t_{ON}/(t_{ON} + t_{OFF})$ in isometrically working muscle, was ~34%. Figure 5 hints at an interesting phenomenon whereby each peak is composed of several harmonic "subpeaks" no doubt reflecting over-simplification of the model.

From the analysis of crystal structures of different forms of myosin, single molecule studies, spectroscopic experiments, and X-ray diffraction of muscle fibers, a picture of the cross-bridge cycle emerges that suggests that in addition to ON and OFF conformations, cross-bridge lever arm assumes an additional—weak binding—conformation (46–48). As a result, polarized intensity of actin monomer could have assumed three different values. For example, the intensity in the OFF state could be 0, the intensity of weakly attached state could be 0.5, and the intensity of the ON state be 1. If the duration of each step were equal, the result would have led to correlation function of the same shape as a two-step process. Because of this degenerative property of the Fourier transform, our results cannot provide evidence for such a state.

The method of extracting kinetic parameters from steady-state data has been applied to the problem of FHC. One of the causes of a malignant phenotype of this disease is D166V point mutation of the ventricular myosin regulatory light chain (RLC) (49–55). It is characterized by ventricular and septal hypertrophy, myofibrillar disarray, cardiac arrhythmias, and frequent sudden cardiac death (SCD). The insidiousness of this disease is that it often leads to death at a younger age (41, 56). Clinical studies have revealed that the D166V mutation in myosin RLC is associated with a malignant FHC disease phenotype (53). It has been proposed that this mutation results in the alteration of cross-bridge kinetics. We show that SMD allowed us to measure both the rate of cross-bridge attachment and detachment in isometrically contracting cardiac myofibrils. We compare rates in wild-type and mutated myofibrils from transgenic (Tg) mice. We show that the rates are significantly decreased by the D166V mutation in RLC. The orientation of a single rhodamine-phalloidin labeled actin protomer was monitored by tracking its fluorescence anisotropy. We suggest that the D166V mutation of RLC leads to alterations in myosin cross-bridge kinetics which affects the interaction of the thick and thin filaments during contraction of cardiac muscle and ultimately leads to inability of heart to pump blood efficiently.

Figure 8a–c shows a typical image of rigor transgenic WT myofibril from right ventricle of mouse heart labeled with Alexa488-phalloidin. The images clearly show that in cardiac muscle the entire I-bands are labeled. In skeletal muscle, by contrast, phalloidin labels only the overlap zone (32). Binding of phalloidin in skeletal muscle is regulated by nebulin (57), and cardiac muscle has no nebulin. The Z-lines are heavily stained because they contain high concentration of actin, and there is no fluorescence from the H-bands because they contain no actin. The orthogonal polarization images show that the fluorescence is highly anisotropic (58) as is expected from aligned array of polar actin filaments. Figure 8g shows the

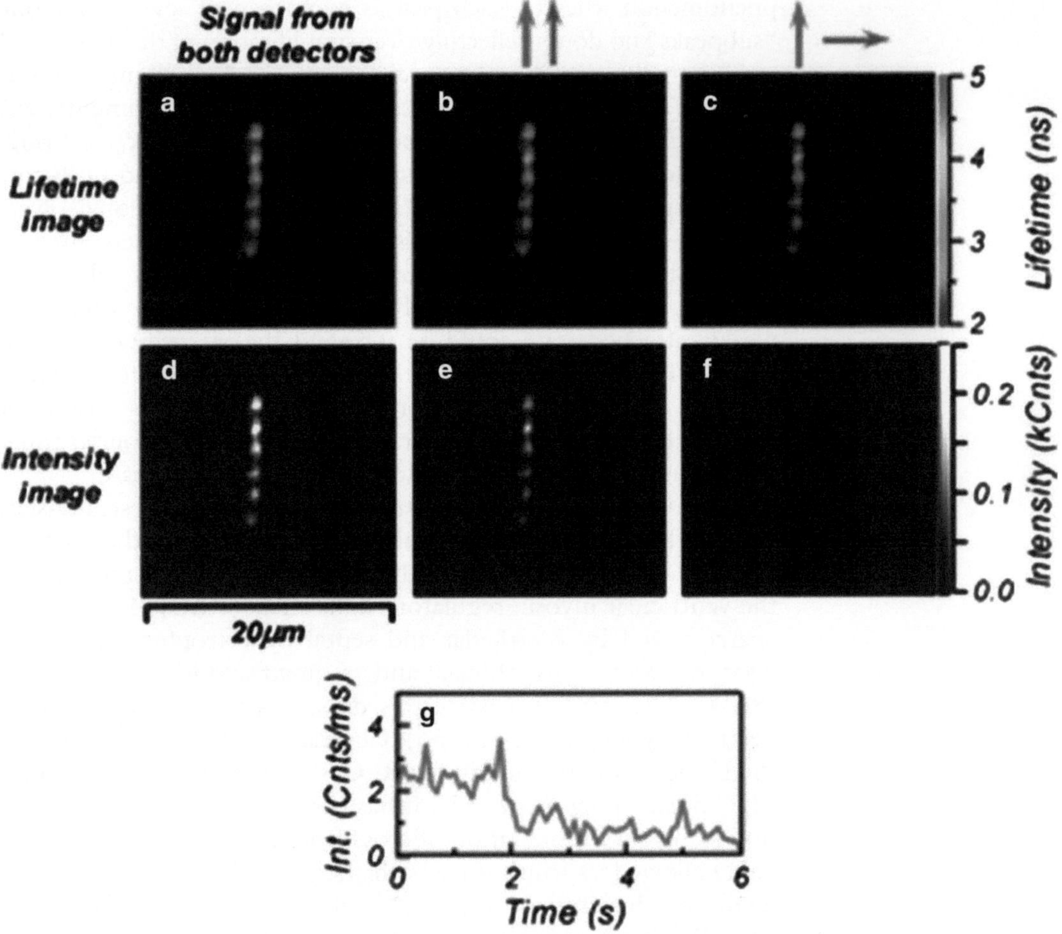


Fig. 8. Lifetime (**a–c**) and polarization (**d–f**) images of rigor transgenic WT myofibril from right ventricle of mouse heart. The *bar* at the right of lifetime images is the lifetime scale, with *maximum* corresponding to 5 ns and *minimum* to 2 ns. The *B/W bar* at the right of polarization images is the intensity scale, with 255 corresponding to *white* and 0 to *black*. *Gray arrows* indicate direction of polarization of exciting and fluorescent light, respectively. Myofibrils were labeled with 0.1 μM Alexa488-phalloidin and cross-linked to avoid shortening. (**g**) Sudden drop of total intensity to the level of the background in rigor muscle—behavior characteristic of single molecule bleaching. The intensity suddenly dropped at ~2 s. The rate of arrival of fluorescent photons before bleaching was 2.4 photons/ms (see text). Data taken on PicoQuant Micro Time 200 confocal lifetime microscope. Excitation with a 470 nm pulse of light, emission through LP500 filter.

time trace of polarized fluorescence intensity in the parallel channel. It shows sudden drop of polarized intensity to the level of the background in rigor muscle—behavior characteristic of single molecule bleaching. The intensity suddenly dropped at ~2 s from the rate of ~3 to ~1 counts/ms. Since we work in a saturated regime, where increasing laser power does not lead to an increase in the number of photons emitted by a fluorophores, we conclude that in cardiac muscle a single chromophore gives rise to fluorescent rate of ~2,000–4,000 photons/s per channel.

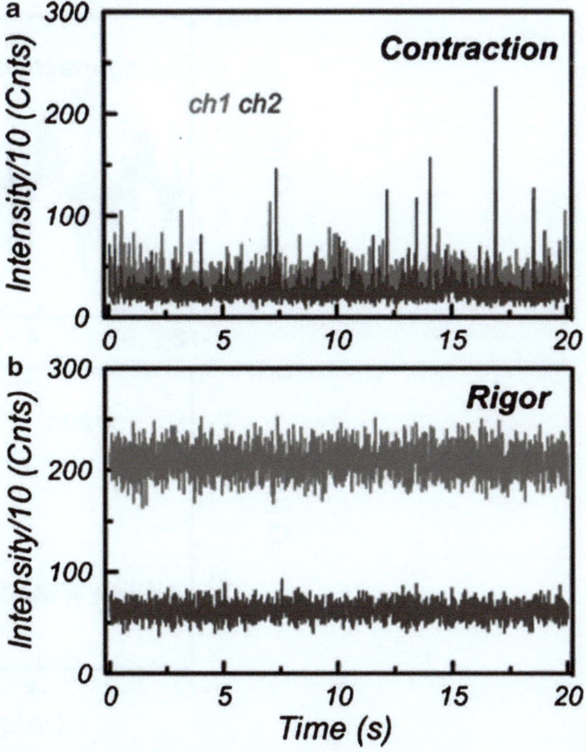


Fig. 9. The time course of polarized intensity of contracting (**a**) and rigor (**b**) myofibril from healthy heart. The original data was collected every 10 μs but 1,000 points are binned together to give 10 ms time resolution. The vertical scale is the number of counts during 10 ms. Note that the scale in both panels is the same. Myofibrillar axis is horizontal on the microscope stage. Laser polarization is vertical on the microscope stage. Ch1 (*red*) and ch2 (*blue*) are the fluorescence intensities polarized perpendicular and parallel to the myofibrillar axis, respectively.

The fluorescence intensity was measured by positioning the laser beam at the center of the I-band. The laser beam was not scanned. After opening the laser shutter, the fluorescence intensity, which was initially high (~50 counts/ms), decayed after several seconds to a steady-state value of ~2–4 counts/ms. Figure 9 shows a typical intensity data collected from healthy WT heart ventricular myofibrils during 20 s of contraction (a) and rigor (b). This fluorescence was due to a single fluorophore in the detection volume (because the rate of fluorescence from a single molecule in myofibrils was ~2–4 photons/ms, see Fig. 8g). As before, in order to decrease the noise, we apply the Fourier transform to compute correlation function of the signal. The autocorrelation function corresponding to the signal from WT heart ventricle during contraction (ch1 of Fig. 9a) is shown in Fig. 10a. The autocorrelation function corresponding to the signal from WT heart ventricle in rigor (ch2 of Fig. 9b) is shown in Fig. 10b.

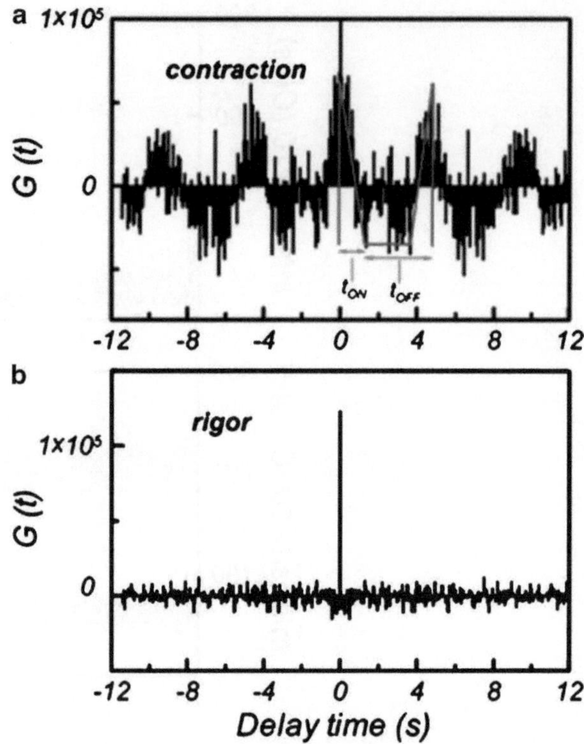


Fig. 10. The correlation function of ch1 data from Fig. 9a (**a**) and ch2 data from Fig. 9b (**b**). Data was fitted to the exponential function, the exponential was subtracted from the data to get zero average value and correlation function was computed. In order to get correlations only between corresponding points, padded zeros have been added before the correlation calculation. Ten points averaging was used. The times are as previously defined.

Figure 11a and b shows power spectra of contracting and rigor myofibril of WT heart ventricle. The spectra show that the signal contains high-frequency components. This reflects the fact that each peak of a correlation function is composed of several harmonic "subpeaks" (Fig. 10a).

Figure 12 shows a typical intensity data collected from a single molecule of a diseased heart ventricular myofibrils during 20 s. The autocorrelation function of the signal from contracting diseased heart ventricle (ch2 of Fig. 12a) is shown in Fig. 13a. Figure 13b is an autocorrelation function of healthy heart (from Fig. 10a) shown here for comparison.

The parameters that are extracted from correlation are $t_{ON}$ and $t_{OFF}$ as defined previously. Values of $t_{ON}$ and $t_{OFF}$ from 45 myofibrils of healthy hearts and 57 myofibrils of diseased hearts are summarized in Table 1 and plotted in Fig. 14. The differences between times were highly statistically significant ($t = 45.2$, $P << 0.005$ for $t_{ON}$) and ($t = 91.1$, $P < < 0.005$ for $t_{OFF}$). The results show that for healthy heart the OFF time was approximately four times as long as the ON time, suggesting the duty ratio,

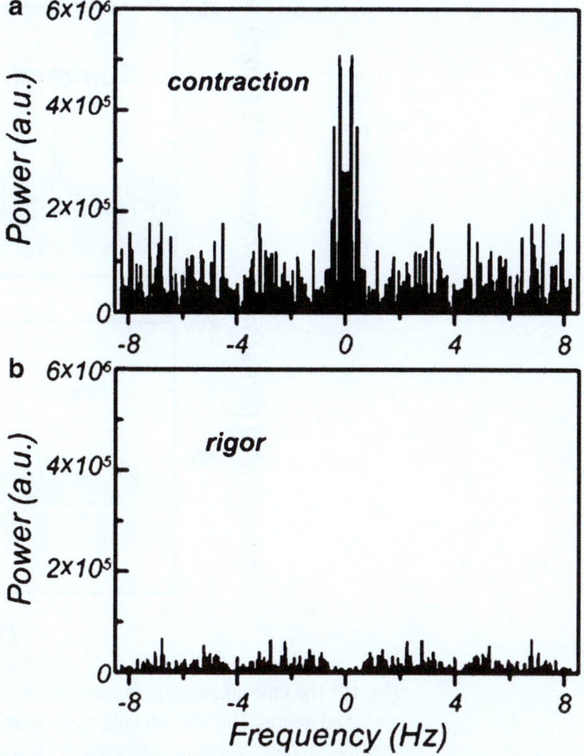


Fig. 11. Power density spectra of contracting (**a**) and rigor (**b**) WT myofibril of Fig. 9.

defined as the fraction of a total cycle time that a cross-bridge remains attached to actin $\Psi = t_{ON}/(t_{ON} + t_{OFF})$ in isometrically working healthy heart, was ~20%. For the diseased heart, the OFF time was a little over twice that of ON time, suggesting the duty ratio in isometrically working diseased heart was ~30%.

## 4. Discussion

We used fluorescence anisotropy to compare kinetics of cross-bridge binding to actin in myofibrils. Not surprisingly, signals were very noisy because they originate from a single molecule among billions present in a myofibril. To reduce noise, we analyzed correlation function of a signal. Kinetic parameters were extracted from correlation function by fitting it to a piecewise linear train of triangles. Such a train of triangles is the correlation function of a train of rectangular waves. Train of rectangular waves is the simplest time course of anisotropy change during contraction. While this scheme is an oversimplification of actual events (59), there is little doubt that

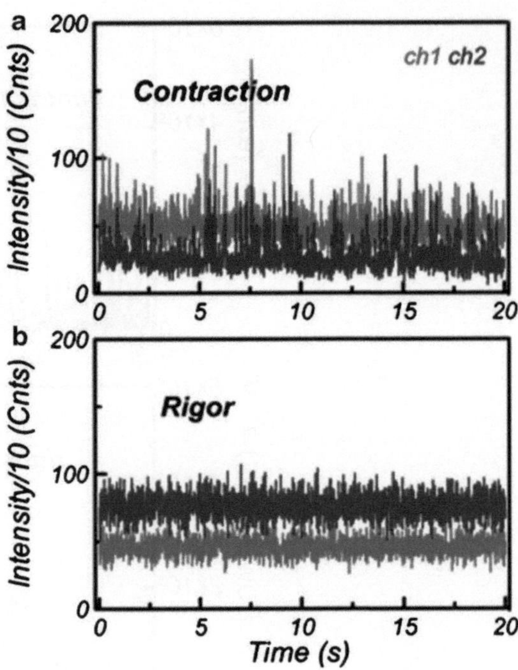


Fig. 12. The time course of polarized intensity of contracting (**a**) and rigor (**b**) myofibril from diseased hearts. The original data was collected every 10 μs but 1,000 points are binned together to give 10 ms time resolution. The vertical scale is the number of counts during 10 ms. Myofibrillar axis is horizontal on the microscope stage. Laser polarization is vertical on the microscope stage. Ch1 (*red*) and ch2 (*blue*) are the fluorescence intensities polarized perpendicular and parallel to the myofibrillar axis, respectively.

this simplified scheme represents correctly the main events occurring during a contractile cycle, i.e., binding dissociation of a cross-bridge. Anisotropy of actin must be high when it is immobilized by binding to myosin and low when it is free from myosin (60). Myosin cross-bridge carrying the products of hydrolysis of ATP binds strongly to actin leading to its immobilization and high anisotropy. Cross-bridge binding probably prevents bending of a whole filament (33, 34) rather than the rotation of actin monomer (Fig. 15a). Dissociation of Pi leads to force generation followed by dissociation of ADP and rigor conformation. Attached state is ended when myosin binds a fresh molecule of ATP. The strongly attached state lasts for $t_{ON}$ seconds. Cross-bridge detachment results in a rapid flutter of F-actin segment to which cross-bridge was bound and leads to a low anisotropy. This state lasts for $t_{OFF}$ seconds. This simplified scheme is shown in Fig. 15b. The rectangular signal was calculated from the fitted triangle train such as experimental train shown in Figs. 5 and 9. It is shown diagrammatically in Fig. 15c.

### 4.1. Duty Ratio

The duty cycle of myosin cross-bridge is another meaningful indicator of the state of muscle. A cross-bridge is in a strongly attached state

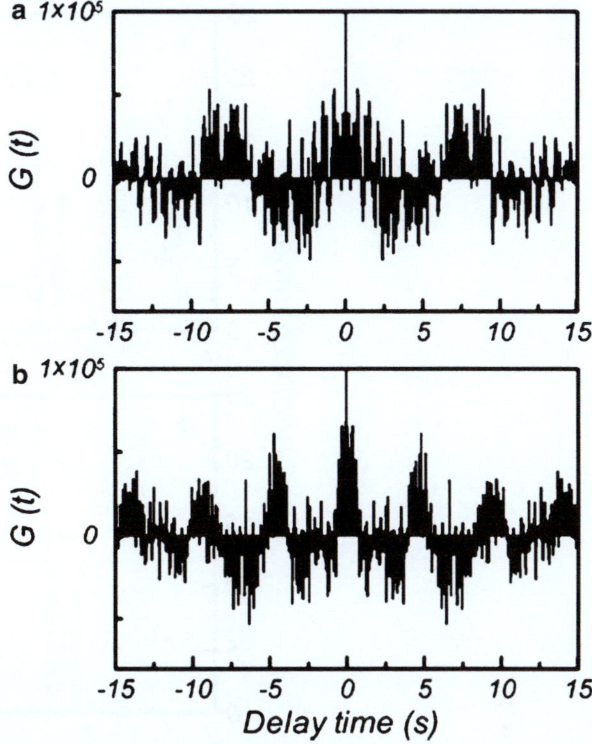


Fig. 13. Side-by-side comparison of typical correlation functions obtained from contracting diseased (**a**) and healthy (**b**) (data from Fig. 9a) heart ventricle.

**Table 1**
**ON and OFF times in WT and diseased right ventricles transgenic mouse**

| Heart | $t_{ON}$ (s) | $t_{OFF}$ (s) | $N$ |
|---|---|---|---|
| WT | 0.7 ± 0.1 | 2.7 ± 0.2 | 45 |
| Mutated | 2.5 ± 0.2 | 5.8 ± 0.2 | 57 |

for a period of time $t_{ON}$, while it remains in a dissociated or a weakly attached state for a period of time, $t_{OFF}$. The duty cycle ratio $\Psi = t_{ON}/(t_{ON} + t_{OFF})$ is equal to the fraction of a total cycle time that a cross-bridge remains attached to actin. To measure $\Psi$ one can follow the changes in anisotropy, as in this work, or the changes in the environment of a cross-bridge while it undergoes a cycle of binding and dissociation from actin, as in (61). The average value of

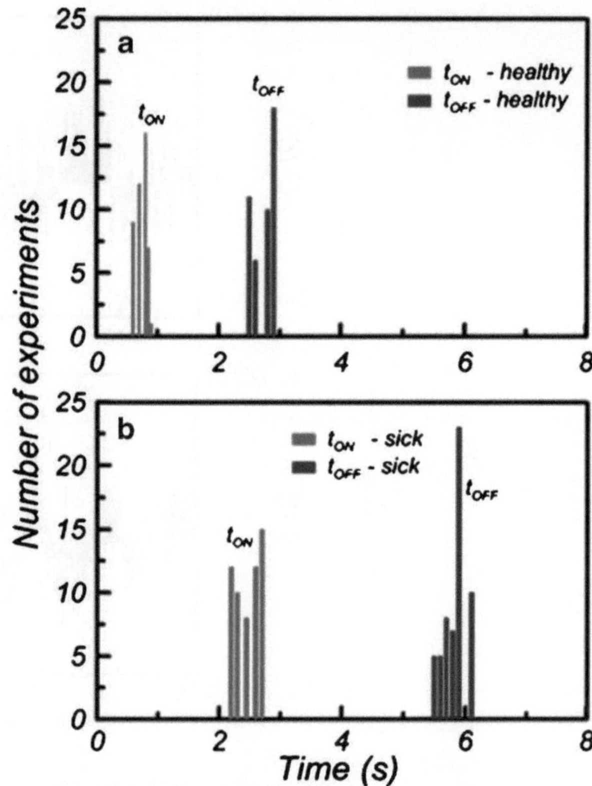


Fig. 14. Comparison of $t_{ON}$ and $t_{OFF}$ from 45 contracting healthy (**a**) and 57 diseased (**b**) heart ventricles.

duty cycle in isometrically working skeletal muscle was 34%. This agrees with recent results which suggested that close to one-third cross-bridges were active during isometric contraction (62). In WT heart myofibrils, $\Psi$ was 21%. This agrees with earlier data obtained by Cooke et al. (44) in isometrically contracting skeletal muscle, and is very similar to the results of Duong and Reisler (63) obtained by measuring protection from tryptic digestion of the 50/20 kDa junction of the myosin heavy chain. The average value of duty cycle in isometrically working diseased heart myofibrils was 30%. The two values (WT and diseased muscle) are statistically significantly different.

### 4.2. Relation Between Cycle Time and ATPase

The average cycle time determined in the present work was 3.5 s in skeletal muscle and 3.4 and 8.3 s in healthy and diseased heart, respectively. This is order-of-magnitude slower than ATPase determined from the maximal ATPase activity measured in skeletal (64) (0.1–0.125 s) and cardiac (65) (0.25–0.27 s) fibers. The reason for discrepancy most likely is the high-frequency substructure seen in the correlation function. The presence of this substructure reflects rapid cycling during the ON time. There is no cycling during the OFF time.

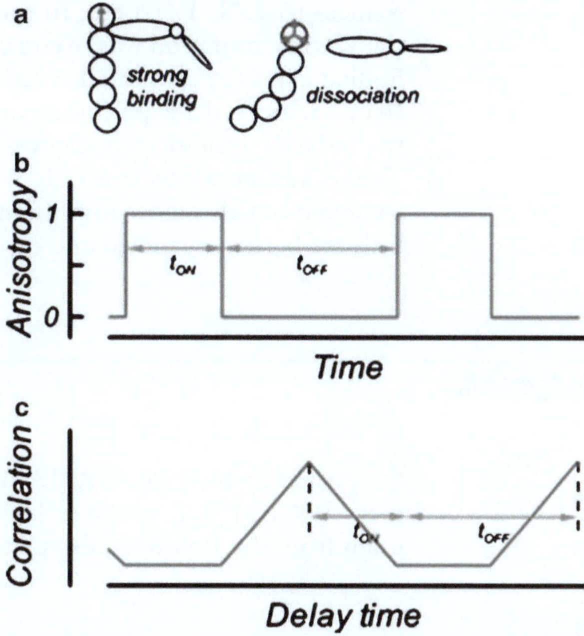


Fig. 15. The simplest scheme for generating anisotropy fluctuations in contracting myofibrils. (**a**) Actin becomes immobilized when a cross-bridge binds to it resulting in a well-defined direction of transition dipole (*red arrow*) and high anisotropy. Dipoles becomes disorganized when a cross-bridge dissociates, leading to low anisotropy (*multiple arrows*). (**b**) A simple rectangular signal reflecting binding dissociation of a cross-bridge. (**c**) Autocorrelation function of a rectangular signal showing duration ($t_{ON}$ seconds) of strongly attached state and detached (or weakly attached) state, lasting for $t_{OFF}$ seconds.

The ON time perhaps reflects the time when a cross-bridge has favorable access to actin monomer. The OFF time perhaps reflects the time when a cross-bridge is unable to interact with a distant monomer; perhaps it is too far to be felt by actin under observation. Not every hydrolysis event leads to the change of phalloidin-actin orientation: hydrolysis by a distant cross-bridge may be not felt by actin molecule under observation. The fact that the nearest cross-bridge on the same actin filament is 38.5 nm away suggests that the signal is not transmitted over such a distance, consistent with the view that signal propagates only over a few actin monomers (66).

### 4.3. Clinical Applications

Application will be harder in human myocardium than in transgenic myofibrils used here because human patients are normally heterozygous for FHC mutations, so 50% of myosin containing thick filaments are composed of wild-type myosin heads interspersed with FHC mutant heads. The results of experiments must reflect the fact that molecule belongs to a healthy or to a diseased subpopulation. If the distribution is 50/50, if we do two experiment there is 1/4 chance that the diseased population will not be revealed, but if we carry out 3, 4 etc experiments, the chance reduces to 1/8, 1/16 etc. In the transgenic mouse myocardium, the D166V mutation was overexpressed at 90%. The probability of finding a healthy actin in diseased sample was therefore quite small. In one respect, the experiments in humans will be easier: in humans the D166V mutation is expressed also in slow skeletal muscle. This is advantageous not only because biopsied samples are easy to obtain but also more importantly because skeletal muscle myofibrils are better organized and easier to work with.

## Acknowledgments

Supported by NIH grant R01AR048622 to J.B. and by Texas ETF grant (CCFT). R.L. is the recipient of the Research Mobility program from the Polish Ministry of Science and Higher Education.

### References

1. Nihei T, Mendelson RA, Botts J (1974) Use of fluorescence polarization to observe changes in attitude of S1 moieties in muscle fibers. Biophys J 14:236–242
2. Borejdo J, Assulin O, Ando T, Putnam S (1982) Cross-bridge orientation in skeletal muscle measured by linear dichroism of an extrinsic chromophore. J Mol Biol 158: 391–414
3. Thomas DD, Cooke R (1980) Orientation of spin-labeled myosin heads in glycerinated muscle fibers. Biophys J 32:891–905
4. Huxley AF, Simmons RM (1971) Proposed mechanism of force generation in striated muscle. Nature 233:533–538
5. Lymn RW, Taylor EW (1971) Mechanism of adenosine triphosphate hydrolysis by actomyosin. Biochemistry 10:4617–4624
6. Siemankowski RF, Wiseman MO, White HD (1985) ADP dissociation from actomyosin subfragment 1 is sufficiently slow to limit the unloaded shortening velocity in vertebrate muscle. Proc Natl Acad Sci U S A 82:658–662
7. Geeves MA, Holmes KC, Bodis E et al (2005) The molecular mechanism of muscle contraction. Adv Protein Chem 71:161–193
8. Magde D, Elson EL, Webb WW (1974) Fluorescence correlation spectroscopy. II. An experimental realization. Biopolymers 13:29–61
9. Qian H, Saffarian S, Elson EL (2002) Concentration fluctuations in a mesoscopic oscillating chemical reaction system. Proc Natl Acad Sci U S A 99:10376–10381
10. Bagshaw CR (1982) Muscle contraction. Chapman & Hall, London
11. Enderlein J, Ambrose WP (1997) Optical collection efficiency function in single-molecule detection experiments. Appl Opt 36: 5298–5302
12. Willets KA, Ostroverkhova O, He M, Twieg RJ, Moerner WE (2003) Novel fluorophores for single-molecule imaging. J Am Chem Soc 125:1174–1175
13. Wang Y, Qin H, Kudaravalli RD et al (2007) Single-molecule structural dynamics of EF-G-ribosome interaction during translocation. Biochemistry 46:10767–10775
14. Taniguchi Y, Karagiannis P, Nishiyama M, Ishii Y, Yanagida T (2007) Single molecule thermodynamics in biological motors. Biosystems 88:283–292
15. Warshaw DM, Hayes E, Gaffney D et al (1998) Myosin conformational states determined by single fluorophore polarization. Proc Natl Acad Sci U S A 95:8034–8039
16. Quinlan ME, Forkey JN, Goldman YE (2005) Orientation of the myosin light chain region by single molecule total internal reflection fluorescence polarization microscopy. Biophys J 89: 1132–1142
17. Forkey JN, Quinlan ME, Shaw MA, Corrie JE, Goldman YE (2003) Three-dimensional

structural dynamics of myosin V by single-molecule fluorescence polarization. Nature 422:399–404

18. Yildiz A, Forkey JN, McKinney SA, Ha T, Goldman YE, Selvin PR (2003) Myosin V walks hand-over-hand: single fluorophore imaging with 1.5-nm localization. Science 300: 2061–2065
19. Toprak E, Enderlein J, Syed S et al (2006) Defocused orientation and position imaging (DOPI) of myosin V. Proc Natl Acad Sci U S A 103:6495–6499
20. Minton AP, Wilf J (1981) Effect of macromolecular crowding upon the structure and function of an enzyme: glyceraldehyde-3-phosphate dehydrogenase. Biochemistry 20:4821–4826
21. Minton AP (1981) Excluded volume as a determinant of macromolecular structure and reactivity. Biopolymers 20:2093–2120
22. Eigen M, Rigler R (1994) Sorting single molecules: application to diagnostics and evolutionary biotechnology. Proc Natl Acad Sci U S A 91:5740–5747
23. Borejdo J, Shepard AA, Akopova I, Grudzinski W, Malicka J (2004) Rotation of the lever-arm of myosin in contracting skeletal muscle fiber measured by two-photon anisotropy. Biophys J 87:3912–3921
24. Levene MJ, Korlach J, Turner SW, Foquet M, Craighead HG, Webb WW (2003) Zero-mode waveguides for single-molecule analysis at high concentrations. Science 299:682–686
25. Lewis A (1991) The optical near-field and cell biology. Semin Cell Biol 2:187–192
26. de Lange F, Cambi A, Huijbens R et al (2001) Cell biology beyond the diffraction limit: near-field scanning optical microscopy. J Cell Sci 114:4153–4160
27. Borejdo J, Talent J, Akopova I, Burghardt TP (2006) Rotations of a few cross-bridges in muscle by confocal total internal reflection microscopy. Biochim Biophys Acta 1763:137–140
28. Vecer J, Kowalczyk AA, Davenport L, Dale RE (1993) Reconvolution analysis in time-resolved fluorescence experiments, an alternative approach: reference-to-excitation-to-fluorescence reconvolution. Rev Sci Instrum 64:3413–3424
29. Bukatina AE, Fuchs F, Watkins SC (1996) A study on the mechanism of phalloidin-induced tension changes in skinned rabbit psoas muscle fibres. J Muscle Res Cell Motil 17:365–371
30. Prochniewicz-Nakayama E, Yanagida T, Oosawa F (1983) Studies on conformation of F-actin in muscle fibers in the relaxed state, rigor, and during contraction using fluorescent phalloidin. J Cell Biol 97:1663–1667
31. Shepard A, Borejdo J (2004) Correlation between mechanical and enzymatic events in contracting skeletal muscle fiber. Biochemistry 43:2804–2811
32. Szczesna D, Lehrer SS (1993) The binding of fluorescent phallotoxins to actin in myofibrils. J Muscle Res Cell Motil 14:594–597
33. Yanagida T, Oosawa F (1980) Conformational changes of F-actin-epsilon-ADP in thin filaments in myosin-free muscle fibers induced by $Ca^{2+}$. J Mol Biol 140:313–320
34. Yanagida T, Oosawa F (1998) Polarized fluorescence from epsilon-ADP incorporated into F-actin in a myosin-free single fiber: conformation of F-actin and changes induced in it by heavy meromyosin. J Mol Biol 126:507–524
35. Borovikov YS, Kuleva NV, Khoroshev MI (1991) Polarization microfluorimetry study of interaction between myosin head and F-actin in muscle fibers. Gen Physiol Biophys 10: 441–459
36. Borejdo J, Shepard A, Dumka D et al (2004) Changes in orientation of actin during contraction of muscle. Biophys J 86:2308–2317
37. Dos Remedios CG, Millikan RG, Morales MF (1972) Polarization of tryptophan fluorescence from single striated muscle fibers. A molecular probe of contractile state. J Gen Physiol 59: 103–120
38. Dos Remedios CG, Yount RG, Morales MF (1972) Individual states in the cycle of muscle contraction. Proc Natl Acad Sci U S A 69: 2542–2546
39. Tregear RT, Mendelson RA (1975) Polarization from a helix of fluorophores and its relation to that obtained from muscle. Biophys J 15:455–467
40. Morales MF (1984) Calculation of the polarized fluorescence from a labeled muscle fiber. Proc Nat Acad Sci U S A 81:145–149
41. Maron BJ, Olivotto I, Spirito P et al (2000) Epidemiology of hypertrophic cardiomyopathy-related death: revisited in a large non-referral-based patient population. Circulation 102: 858–864
42. Elson EL (2008) Introduction to FCS. Course Cell Mol Fluorescence 2:8
43. Huxley AF (1957) A hypothesis for the mechanism of contraction of muscle. Prog Biophys Biophys Chem 7:255–318
44. Cooke R, Crowder MS, Thomas DD (1982) Orientation of spin labels attached to cross-bridges in contracting muscle fibres. Nature 300:776–778

45. Thomas DD, Ramachandran S, Roopnarine O, Hayden DW, Ostap EM (1995) The mechanism of force generation in myosin: a disorder-to-order transition, coupled to internal structural changes. Biophys J 68:135S–141S
46. Reedy MC (2000) Visualizing myosin's power stroke in muscle contraction. J Cell Sci 113: 3551–3562
47. Sweeney HL, Houdusse A (2004) The motor mechanism of myosin V: insights for muscle contraction. Philos Trans R Soc Lond B Biol Sci 359:1829–1841
48. Takagi Y, Shuman H, Goldman YE (2004) Coupling between phosphate release and force generation in muscle actomyosin. Philos Trans R Soc Lond B Biol Sci 359:1913–1920
49. Poetter K, Jiang H, Hassanzadeh S et al (1996) Mutations in either the essential or regulatory light chains of myosin are associated with a rare myopathy in human heart and skeletal muscle. Nat Genet 13:63–69
50. Flavigny J, Richard P, Isnard R et al (1998) Identification of two novel mutations in the ventricular regulatory myosin light chain gene (MYL2) associated with familial and classical forms of hypertrophic cardiomyopathy. J Mol Med 76:208–214
51. Andersen PS, Havndrup O, Bundgaard H et al (2001) Myosin light chain mutations in familial hypertrophic cardiomyopathy: phenotypic presentation and frequency in Danish and South African populations. J Med Genet 38:E43
52. Kabaeva ZT, Perrot A, Wolter B et al (2002) Systematic analysis of the regulatory and essential myosin light chain genes: genetic variants and mutations in hypertrophic cardiomyopathy. Eur J Hum Genet 10:741–748
53. Richard P, Charron P, Carrier L et al (2003) Hypertrophic cardiomyopathy: distribution of disease genes. Spectrum of mutations, and implications for a molecular diagnosis strategy. Circulation 107:2227–2232
54. Morner S, Richard P, Kazzam E et al (2003) Identification of the genotypes causing hypertrophic cardiomyopathy in northern Sweden. J Mol Cell Cardiol 35:841–849
55. Hougs L, Havndrup O, Bundgaard H et al (2005) One third of Danish hypertrophic cardiomyopathy patients have mutations in MYH7 rod region. Eur J Hum Genet 13:161–165
56. Maron BJ (2002) The young competitive athlete with cardiovascular abnormalities: causes of sudden death, detection by preparticipation screening, and standards for disqualification. Card Electrophysiol Rev 6:100–103
57. Ao X, Lehrer SS (1995) Phalloidin unzips nebulin from thin filaments in skeletal myofibrils. J Cell Sci 108:3397–3403
58. Borovikov YS, Chernogriadskaia NA (1979) Studies on conformational changes in F-actin of glycerinated muscle fibers during relaxation by means of polarized ultraviolet fluorescence microscopy. Microsc Acta 81:383–392
59. Houdusse A, Sweeney HL (2001) Myosin motors: missing structures and hidden springs. Curr Opin Struct Biol 11:182–194
60. Pesce AJ, Rosen CG, Pasby TL (1971) Fluorescence spectroscopy. Marcel Dekker, New York
61. Mettikolla P, Luchowski R, Gryczynski I, Gryczynski Z, Szczesna-Cordary D, Borejdo J (2009) Fluorescence lifetime of actin in the FHC transgenic heart. Biochemistry 48(6): 1264–1271
62. Cooper WC, Chrin LR, Berger CL (2000) Detection of fluorescently labeled actin-bound cross-bridges in actively contracting myofibrils. Biophys J 78:1449–1457
63. Duong AM, Reisler E (1989) Binding of myosin to actin in myofibrils during ATP hydrolysis. Biochemistry 28:1307–1313
64. Hilber K, Sun YB, Irving M (2001) Effects of sarcomere length and temperature on the rate of ATP utilisation by rabbit psoas muscle fibres. J Physiol 531:771–780
65. Kerrick WG, Kazmierczak K, Xu Y, Wang Y, Szczesna-Cordary D (2009) Malignant familial hypertrophic cardiomyopathy D166V mutation in the ventricular myosin regulatory light chain causes profound effects in skinned and intact papillary muscle fibers from transgenic mice. FASEB J 23:855–865
66. Ando T (1987) Propagation of Acto-S-1 ATPase reaction-coupled conformational change in actin along the filament. J Biochem (Tokyo) 105:818–822

# Chapter 18

# Single-Molecule Optical-Trapping Measurements with DNA Anchored to an Array of Gold Nanoposts

**D. Hern Paik and Thomas T. Perkins**

## Abstract

Gold–thiol chemistry is one of the most successful chemistries for conjugating biomolecules to surfaces, but such chemistry has not been exploited in optical-trapping experiments because of laser-induced ablation of gold. In this work, we describe a method to combine these two separate technologies without undue heating using DNA anchored to gold nanostructures ($r = 50$–$250$ nm; $h \approx 20$ nm). Moreover, we demonstrate a quantitative and mechanically robust (>100 pN) optical-trapping assay. By using three dithiol phosphoramidites (DTPAs) incorporated into a polymerase chain reaction (PCR) primer, the gold–DNA bond remained stable in the presence of excess thiolated compounds. This chemical robustness allowed us to reduce nonspecific sticking by passivating the unreacted gold with methoxy-(polyethylene glycol)-thiol (mPEG-SH). Overall, this surface conjugation of biomolecules onto an ordered array of gold nanostructures by chemically and mechanically robust bonds provides a unique way to carry out spatially controlled, repeatable measurements of single molecules.

**Key words:** Single molecule, Optical trap, Optical tweezers, Gold–thiol bond, Gold–DNA bond, DNA, Force spectroscopy

## 1. Introduction

Optical traps are widely used for studying biological systems (1–3), including stretching DNA (4), characterizing molecular motors (5), and unfolding RNA structures (6). These and other single-molecule force spectroscopy assays (7) require coupling of biomolecules to surfaces through mechanically strong bonds. The challenge is to minimize nonspecific sticking (3) while maintaining unaltered biological activity and/or a native structure. Single-molecule assays are usually assembled by random absorption of biomolecules onto bare glass or protein-coated surfaces. Since different single-molecule assays have different geometric requirements, one would like to pattern biomolecules onto an otherwise nonstick surface. Gold–thiol chemistry provides an important

Wlodek M. Bujalowski (ed.), *Spectroscopic Methods of Analysis: Methods and Protocols*, Methods in Molecular Biology, vol. 875, DOI 10.1007/978-1-61779-806-1_18, © Springer Science+Business Media New York 2012

complementary conjugation technique to the commonly used, but mechanically weaker, ligand–receptor couplings (e.g., streptavidin–biotin, digoxigenin–antidigoxigenin). Gold–thiol bonds are covalent and therefore mechanically strong (~1.4 nN) (8). Gold is also easy to pattern at the nanometer scale (9). However, gold–thiol chemistry has not been integrated into optical-trapping assays due to laser-induced ablation of the gold.

In addressing this challenge, we present a chemically and mechanically robust means for coupling biological molecules to gold nanostructures for use in high-force optical-trapping assays (see Fig. 1a). Our method also minimizes nonspecific sticking—a ubiquitous problem when trying to anchor biomolecules to surfaces. By using multiple gold–thiol bonds (10) (see Fig. 1b), we stably anchored DNA to gold in the presence of excess thiol compounds [including dithiothreitol (DTT)]. This stability allowed us to prevent unwanted sticking by passivating the unreacted gold with methoxy-(polyethylene glycol)-thiol (mPEG-SH), a reducing compound that would displace DNA anchored by a single gold–thiol bond. The mechanical strength of the gold–thiol bond allowed us to perform high-force optical-trapping assays and to rupture the strongest biological interactions (i.e., the streptavidin–biotin bond) (11). Since the DNA remained attached to the gold nanopost, we could bring the optically trapped bead back to the surface-anchored DNA and repeatedly measure the same individual DNA molecule.

To achieve efficient patterning of the gold nanoposts onto glass cover slips, we developed a set of protocols for making and using shadow masks (12). Use of shadow masks avoids exposing the unreacted gold to the chemical solution normally used to remove a polymeric mask. Additionally, the shadow masks are reusable and can be used in parallel to pattern multiple cover slips. We expect that spatial patterning of single molecules via gold nanoposts with very low nonspecific sticking to be useful for other single-molecule techniques, including single-molecule fluorescence and atomic force microscopy.

## 2. Materials

### 2.1. Spin-Coating Resist onto a Silicon Nitride Window

1. Positive resist for electron beam (e-beam) lithography (ZEP520A, Zeon, stored at 4°C).
2. Silicon nitride ($Si_3N_4$) membrane window (membrane size = $1.5 \times 1.5\ mm^2$; 200-nm thickness; frame size = $7.5 \times 7.5\ mm^2$; frame thickness = 300–500 μm; Silson) (see Fig. 2).
3. Microscope cover slip ($22 \times 40\ mm^2$, #1.5 thickness, Corning).
4. Double-sided tape (3 M).

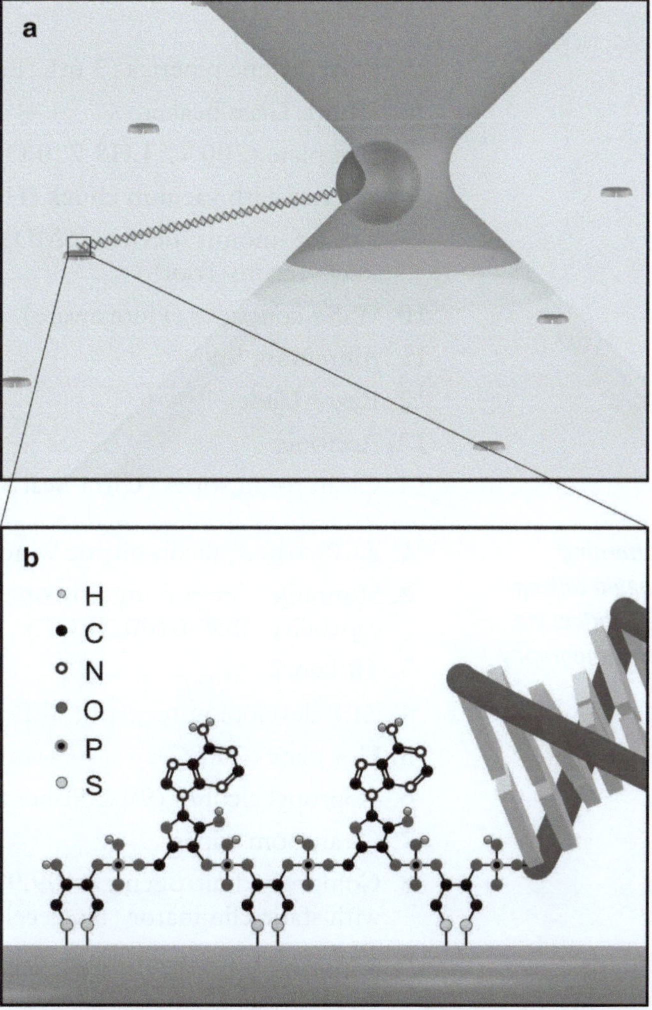


Fig. 1. (**a**) Schematic of an optical-trapping assay integrated with an array of gold nanoposts (*not to scale*). The DNA was anchored at one end via a gold–DTPA bond while the opposite end was attached to an optically trapped streptavidin-coated bead via biotin. (**b**) Chemical structure of the triple repeat of the DTPA linker attached to gold surface. Hydrogen atoms on carbon atoms are not shown (Reprinted with permission from (11); © 2009 American Chemical Society).

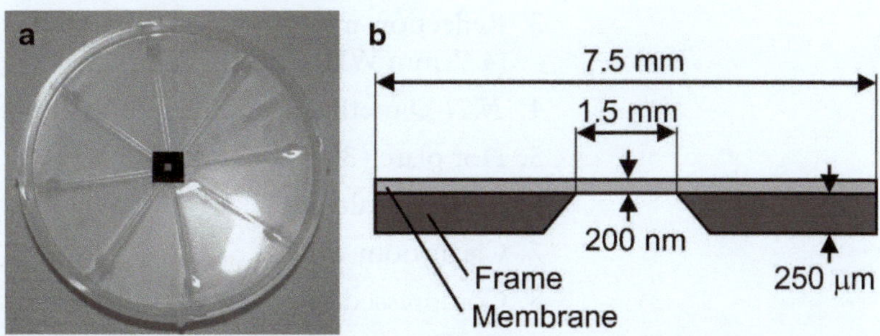


Fig. 2. (**a**) Picture of a 200-nm thick silicon nitride membrane window (*inner clear area*) suspended from a larger silicon frame resting in a 3-in. wafer holder. (**b**) Side view of the silicon nitride window (*not to scale*).

5. Polyethylene pipettes (3 mL, Becton Dickinson).
6. 10-mL Glass beaker.
7. Hot plate (200°C, LHS-720 Omega).
8. Spinner with vacuum chuck (Headway Research).
9. Surface mount device (SMD) handling tweezers (Stainless steel, Techni-Tool).
10. Wafer container (Fluoroware).
11. Aluminum foil.
12. Razor blade.
13. Acetone.
14. Clean room wipes (Ultra-Seal 3000, Berkshire).

### 2.2. Patterning ZEP-Coated Silicon Nitride Window via E-Beam Lithography

1. ZEP-coated silicon nitride window from Subheading 2.1.
2. Scanning electron microscope (SEM) with e-beam writing capability (JSM-6400, JEOL).
3. 4B Pencil.
4. ZEP developing reagent (ZEP-RD, Zeon).
5. Hot plate (200°C).
6. Isopropyl alcohol (99%, Fisher Scientific).
7. Cleanroom wipes.
8. Compressed nitrogen gas (99.9%, separated from air) equipped with static eliminator (Nuclecel, NRD Inc.).
9. Two Pyrex crystallization dish (dia. = 80 mm; $h$ = 40 mm; Sigma).
10. SMD handling tweezers.
11. Aluminum foil.

### 2.3. Etching Silicon Nitride Membrane with Reactive Ion Etch

1. E-beam-patterned silicon nitride window from Subheading 2.2.
2. Reactive ion etch (RIE) (Plasmathern) equipped with $N_2$ and $CF_4$ gas.
3. Reflection microscope (DM LM, Leica) with 100× objective (4.7-mm WD, Leica).
4. *N*,*N*-Dimethyl actamide (99%, Sigma).
5. Hot plate (35°C).
6. Isopropyl alcohol.
7. Cleanroom wipes.
8. Compressed nitrogen gas with static eliminator.
9. Two Pyrex crystallization dishes (dia. = 80 mm; $h$ = 40 mm tall).
10. SMD handling tweezers.
11. Acetone.

#### 2.4. Adding Stand-Off Spacers to the Frame of Silicon Nitride Window

1. RIE-processed silicon nitride window from Subheading 2.3.
2. SU-8 photoresist (SU-8 2005, MicroChem).
3. Microscope cover slips.
4. Double-sided tape (3 M).
5. Polyethylene pipettes.
6. 10-mL Beaker.
7. Two hot plates (65 and 95°C).
8. SU-8 Developer (Methoxy isopropyl acetate, MicroChem).
9. Isopropyl alcohol.
10. Ultrapure water (18.2 MΩ cm, Millipore).
11. Maskaligner (MJB3, Karl Suss).
12. Lithographic photomask (Mylar, Infinite Graphics Inc.). Mask has three (equilateral triangle) or four (square) openings that expose 1-mm dia. circles located along the edge of the silicon frame ($7.5 \times 7.5\ mm^2$).
13. Oxygen plasma clean (PlasmaStar, Axic).
14. Three Pyrex crystallization dishes (dia. = 80 mm; $h = 40$ mm).
15. Pyrex crystallization dish (dia. = 125 mm; $h = 65$ mm).
16. SMD handling tweezers.
17. Acetone.
18. Razor blade.
19. Compressed nitrogen gas (99.9%, separated from air) equipped with static eliminator.
20. Reflection microscope with 100× objective.
21. RIE with fused silica plate (dia. = 75 mm; thickness 1 mm).

#### 2.5. Cleaning Glass Cover Slips

1. KOH pellets (Mallinckrodt).
2. Ethyl alcohol, completely denatured (Mallinckrodt Chemicals).
3. Ultrasonic bath.
4. Microscope cover slips.
5. Squirt bottles with ultraclean $H_2O$ and denatured ethyl alcohol.
6. Custom-made cover-slip holder machined from Teflon.
7. Three 1-L glass beakers.
8. Microwave.
9. Container for cleaned cover slips (e.g., an empty 1-mL pipette tip box).
10. Parafilm.

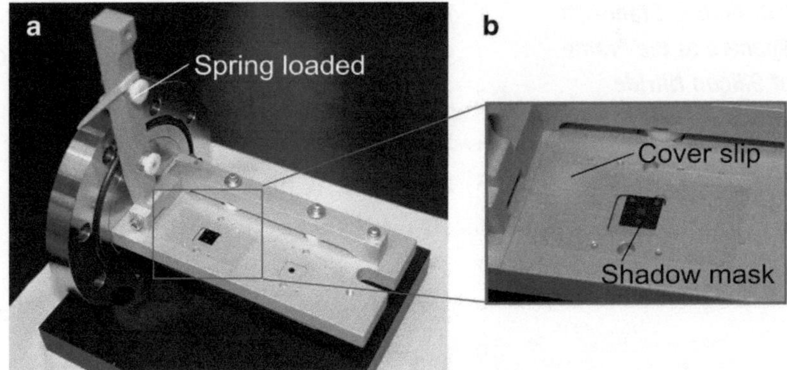


Fig. 3. (**a**) Shadow-mask loading platform, which holds four cover slips so gold can be deposited onto multiple cover slips in parallel. (**b**) Zoomed-in view of the sample loading area. The shadow mask sits in the grove, membrane side up. A cleaned cover slip sits on top of the mask spaced by SU-8 stand-offs (~3 μm). The stacked shadow mask cover slip is gently pressed by the spring-loaded lever arm. Both the shadow mask and the cover slip are registered in the upper right corner to ensure that gold deposition occurs at the center of cover slip.

### 2.6. Depositing Gold Posts onto Cover Slips via Physical Deposition

1. Silicon nitride shadow mask from Subheading 2.4.
2. Chromium pellets (99.99%, Alfa Aesar).
3. Gold pellets (99.999%, Garfield Refining Company).
4. Cleaned cover slip from Subheading 2.5.
5. Custom-made shadow mask loading platform (see Fig. 3).
6. Evaporation chamber with e-beam heater (FL400, Edwards).
7. SMD handling tweezers.
8. Crucible liner (Fabmate Graphite, 1.8 cc, Kurt-Lesker).

### 2.7. Cleaning Shadow Mask

1. Aqua regia (3:1 (v/v) of $HNO_3/HCl$).
2. Nitric-acid based chromium etchant (TFN, Trensene, Inc.).
3. Custom-made shadow-mask holder machined from Teflon (see Fig. 4).
4. Glass Pasteur pipettes [5¾″ (14.6 cm), VWR].
5. Two glass Petri dishes ($r = 50$ mm; $h = 15$ mm).
6. Carbon-fiber tweezers (polyvinylidene fluoride reinforced, Techni-Tool).
7. Two 50-mL beakers.
8. 1-L Beaker.
9. Ultrapure water.
10. Isopropyl alcohol.
11. Cleanroom wipes.
12. Compressed nitrogen gas with static eliminator.
13. Two Pyrex crystallization dishes (dia. = 80 mm; $h = 40$ mm).

Fig. 4. Shadow-mask holder for etching off gold and chromium films. The holder is machined from Teflon for use in acid.

### 2.8. Assembling Flow Cell

1. Microscope slides (75 × 25 × 1 mm$^3$, Corning).
2. Gold-post cover slips from Subheading 2.6.
3. Double-sided tape (3 M).
4. Five-minute epoxy (Devcon).
5. Razor blade.

### 2.9. Assembling Bead: DNA:Gold-Post Complexes While Reducing Nonspecific Sticking

1. Flow cell from Subheading 2.8.
2. DTPA/Biotin-DNA [6,230 base pair (bp) = 2.1 μm in length], where DTPA is an abbreviation for dithiol phosphoramidite. The end-labeled DNA is made via polymerase chain reaction (PCR) (GeneAmp XL Kit, Applied Biosystems) using 5′-labeled primers (IDT) and M13mp18 plasmid (Bayou BioLabs) linearized with BspH1 (NEB). DNA is stored in TE Buffer [10 mM Tris–Cl (pH = 7.4), 1 mM EDTA] at 4°C. The DTPA primer has three DTPAs separated by adenine bases to promote binding to the gold.
3. 92 pM streptavidin-coated polystyrene beads [690-nm dia. (Spherotech); 1.0 % (w/v)].
4. Wash Buffer: 100 mM sodium phosphate (pH = 7.4), 0.4% Tween-20 (biological studies grade, Sigma).
5. TBS Buffer: 20 mM Tris–Cl (pH 7.4), 150 mM NaCl.

6. 100 μM mPEG-SH (Nektar) in TBS Buffer.
7. Tethering Buffer: 20 mM Tris–Cl (pH = 7.4), 50 mM NaCl, 3 mg/mL bovine serum albumin (BSA, Sigma), and 0.4% Tween-20, filtered through 0.22-μm Durapore PVDF membrane (Steriflip, Millipore).
8. Cup-horn sonicator (Sonics Vibracell).
9. Humidity chamber, which can be made using an empty pipette tip box, two short portions of plastic 1-mL pipettes placed about ~6 cm apart, and double-sided tape.
10. 1.5 mL Eppendorf tube.
11. Parafilm.
12. Microcentrifuge (Beckman Coulter).
13. Vortex (Scientific Industries).
14. Research-grade microscope with 100× objective (Nikon).

#### 2.10. Measuring a Single DNA Anchored to Gold with an Optical Trap

1. Flow cell with bead:DNA:gold-post complexes from Subheading 2.9.
2. Flow cell with unpatterned cover slip.
3. RXN Buffer: 20 mM Tris–Cl (pH = 7.4), 150 mM NaCl, 3 mg/mL BSA, and 0.4% Tween-20.
4. High-resolution computer-controlled optical-trapping microscope (13).

## 3. Methods

Nanometer-scale gold posts are efficiently patterned onto glass cover slips via physical deposition through shadow masks. As illustrated in Fig. 5, shadow masks are made from commercially available silicon nitride membrane windows in a sequential process consisting of: coating with a resist (Subheading 3.1), patterning the resist with an electron beam (Subheading 3.2), etching the silicon nitride membrane window using a RIE (Subheading 3.3), and adding spacers to the window (Subheading 3.4).

#### 3.1. Spin-Coating Resist onto a Silicon Nitride Window

1. Bring ZEP520A to room temperature (RT ~22°C) in a nitrogen box (see Note 1).
2. Place double-sided tape at the center of a cover slip. Using tweezers, affix a silicon nitride membrane window (membrane side up) onto the center of the tape-coated cover slip. The silicon nitride membrane window is extremely fragile, so handle with care.
3. Place the cover slip on the vacuum chuck mounted in the spinner (see Fig. 6).

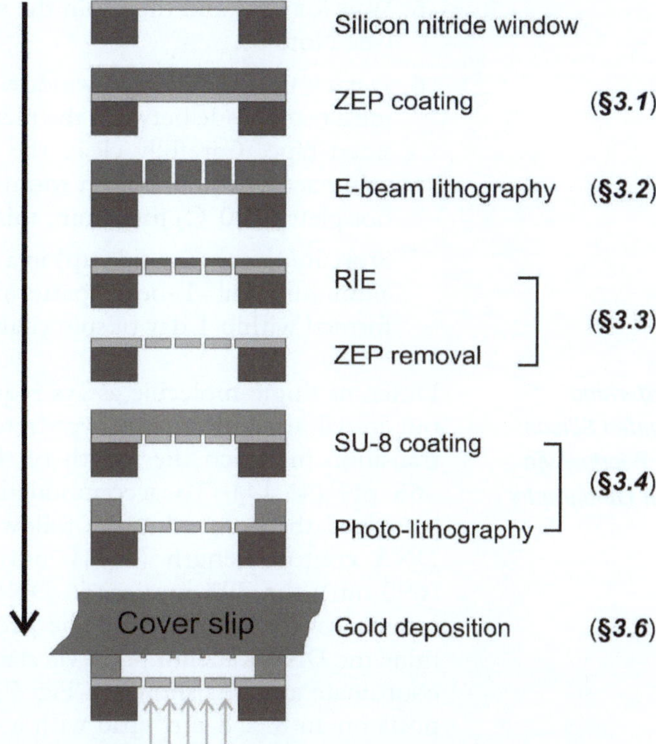


Fig. 5. Shadow-mask fabrication process with corresponding section numbers.

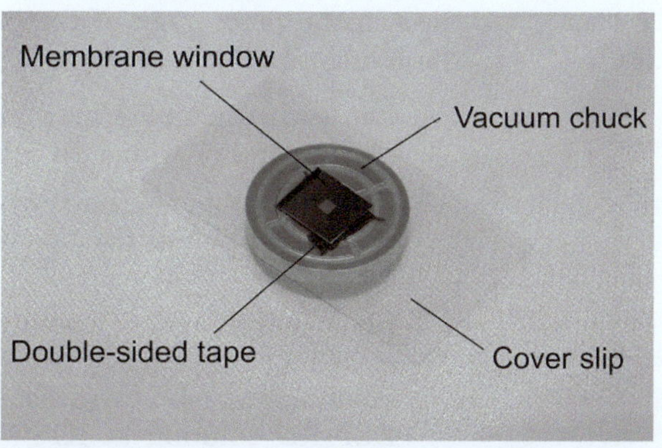


Fig. 6. Spin-coating setup. Membrane window is attached to a cover slip by double-sided tape. The cover slip is placed at the center of spinner chuck and held by a vacuum. The membrane window cannot be placed directly on the spinner chuck because the vacuum will break the membrane.

4. Dispense ~5 mL of ZEP520A to a 10-mL beaker and cover it with aluminum foil (see Note 2). Take ~0.3 mL of ZEP520A from the beaker using a polyethylene pipette and recover the beaker with aluminum foil. Put 2–3 drops of ZEP520A onto the window. Make sure that the entire window is covered and no bubbles are formed.

5. Wait for ~5 s and then spin the window at 2,500 rpm for 60 s (see Note 3).
6. Detach the ZEP-coated window from the cover slip by wedging a razor blade between the membrane frame and the double-sided tape. Carefully clean the tape residue from the frame using acetone and a clean room wipe. Place the window on a hot plate (200°C) for 2 min; this is called a "prebake."
7. Store it (membrane side up) in a wafer container wrapped with aluminum foil. E-beam patterning (Subheading 3.2) is performed within 1 day of spin coating.

### 3.2. Patterning ZEP-Coated Silicon Nitride Window via E-Beam Lithography

Different single-molecule assays require different geometries. For our initial application, we overstretched the DNA, a remarkable transition in which the length of the DNA increases by 70% at ~65 pN (4, 14). To accommodate this increase in length, we patterned the gold using the following considerations: the initial DNA contour length (2,081 nm), the diameter of the bead (690 nm), the 70% increase in DNA extension during the force-induced overstretching, and the geometric area needed to determine the DNA's anchor point via elasticity measurements along the coordinate axes. As shown in Fig. 7, our optimized geometry has posts on an $8 \times 8\text{-}\mu m^2$ grid with a 4-μm offset between adjacent rows. The size of the holes in the shadow mask dictates the size of the resulting gold posts and ranged from 50 to 500 nm in radius. Larger posts lead to more DNA bound per post but with a reduction in the percentage of the beads anchored by a single DNA molecule.

1. Put a small graphite mark on the silicon frame of the ZEP-coated window with a 4B pencil (see Note 4).
2. Load the sample in the SEM. E-beam write with the following settings: e-beam acceleration voltage = 38 kV; e-beam current = 15 pA; dose = $50\ \mu C/cm^2$.
3. Remove the window from the SEM. Develop the e-beam patterned ZEP coating by placing the window (facing the membrane side up) in a crystallization dish containing 30–40 mL of ZEP-RD solution at RT for 5 min. Rinse by placing in a crystallization dish containing 30–40 mL of isopropyl alcohol at RT for 1 min. Dry the window by putting it (membrane facing up) onto a clean room wipe to wick off liquid and then further drying with a gentle flow of nitrogen gas (see Note 5).
4. Place the window on a hot plate (200°C ) for 1 min; this is called a "postbake."
5. Store the window membrane side up in a wafer container wrapped with aluminum foil.

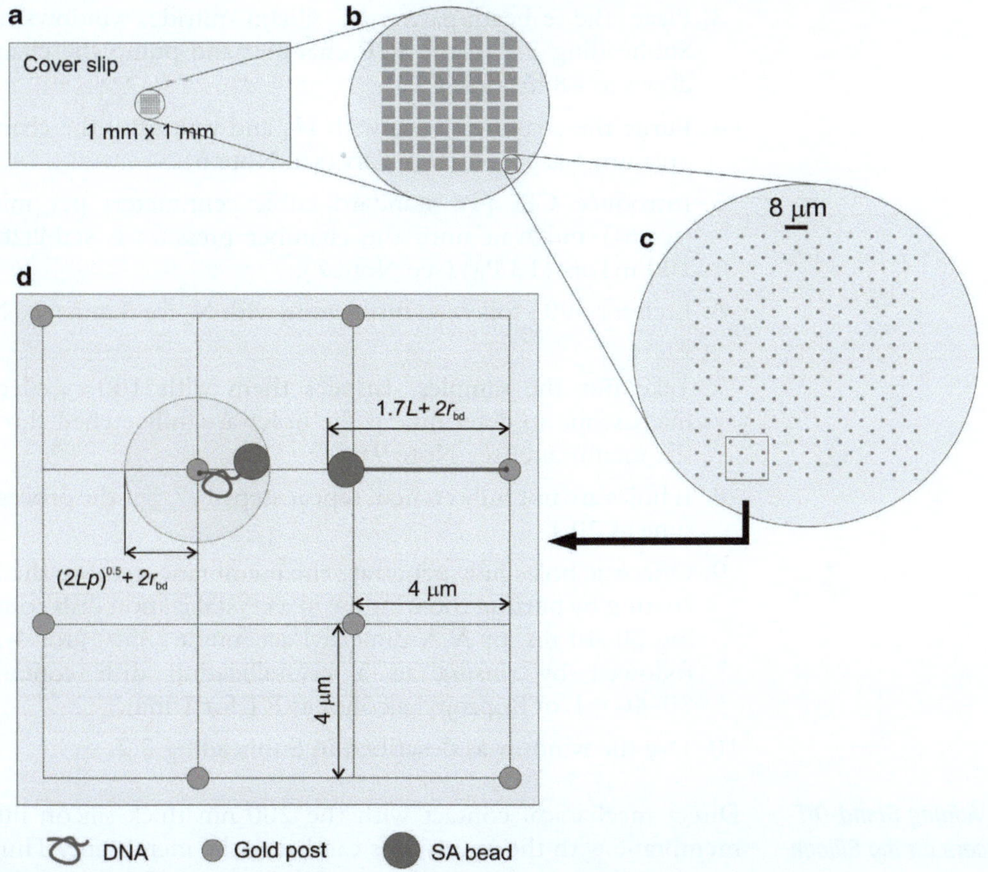


Fig. 7. Array of gold posts patterned for studying a DNA overstretching transition. (**a**) Gold posts are deposited on the glass cover slip through a shadow mask. (**b**) The $1 \times 1$ mm$^2$ area consists of $8 \times 8$ matrices. Each matrix is $80 \times 80$ μm$^2$, and the gap between matrices is 20 μm. (**c**) Each matrix consists of 221 ($10 \times 10 + 11 \times 11$) posts. (**d**) For the DNA overstretching transition, we choose the distance between the posts to be 8 μm. To avoid interference with neighboring tethered beads, spacing between the posts should be large enough to accommodate the bead diameter (690 nm), the overstretched DNA length ($1.7 \times 2{,}100$ nm $=$ 3,600 nm), and the radius of gyration of a neighboring DNA molecule plus the diameter of the bead ($\sqrt{2Lp} + 2r_{\mathrm{bd}}=$ 1,250 nm) (Reprinted with permission from (11); © 2009 American Chemical Society).

### 3.3. Etching Silicon Nitride Membrane Windows with RIE

RIE of a silicon nitride membrane is very sensitive to etching conditions. Different RIE apparatuses will have different etching conditions. Once the conditions are optimized, it is important to place the windows at the same distance from the center of the RIE chamber, since there is often a radial-gradient etching rate. Nominal etching rates for the conditions used are 50–70 nm/min for the ZEP and 60 nm/min for the SiN.

1. Vent RIE chamber to atmosphere and open.
2. Clean inside of the RIE chamber by wiping with acetone and isopropyl alcohol.

3. Place the e-beam-patterned silicon nitride windows from Subheading 3.2 in the RIE chamber and pump the chamber down to ~8 mTorr (1 Pa).
4. Purge the chamber twice with $N_2$ and wait until the chamber pressure reaches < 10 mTorr (see Note 6).
5. Introduce $CF_4$ [16 standard cubic centimeters per minute (sccm)] and wait until the chamber pressure is stabilized at 100 mTorr (13 Pa) (see Note 7).
6. Etch for ~90 s followed by purging with $N_2$ for 3 min (see Note 8).
7. Take out the samples. Inspect them with 100× reflection microscope to determine if the holes are fully etched through the membrane (see Note 9).
8. If holes are not fully etched, repeat steps 1–7. Set the processing time at 20 s.
9. Once the holes fully penetrate the membrane, remove the ZEP coating by putting the window in a crystallization dish containing 30–40 mL of *N*,*N*-dimethyl acetamide (35°C) for 4 min; followed by rinsing in a crystallization dish containing 30–40 mL of isopropyl alcohol at RT for 1 min.
10. Dry the window as described in Subheading 3.2.

### 3.4. Adding Stand-Off Spacers on the Silicon Nitride Window Frame

Direct mechanical contact with the 200-nm thick silicon nitride membrane with the cover glass can break the membrane. Thus, to use the silicon nitride window as a shadow mask, we put micron-sized stand-offs (~3 μm) onto the more mechanically robust silicon frame of the silicon nitride window. These spacers are preserved through multiple rounds of patterning cover slips (see Subheading 3.6) and cleaning the window (see Subheading 3.7), because the material of the spacers (SU-8) is only very slowly degraded by the solvents used in Subheading 3.7.

1. As described in Subheading 3.1, place double-sided tape at the center of a cover slip. Using tweezers, affix the patterned window from Subheading 3.3 onto the center of the tape-coated cover slip. Place this assembly into the spinner.
2. Dispense ~5 mL of SU-8 into a 10-mL beaker and cover the beaker with aluminum foil. Take 0.2–0.3 mL of SU-8 from the beaker using a polyethylene pipette and recover the beaker (see Note 10). Put 2–3 drops of SU-8 onto the window. Make sure that the entire window is covered and no bubble is formed. If a bubble is formed, clean with acetone by placing in a crystallization dish containing 30–40 mL of acetone at RT for 1 min followed by rinsing in a crystallization dish containing 30–40 mL of isopropyl alcohol at RT for 1 min. Dry as described in Subheading 3.2, and attempt to re-coat with SU-8.

3. Wait for 5 s. Spin the window at 500 rpm for 10 s and then quickly increase the spin speed to 3,000 rpm for 30 s (see Note 11).
4. Detach the SU-8-coated window from the cover slip with a razor blade as described in Subheading 3.1. Place the coated window on a hot plate (65°C) for 1 min followed by 2 min on a second hot plate (95°C). This is sometimes referred to as a "soft-bake," or a "prebake."
5. Place the sample into the mask aligner and align it with the UV-lithographic photomask.
6. Irradiate with UV light for 30 s.
7. Place the exposed samples onto a hot place (65°C) for 1 min followed by 1 min on a second hot plate (95°C). This is referred to as a "postbake."
8. Develop the SU-8 coating by putting the sample into a crystallization dish (80-mm dia.) filled with 30–40 mL of SU-8 developer (methoxy isopropyl acetate) for 50 s with slight agitation. Rinse by placing the sample for 1 min in a crystallization dish (125-mm dia.) that is half filled with $H_2O$. Dry it as described in Subheading 3.2.
9. Inspect the sample with 100× reflection microscope.
10. If there is residue on the mask (likely from the double-sided tape), clean the sample with a dry etch in an $O_2$ plasma. Load the mask into the plasma cleaner, placing it onto a silica plate—this plate buffers the sample from the electrode and prevents deposition of metallic electrode material onto the sample. Clean for 600 s using 150 sccm $O_2$ at 275 mTorr (37 Pa) and a forward power of 500 W. Remove and place back into the wafer container.

### 3.5. Cleaning Glass Cover Slips

1. Prepare ethanolic KOH solution in a 1-L beaker by dissolving 80 g of potassium hydroxide pellets in 250 mL ethyl alcohol (completely denatured) (see Note 12).
2. Place this beaker and two 1-L beakers, which are half filled with a $H_2O$, into an ultrasonic bath.
3. Load cover slips into the cover-slip holder.
4. Place the cover-slip holder into the ethanolic KOH solution and sonicate for 3 min. Occasionally, agitate the solution by manually moving the holder up and down.
5. Take the holder out and remove most of KOH solution by rinsing with a $H_2O$ squirt bottle.
6. Place the holder into one of the $H_2O$-filled 1-L beakers. Sonicate for 3 min.
7. Take the holder out and briefly rinse the cover slips using the $H_2O$ squirt bottle.

8. Place the cover-slip holder into the second $H_2O$-filled 1-L beakers. Sonicate for 3 min.
9. Take the holder out and rinse with ethyl alcohol (completely denatured).
10. Dry the cover slips (with holder) in a microwave for 2 min.
11. Place the cover slips (with holder) into a sealed container (see Note 13).

### 3.6. Depositing Gold Posts onto Cover Slips via Evaporative Physical Deposition

Our custom-made shadow-mask loading platform can hold four silicon nitride membrane windows. Thus, four cover slips are patterned in parallel with gold posts during each round of evaporation. The evaporative source is ~1 m from the cover slips, while the silicon nitride mask is only 3 μm above the cover slips. This geometry leads to gold posts whose size is determined by the holes in the silicon membrane, and therefore the term shadow mask.

1. Place shadow masks and KOH-cleaned cover slips onto the custom-made shadow-mask loading platform, as illustrated in Fig. 3.
2. Load the shadow mask holder into the evaporation chamber.
3. Put chromium and gold pellets into the crucibles.
4. Pump the evaporation chamber down to mid-$10^{-7}$ Torr ($10^{-6}$ Pa) (see Note 14).
5. Evaporate chromium using an e-beam. Set the beam current such that chromium is deposited at the speed of ~0.2 nm/s. Deposit 5 nm total (see Note 15).
6. Deposit gold using an e-beam evaporator. Set the deposition rate to 0.2–0.4 nm/s. Deposit a total of 20-nm gold.
7. Cool the crucibles and vent the chamber to the atmosphere. Unload the cover slips and put them back in the dust-free container. For best results, proceed with sample cell assembly (see Subheading 3.8) and DNA coupling (see Subheading 3.9) within 1 h of the evaporation.

### 3.7. Cleaning Shadow Mask

Each round of evaporation closes the holes in the shadow mask, due to deposition onto the edges of the holes in the silicon nitride membrane. Repeated deposition without cleaning will eventually lead to complete closure of the holes. Deposition of two metals in alternating layers complicates cleaning of the shadow masks. Thus, we clean the masks after each deposition. All mask handling steps in Subheading 3.7 use carbon-fiber tweezers to prevent degradation of the tweezers by the strong acids.

1. Load four shadow masks on the custom-made shadow mask holder (Fig. 4). Place the holder into a glass Petri dish.

2. Prepare aqua regia solution in a 50-mL beaker by adding 10 mL of HCl to 30 mL of $HNO_3$. Wait a few minutes until the solution turns to orange. Pour 30 mL of chromium etchant into another 50-mL beaker.
3. With the holder aligned vertically and resting on a glass petri dish, deliver aqua regia (~5 mL) dropwise to the membrane window (front and back) using a Pasteur pipette. Next, place the holder on its side (membrane widows up) and deliver one drop of aqua regia to each window. Let sit for ~1 min. Repeat this 1-min incubation on the other side.
4. Rinse with a large volume of ultraclean water by immersing the shadow mask holder into the 1-L $H_2O$-containing beaker for 1 min.
5. Remove the holder from the $H_2O$. Place it into a different glass Petri dish. Clean the membrane window with chromium etchant using the process described in step 3.
6. Rinse with a large volume of ultraclean water.
7. Repeat steps 3–6 three times (see Note 16). Discard waste liquid from Petri dishes after each cycle.
8. Take the shadow masks out from the holder. Rinse in a crystallization dish containing 30–40 mL of isopropyl alcohol at RT for 1 min.
9. Dry them as described in Subheading 3.2.

### 3.8. Assembling Flow Cell

Epoxy-rigidified flow cells provide increased mechanical rigidity during single-molecule assays. The total volume of the flow cell is ~15 μL.

1. Place microscope slide on clean bench surface.
2. Carefully cut ~5 cm of double-sided tape in half lengthwise with a razor blade. Place each half perpendicular to the long axis of the microscope slide, but centered on the slide. The interior gap between the tapes should be 5–8 mm.
3. Using a gloved hand, place the cover slip from Subheading 3.6 with the gold-post side face down and centered with respect to the microscope slide so that the cover slip overhangs on both sides of the short axis of the slide. Gently press down on the cover slip to seal the tape to the cover slip.
4. Repeat steps 1–3 until the appropriate number of slides is made (typically 4–8 per day).
5. Apply 5-min epoxy to the long edges of cover slip first and wait a few seconds for the epoxy to flow into the gap between the tape and slide. Then apply a small dab of epoxy on each side of the fluid channel, but slightly removed from the channel itself.

### *3.9. Assembling Bead: DNA:Gold-Post Complexes While Reducing Nonspecific Sticking*

The goal is to link the DNA to the gold post via a gold–thiol bond while minimizing nonspecific sticking of both the protein-coated beads and the DNA to the surface. mPEG-SH and BSA are used to passivate the gold and bare glass, respectively. In addition, a low concentration (0.4%) of a nonionic detergent (e.g., Tween-20) significantly reduces unwanted sticking. As illustrated in Fig. 8, the tethered bead assay was formed by sequentially adding the DTPA–DNA, mPEG-SH, BSA, and then streptavidin-coated (SA) beads to an array of gold nanoposts on glass cover slips.

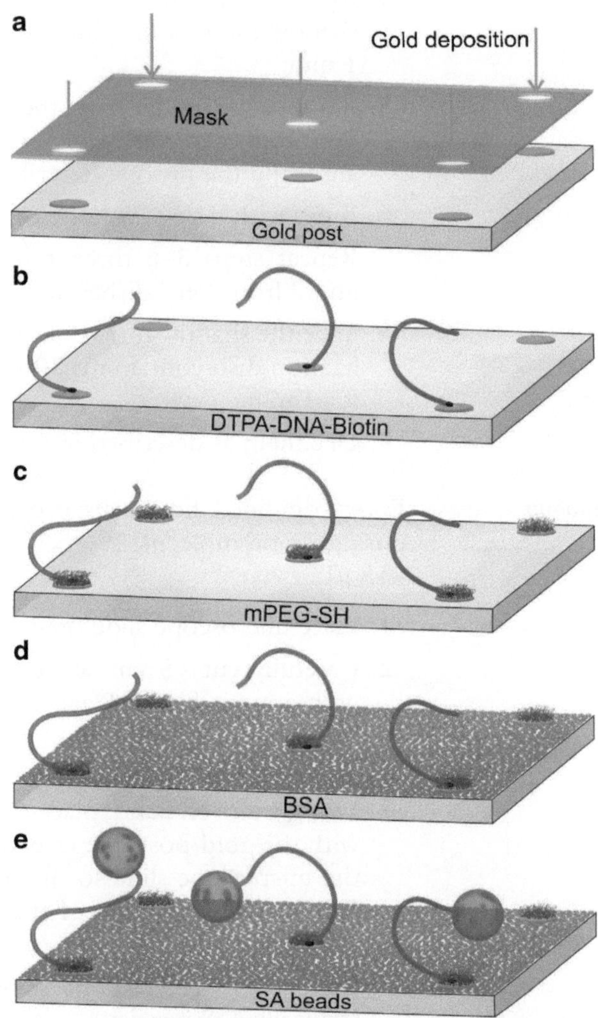


Fig. 8. Steps for making gold nanoposts and attaching bead–DNA complexes to them (*not to scale*). (**a**) Gold is deposited onto a glass cover slip through a shadow mask. (**b**) DNA, end-labeled with DTPA (*black dot*), is incubated with freshly deposited gold. (**c**) The unreacted gold is passivated with mPEG-SH. (**d**) The cover glass is passivated with bovine serum albumin (BSA). (**e**) Streptavidin-coated beads are attached to biotinylated end of the DNA (Reprinted with permission from (11); © 2009 American Chemical Society).

1. Dilute DTPA/Biotin-DNA to 100 pM using TE Buffer, and then a second dilution to 10 pM using TBS Buffer. Briefly vortex on medium setting.
2. Flow 15 μL of DTPA/Biotin-DNA into the flow cell and incubate for 10–12 h at RT in a humidity chamber (see Note 17).
3. Rinse with 600 μL of TBS Buffer (see Note 18) and then flow through 50–100 μL of 100 μM mPEG-SH and let sit for 2 h at RT in humidity chamber.
4. Meanwhile, wash the streptavidin (SA)-coated beads. First, vortex the stock solution for 20 s at high and put 55 μL of the stock solution per slide to be studied into a 1.5 mL Eppendorf tube and add 45 μL of Wash Buffer per slide. Vortex for 30 s.
5. Centrifuge for 4 min at 5,000 × *g* (or until a pellet is formed, which depends on volume). Remove supernatant. Resuspend the beads in 200 μL of Wash Buffer by vortexing and pipetting, while trying to minimize bubble formation.
6. Repeat step 5 two more times. In the final wash, resuspend the beads in 50 μL of Tethering Buffer per slide to be studied. Final bead concentration is ~100 pM.
7. Seal Eppendorf tube with parafilm. Sonicate the beads in the chilled (15°C) cup-horn sonicator for 30–40 min.
8. Under microscope, check whether beads are mono-dispersed (i.e., ~9 out of 10 not clumped).
9. Rinse flow cell with 600 μL of TBS Buffer. Rinse with 600 μL of Tethering Buffer.
10. Incubate the Tethering Buffer for 1 h to passivate the glass surface with BSA.
11. Flow 50 μL of 100 pM SA into the flow chamber and incubate for 1 h at RT, followed by rinsing with 600 μL of Tethering Buffer.

### 3.10. Measuring a Single DNA Anchored to Gold with an Optical Trap

The conventional optical-trapping procedures for DNA elasticity measurement (15) are modified to avoid direct irradiation of gold posts by an intense trapping laser (see Fig. 9a). We use two lasers [one for trapping (100–600 mW) and one for position detection (<1 mW)], but the whole experiment could be done with one laser as long as the trap laser can be modulated down to both very low levels (≈1 mW) to avoid damage to the gold and completely off.

1. Turn on the optical-trapping apparatus (13). Mount a flow cell made with an unpatterned cover slip and filled with Tethering Buffer. Align microscope for Köhler illumination.
2. Warm up the instrument for > 2 h with the trapping laser at a nominal trapping stiffness (0.08 pN/nm).

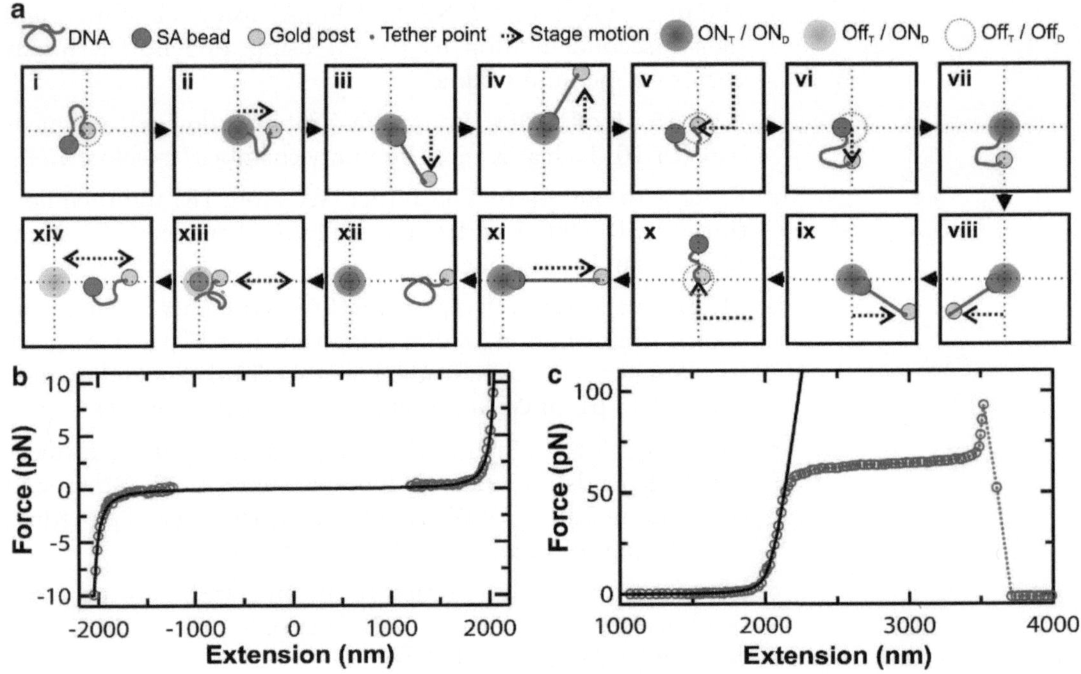


Fig. 9. (**a**) Procedure for measuring DNA anchored to a gold nanopost with an optical trap (*not to scale*). The trap and the co-linear detection lasers are stationary during the experiment (intersection of the *dotted lines*), while the sample is moved in 3D using a three-axis, closed-loop piezoelectric stage. The on/off states of the trap and detection lasers are indicated by shading. A 2D elasticity-centering routine (**i–x**) determines the anchor point of the DNA along the *y*- and *x*-axes sequentially, but with a substantial (~1 μm) lateral offset. By moving the stage at constant velocity (**xi**), the trap exerts enough force to rupture the biotin–streptavidin bond anchoring the DNA to the bead (**xii**). To reattach the bead to the same individual DNA molecule, the disconnected bead is trapped by a weak detection beam (<1 mW), positioned near the post, and then slowly moved back and forth (**xiii**) until the bead rebinds to the DNA and this stage motion pulls the bead out of the weak trap (**xiv**). (**b**) Elasticity measurement of a DNA molecule (*circle*) stretched in the positive and negative directions. Using an analysis routine that takes into account the substantial lateral offset (11), the data is fit to the WLC model (17) (*line*) to determine three values, *p*, *L*, and Δ. For the trace shown, the fitted values were 48.9 (±1.6) nm, 2,093 (±2) nm, and 0.2 (±0.5) nm, respectively. (**c**) Force-extension record for DNA when moving the stage at a uniform velocity (250 nm/s) (*circles*). The plateau at ~65 pN corresponds to the overstretching transition of the DNA (4, 14). The cusp at 90 pN corresponds to the rupture of the biotin–streptavidin bond. The force-extension record below 50 pN is well fit by a WLC model that includes enthalpic stretching of the DNA backbone (*line*) (18) (Reprinted with permission from (11); © 2009 American Chemical Society).

3. Turn off the trapping laser. Mount the flow cell with DNA–gold-post tethers from Subheading 3.9. Adjust the flow cell position such that the gold-post array is in the field of view.
4. Verify through the Bertrand lens that there are no bubbles in the immersion oil.
5. Exchange the flow cell fluid from Tethering Buffer to RXN Buffer by flowing 200 μL RXN Buffer through the flow cell. Turn trap on. Let it sit for 5 min.
6. Turn trap off. Move sample using a three-axis piezoelectric translation stage and/or a coarse translation stage to locate

a tethered bead on a gold post. Visually align the gold post with the location of the stationary trap.

7. Move the post laterally ~1 μm along the $x$-axis and turn both lasers on, which leads to rapid trapping of the tethered bead [see Fig. 9a(ii)].
8. At this position, scan the stage vertically until the trapped bead is in contact with the glass surface (1) and move the surface 300 nm below the contact point.
9. Measure the elasticity of the DNA along both the negative and positive $y$ directions and fit the resulting data to the worm-like chain (WLC) model to determine $p$, $L$, and $\Delta$, where $p$ is the persistence length, $L$ is the contour length, and $\Delta$ is the distance to the anchor point [see Fig. 9a(iii–iv), b, Note 19].
10. Turn off both lasers, and correct for the $y$-lateral offset ($\Delta$) by a stage movement [see Fig. 9a(v)].
11. Repeat the elasticity measurement along the $x$-axis [see Fig. 9a (vi–x)].
12. Repeat steps 7–11 one or two more times to determine the anchor point more accurately.
13. Use the DNA tether for the desired single-molecule assay. For instance, one could overstretch the DNA by moving the stage in the positive $x$ direction [see Fig. 9a(xi), c].
14. Upon rupture, one can either reconnect the trapped bead to the DNA [see Fig. 9a(xii–xiv), Note 20] or discard the bead and repeat steps 6–13 on a new tether.

## 4. Notes

1. The quality (i.e., the uniformity) of the spin coating of ZEP520A depends on humidity. Good results with our protocol were achieved when the humidity of clean room was < 20%. Different environmental conditions will most likely require variations on the exact protocol presented. All mask making and cleaning steps were done in a clean room.
2. ZEP520A is photosensitive and volatile. Completely wrap the beaker with aluminum foil to prevent evaporation of the ZEP520.
3. The thickness of ZEP film is approximately 500 nm at 2,500 rpm.
4. The graphite mark is to facilitate focusing e-beam.
5. Never put the membrane in direct contact with a surface (e.g., membrane side down). Such contact can break the membrane.

Unless otherwise stated, always handle the window with tweezers. For drying, hold the silicon frame by its edges (with a powder-free latex-gloved hand), and orient the frame so that the nitrogen gas flow is parallel to the membrane (i.e., the flow hits the edge of the frame) not perpendicular. In a perpendicular orientation, the flow can exert enough force to break the membrane.

6. The cleanliness of the RIE chamber is crucial for consistent results. If necessary, more thoroughly wipe inside of the chamber with acetone and isopropyl alcohol to reduce the base pressure.
7. Chamber pressure should be set to ~100 mTorr, so in practice we set the pressure to 99 mTorr due to significant digits on our vacuum gauge.
8. The forward incident power reading is about 100 W at these conditions.
9. In a reflection microscope, partially etched holes appear brighter compared to holes that are fully etched through the membrane.
10. SU-8 is photosensitive and dissolved in volatile solvent.
11. The resulting thickness of the SU-8 layer near the edge of the window frame is ~3 μm at 3,000 rpm.
12. KOH solution needs to be clear not cloudy and approximate color of apple juice. This is very sensitive to the source and the exact denatured ethyl alcohol used. We use catalog #7018.
13. Clean cover slips are usable for 2 weeks, if properly stored. Passive absorption of the BSA, as is used here, is sensitive to surface cleanliness, but not nearly as sensitive as PEG-silanization of glass cover slips.
14. The reactivity of DTPA-labeled DNA with the gold post depends on the cleanliness of the surface and hence the base pressure during the gold evaporation. We found significantly improved results if the base pressure was reduced to mid-$10^{-7}$ torr (mid-$10^{-6}$ Pa). In general, lower base pressures yield more reactive gold surfaces.
15. Chromium promotes adhesion of the gold to the glass.
16. The SU-8 stand-offs are slowly removed by aqua regia/chromium etchant. SU-8 stand-offs are reapplied after 20–25 depositions.
17. The number of DNA molecules per gold post depends on the concentration of DTPA-labeled DNA and incubation time. At the conditions presented, one DNA molecule per gold post was achieved, but only ~20% of gold posts were tethered.
18. To facilitate buffer exchange, fluid is added dropwise to one side of the flow cell (the "clean inlet") while aspirating off the excess

fluid from the distal side. To reduce unwanted mechanical agitation, gel-loading tips are used on the aspirator. Take care not to suck the flow cell dry; setting a low aspiration rate and practice are critical. If a bubble starts to enter the flow cell, stop aspiration, change the current aspiration tip for a clean one, and then add a drop of the fluid to the outlet while gently aspirating from the "clean inlet" until the bubble is removed.

19. Single DNA molecules are selected based on their mechanical properties. A single DNA molecule has a characteristic stiffness that corresponds to a persistence length ($p$) of ~50 nm within the WLC model, though $p$ depends slightly on the contour length (16).
20. Since the DNA–gold bond is covalent, it is stronger than biotin–streptavidin coupling. Repeated measurements of this rupture force are consistent with this supposition (11). Thus, disconnected beads can be reattached to the DNA that remains bound to the gold post by holding the bead with the weak detection laser near the gold post.

## Acknowledgments

We thank Gavin King for a careful reading of this manuscript. This work was supported by a National Research Council Research Associateship (D.H.P.), a W.M. Keck Grant in the RNA Sciences, the National Science Foundation (NSF Phys-0404286 to T.T.P.), and National Institute of Standards and Technology (NIST). Mention of commercial products is for information only; it does not imply NIST recommendation or endorsement, nor does it imply that the products mentioned are necessarily the best available for the purpose. T.T.P. is a staff member of NIST's Quantum Physics Division.

### References

1. Neuman KC, Block SM (2004) Optical trapping. Rev Sci Instrum 75:2787–2809
2. Moffitt JR, Chemla YR, Smith SB, Bustamante C (2008) Recent advances in optical tweezers. Annu Rev Biochem 77:205–228
3. Perkins TT (2009) Optical traps for single molecule biophysics: a primer. Laser Photon Rev 3:203–220
4. Smith SB, Cui Y, Bustamante C (1996) Overstretching of B-DNA: the elastic response of individual double-stranded and single stranded DNA molecules. Science 271:795–799
5. Svoboda K, Schmidt CF, Schnapp BJ, Block SM (1993) Direct observation of kinesin stepping by optical trapping interferometry. Nature 365:721–727
6. Liphardt J, Onoa B, Smith SB, Tinoco IJ, Bustamante C (2001) Reversible unfolding of single RNA molecules by mechanical force. Science 292:733–737
7. Fernandez JM, Li H (2004) Force-clamp spectroscopy monitors the folding trajectory of a single protein. Science 303:1674–1678

8. Grandbois M, Beyer M, Rief M, Clausen-Schaumann H, Gaub HE (1999) How strong is a covalent bond? Science 283:1727–1730
9. Zhao XM, Xia YN, Whitesides GM (1997) Soft lithographic methods for nano-fabrication. J Mater Chem 7:1069–1074
10. Liepold P, Kratzmueller T, Persike N, Bandilla M, Hinz M, Wieder H, Hillebrandt H, Ferrer E, Hartwich G (2008) Electrically detected displacement assay (EDDA): a practical approach to nucleic acid testing in clinical or medical diagnosis. Anal Bioanal Chem 391:1759–1772
11. Paik DH, Seol Y, Halsey WA, Perkins TT (2009) Integrating a high-force optical trap with gold nanoposts and a robust gold-DNA bond. Nano Lett 9:2978–2983
12. Kohler J, Albrecht M, Musil CR, Bucher E (1999) Direct growth of nanostructures by deposition through an Si3N4 shadow mask. Physica E 4:196–200
13. Carter AR, Seol Y, Perkins TT (2009) Precision surface-coupled optical-trapping assays with 1 base-pair resolution. Biophys J 96:2926–2934
14. Cluzel P, Lebrun A, Heller C, Lavery R, Viovy JL, Chatenay D, Caron F (1996) DNA: an extensible molecule. Science 271: 792–794
15. Wang MD, Yin H, Landick R, Gelles J, Block SM (1997) Stretching DNA with optical tweezers. Biophys J 72:1335–1346
16. Seol Y, Li J, Nelson PC, Perkins TT, Betterton MD (2007) Elasticity of short DNA molecules: theory and experiment for contour lengths of 0.6-7 μm. Biophys J 93:4360–4373
17. Marko JF, Siggia ED (1995) Stretching of DNA. Macromolecules 28:8759–8770
18. Odijk T (1995) Stiff chains and filaments under tension. Macromolecules 28:7016–7018

# Chapter 19

# Electro-optical Analysis of Macromolecular Structure and Dynamics


## Dietmar Porschke



## Abstract

Electro-optical effects are induced by external electric field pulses applied to solutions or suspensions and are recorded by various optical techniques. These effects are very useful for the characterization of macromolecular structures and their dynamics in solution. One of the field-induced effects is alignment of molecular dipoles, which can be detected at a very high sensitivity by measurements of the dichroism or the birefringence. Stationary values of these optical parameters recorded at different electric field strengths can be used to characterize dipole moments and to determine the orientation of chromophores with respect to the dipole vector. The transients reflect rotational diffusion, providing a particularly accurate measure of size and shape. The internal flexibility is also reflected in these transients. Another type of field-induced effect is chemical relaxation, which can be detected selectively and is very useful for the characterization of reactions, like ligand binding and conformation changes. The techniques based on electric field effects are unique in the sense that problems can be solved, which are difficult or even impossible to be sorted out by other techniques.



**Key words:** Electric dichroism, Electric birefringence, Rotational diffusion, Chemical relaxation, Stopped flow


## 1. Introduction

Electrostatic interactions are known to be essential for most biological structures and their functions. This is simply due to the large number of charged residues in biological macromolecules and the fact that electrostatic interactions operate over long distances. Electrostatics is not only important for internal interactions but also for the molecular identity exposed to the environment, i.e., for interactions with other molecules. The electrostatic “character” is reflected by the response of these molecules to external electric fields. Studies of molecules by measurements of the response to


Wlodek M. Bujalowski (ed.), *Spectroscopic Methods of Analysis: Methods and Protocols*, Methods in Molecular Biology, vol. 875, DOI 10.1007/978-1-61779-806-1_19, © Springer Science+Business Media New York 2012

external electric field have a long history. One of the pioneers was Peter Debye, the "master of molecules," who provided the basis for characterization of molecules by dipole moments (1). The response of molecules to an external electric field by a sensitive optical procedure was first demonstrated by John Kerr (2). This approach has been developed to a powerful method (3–6) providing electrical, optical, and hydrodynamic parameters at a high accuracy. Another closely related technique was developed by Manfred Eigen and coworkers, who used electric field pulses for perturbation of molecular interactions and established chemical relaxation for studies of fast reactions (7).

The methods based on effects induced by external electric fields are simple in principle, but not always simple in the technical implementation. Some of the instruments were developed to a high degree of sophistication. Unfortunately, the special instruments are currently not available commercially as a whole and must be assembled from components. The experiments described below require two separate functions: a generator for the electric field pulses and a spectro-photometric detection system. The details are very much dependent on the goals of the experiments.

## 2. General

### 2.1. Orientation and Reaction Effects Induced by Electric Fields

As indicated already above, electric fields may induce two different types of effects:

1. Orientation corresponding to alignment of the molecules in the direction of the field vector, determined by an induced, permanent, or other dipole moment associated with the molecules. When the electric field is applied, the molecules rotate from a random distribution in space toward a state with their dipoles being partially aligned in the direction of the field vector (cf. Fig. 1). Upon pulse termination, the molecules turn back to the random distribution by rotational diffusion with a characteristic time constant or a spectrum of time constants. Because rotational diffusion is very much dependent on the size and the shape, the decay of the orientation provides accurate information on the structure. Moreover, the optical anisotropy may be studied at different field strengths corresponding to different degrees of orientation. Such data can be used to evaluate the nature and the magnitude of the electric dipole. Furthermore, these data tell the orientation of the chromophore(s) used for the measurements with respect to the dipole vector. The orientation of molecules does not change their conformation or state of aggregation and thus may be classified as a "physical" effect.

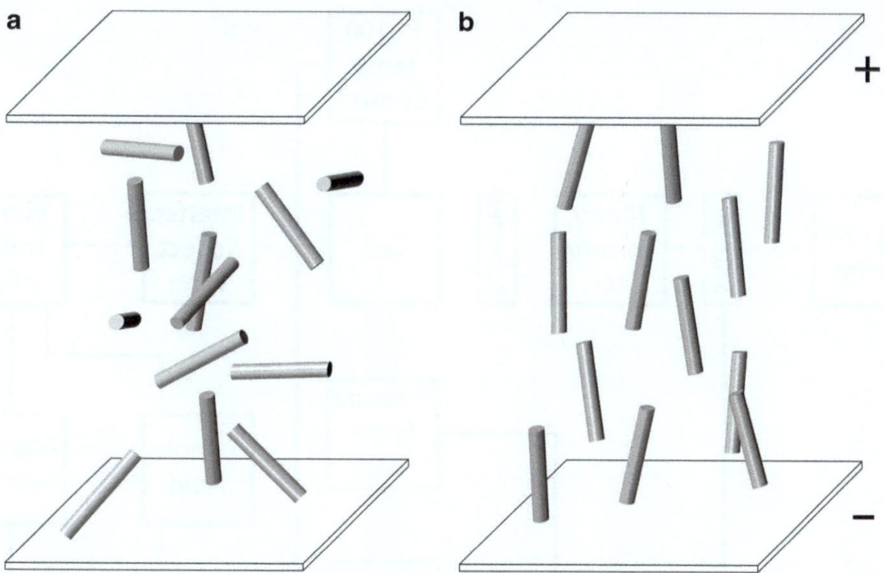


Fig. 1. Scheme of the alignment of rod-like molecules under the action of an electric field. (**a**) Random orientation in the absence of an external field; (**b**) partial alignment in the presence of an external electric field.

2. Examples for reactions induced by electric field pulses are changes of conformation, changes of ligand binding, or changes in the degree of association. In most cases, changes of noncovalent interactions are induced, whereas chemical bonds usually are not affected. Thus, field-induced reactions usually are in the domain of "soft" chemistry. These effects are very useful to study the dynamics of the molecules under investigation. When the perturbations induced by the field remain small, the transients are described by relaxation time constants. Measurements at different concentrations can be used to derive rate constants and also equilibrium constants.

### 2.2. Electric Dichroism

The field-induced orientation is reflected by changes in the absorbance of polarized light. The absorbance change $\Delta A(\varphi)$ at an angle $\varphi$ of the polarization plane with respect to the vector of a given external electric field is dependent on $\varphi$ according to

$$\Delta A(\varphi) = \Delta A_{\|}(3 \times \cos^2\varphi - 1)/2. \quad (1)$$

The maximal change of absorbance, $\Delta A_{\|}$, is observed, when the plane of polarization and the field vector are parallel ($\varphi = 0$). For perpendicular alignment of polarization plane and field vector ($\varphi = 90$), an effect in the opposite direction, $\Delta A_{\perp}$, appears as described by

$$\Delta A_{\|} = -2\Delta A_{\perp}. \quad (2)$$

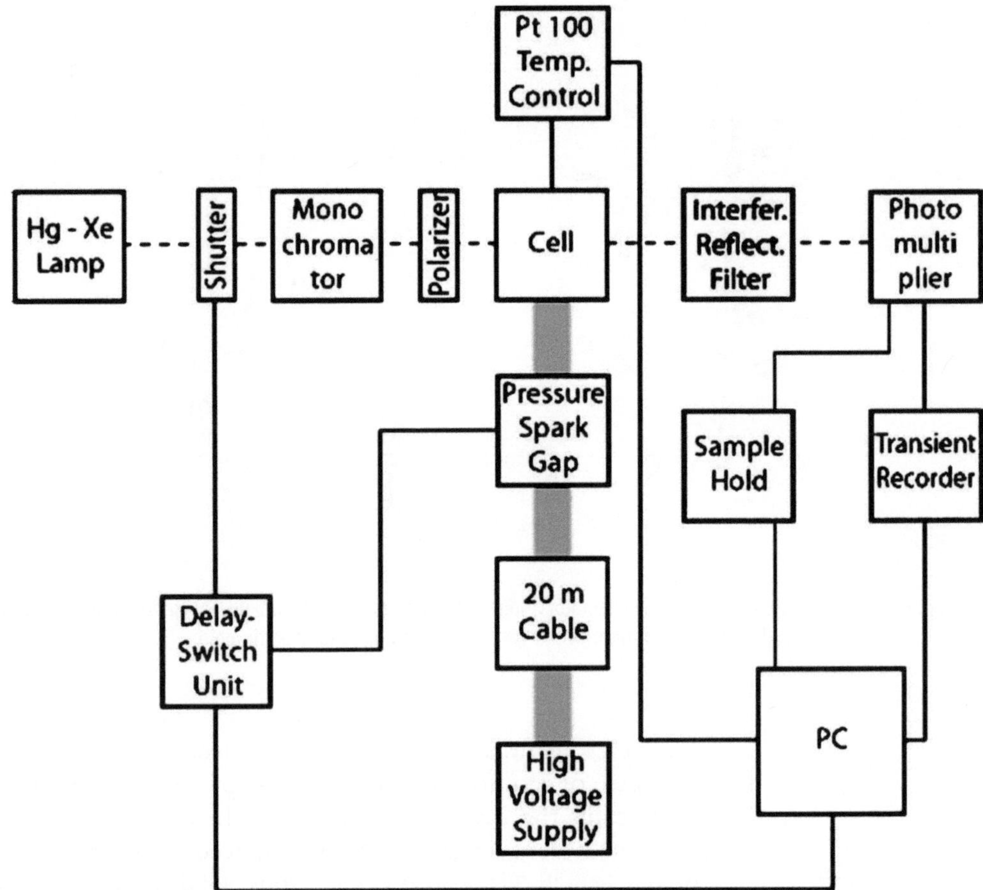


Fig. 2. Scheme of an instrument for automatic measurements of the electric dichroism. A high-pressure Hg–Xe lamp is used as a light source. The interference reflection filter is used to suppress electro-luminescence. Electric field pulses are generated by discharge of a high voltage cable, using a pressure spark gap as a fast high voltage switch. The instrument is used for automatic data collection: at fixed time intervals and provided that the temperature in the cell is within preset limits, an initiation signal is sent by the PC, which opens the shutter and activates the pressure spark gap. A "Sample Hold" records the total light intensity before pulse application. The time sequence of these events is optimized to reduce photo-damage as much as possible and to apply the field pulse as soon as the detection system is stationary. Most of the high voltage energy is dissipated in a compensation cell parallel to the measuring cell (Figure adapted from ref. (13)), reprinted with permission from American Institute of Physics.

The relation between $\Delta A_{\parallel}$ and $\Delta A_{\perp}$ is useful as a control (cf. below). The optical effect is presented in most cases in the form of the reduced dichroism

$$\xi = (\Delta A_{\parallel} - \Delta A_{\perp})/\bar{A}, \tag{3}$$

where $\bar{A}$ is the isotropic absorbance (in the absence of the external electric field). A scheme including the optical components required for measurements of the electric dichroism is shown in Fig. 2: a high intensity light source, a monochromator, a polarizer, and a photomultiplier detector (for more details, cf. Subheading 2.6).

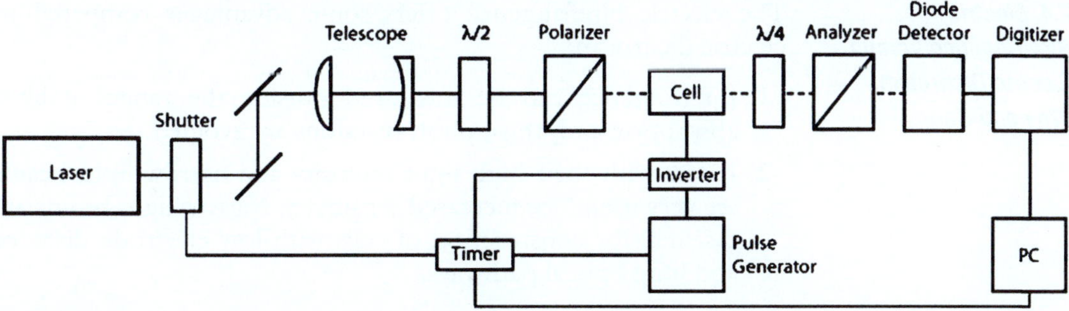


Fig. 3. Scheme of an instrument for measurements of the electric birefringence. The telescope is used to reduce the beam diameter. The PC controls automatic data accumulation: (1) The digitizer is armed; (2) The timer is activated and gives signals to open the shutter, to invert the pulse polarity, and finally to trigger the pulse generator.

### 2.3. Electric Birefringence

The orientation of molecules under electric field pulses is not only reflected by changes of absorbance but also by changes in the refraction index. These optical signals are complementary. The electric birefringence is the difference in the refraction index for light polarized parallel and perpendicular to the field vector. A standard setup for measurements of the electric birefringence (3) has the following optical components (Fig. 3): (1) Lasers are used as a convenient light source. (2) The light must be polarized at an angle of 45° with respect to the vector of the applied external electric field. If the laser light is polarized, the plane of polarization may be rotated to the required 45° by a $\lambda/2$ plate, and finally the degree of polarization is increased by a polarizer. The polarized light passes the cell, a $\lambda/4$ plate with its slow axis parallel to the polarization plane of the polarizer, an analyzer in almost crossed position and finally arrives at the diode detector (see Fig. 3). A procedure for proper alignment is presented below (Subheading 4.3.1).

The change of light intensity $\Delta I_\delta$ induced by an electric field pulse is given (3) by

$$\Delta I_\delta / I_\alpha = \left[\sin^2(\alpha + \delta/2) - \sin^2(\alpha)\right] / \sin^2(\alpha). \tag{4}$$

$I_\alpha$ is the total light signal measured in the absence of an external electric field at the analyzer position rotated by an angle $\alpha$ from the crossed position. $\delta$ is the phase shift or optical retardation:

$$\delta = 2\pi \ell \Delta n / \lambda, \tag{5}$$

where $\Delta n$ is the difference in the refractive index, $\ell$ is the optical path length, and $\lambda$ is the wavelength of light used for the measurement.

### 2.4. Electric Birefringence Versus Electric Dichroism: Pros and Cons

The electric birefringence offers some advantages compared to electric dichroism:

1. Birefringence can be measured outside the range of light absorption and, thus, photoreactions are avoided.
2. Lasers with their high light intensity and narrow light beams are very useful for increased sensitivity. Narrow light beams are essential for construction of cells with low electrode distance and long optical pathways.

However, the electric dichroism also has some advantages:

1. Transients of the electric dichroism are not perturbed by contributions of the solvent. (In aqueous solutions the birefringence of water may be relatively large.)
2. Field-induced reactions can be detected more easily: measurements at the magic angle demonstrate directly and selectively the existence of reactions (see Subheadings 2.5 and 4.2.3).
3. Calculations of the electric dichroism from known structures or from models are more simple and straightforward.

### 2.5. Selective Study of Reaction Effects in the Presence of Orientation Effects

In many polymer systems, electric fields induce orientation of molecules without changes of the conformation or other reaction effects, but it is always advisable to check for the existence of any field-induced reaction. Fortunately, there is a simple approach for selective detection of field-induced reactions by measurements of the absorbance or fluorescence under magic angle conditions (8). This approach is based on equation (1), which implies that the absorbance change due to orientation disappears at the "magic angle" $\varphi = 55.74^{\circ}$. Thus, any change of absorbance observed at this orientation of the polarized light must be due to some reaction. A corresponding relation holds for fluorescence (9).

The theoretical basis of the magic angle technique is relatively simple, whereas clean experiments under magic angle conditions require special caution. This is due to the fact that absorbance changes resulting from orientation are usually very large, whereas the changes from reaction effects often are much smaller. Thus, the magic angle has to be adjusted very carefully. Furthermore, optical windows of standard observation chambers may change the plane of polarization, whenever these windows experience some strain. For this reason, the optical windows must be absolutely strain-free and must be inserted into the cell without inducing any strain.

In most cases, electric field pulses induce only relatively small changes of chemical reaction equilibria. Under these conditions, the approximations used for the analysis of chemical relaxation are valid: rate equations may be simplified by linearization (7, 10, 11) and, thus, the evaluation of data is facilitated considerably. An obvious drawback of small perturbations is small amplitudes of observed effects. However, this problem may be overcome by repeating the relaxation experiment many times with signal averaging, until the signal-to-noise ratio is sufficiently large.

### *2.6. Components for Instruments*

Studies of electric field effects require special instruments, which are not available commercially but must be assembled from components. An extensive choice of optical components is provided by various companies. Currently, it is more difficult to find suitable generators for high voltage pulses. A wide range of different generators has been used for applications starting from relatively low field strengths in the range around 1 kV/cm up to very high field strengths close to the limit of breakdown voltages around 100 kV/cm. The required technical efforts are not only dependent on the voltage range but also on the limit of time resolution. Among the advantages of electric field techniques is the possibility of a high time resolution. Pulse generators based on the cable discharge technique with a time resolution of a few nanoseconds have been described (12, 13). An example of an instrument for automatic collection of dichroism data with continuous supervision of operating conditions is illustrated in Fig. 2. Fast analog–digital converters and automatic sampling together with signal averaging for increased signal-to-noise ratios are required for optimal processing of experimental data.

A very sensitive detection of field-induced orientation effects is possible by birefringence measurements (cf. Fig. 3). Using lasers with a high light intensity and a sufficiently high stability, the sensitivity of birefringence measurements could be extended beyond the standard level used in the past. Because of narrow light beams, it is possible to construct cells with short electrode distances and high optical pathlengths. A major challenge remains the construction of cells with strain-free optical windows. A simple cell construction used at low field strengths is given in Fig. 4a. A novel type of cell window construction (Fig. 4b) proved to be very useful for minimization of strain and reduction of cell window birefringence (27).

## 3. Sample Preparation

### *3.1. Dialysis*

The signals of most biological macromolecules induced by electric fields are very strongly dependent on the salt concentration. This is mainly due to ion binding at various sites on these macromolecules, affecting their electric parameters. In order to avoid problems with reproducibility, the samples should be transferred into a well-defined ionic environment by extensive dialysis against the buffer used for measurements. Bivalent ions like $Mg^{2+}$ or $Ca^{2+}$ often have a particularly high affinity. Thus, measurements to be conducted in the absence of these ligands require extensive dialysis of solutions

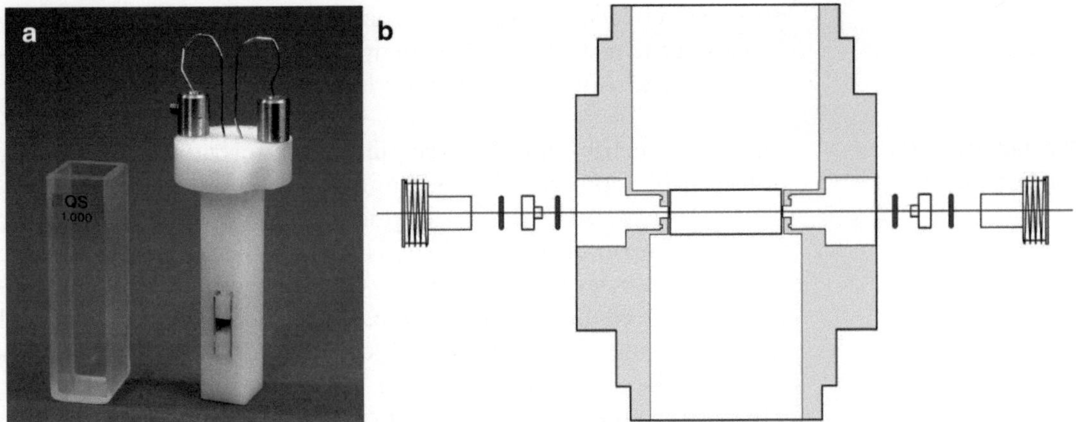


Fig. 4. Cells for electro-optical measurements. (**a**) The quartz cuvette (with an optical pathlength of 1 cm) is selected for minimal strain. The insert, made from Teflon, holds the Pt-electrodes at a given distance and the connection wires for pulse application. (**b**) Construction of optical windows inserted in cell bodies machined from dynal, polyacrylate, or any other synthetic polymer: silicon O-rings hold the quartz windows in a state with minimal strain and, thus, birefringence of the windows is reduced to a minimum (27). Pt-electrodes are inserted from the bottom and from the top.

against buffers containing EDTA; for complete exclusion of bivalent ions, some initial dialysis cycles in the presence of high concentrations of monovalent salt are recommended.

1. Select dialysis tubing with a pore size smaller than the size of the macromolecule.
2. Boil the tubing in water for ~5 min; exchange the water and boil again—during this procedure impurities are washed out, the tubing material is converted to a swollen and sterile state.
3. Seal one end of the tubing by a double knot—use gloves whenever handling the tubing.
4. Fill in the solution of your macromolecule and seal the open end of the tubing by a double knot.
5. Put the tubing into a reservoir of buffer with a volume approximately 100× larger than the volume inside the tubing. The buffer in the reservoir should be stirred. Removing of multivalent ions is supported by adding a high concentration of monovalent salt (e.g., 1 M NaCl) and by adding EDTA during initial dialysis cycles. The dialysis time depends on the pore size—usually a few hours for one equilibration. The number of repetitions depends on the required purity—usually several times. If the sample requires the presence of multivalent ions for remaining in its native state, the dialysis against EDTA and probably also against high salt should be omitted.
6. The procedure is completed by extensive dialysis against the final buffer—with at least several exchange cycles.

For safety of the sample, the dialysis should be conducted in a cold room. If ions like $Ca^{2+}$ or $Mg^{2+}$ must be removed and ETDA cannot be a component of the final dialysis buffer (e.g., for studies of bivalent ion binding), dialysis and storage of the samples should be in vessels from, e.g., polypropylene or quartz.

### 3.2. Degassing

Application of electric field pulses may cause cavitation effects, perturbing the measurements and even leading to cell destruction in the worst case. This type of problem may be avoided by careful degassing of solutions within the measuring cell. A standard procedure involves filling the solution into the measuring cell, putting the cell into a desiccator, which is then evacuated by a simple pump (water jet or one stage oil pump). The final pressure should be only slightly below the boiling pressure, in order to avoid loss of solution by boiling retardation. It is advisable to reduce the pressure slowly and watch carefully the appearance of bubbles. Gentle mechanical impacts may be applied to support detachment of bubbles from surfaces. When the solution is at its boiling point, bubbles tend to appear in pulses. It is recommended to return to standard pressure quickly after appearance of such pulse by stripping the vacuum tube from the desiccator.

Some protein solutions tend to precipitate upon degassing. In such cases, the cell may be filled with buffer and degassed separately. Subsequently, the buffer is removed from the cell, keeping a thin layer of the solvent at the surfaces and in the corners. The protein solution is degassed in another vessel separately and then filtrated directly into the cell.

It is always useful to check the concentration and the state of the sample by taking a spectrum in the measuring cell. Any changes of the sample during the measurements may be detected by a comparison of spectra before and after measurements.

#### 3.2.1. Resistance

Whenever electric fields are applied to samples, the resistance of the sample in the measuring cell is essential. Thus, a control of the resistance of each sample before start of the measurements is important—using an instrument for recording of ionic conductivities. The resistance may also be used to determine the increase of the temperature induced during pulse application.

## 4. Measurements

### 4.1. General

#### 4.1.1. Temperature Control

Measurements of the electric dichroism/birefringence may provide rotational time constants at a very high accuracy. For quantitative evaluations, the temperature of the solutions must be under careful control, because the rate of rotational diffusion is directly dependent on the viscosity and the viscosity of water is strongly dependent on the temperature. A change of the temperature from 20 to 19°C induces a change of the water viscosity by 2.5%, which is a very large effect in view of the high accuracy routinely accessible by the electric dichroism/birefringence technique (±1% for the time constants of rigid rods like DNA fragments).

#### 4.1.2. Field Strengths

For various reasons, it is useful to measure experimental data for a range of electric field strengths. Any change of, e.g., rotational time constants with the field strength may indicate some special effect, e.g., a field-induced change of the conformation. However, there may be also other special phenomena from superposition of different relaxation effects associated with opposite amplitudes, resulting in an unusual relaxation effect at a given field strength. As demonstrated by recent model calculations (14), singular points with a relaxation effect suggesting an unusual conformation may appear. Erroneous conclusions from this type of effect may be avoided easily by measurements over a sufficiently wide range of electric field strengths.

#### 4.1.3. Control of Dichroism

For convenience, the electric dichroism is often measured only at parallel orientation of the polarized light with respect to the field vector, because the effect is maximal under these conditions. It is very useful, however, to check the magnitude of the dichroism also for perpendicular orientation, because deviations from the expected ratio $\Delta A_{\|}/\Delta A_{\perp} = -2$ may indicate field-induced reaction effects. A direct control for the existence of field-induced reaction effects is possible by measurements at the magic angle (cf. Subheading 4.2.3).

#### 4.1.4. Spectroscopic Parameters

Selection of the optimal wavelength is crucial for any spectroscopy investigation. Usually, a wavelength is selected close to the maximum of the absorbance of the molecules under investigation. A major advantage of this choice is the fact that measurements can be conducted at concentrations where aggregation effects are minimal. However, there may be other criteria for selection. One of them is a maximal light intensity for optimal signal-to-noise ratios. Among the available light sources, mercury–xenon arc lamps provide particularly high light intensities at special mercury emission lines: 248.2, 265.2, 280.4, 289.4, 302.2, 313.2, 334.2, 366.3, 404.7, 435.8, and 546.1. Because of the relatively close spacing, useful lines may be found for most applications.

In some cases, a wavelength outside the range of maximal absorbance is favorable. The electric dichroism strongly depends on the orientation of the transition dipole: usually ($\pi^* \leftarrow \pi$) transitions are polarized in the plane of aromatic chromophores, whereas ($\pi^* \leftarrow n$) transitions are usually polarized in perpendicular direction to the aromatic plane. The difference in orientation leads to large changes in the electric dichroism and, thus, it is very useful to do test measurements in different wavelength ranges. In close analogy, it is crucial to check spectral changes before starting measurements of chemical relaxation. Here, the relative change may be more important than the absolute one. In some cases, these relative changes are much higher at spectral shoulders than at the absorbance center.

#### 4.1.5. Time Resolution and Signal-to-Noise Ratio

The quality of relaxation transients is determined by the signal-to-noise ratio, which is affected by many different factors. This ratio is increased by increasing the light intensity and also by limiting the bandwidth, i.e., the time constant of the detector. In both cases, there are practical limits given by the photo-stability of the sample and the required time resolution. The photo-stability must be checked experimentally. A simple rule of thumb for the time resolution indicates that the time constant of the detector should be adjusted to a value smaller by a factor of ~10 than the time constant of the fastest process to be detected. When this rule cannot be implemented, the experimental data may be evaluated by numerical deconvolution procedures. Because there are quite reliable deconvolution procedures (15), it may be even useful to limit the bandwidth of the detector beyond the usual rule of thumb with the benefit of an increased signal-to-noise ratio and subsequently eliminate the detector effect on the transients by deconvolution.

#### 4.1.6. Averaging

When all the spectroscopic parameters are optimized, the quality of the experimental data may still be improved by signal averaging. Because the signal-to-noise ratio increases with the square root of the number of shots included in the average, an increase of the signal quality by a factor of 10 requires 100 shots. Thus, there are practical limits in averaging. Constraints may also be given by insufficient sample stability. This stability may be checked by taking spectra before and after measurements.

### 4.2. Dichroism

#### 4.2.1. Dichroism Amplitude

1. Adjust polarization plane parallel to field vector.
2. Adjust wavelength and slit width of monochromator and bandwidth of detector.
3. Measure the total light intensity $I_{\|}^{0}$ for light polarized parallel to the field vector in the absence of an external electric field; take as much light as possible, but ensure that all parts of your transients are recorded in the linear range of your detector.
4. Apply an electric field pulse and record the transient; the electric field pulse should be long enough to induce a stationary level of the light intensity at the end of the pulse, i.e., the intensity should not change anymore, when the pulse length

is increased. For evaluations of accurate time constants, it is important to record signals for a sufficiently long time period such that the final level of the signal is defined precisely.

5. Read the difference $\Delta I_{\|}$ between the light intensities before application of the pulse and the stationary level at the end of the pulse.
6. Calculate the absorbance change according to $\Delta A_{\|} = -\log[(I_{\|}^{o} + \Delta I_{\|})/I_{\|}^{o}]$.
7. Calculate the reduced dichroism amplitude $\xi = 1.5 \times \Delta A_{\|}/\bar{A}$, where $\bar{A}$ is the isotropic absorbance at the given wavelength.

*4.2.2. Control of Dichroism Optics*

1. Strain in the windows of your cell may be detected by a simple device with a bright light source and two crossed polarizers; put the cell between the crossed polarizers, and look through the windows toward the light source; if there is no strain, light does not pass through the windows; strain is indicated by bright areas appearing in the windows.
2. A solution with a high electric dichroism, e.g., DNA double helices, in a low salt buffer is filled into the cell and the solution is degassed (field-induced denaturation is avoided by, e.g., 100 μM $Mg^{2+}$ in the buffer).
3. Measure dichroism amplitude $\Delta A_{\|}$ with light parallel to the field vector (see Subheading 4.2.1, steps 1–6).
4. Measure dichroism amplitude $\Delta A_{\perp}$ with light perpendicular to the field vector (see Subheading 4.2.1, steps 1–6 but polarizer perpendicular to the field vector).
5. The values $\Delta A_{\|}$ and $\Delta A_{\perp}$ should be consistent with $\Delta A_{\|} = -2\Delta A_{\perp}$ within a few percent.
6. As an extension, the polarizer may be adjusted to the magic angle 55.7° with respect to the field vector; the field-induced amplitude $\Delta A_{55.7}$ should be zero.
7. Deviations may be due to one the following reasons:
   (a) If the amplitude $\Delta A_{\|}$ is larger than ~10%, the dichroism relations are usually not fulfilled; then the test should be repeated at a lower electric field strength or at a lower concentration of the polymer.
   (b) The windows of the measuring cell may be under strain.
   (c) There may be a field-induced reaction—see next procedure.

*4.2.3. Reactions Effects/ Field-Induced Conformation Changes*

1. Measure the field-induced changes of light intensity at the polarizer angles 0, 55.7, and 90° and get the absorbance changes $\Delta A_{\|}$, $\Delta A_{55.7}$, and $\Delta A_{\perp}$.
2. Given a significant value $\Delta A_{55.7}$ check whether $[\Delta A_{\|} - \Delta A_{55.7}] = -2 \times [\Delta A_{\perp} - \Delta A_{55.7}]$; this is a test for internal consistency.

3. Additional evidence may be obtained by an analysis of time constants; usually the time constants of orientation are different from those of reactions.

### 4.3. Birefringence

#### 4.3.1. Adjustment of Optics for Birefringence Measurements

1. The polarizer between the light source and the measuring cell should be rotated to an angle of 45° with respect to the field vector in clockwise direction viewed from the direction of the light source.
2. In the absence of the $\lambda/4$ plate, the analyzer, located between the cell and the detector, should be turned to the crossed position, corresponding to the minimum of light intensity passing from the light source to the detector.
3. The $\lambda/4$ plate is inserted and rotated until the light intensity arriving at the detector is minimal again; in this angular orientation, the slow axis of the $\lambda/4$ plate is parallel to the polarization plane of the polarizer.
4. The analyzer is rotated from the crossed position toward the polarization plane of the polarizer by an angle $\alpha$; in this arrangement a negative birefringence of the sample results in a decrease of the light intensity, whereas a positive birefringence induces an increase of the light intensity; the angle $\alpha$ depends on the available light intensity and the sensitivity of the detector; for measurements of negative birefringence $\alpha$ is selected such that the total light intensity is close to the upper end of the linear response domain of the detector (e.g., 10% below the maximal value).

#### 4.3.2. Measure Birefringence

1. Read the total light intensity $I_\alpha$ at the angle $\alpha$ from the crossed position.
2. Apply an electric field pulse and record the transient induced by the field; read the stationary change of the light intensity $\Delta I_\delta$ (in the time range where the change arrives at a constant level).
3. The phase shift or optical retardation is obtained by

$$\delta = 2 \times \left(\arcsin\left[(\Delta I_\delta / I_\alpha) \times \sin^2(\alpha) + \sin^2(\alpha)\right]^{1/2} - \alpha\right). \quad (6)$$

4. The birefringence corresponding to the difference of the refractive index of the solution parallel and perpendicular to the applied field is given by $\Delta n = \delta\lambda/(2\pi\ell)$; $\lambda$ is the wavelength used for the measurement and $\ell$ is optical path length of the cell.

## 5. Data Analysis

### 5.1. Relaxation Components

The field-induced change of the absorbance, dichroism, or birefringence may be composed of several different individual relaxation components, which are identified by a detailed analysis of the transient(s) in terms of exponential functions. There are professional fitting programs (16) for this purpose, but the final decision on the number of contributing exponential terms is not always simple. One of the criteria is of course a direct visual inspection of the quality of the fit(s). A decision may be supported by inspection of plots of the difference between the data and the fit. Moreover, the autocorrelation of this difference may be useful. Finally, the decision must be often based on pragmatic considerations: when admission of an additional exponential does not decrease the error sum by, e.g., more than 20% reproducibly, then there is not enough information on such additional exponential and therefore the parameters of an additional exponential cannot be defined at a sufficient accuracy.

### 5.2. Amplitudes

Electro-optical amplitudes corresponding to stationary changes of the dichroism or the birefringence measured over a range of different electric field strengths are useful for the determination of dipole moments and optical anisotropies. For this purpose, the data must be analyzed by orientation functions (3), which are available for the basic cases of induced and permanent dipole moments. Biological macromolecules usually have many charged residues, which may lead to more complex cases of dipolar orientation and require special orientation functions (17). Moreover, electric fields may induce stretching of bent conformations, which has to be considered explicitly (18).

Amplitudes obtained for reaction effects (19) may be used to derive detailed thermodynamic parameters for reactions, provided that the amplitudes are measured over a sufficiently wide range of concentrations.

### 5.3. Time Constants

#### 5.3.1. Electric Dichroism and Electric Birefringence

The time constants derived from decay curves of the electric dichroism or the electric birefringence reflect rotational diffusion of the particles under investigation. These time constants provide detailed information on the size, shape, and also on the flexibility of the particles. The interpretation for particles of simple shape like spheres or rods is based on analytical expressions.

Data for particles of more complex shape can be analyzed by numerical procedures, which proved to be very reliable. Using the procedures of "Quantitative Molecular Electro-Optics", the electro-optical response of rigid macromolecular structures can be calculated (20). This is very useful not only for comparisons of structures in crystals and in solutions but also for control of structures designed by molecular modeling (21).

The theory predicts for rigid structures of arbitrary shape the existence of five different exponential terms in the decay curves (22, 23). In practice, the amplitudes for most of these terms are negligible and usually not more than two components can be resolved.

Electro-optical transients have a great potential for analysis of internal mobilities: However, quantitative interpretations require extensive simulations (24, 25).

#### *5.3.2. Reaction Effects*

In the limit of small perturbations, the interpretation of reaction time constants is relatively simple, because the differential equations can be linearized and the general results obtained for relaxation kinetics can be applied (7, 10, 11). Under these conditions, the number of relaxation effects directly indicates the number of independent reactions for the system under investigation. The concentration dependence of the time constants demonstrates reaction orders and can be used to determine rate constants.

## 6. Stopped-Flow Electro-Optics

Electro-optical transients provide information on macromolecular structures in a rather compact form. Because these transients can be measured within short time intervals—usually within a few microseconds, it is possible to utilize the electro-optical approach for analysis of macromolecular kinetics by the stopped flow technique. The main requirement for this combination is construction of a simple and reliable stopped-flow electric-field jump instrument. This was enabled by integration of Pt-electrodes into standard quartz cells used for stopped flow measurements (26). Cells with different optical pathlengths have been constructed and proved to be reliable. The information on the reaction is obtained by recording transients measured by application of electric field pulses at different times after the flow is stopped. The method is very sensitive: for the case of ethidium intercalation into DNA double helices (26) the increase in the length of a DNA fragment with 95 base pairs resulting from intercalation of single ethidium molecules could be measured down to reaction times of less than a millisecond. Stopped-flow electro-optics proved to be very useful also for the analysis of structures during the reaction of the cAMP receptor with promoter DNA (28).

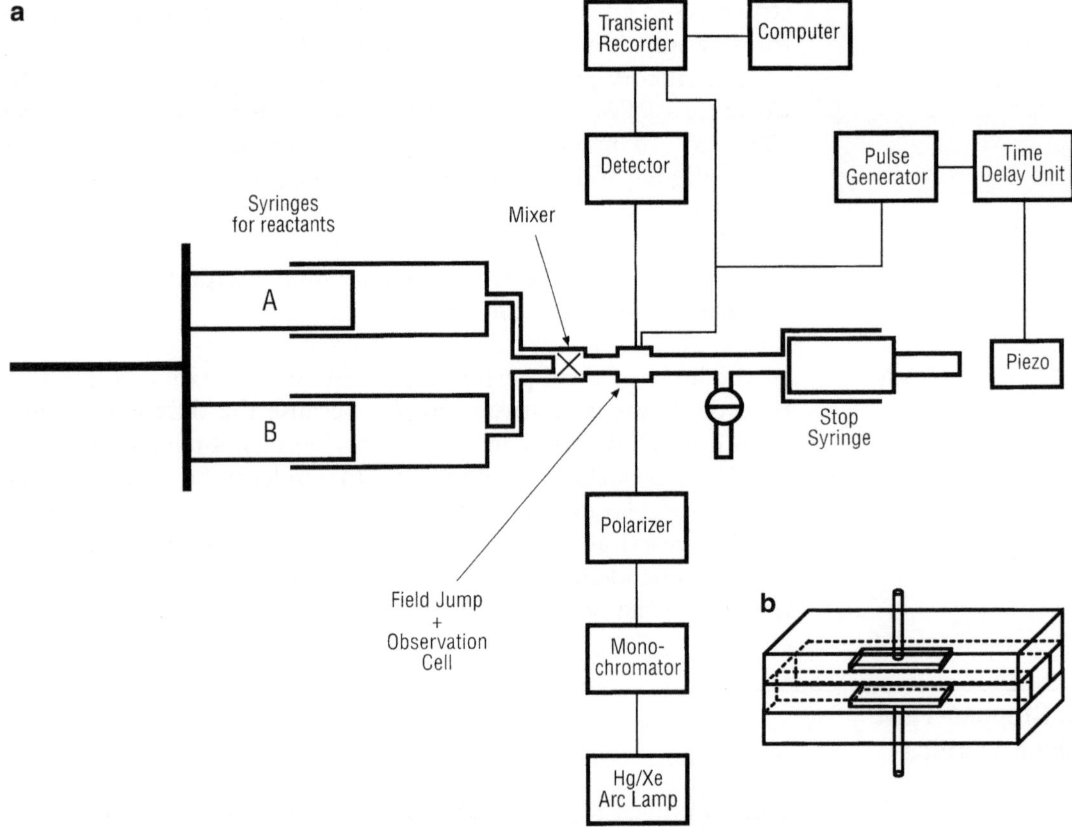


Fig. 5. Stopped flow electric field jump used for measurements of electric dichroism transients. (**a**) Reactants A and B are pushed from syringes through a mixing chamber via the observation cell into a stop syringe. The flow is stopped, when the piston of the stop syringe hits the piezo. The piezo-signal initiates a time delay unit, which triggers after a predefined delay time an electric field pulse applied to the cell. The optical components for measurement of the electric dichroism correspond to those shown in Fig. 2. The information on kinetics is obtained from a sequence of transients measured at different delay times. (**b**) The measuring cell is a quartz channel (cross section 2 × 2 mm) with inserted Pt-electrodes (adapted from ref. (26)), reprinted with permission from Elsevier.

## References


1. Debye P (1929) Polare Molekeln. Hirzel, Leipzig
2. Kerr J (1875) A new relation between electricity and light: dielectrified media birefringent. Philos Mag 50:337–348
3. Fredericq E, Houssier C (1973) Electric dichroism and electric birefringence. Clarendon, Oxford
4. O'Konski CT (ed) (1976) Molecular electro-optics: part 1—theory and methods. Marcel Dekker, New York
5. O'Konski CT (ed) (1978) Molecular electro-optics: part 2—applications to biopolymers. Marcel Dekker, New York
6. Stoylov SP (1991) Colloid electro-optics: theory, techniques, applications. Academic, London
7. Eigen M, DeMaeyer L (1963) Relaxation methods. In: Friess SL, Lewis ES, Weissberger A (eds) Investigation of rates and mechanisms of reactions, vol. VIII, part II. Interscience, New York, pp 895–1054.
8. Labhart H (1961) Methoden der Zuordnung von Absorptionsbanden von Farbstoffen zu berechneten Übergängen. Chimia 15:20–27
9. Meyer-Almes FJ, Porschke D (1993) Mechanism of intercalation into the DNA double helix by ethidium. Biochemistry 32:4246–4253

10. Bernasconi CF (1976) Relaxation kinetics. Academic, New York
11. Strehlow H (1995) Rapid reactions in solution. Weinheim, VCH
12. Grunhagen HH (1974) High-power square-wave pulse-generator for investigation of fast electric-field effects in solution. Messtechnik 82:19–23
13. Porschke D, Obst A (1991) An electric-field jump apparatus with ns time resolution for electrooptical measurements at physiological salt concentrations. Rev Sci Instrum 62:818–820
14. Antosiewicz JM, Porschke D (2009) Effects of hydrodynamic coupling on electro-optical transients. J Phys Chem B 113:13988–13992
15. Porschke D, Jung M (1985) The conformation of single stranded oligonucleotides and of oligonucleotide-oligopeptide complexes from their rotation relaxation in the nanosecond time range. J Biomol Struct Dyn 2:1173–1184
16. Provencher SW (1976) Fourier method for analysis of exponential decay curves. Biophys J 16:27–41
17. Shah MJ (1963) Electric birefringence of bentonite. 2. An extension of saturation birefringence theory. J Phys Chem 67:2215–2219
18. Porschke D (1989) Electric dichroism and bending amplitudes of DNA fragments according to a simple orientation function for weakly bent rods. Biopolymers 28:1383–1396
19. Winkler-Oswatitsch R, Eigen M (1979) Art of titration—from classical end-points to modern differential and dynamic analysis. Angew Chem 18:20–49
20. Porschke D, Antosiewicz JM (2007) Quantitative molecular electro-optics: Macromolecular structures and their dynamics in solution. In: Stoylov SP, Stoimenova MV (eds) Molecular and colloidal electro-optics. CRC, Boca Raton, FL, pp 55–1007
21. Porschke D (2010) Allosteric control of promoter DNA bending by cyclic AMP receptor and cyclic AMP. Biochemistry 49:5553–5559
22. Wegener WA, Dowben RM, Koester VJ (1979) Time-dependent birefringence, linear dichroism, and optical-rotation resulting from rigid-body rotational diffusion. J Chem Phys 70: 622–632
23. Wegener WA (1986) Transient electric birefringence of dilute rigid-body suspensions at low field strengths. J Chem Phys 84:5989–6004
24. Hagerman PJ, Zimm BH (1981) Monte-Carlo approach to the analysis of the rotational diffusion of wormlike chains. Biopolymers 20: 1481–1502
25. Antosiewicz J, Nolte G, Porschke D (1992) Modes of rotational motion of wormlike chains and the effect of charges on electrooptical transients. Macromolecules 25:6500–6504
26. Porschke D (1998) Time-resolved analysis of macromolecular structures during reactions by stopped-flow electrooptics. Biophys J 75: 528–537
27. Porschke D (2011) Electric birefringence at small angles from cross position: enhanced sensitivity and special effects. J Phys Chem 115: 4177–4183
28. Porschke D (2012) Structures during binding of cAMP receptor to promoter DNA: promoter search slowed by non-specific sites. Eur Biophys J 41:415–424

# Chapter 20

# Cryoradiolysis and Cryospectroscopy for Studies of Heme-Oxygen Intermediates in Cytochromes P450

**I.G. Denisov, Y.V. Grinkova, and S.G. Sligar**

## Abstract

Cryogenic radiolytic reduction is one of the most straightforward and convenient methods of generation and stabilization of reactive iron–oxygen intermediates for mechanistic studies in chemistry and biochemistry. The method is based on one-electron reduction of the precursor complex in frozen solution via exposure to the ionizing radiation at cryogenic temperatures. Such approach allows for accumulation of the fleeting reactive complexes which otherwise could not be generated at sufficient amount for structural and mechanistic studies. Application of this method allowed for characterizing of peroxo-ferric and hydroperoxo-ferric intermediates, which are common for the oxygen activation mechanism in cytochromes P450, heme oxygenases, and nitric oxide synthases, as well as for the peroxide metabolism by peroxidases and catalases.

**Key words:** Radiolysis, Radiolytic reduction, Peroxo-ferric complex, Hydroperoxo-ferric complex, Cytochrome P450, Low-temperature methods

## 1. Introduction

The detailed understanding of the chemical mechanism implies the knowledge of the most important intermediates of the reaction, their structure, and properties. This information can be experimentally obtained by means of fast kinetic methods, slowing down the progress of reaction by varying experimental conditions, trapping of intermediates, or using a combination of all these methods. Usually significant results can be achieved by using low temperatures, just because rates of almost all chemical reactions strongly depend on temperature. Thus, data collection at low temperatures can often provide more details on chemical kinetics and on the properties of intermediates at an experimentally accessible time range.

Wlodek M. Bujalowski (ed.), *Spectroscopic Methods of Analysis: Methods and Protocols*, Methods in Molecular Biology, vol. 875, DOI 10.1007/978-1-61779-806-1_20, © Springer Science+Business Media New York 2012

This review centers on the application of optical spectroscopy to studies of unstable iron–oxygen intermediates in heme enzymes in frozen glassy solutions. Systematic use of cryogenic temperatures is also essential in many other applications of spectroscopic methods in biochemistry and biophysics. A major part of EPR and Mössbauer data is collected at cryogenic temperatures. Other examples of cryogenic spectroscopic methods include most of the X-ray spectroscopic methods such as XAS, EXAFS, etc., and temperature-dependent magnetic circular dichroism, which provides indispensable information on the EPR-silent systems (1, 2). Moreover, most of X-ray crystallographic data on proteins are also obtained at 100 K to minimize radiation damage caused by modern bright synchrotron sources. Thus, understanding the general aspects of protein behavior at low temperatures is important, including dealing with possible effects of the frozen aqueous or mixed aqueous–organic solutions on the physical and chemical properties of proteins.

Historically, cryogenic optical spectroscopy was used in heme protein studies as long as in 1920, reviewed by Keilin and Hartree (3). More recent applications include resolution of the vibrational structure and conformational assignment of circular dichroism spectra of aromatic amino acids (4, 5), MCD studies of heme enzymes (6, 7), nonheme metalloproteins (8) and comprehensive characterization of photocycle in bacteriorhodopsin (9, 10). Various experimental methods have been developed to study protein dynamics and relaxation using cryogenic optical spectroscopy, such as hole burning spectroscopy (11), photolysis and recombination of CO, NO. and $O_2$ in heme proteins (12–15), and deconvolution of temperature-dependent absorption spectra (16). Optical spectroscopy is also used to monitor the redox state of metalloproteins directly during data collection in cryogenic X-ray crystallographic experiments (17, 18).

### 1.1. Solvents for Low-Temperature and Cryogenic Spectroscopy

The classic work of Meyer (19) provides a comprehensive and detailed description of application of cryogenic techniques for optical spectroscopy, including different designs of low-temperature cells and cryostats, as well as an indispensable collection of physical and chemical properties of mixed cryosolvents which are used at cryogenic temperatures. Several other books (20–22) contain a thorough analysis of fundamental theoretical and experimental aspects of chemistry and biochemistry at low temperatures, particularly in the frozen solutions. Many review articles (23–27) provide an excellent introduction into the field of cryobiochemistry and cryoenzymology and offer both general theoretical concepts and valuable experimental information.

For many spectroscopic methods such as EPR, NMR, XAS, Mössbauer, the homogeneity of the frozen solution is not an essential condition, and microcrystalline or powder samples can be studied with no experimental difficulties. This is not the case for optical absorption spectroscopy, where light scattering in the highly turbid samples can completely obscure absorption spectra even in thin layer cells, especially in the UV and visible range, 200–800 nm. Thus, it is important to use solvents which form optically transparent clear glass when frozen at low temperatures. Many organic and inorganic solvents, such as methylcyclopentane, isooctane, ethanol, isopropanol, glycerol, triethanolamine, boric, phosphoric, and sulfuric acid, can form transparent glasses at 77 K. In addition, binary or ternary solvent mixtures are often used in cryospectroscopy because they may form glasses when cooled, even despite their constituents crystallize with phase separation of pure solvent and freeze concentrated solute. A table of optically clear glassy solvents is given in (19), Chapter 8. However, the choice of solvents for biopolymers is usually limited to mixtures of water with selected alcohols (28, 29), concentrated solutions of sugars (30, 31) or salts (23), because most common organic solvents unfold or precipitate proteins and inactivate enzymes. For proteins the most popular cryosolvents are binary mixtures of water with ethylene glycol or glycerol, although methanol (32, 33), dimethyl sulfoxide, and dimethylformamide (32), as well as ternary mixtures of water with combinations of these solvents (29) also can be used. To prevent crystallization of water upon freezing, 50–70% w/w fractions of organic solvent are used in cryospectroscopic studies. For instance, FTIR study of water–glycerol mixtures at low temperatures indicated that phase separation and crystallization of water is fully suppressed at 65% w/w (60% v/v) fraction of glycerol (34). Fortunately, in most cases, enzymes remain folded and active in such mixed cryosolvents even at extremely low temperatures (32, 33, 35–38) despite the very well-known phenomenon of "cold denaturation" (39, 40).

An important aspect of biochemical studies at low temperatures is the necessity to control the pH of aqueous–organic solutions and to avoid large shifts of pH while changing temperature. The methods of pH measurements and the properties of the most popular buffer systems in mixtures of water with alcohols at subzero temperatures have been studied and reviewed by Douzou and colleagues (29, 36, 41). Despite the fact that some buffers, such as acetate and phosphate have much smaller $pK_a$ temperature dependence than amine buffers, a general recommendation for the most stable buffer system is difficult because of the changes of ionization enthalpies of these groups in organic and mixed solvents. Several reviews by Douzou provide tabulated reference data for buffers in mixed aqueous–organic solvents (24, 29, 36).

Another source of large pH changes at low temperature is phase separation and partial crystallization of water in frozen solvents. Orii have shown that pH may drastically change in the most commonly used buffer systems upon freezing of the solutions (42). For instance, pH in phosphate buffer solutions could drop by 2–3 units during freezing (43). This effect was correctly attributed to the sharp increase of the buffering salt concentration along with the progress of the sample freezing and formation of the grains of pure aqueous ice microcrystals (42, 43). The same phenomenon was described in an optical spectroscopic study of protonation behavior of the dye Cresol Red in the frozen aqueous solutions in the presence of organic and inorganic acids and bases (44). The drastic change of pH during freezing of aqueous solutions represents a very important aspect in cryogenic storage of biochemical samples, and the presence of 10–15% glycerol or other cryoprotectant is important to prevent the damage which may be caused by such changes. The systematic study of pH changes in acetate, phosphate, and cacodylate buffer solutions as a function of pH was published by Jung and coauthors (45). They described in detail an indirect method of pH measurements below 0°C using several acid–base indicator dyes. In general, an increase of pH by 0.6–1.1 units has been detected at low temperatures. To avoid such changes, a new method was developed recently (46), which allows to create buffers with temperature-independent pH by mixing of two buffers, one with positive and another with negative temperature coefficients of their pH. This approach can be very useful if constant pH has to be maintained in the broad temperature range.

### 1.2. Temperature Dependence of Chemical Kinetics

For many chemical reactions, the dependence of their rates on the temperature obeys the Arrhenius equation:

$$k = A \exp(-E_a/RT).$$

This equation holds for the reactions with the strong temperature coefficient and hence relatively high activation energies, i.e., much higher than $RT$ (0.6 kcal/mol at $T = 300$ K). For such reactions, the Arrhenius plots are linear in the accessible temperature intervals, and activation energy $E_a$, which is approximately equal to the enthalpy of activation (47), can be estimated from the slope of $\ln(k)$ vs. $1/T$.

As an example, consider the stability of the oxy-ferrous complex in cytochromes P450, or any other heme protein against autoxidation at different temperatures. Concentration of the oxy-complex is determined by the rates of oxygen binding to the ferrous heme protein and of oxidative decomposition of this complex with dissociation of superoxide anion:

$$Fe^{2+} + O_2 \xrightarrow[{[k_{binding}]}]{} [Fe^{2+}O_2] \xrightarrow[{[k_{autox}]}]{} Fe^{3+} + O_2^{-}. \quad (1)$$

For many heme proteins, the experimentally observed apparent rates of oxygen binding are in the range $10^2$–$10^3$ $s^{-1}$ at ambient oxygen concentrations. Thus, oxygen binding is fast at air-saturated buffers and usually does not limit complete oxygenation of the reduced heme proteins. However, for many cytochromes P450 and nitric oxide synthases, the measured apparent autoxidation

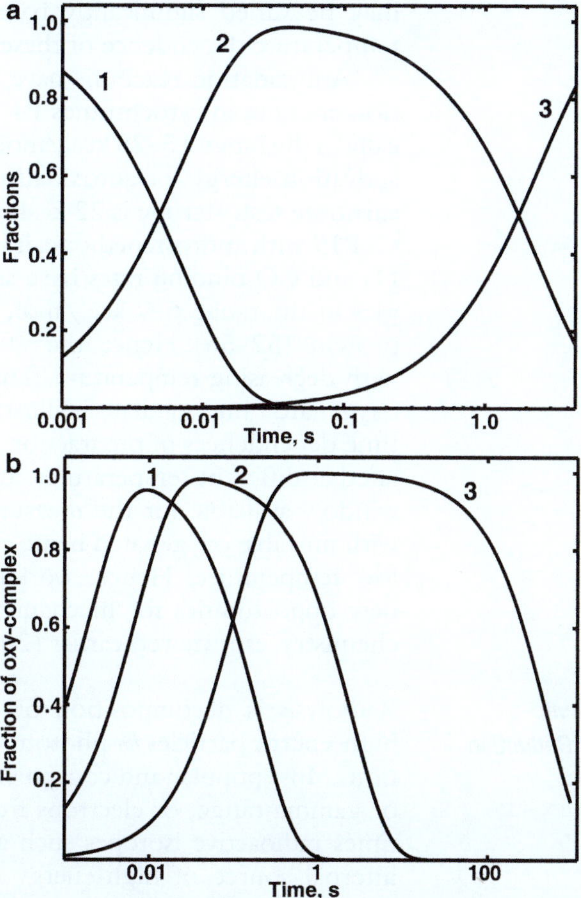


Fig. 1. The kinetics of formation and decay of oxyferrous complex in cytochrome P450 as described by (1). (**a**) Fractions of ferrous (*1*), oxy-ferrous (*2*), and ferric (*3*) enzyme at $T = 279$ K as a function of time. (**b**) Fractions of oxy-ferrous complex as a function of time at three different temperatures: (*1*) 300 K; (*2*) 279 K; (*3*) 243 K. Parameters used for calculations: rates $k_1 = 500\ \mathrm{s}^{-1}$ and $k_2 = 5\ \mathrm{s}^{-1}$ at $T = 300$ K, corresponding activation enthalpies $\Delta H_1^{\neq} = 10\,\mathrm{kcal/mol}$; $\Delta H_2^{\neq} = 20\,\mathrm{kcal/mol}$.

rates are also very high, and may reach the same order of magnitude, up to $10^2\ \mathrm{s}^{-1}$ and higher (48, 49). For these enzymes, the maximum concentration of oxy-ferrous complex, which can be reached by fast mixing of the reduced protein solution with aerated or oxygen saturated buffer, is determined by the ratio of the rates of oxygen binding and oxidative decomposition of oxyferrous complex, as shown in Fig. 1. To increase the population of the oxy-complex, it is necessary to maximize this ratio by the appropriate choice of the experimental conditions, as shown in the Fig. 1a. Acceleration of oxygen binding is limited by oxygen solubility and cannot be significantly improved without using high-pressure technique (see (50), for an example). However, the ratio $k_{\mathrm{binding}}/k_{\mathrm{autox}}$

may be varied significantly based on the substantially different temperature dependence of these rates.

Autoxidation reactions have been shown to have high activation energies in cytochromes P450 and other heme proteins, typically in the range 15–25 kcal/mol (48). For example, the Arrhenius activation energy for autoxidation of CYP3A4 saturated with the substrate testosterone is 22 kcal/mol (48), and for autoxidation of CYP19 with androstenedione, it is 18 kcal/mol (51). Contrary, the $O_2$ and CO binding rates have significantly lower activation energies in the range 6–9 kcal/mol, as documented for several heme proteins (52–55). Hence, the ratio $k_{binding}/k_{autox}$ increases notably with decreasing temperature. The resulting increase in stability of oxygenated intermediate is illustrated in Fig. 1b, where the same time dependences of the fraction of oxy-ferrous complex are calculated at different temperatures. In addition, the characteristic time window available for the measurements and other manipulations with unstable oxygenated heme enzyme is significantly increased at low temperature. Hence, working at low temperatures provides new opportunities for mechanistic studies in chemistry and biochemistry, as reviewed earlier (20, 22, 56).

### 1.3. Cryogenic Radiolytic Reduction

Radiolysis is decomposition or fragmentation of a molecule by high-energy particles or photons commonly termed ionizing radiation. Most popular and convenient to use are photons in the X-ray or gamma-range, or electrons from the linear accelerators. Sometimes radioactive isotopes such as $^{3}H$, $^{32}P$, $^{35}S$, also can serve as internal source of high-energy electrons (57–59). The yield of radiolysis is usually measured as an average number of corresponding products per 100 eV absorbed by the sample. The main products of the water radiolysis are hydroxyl radicals, hydrated electrons, hydrogen atoms, and several other highly reactive intermediate species which quickly react with solutes or recombine to produce more stable compounds such as hydrogen peroxide. Radiolysis of alcohols, which include ethylene glycol and glycerol, results in the formation of hydrogen atoms, CO, aldehydes, ketones, alkanes, and alkenes. In all cases, the final radiolytic yield depends on the secondary reactions between primary radicals and on the presence of other compounds (60, 61). However, at cryogenic temperatures almost all reactive products of radiolysis are immobilized in the frozen solvent matrix or trapped by radical quenchers. Only solvated electrons are mobile enough to escape spurs and to react with the heme protein complexes even in frozen solutions at 77 K.

A large number of published works utilized $\gamma$-irradiation from $^{60}Co$ source for cryoradiolytic reduction of cytochromes P450 and other heme proteins (59, 62, 63). Photons from this source, with energies 1.13 and 1.3 MeV, interact with atoms with low $Z$ (atomic mass) mostly through inelastic Compton scattering with emission of lower energy photon, and generation of free electron and cation radical as primary products of this scattering (60). The scattered photon and electron generate more electrons with the energy of the initial γ-photon gradually dissipated in the cascade of scattering events. According to the experimentally measured yields (60), one γ-photon with an energy of 1 MeV absorbed in water will generate approximately 27,000 electrons, which means that approximately 2 mM average concentration of free electrons will be generated by 1 Mrad of absorbed dose from a $^{60}$Co source (1 rad is defined as $6.24 \times 10^{13}$ eV/g, or $10^{-5}$ J/g of absorbed energy). The actual concentration of generated electrons in ice at 77 K may be several times lower because of recombination of electrons with parent radicals, and the typical doses used in cryoradiolytic reduction range from 2 to 10 Mrad. Recombination of electrons with radicals also results in a saturation behavior of cryoradiolytic reduction with increasing dose, and the yield for metalloproteins usually is not improved by prolonged irradiation with the doses higher than 4–6 Mrad (63, 64).

Recently, the method of cryoradiolytic reduction was established as an indispensable tool in the studies of unstable iron–oxygen intermediates in cytochromes P450 and other heme enzymes (59, 62, 65–72). In these works, the oxy-ferrous complexes of heme proteins have been prepared by direct oxygenation of the reduced ferrous protein and then frozen in liquid nitrogen. One-electron reduction of these samples by radiolytic electrons generated using irradiation with high energy photons at 77 K (X-rays or γ-rays) results in the formation of peroxo-ferric or hydroperoxo-ferric complexes. These fleeting intermediates of heme enzyme catalysis could not be isolated in ambient conditions, despite numerous attempts. However, being immersed in liquid nitrogen, they are stable for many months and can be studied using various spectroscopic methods, such as EPR (62, 65, 67, 73–76), UV–vis absorption spectroscopy (58, 59, 69, 70, 76–79), resonance Raman (69, 71, 72, 79, 80), Mössbauer (67), and EXAFS (81).

## 2. Materials

1. Methacrylate UV–vis enhanced semimicro cells for optical absorption (Fisher Scientific, Cat 14-955-128). Polystyrene semimicro cells can be also used at wavelengths longer than 320 nm (Cat 14-955-127).
2. NMR (5 mm outer diameter) or EPR (4 mm outer diameter) tubes, glass or quartz, for resonance Raman or EPR samples. For EPR, quartz is preferred for higher purity and much lower contamination with iron.

3. Glycerol, ultrahigh purity (Fluka), dithionite (Sigma), phosphate buffer for appropriate pH.
4. Dewar flasks, including one for cryogenic irradiation, 350 ml, outer diameter 8.5 cm, height 15 cm to fit the irradiation compartment of Gammacell-220 $^{60}Co$ source, which has the shape of cylinder with 20 cm height and 15 cm diameter.
5. Dry ice and ethanol for two cold baths (see Subheading 3.1, step 3).
6. Long thongs for sample manipulation when fully immersed in liquid nitrogen.
7. Protective eye goggles.
8. Large storage Dewar for prolonged storage of samples at 77 K fully immersed in liquid nitrogen.
9. Equipment for the cryospectroscopy. (1) Commercially available optical cryostat, or home made, as described (our papers). (2) Thermocouple and thermometer for the temperature control in the broad range, from 77 to 300 K (Fisher). (3) Liquid nitrogen to use as a cooling agent. *Warning: work with liquid gases has to be performed according to the safety instructions*!
10. Spectrophotometer with a large enough sample compartment to accommodate the optical cryostat. Sample compartment has to be purged by a constant flow of dry air or nitrogen to prevent condensation of atmospheric water vapor on the cryostat to avoid light scattering during spectroscopic measurements.
11. Access to the large $^{60}Co$ source or alternative source of ionizing radiation.

## 3. Methods

### *3.1. Preparation of Oxy-Ferrous Complex in Cytochromes P450*

1. Solutions of cytochromes P450 are deoxygenated for several minutes under flow of humidified nitrogen or argon of high purity and transferred to the anaerobic chamber. Deoxygenation of glycerol (ultra-high purity glycerol is used to minimize contamination with organic peroxides) is achieved by multiple freeze–pump–thaw cycles under pure nitrogen or argon.
2. After addition of glycerol (15% v/v), cytochrome P450 is reduced by twofold molar excess of dithionite, added from the freshly prepared solution at the same phosphate buffer. The reduced sample is loaded into the gas tight Hamilton syringe and kept in the anaerobic chamber till oxygenation.
3. To provide the maximum stability of the oxy-ferrous complex, two cold baths with different ratios of ethanol/dry ice have

been used, Bath A with excess ethanol for −30°C, and Bath B with excess of dry ice at −70°C. Note that temperature of Bath A is important for the stability of oxy-complex, and has to be maintained at the desired level by gradual addition of dry ice.

4. The methacrylate optical cell, or the 4-mm EPR tube with 85% glycerol/buffer with appropriate concentration of substrate and saturated with oxygen gas is equilibrated at −30°C by immersion into the first bath.
5. The concentrated solution of the reduced cytochrome P450 is quickly inserted into this glycerol/buffer using gas tight syringe and stirred for 25–30 s using a cold metal stirrer. As soon as the sample looks homogeneous, it is transferred to the second bath at −70°C to quench the oxy-ferrous complex.
6. The good thermal contact of the cell with the sample and the bath provides fast cooling rate and ensures longer life-time of oxyferrous complex. After cooling at −70°C for 2 min, the cell is quickly transferred to the optical cryostat which is precooled at −100°C, or 170 K. The sample is gradually cooled down from this temperature to 77 K, and the absorption spectra are taken at several intermediate temperatures to ensure the formation and stability of oxy-ferrous complex.

After cooling at cryogenic temperatures, the solvent solidifies and forms optically transparent glass, where diffusion of even smallest molecules is severely impaired, and no autoxidation is observed at the temperatures below 200 K. Stabilization of oxy-ferrous intermediates in cytochromes P450 at cryogenic temperatures makes possible application of slow spectroscopic and structural methods for comprehensive studies of both oxy-ferrous and peroxo-ferric complexes, which can be obtained by cryoradiolytic reduction.

Another approach can be used if oxygen isotope $^{18}O$ is used for vibrational spectroscopy (69, 71, 72, 79). In such case, aerobic oxygenation has to be replaced by application of isotope-enriched oxygen using gas tight syringe and inserting the needle through the septum or on the Schlenk line (69). Deoxygenation of concentrated P450 solution and ultra-pure glycerol is accomplished as described above. Next, they are mixed in appropriate proportions in the anaerobic chamber and ferric cytochrome P450 is reduced by addition of small molar excess of dithionite solution of known concentration. To maintain anaerobic conditions, the samples in NMR or EPR tubes with rubber septa are packed into the second closed container and only after that transferred from anaerobic chamber into the −80°C refrigerator. After cooling the samples can be safely connected to the Schlenk line and purged with nitrogen or argon before application of isotope-enriched oxygen $^{18}O_2$ or other gas.

### *3.2. Cryogenic Radiolytic Reduction*

1. Samples are placed into the Dewar flask filled with liquid nitrogen and covered with a styrofoam stopper to minimize evaporation of the coolant during irradiation. The Dewar flask must be made of glass, possibly silvered, but not stainless steel or other metal, to avoid the absorption of $\gamma$-rays. The styrofoam cover can be easily cut from the styrofoam packing material to match the size of the Dewar flask and to allow holes for the NMR tubes, if necessary.
2. The Dewar containing samples and filled with liquid nitrogen is placed into the irradiation compartment of $^{60}$Co source (Gammacell 220 has a cylindrical irradiation compartment with approximately 20 cm height and 15 cm diameter). Liquid nitrogen constantly evaporates during irradiation and has to be added every 90–100 min.
3. After irradiation all samples contain high concentrations of trapped electrons, which have strong absorption band in the visible region. Thus, the samples for optical absorption spectroscopy before measurements have to be illuminated for 10–15 min with the visible light ($\lambda > 450$ nm) from the Xe lamp or some analogue, to photobleach these electrons and to make the samples transparent (59, 63). During photobleaching samples are fully immersed in liquid nitrogen.

## 4. Notes

Making oxy-complex in aromatase and CYP3A4 by quick mixing of the concentrated anaerobic solutions of reduced P450 in 15% glycerol with sixfold volume excess of 70% glycerol/phosphate buffer saturated with oxygen and precooled at −30°C. This approach is preferable for the enzymes with very fast autoxidation rates, because oxygen binding happens at low temperature, where the oxy-ferrous complex has a longer half-life. Typical Arrhenius activation free energies for autoxidation of heme enzymes are in the range of 15–25 kcal/mol, corresponding to approximately 2.2–4.3-fold increase of the half-life of oxy-complex with every 10°C decrease of the temperature. This strong temperature dependence of autoxidation rate implies an increase of the half-life of unstable oxy-ferrous complexes in cytochromes P450 from tens of milliseconds at 37°C to hundreds of seconds at −30°C, Fig. 2. We utilized this method to obtain oxycomplexes of CYP19 and CYP3A4 for the optical absorption and EPR studies of peroxo-intermediates obtained by radiolytic reduction of oxy-ferrous precursors (82).

Making unstable oxy-ferrous complexes—mixing at low temperatures in cryosolvents or bubbling cold gas though the

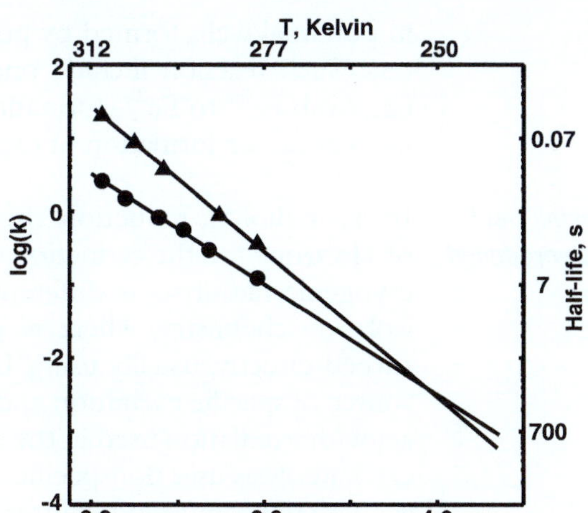


Fig. 2. Arrhenius plot of the temperature dependences of autoxidation rate constants for CYP3A4. Experimental data (*symbols*) and linear fits (*lines*) are shown for CYP3A4 saturated with testosterone (*triangles*) and bromocriptine (*circles*). For clarity, corresponding absolute temperatures are shown on *top*, and calculated half-life times for the oxy-complex are listed on the *right*.

anaerobic solution of reduced heme protein. Oxy-ferrous complex in cytochromes P450 is usually only marginally stable at ambient conditions, and irreversibly decomposes with the escape of superoxide from the remaining ferric hemoprotein with typical rates 0.01–10 $s^{-1}$ (48). However, high activation energies of autoxidation reactions, 15–25 kcal/mol, provide very steep increase of oxyferrous complex stability with the decrease of temperature, approximately by three orders of magnitude on cooling from the room temperature to −30°C, Fig. 2. Thus, oxygenation of cytochromes P450 can be accomplished by manual mixing at −30°C in aqueous glycerol buffer.

#### 4.1. Radiolytic Reduction

Ionizing radiation, i.e., high-energy particles or photons such as X-rays or gamma-rays, interacts with matter nonselectively, generating electrons and radicals. Using electrons generated by radiolysis of the solvent radiation, one can initiate the desired chemical reactions, both in solution and in solid state. While in solution multiple reactive radicals generated by radiolysis can interfere with the process under study through side reactions, and the overall radiochemical effect is in most cases oxidative transformation of the solute. Contrary, in the solid state the diffusion of all products of radiolysis with the exception of electrons is impaired, and side reactions of the target compound with nonspecific radicals can be prevented by keeping the sample well below the melting temperature. Radiolytic electrons, which escape recombination and lose their kinetic energy through inelastic scattering on the solvent atoms, are either trapped in potential wells formed by polar groups or react with redox centers. Such reaction include one-electron reduction of metal ions, i.e., from $Fe^{3+}$ to $Fe^{2+}$, formation of hydrogen atoms from protons in solution, or formation of organic radicals (83).

### 4.2. Mechanism, Yield, Stability of Cryoreduced Intermediates

In cryoradiolytic reduction, the solvent serves as the primary source of electrons for the reduction of target compound. In this sense, cryogenic radiolysis is different from other approaches in matrix isolation chemistry, where reagents or reactive complex are perturbed directly, usually using UV-photolysis. The latter requires a source of specific excitation and inert matrix, which does not interact with irradiation used in the experiments. In turn, the method of cryoradiolysis uses nonspecific ionizing radiation, which may interact with the compound of interest as well as with the solvent matrix. For the dilute solutions of the reactive complexes, the volume fraction of the solvent is much higher, and the effect of direct radiolysis of the target compound can be neglected. Note that the solvent can also serve as a selective quencher of undesired radiolysis products. For example, glycerol or ethylene glycol efficiently trap and immobilize hydroxyl radicals in experiments on cryogenic radiolytic reduction of metalloproteins and greatly improve the yield of solvated electrons, which in turn results in higher yield of reduced protein (64).

## 5. Conclusion

Cryogenic stabilization in frozen glassy solvents is a very useful method for the studies of unstable iron–oxygen intermediates in heme enzymes. Combining this method with cryogenic radiolytic reduction allows for accumulation of one-electron reduced transient peroxo-ferric and hydroperoxo-ferric complexes and for detailed spectroscopic and structural characterization of such complexes, which usually cannot be isolated and stabilized by other methods.

## 6. Additional Reading

Different aspects of low-temperature spectroscopy have been described in many useful reviews and original papers. Elaborated and systematic development of low-temperature methods for biochemistry by Pierre Douzou and his colleagues produced dozens of seminal publications, summarized in many excellent reviews (24–26, 35, 84–87) and a book (20), which describe in detail theoretical basis and numerous indispensable details of experimental methods. The works of Roy Daniel and collaborators also provide a useful insight into the role of enzyme dynamics in the catalytic activity by using low-temperature methods and cryosolvents (27, 28, 88). Absorption spectra and MCD of unstable high-valent intermediates in heme enzymes have been described by Gasyna, Stillman, and coworkers (76, 89–91). Application of cryogenic optical spectroscopy to biophysical chemistry of proteins, including heme enzymes, has been described by Vanderkooi, Friedrich, and colleagues (11, 34, 92–98). Frauenfelder and colleagues systematically applied various spectroscopic methods at cryogenic temperatures to study fundamental aspects of protein structure and dynamics (12, 99–101).

A number of analytical and preparative biophysical and biochemical methods other than spectroscopic, also can be modified to take advantage of improved sample stability at subzero temperatures. Application of aqueous–organic solvents for extending the available temperature range down to −25°C to −100°C and appropriate experimental modifications for work at such low-temperatures have been described for isoelectric focusing and electrophoresis (102–104), potentiometry (41), size-exclusion and affinity chromatography (84, 105, 106), and direct assays of enzymatic catalysis (28, 86, 88, 107).

## Acknowledgments

We gratefully acknowledge the contribution into development of these methods and collaboration with Drs. T.M. Makris, I. Schlichting, B.M. Hoffman, R.M. Davydov, M. Ikeda-Saito, J.R. Kincaid, and P.J. Mak, much of which resulted in the cited works. We appreciate the help provided by Dr. S. Toshkov at the Nuclear Radiation Lab, University of Illinois, Urbana-Champaign, and Dr. J. Bentley while using the $^{60}$Co source in the Notre Dame Radiation Laboratory (Notre Dame University, IN). Irradiations were conducted partly at the Notre Dame Radiation Laboratory, which is a facility of the U.S. Department of Energy, Office of Basic Energy Sciences. This work is supported by NIH grants GM31756 and GM33775 to S.G.S.

### References

1. Sharonov YA (1986) The heme electronic structure of reduced cytochromes P 450 and P 420 as studied by low-temperature magnetic circular dichroism. Mol Biol (Moscow) 20:440–450
2. Solomon EI, Hanson MA (1999) Bioinorganic spectroscopy. In: Solomon EI, Lever ABP (eds) Inorganic electronic structure and spectroscopy, vol 2. Wiley, New York, pp 1–129

3. Keilin D, Hartree EF (1949) Effect of low temperature on the absorption spectra of hemoproteins, with observations on the absorption spectrum of oxygen. Nature 164:254–259
4. Strickland EH (1974) Aromatic contributions to circular dichroism spectra of proteins. CRC Crit Rev Biochem 2:113–175
5. Strickland EH, Horwitz J, Billups C (1969) Fine structure in the near-ultraviolet circular dichroism and absorption spectra of tryptophan derivatives and chymotrypsinogen A at 77 K. Biochemistry 8:3205–3213
6. Sharonov YA (1992) Substrate induced electronic-conformational interactions in active site of reduced bacterial cytochrome P 450CAM and analysis of the heme electronic structure. Mol Biol (Moscow) 26:1251–1262
7. Sharonov YA (2001) The energy level scheme for the ferryl heme in compound II of the peroxidase-catalase family as determined from analysis of low-temperature magnetic circular dichroism. Biochim Biophys Acta 1504:444–451
8. Solomon EI, Pavel EG, Loeb KE, Campochiaro C (1995) Magnetic circular dichroism spectroscopy as a probe of the geometric and electronic structure of non-heme ferrous enzymes. Coord Chem Rev 144:369–460
9. Honig B, Ebrey TG (1974) The structure and spectra of the chromophore of the visual pigments. Annu Rev Biophys Bioeng 3:151–177
10. Balashov SP, Ebrey TG (2001) Trapping and spectroscopic identification of the photointermediates of bacteriorhodopsin at low temperatures. Photochem Photobiol 73:453–462
11. Ponkratov VV, Friedrich J, Vanderkooi JM, Burin AL, Berlin YA (2004) Physics of proteins at low temperature. J Low Temp Phys 137:289–317
12. Frauenfelder H, Alberding NA, Ansari A, Braunstein D, Cowen BR, Hong MK, Iben IET, Johnson JB, Luck S et al (1990) Proteins and pressure. J Phys Chem 94:1024–1037
13. Miller LM, Chance MR (1995) Structural and electronic factors that influence oxygen affinities: A spectroscopic comparison of ferrous and cobaltous oxymyoglobin. Biochemistry 34:10170–10179
14. Nienhaus K, Lamb DC, Deng P, Nienhaus GU (2002) The effect of ligand dynamics on heme electronic transition band III in myoglobin. Biophys J 82:1059–1067
15. Tetreau C, Mouawad L, Murail S, Duchambon P, Blouquit Y, Lavalette D (2005) Disentangling ligand migration and heme pocket relaxation in cytochrome P450cam. Biophys J 88:1250–1263
16. Cupane A, Leone M, Vitrano E, Cordone L (1995) Low temperature optical absorption spectroscopy: an approach to the study of stereodynamic properties of hemeproteins. Eur Biophys J 23:385–398
17. Unno M, Chen H, Kusama S, Shaik S, Ikeda-Saito M (2007) Structural characterization of the fleeting ferric peroxo species in myoglobin: Experiment and theory. J Am Chem Soc 129:13394–13395
18. Beitlich T, Kuehnel K, Schulze-Briese C, Shoeman RL, Schlichting I (2007) Cryoradiolytic reduction of crystalline heme proteins: analysis by UV-vis spectroscopy and X-ray crystallography. J Synchrotron Radiat 14:11–23
19. Meyer B (1971) Low temperature spectroscopy. American Elsevier Publishing Co., New York, 653 pp
20. Douzou P (1977) Cryobiochemistry: an introduction. Academic, London, 286 pp
21. Sergeev GB, Batyuk VA (1981) Cryochemistry. Mir Publishers, Moscow, 298 pp
22. Franks F (1985) Biophysics and biochemistry at low temperatures. Cambridge University Press, Cambridge, 210 pp
23. Auld DS (1993) Low-temperature stopped-flow rapid-scanning spectroscopy: performance tests and use of aqueous salt cryosolvents. Meth Enzymol 226:553–565
24. Douzou P (1980) Cryoenzymology in aqueous media. Adv Enzymol Relat Areas Mol Biol 51:1–74
25. Douzou P, Balny C (1977) Cryoenzymology in mixed solvents without cosolvent effects on enzyme specific activity. Proc Natl Acad Sci U S A 74:2297–2300
26. Douzou P, Petsko GA (1984) Proteins at work: "stop-action" pictures at subzero temperatures. Adv Protein Chem 36:245–361
27. Daniel RM, Dunn RV, Finney JL, Smith JC (2003) The role of dynamics in enzyme activity. Annu Rev Biophys Biomol Struct 32:69–92
28. Bragger JM, Dunn RV, Daniel RM (2000) Enzyme activity down to −100 degrees C. Biochim Biophys Acta 1480:278–282
29. Douzou P, Hui Bon Hoa G, Maurel P, Travers F (1976) Physical chemical data for mixed solvents used in low temperature biochemistry. In: Fasman GD (ed) Physical and chemical data, vol 1, CRC handbook of biochemistry and molecular biology. CRC, Cleveland, pp 520–539
30. Wright WW, Guffanti GT, Vanderkooi JM (2003) Protein in sugar films and in glycerol/water as examined by infrared

spectroscopy and by the fluorescence and phosphorescence of tryptophan. Biophys J 85:1980–1995

31. Nienhaus K, Nienhaus GU (2008) Ligand dynamics in heme proteins observed by Fourier transform infrared spectroscopy at cryogenic temperatures. Meth Enzymol 437:347–378
32. Fink AL, Cartwright SJ (1981) Cryoenzymology. CRC Crit Rev Biochem 11:145–207
33. Reat V, Finney JL, Steer A, Roberts MA, Smith J, Dunn R, Peterson M, Daniel R (2000) Cryosolvents useful for protein and enzyme studies below −100 degrees C. J Biochem Biophys Meth 42:97–103
34. Dashnau JL, Nucci NV, Sharp KA, Vanderkooi JM (2006) Hydrogen bonding and the cryoprotective properties of glycerol/water mixtures. J Phys Chem B 110:13670–13677
35. Douzou P (1973) Enzymology at sub-zero temperatures. Mol Cell Biochem 1:15–27
36. Douzou P (1977) Enzymology at subzero temperatures. Adv Enzymol Relat Areas Mol Biol 45:157–272
37. Cox RP (1978) Cryoenzymology: the use of fluid solvent mixtures at subzero temperatures for the study of biochemical reactions. Biochem Soc Trans 6:689–697
38. Fink AL (1986) Protein folding in cryosolvents and at subzero temperatures. Methods Enzymol 131:173–185
39. Privalov PL (1990) Cold denaturation of proteins. CRC Crit Rev Biochem Mol Biol 25:281–305
40. Prabhu NV, Sharp KA (2005) Heat capacity in proteins. Annu Rev Phys Chem 56:521–548
41. Larroque C, Maurel P, Balny C, Douzou P (1976) Practical potentiometric determinations of proton activity in hydro organic solvents at subzero temperatures. Anal Biochem 73:9–19
42. Orii Y, Morita M (1977) Measurement of the pH of frozen buffer solutions by using pH indicators. J Biochem Biophys Meth 81:163–168
43. Williams-Smith DL, Bray RC, Barber MJ, Tsopanakis AD, Vincent SP (1977) Changes in apparent pH on freezing aqueous buffer solutions and their relevance to biochemical electron-paramagnetic-resonance spectroscopy. Biochem J 167:593–600
44. Heger D, Klanova J, Klan P (2006) Enhanced protonation of cresol red in acidic aqueous solutions caused by freezing. J Phys Chem B 110:1277–1287
45. Schulze H, Ristau O, Jung C (1994) The proton activity at cryogenic temperatures–a possible influence on the spin state of the heme iron of cytochrome P-450cam in supercooled buffered solutions. Biochim Biophys Acta 1183:491–498
46. Sieracki NA, Hwang HJ, Lee MK, Garner DK, Lu Y (2008) A temperature independent pH (TIP) buffer for biomedical biophysical applications at low temperatures. Chem Commun: 823–825
47. Laidler KJ (1996) A glossary of terms used in chemical kinetics, including reaction dynamics. Pure Appl Chem 68:149–192
48. Denisov IG, Grinkova YV, Baas BJ, Sligar SG (2006) The ferrous-dioxygen intermediate in human cytochrome P450 3A4: Substrate dependence of formation and decay kinetics. J Biol Chem 281:23313–23318
49. Lefevre-Groboillot D, Boucher JL, Mansuy D, Stuehr DJ (2006) Reactivity of the heme-dioxygen complex of the inducible nitric oxide synthase in the presence of alternative substrates. FEBS J 273:180–191
50. Schlichting I, Berendzen J, Chu K, Stock AM, Maves SA, Benson DE, Sweet RM, Ringe D, Petsko GA, Sligar SG (2000) The catalytic pathway of cytochrome P450cam at atomic resolution. Science 287:1615–1622
51. Grinkova YV, Denisov IG, Waterman MR, Arase M, Kagawa N, Sligar SG (2008) The ferrous-oxy complex of human aromatase. Biochem Biophys Res Commun 372:379–382
52. Van Leeuwen JW, Butler J, Swallow AJ (1981) A non-equilibrium state of deoxyhaemoglobin. Temperature-dependence and oxygen binding. Biochim Biophys Acta 667:185–196
53. Sato F, Shiro Y, Sakaguchi Y, Iizuka T, Hayashi H (1990) Thermodynamic study of protein dynamic structure in the oxygen binding reaction of myoglobin. J Biol Chem 265:18823–18828
54. Filiaci M, Nienhaus GU (1997) The role of entropy in the discrimination between CO and O2 in myoglobin. Eur Biophys J 26: 209–214
55. Tetreau C, Di Primo C, Lange R, Tourbez H, Lavalette D (1997) Dynamics of carbon monoxide binding with cytochromes P-450. Biochemistry 36:10262–10275
56. Barman T, Travers F, Balny C, Hui Bon Hoa G, Douzou P (1986) New trends in cryoenzymology: probing the functional role of protein dynamics by single-step kinetics. Biochimie 68:1041–1051
57. Denisov IG, Hung S-C, Weiss KE, Mclean MA, Shiro Y, Park S-Y, Champion PM, Sligar SG (2001) Characterization of the oxygenated intermediate of the thermophilic cytochrome P450 CYP119. J Inorg Biochem 87:215–226

58. Denisov IG, Ikeda-Saito M, Yoshida T, Sligar SG (2002) Cryogenic absorption spectra of hydroperoxo-ferric heme oxygenase, the active intermediate of enzymatic heme oxygenation. FEBS Lett 532:203–206
59. Denisov IG, Makris TM, Sligar SG (2002) Cryoradiolysis for the study of P450 reaction intermediates. Meth Enzymol 357:103–115
60. Spinks JWT, Woods RJ (1990) An introduction to radiation chemistry, 3rd edn. Wiley, New York, 574 pp
61. Woods RJ, Pikaev AK (1994) Applied radiation chemistry, radiation processing. Wiley, New York, 535 pp
62. Davydov R, Makris TM, Kofman V, Werst DE, Sligar SG, Hoffman BM (2001) Hydroxylation of camphor by reduced oxy-cytochrome P450cam: Mechanistic implications of EPR and ENDOR studies of catalytic intermediates in native and mutant enzymes. J Am Chem Soc 123:1403–1415
63. Denisov IG, Victoria DC, Sligar SG (2007) Cryoradiolytic reduction of heme proteins: Maximizing dose-dependent yield. Radiat Phys Chem 76:714–721
64. Davydov R, Kuprin S, Graeslund A, Ehrenberg A (1994) Electron paramagnetic resonance study of the mixed-valent diiron center in *Escherichia coli* ribonucleotide reductase produced by reduction of radical-free protein R2 at 77 K. J Am Chem Soc 116:11120–11128
65. Davydov R, Ledbetter-Rogers A, Martasek P, Larukhin M, Sono M, Dawson JH, Siler Masters BS, Hoffman BM (2002) EPR and ENDOR characterization of intermediates in the cryoreduced oxy-nitric oxide synthase heme domain with bound L-arginine or N-hydroxyarginine. Biochemistry 41:10375–10381
66. Davydov R, Kofman V, Fujii H, Yoshida T, Ikeda-Saito M, Hoffman BM (2002) Catalytic mechanism of heme oxygenase through EPR and ENDOR of cryoreduced oxy-heme oxygenase and its Asp 140 mutants. J Am Chem Soc 124:1798–1808
67. Garcia-Serres R, Davydov RM, Matsui T, Ikeda-Saito M, Hoffman BM, Huynh BH (2007) Distinct reaction pathways followed upon reduction of oxy-heme oxygenase and oxy-myoglobin as characterized by Mossbauer spectroscopy. J Am Chem Soc 129:1402–1412
68. Davydov R, Osborne RL, Kim SH, Dawson JH, Hoffman BM (2008) EPR and ENDOR studies of cryoreduced Compounds II of peroxidases and myoglobin. Proton-coupled electron transfer and protonation status of ferryl hemes. Biochemistry 47:5147–5155
69. Denisov IG, Mak PJ, Makris TM, Sligar SG, Kincaid JR (2008) Resonance Raman characterization of the peroxo and hydroperoxo intermediates in cytochrome P450. J Phys Chem A 112:13172–13179
70. Denisov IG, Makris TM, Sligar SG (2002) Formation and decay of hydroperoxo-ferric heme complex in horseradish peroxidase studied by cryoradiolysis. J Biol Chem 277: 42706–42710
71. Mak PJ, Denisov IG, Victoria D, Makris TM, Deng T, Sligar SG, Kincaid JR (2007) Resonance Raman detection of the hydroperoxo intermediate in the cytochrome P450 enzymatic cycle. J Am Chem Soc 129:6382–6383
72. Mak PJ, Kincaid JR (2008) Resonance Raman spectroscopic studies of hydroperoxo derivatives of cobalt-substituted myoglobin. J Inorg Biochem 102:1952–1957
73. Davydov R, Kappl R, Huettermann J, Peterson JA (1991) EPR-spectroscopy of reduced oxyferrous-P450cam. FEBS Lett 295:113–115
74. Makris TM, Davydov R, Denisov IG, Hoffman BM, Sligar SG (2002) Mechanistic enzymology of oxygen activation by the cytochromes P450. Drug Metab Rev 34:691–708
75. Makris TM, Von Koenig K, Schlichting I, Sligar SG (2007) Alteration of P450 distal pocket solvent leads to impaired proton delivery and changes in heme geometry. Biochemistry 46:14129–14140
76. Gasyna Z (1979) Intermediate spin-states in one-electron reduction of oxygen-hemoprotein complexes at low temperature. FEBS Lett 106:213–218
77. Denisov IG, Makris TM, Sligar SG (2001) Cryotrapped reaction intermediates of cytochrome P450 studied by radiolytic reduction with phosphorus-32. J Biol Chem 276: 11648–11652
78. Denisov IG, Dawson JH, Hager LP, Sligar SG (2007) The ferric-hydroperoxo complex of chloroperoxidase. Biochem Biophys Res Commun 363:954–958
79. Ibrahim M, Denisov IG, Makris TM, Kincaid JR, Sligar SG (2003) Resonance Raman spectroscopic studies of hydroperoxo-myoglobin at cryogenic temperatures. J Am Chem Soc 125:13714–13718
80. Ibrahim M, Kincaid JR (2004) Spectroscopic studies of peroxo/hydroperoxo derivatives of heme proteins and model compounds. J Porphyrins Phthalocyanines 8:215–225
81. Sligar SG, Makris TM, Denisov IG (2005) Thirty years of microbial P450 monooxygenase research: Peroxo-heme intermediates: the central bus station in heme oxygenase

catalysis. Biochem Biophys Res Commun 338:346–354

82. Gantt SL, Denisov IG, Grinkova YV, Sligar SG (2009) The critical iron–oxygen intermediate in human aromatase. Biochem Biophys Res Commun 387:169–173
83. Schuler RH (1994) Three decades of spectroscopic studies of radiation produced intermediates. Radiat Phys Chem 43:417–423
84. Douzou P, Balny C (1978) Protein fractionation at subzero temperatures. Adv Protein Chem 32:77–189
85. Bonfils C, Saldana JL, Debey P, Maurel P, Balny C, Douzou P (1979) Fast photochemical reactions of cytochrome P450 at subzero temperatures. Biochimie 61:681–687
86. Douzou P (1982) Developments in low-temperature biochemistry and biology. Proc R Soc Lond B 217:1–28
87. Douzou P (1983) Cryoenzymology. Cryobiology 20:625–635
88. Daniel RM, Smith JC, Ferrand M, Hery S, Dunn R, Finney JL (1998) Enzyme activity below the dynamical transition at 220 K. Biophys J 75:2504–2507
89. Gasyna Z (1980) Unusual spin-state transitions in the reduction of ferrylmyoglobin at low temperature. Biochem Biophys Res Comm 93:637–644
90. Gasyna Z, Browett WR, Stillman MJ (1988) Low-temperature magnetic circular dichroism studies of the photoreaction of horseradish peroxidase compound I. Biochemistry 27: 2503–2509
91. Browett WR, Gasyna Z, Stillman MJ (1988) Temperature dependence and electronic transition energies in the magnetic circular dichroism spectrum of horeseradish peroxidase compound I. J Am Chem Soc 110: 3633–3640
92. Zollfrank J, Friedrich J, Vanderkooi JM, Fidy J (1991) Conformational relaxation of a low-temperature protein as probed by photochemical hole burning. Horseradish peroxidase. Biophys J 59:305–312
93. Manas ES, Vanderkooi JM, Sharp KA (1999) The effects of protein environment on the low temperature electronic spectroscopy of cytochrome c and microperoxidase-11. J Phys Chem 103B:6334–6348
94. Wright WW, Carlos Baez J, Vanderkooi JM (2002) Mixed trehalose/sucrose glasses used for protein incorporation as studied by infrared and optical spectroscopy. Anal Biochem 307:167–172
95. Khajehpour M, Rietveld I, Vinogradov S, Prabhu NV, Sharp KA, Vanderkooi JM (2003) Accessibility of oxygen with respect to the heme pocket in horseradish peroxidase. Proteins 53:656–666
96. Zelent B, Nucci NV, Vanderkooi JM (2004) Liquid and ice water and glycerol/water glasses compared by infrared spectroscopy from 295 to 12 K. J Phys Chem A 108: 11141–11150
97. Nibbs J, Vinogradov SA, Vanderkooi JM, Zelent B (2004) Flexibility in proteins: tuning the sensitivity to O2 diffusion by varying the lifetime of a phosphorescent sensor in horseradish peroxidase. Photochem Photobiol 80:36–40
98. Ponkratov VV, Wiedersich J, Friedrich J, Vanderkooi JM (2007) Experiments with proteins at low temperature: what do we learn on properties in their functional state? J Chem Phys 126:16510–16514
99. Austin RH, Beeson KW, Eisenstein L, Frauenfelder H, Gunsalus IC (1975) Dynamics of ligand binding to myoglobin. Biochemistry 14:5355–5373
100. Frauenfelder H, Sligar SG, Wolynes PG (1991) The energy landscapes and motions of proteins. Science 254:1598–1603
101. Chen G, Fenimore PW, Frauenfelder H, Mezei F, Swenson J, Young RD (2008) Protein fluctuations explored by inelastic neutron scattering and dielectric relaxation spectroscopy. Phil Mag 88:3877–3883
102. Perrella M, Heyda A, Mosca A, Rossi-Bernardi L (1978) Isoelectric focusing and electrophoresis at subzero temperatures. Anal Biochem 88:212–224
103. Perrella M, Benazzi L, Cremonesi L, Vesely S, Viggiano G, Berger RL (1983) Subzero temperature quenching and electrophoretic methods for the isolation of protein reaction intermediates. J Biochem Biophys Meth 7: 187–197
104. Perrella M, Denisov I (1995) Low-temperature electrophoresis methods. Meth Enzymol 259:468–487
105. Balny C, Le Peuch C, Debey P (1975) Low temperature column chromatography: application to microsomal hydroxylating system. Anal Biochem 63:321–330
106. Balny C, Debey P, Douzou P (1976) The subzero temperature chromatographic isolation of transient intermediates of a multi-step cycle: preparation of the substrate-free oxy-ferrous cytochrome P450. FEBS Lett 69:236–239
107. Debey P, Balny C, Douzou P (1973) Enzyme assay in microsomes below zero degrees. Proc Natl Acad Sci U S A 70:2633–2636

# Index



Wlodek M. Bujalowski (ed.), *Spectroscopic Methods of Analysis: Methods and Protocols*, Methods in Molecular Biology, vol. 875, DOI 10.1007/978-1-61779-806-1, © Springer Science+Business Media New York 2012

## P



## Q



## R



## S

## T



## U



## V



## X



## Z

MIX
Papier aus verantwortungsvollen Quellen
Paper from responsible sources
FSC® C105338

If you have any concerns about our products,
you can contact us on
**ProductSafety@springernature.com**

In case Publisher is established outside the EU,
the EU authorized representative is:
**Springer Nature Customer Service Center GmbH**
**Europaplatz 3, 69115 Heidelberg, Germany**

Printed by Libri Plureos GmbH
in Hamburg, Germany